Lecture Notes in Computer Science 12942

More information about this subseries at http://www.springer.com/series/7407

Victor Malyshkin (Ed.)

Parallel Computing Technologies

16th International Conference, PaCT 2021
Kaliningrad, Russia, September 13–18, 2021
Proceedings

 Springer

Editor
Victor Malyshkin 🆔
Institute of Computational Mathematics
and Mathematical Geophysics SB RAS
Novosibirsk, Russia

ISSN 0302-9743   ISSN 1611-3349   (electronic)
Lecture Notes in Computer Science
ISBN 978-3-030-86358-6   ISBN 978-3-030-86359-3   (eBook)
https://doi.org/10.1007/978-3-030-86359-3

LNCS Sublibrary: SL1 – Theoretical Computer Science and General Issues

This Springer imprint is published by the registered company Springer Nature Switzerland AG
The registered company address is: Gewerbestrasse 11, 6330 Cham, Switzerland

Preface

The 16th International Conference on Parallel Computing Technologies (PaCT 2021) was a four-day event held in Kaliningrad, Russia. It was organized by the Institute of Computational Mathematics and Mathematical Geophysics of the Russian Academy of Sciences (Novosibirsk) in cooperation with the Immanuel Kant Baltic Federal University (Kaliningrad), Novosibirsk State University, and Novosibirsk State Technical University.

Previous conferences of the PaCT series were held in various Russian cities every odd year beginning with PaCT 1991, which took place in Novosibirsk (Akademgorodok), whilst the 15th Conference took place in Almaty, Kazakhstan. Since 1995, all the PaCT proceedings have been published by Springer in the LNCS series.

The aim of the PaCT 2021 conference was to provide a forum for an exchange of views among the international community of researchers in the field of the development of parallel computing technologies. The PaCT 2021 Program Committee selected papers that contributed new knowledge in methods and tools for parallel solution of topical large-scale problems. The papers selected for PaCT 2021

- propose and study tools for parallel program development such as languages, performance analysers, and automated performance tuners,
- examine and optimize the processes related to management of jobs, data, and computing resources at high performance computing centers,
- propose new computer simulation models and algorithms specifically targeted to parallel computing architectures, and
- theoretically study practically relevant properties of parallel programming models and parallel algorithms.

Authors from 15 countries submitted 62 papers. The submitted papers were subject to a single blind reviewing process, with papers receiving an average of 2.8 reviews. The Program Committee selected 24 full papers and 12 short papers for presentation at PaCT 2021.

Many thanks to our sponsors: the Ministry of Science and Higher Education of the Russian Federation, the Russian Academy of Sciences, and the RSC Group.

September 2021   Victor Malyshkin

Organization

The PaCT 2021 was organized by the Institute of Computational Mathematics and Mathematical Geophysics, Siberian Branch of Russian Academy of Sciences (Novosibirsk, Russia) in cooperation with the Immanuel Kant Baltic Federal University, Novosibirsk State University, and Novosibirsk State Technical University.

Organizing Committee

Conference Co-chairs

V. E. Malyshkin	ICMMG SB RAS, NSU, NSTU, Russia
M. V. Demin	IKBFU, Russia
G. N. Erokhin	IKBFU, Russia

Conference Secretary

M. A. Gorodnichev	ICMMG SB RAS, NSU, NSTU, Russia

Organizing Committee

S. M. Achasova	ICMMG SB RAS, Russia
S. B. Arykov	ICMMG SB RAS, NSTU, Russia
A. V. Belova	IKBFU, Russia
E. G. Danilov	IKBFU, Russia
V. M. Filatova	IKBFU, Russia
M. A. Gorodnichev	ICMMG SB RAS, NSU, NSTU, Russia
A. I. Kamyshnikov	IKBFU, Russia
S. E. Kireev	ICMMG SB RAS, NSU, Russia
A. E. Kireeva	ICMMG SB RAS, Russia
T. V. Makhneva	IKBFU, Russia
V. P. Markova	ICMMG SB RAS, NSU, NSTU, Russia
Yu. G. Medvedev	ICMMG SB RAS, Russia
V. A. Perepelkin	ICMMG SB RAS, NSU, Russia
L. N. Pestov	IKBFU, Russia
I. G. Samusev	IKBFU, Russia
G. A. Schukin	ICMMG SB RAS, NSTU, Russia
R. V. Simonov	IKBFU, Russia
Yu. N. Svirina	IKBFU, Russia
V. S. Timofeev	NSTU, Russia

Program Committee

Victor Malyshkin (Co-chair)	ICMMG SB RAS, NSU, NSTU, Russia
Gennady N. Erokhin (Co-chair)	Immanuel Kant Baltic Federal University, Russia
Sergey Abramov	Program Systems Institute, Russian Academy of Sciences, Russia
Darkhan Akhmed-Zaki	Astana IT University and al-Farabi Kazakh National University, Kazakhstan
Farhad Arbab	Leiden University, The Netherlands
Jan Baetens	Ghent University, Belgium
Stefania Bandini	University of Milano-Bicocca, Italy
Thomas Casavant	University of Iowa, USA
Pierpaolo Degano	University of Pisa, Italy
Dominique Désérable	National Institute for Applied Sciences, Rennes, France
Hugues Fauconnier	IRIF, Paris Diderot University, France
Thomas Fahringer	University of Innsbruck, Austria
Victor Gergel	Lobachevsky State University of Nizhni Novgorod, Russia
Juan Manuel Cebrián González	University of Murcia, Spain
Bernard Goossens	University of Perpignan, France
Sergei Gorlatch	University of Münster, Germany
Yuri G. Karpov	Peter the Great St. Petersburg State Polytechnic University, Russia
Alexey Lastovetsky	University College Dublin, Ireland
Jie Li	University of Tsukuba, Japan
Thomas Ludwig	University of Hamburg, Germany
Giancarlo Mauri	University of Milano-Bicocca, Italy
Igor Menshov	Keldysh Institute for Applied Mathematics, Russian Academy of Sciences
Nikolay Mirenkov	University of Aizu, Japan
Marcin Paprzycki	Polish Academy of Sciences, Poland
Dana Petcu	West University of Timisoara, Romania
Viktor Prasanna	University of Southern California, USA
Michel Raynal	Research Institute in Computer Science and Random Systems, Rennes, France
Bernard Roux	National Center for Scientific Research, Aix-Marseille University, France
Uwe Schwiegelshohn	Technical University of Dortmund, Germany
Waleed W. Smari	Ball Aerospace & Technologies Corp., Ohio, USA
Victor Toporkov	National Research University "Moscow Power Engineering Institute", Russia
Carsten Trinitis	University of Bedfordshire, UK, and Technical University of Munich, Germany
Roman Wyrzykowski	Czestochowa University of Technology, Poland

Additional Reviewers

Oleg Bessonov
Carole Delporte-Gallet
Maxim Gorodnichev
Rolf Hoffmann
Evgeny Ivashko
Ivan Kholod
Sergey Kireev
Anastasia Kireeva

Yuri Medvedev
Pavel Pavlukhin
Vladislav Perepelkin
Anastasia Perepelkina
Georgy Schukin
Aleksey Snytnikov
Oleg Sukhoroslov

Sponsoring Institutions

Ministry of Education and Science of the Russian Federation
Russian Academy of Sciences
RSC Group

Contents

Applications

Memory-Efficient Data Structures

Experimental Studies

Cellular Automata

Parallel Programming Methods
and Tools

Trace-Based Optimization of Fragmented Programs Execution in LuNA System

Victor Malyshkin[1,2,3] and Vladislav Perepelkin[1,2(✉)]

[1] Institute of Computational Mathematics and Mathematical Geophysics SB RAS,
Novosibirsk, Russia
perepelkin@ssd.sscc.ru
[2] Novosibirsk State University, Novosibirsk, Russia
[3] Novosibirsk State Technical University, Novosibirsk, Russia

Abstract. Automatic construction of high performance distributed numerical simulation programs is used to reduce complexity of distributed parallel programs development and to improve code efficiency as compared to an average manual development. Development of such means, however, is challenging in general case, that's why a variety of different languages, systems and tools for parallel programs construction exist and evolve. Program tracing (i.e. journaling execution acts of the program) is a valuable source of information, which can be used to optimize efficiency of constructed programs for particular execution conditions and input data peculiarities. One of the optimization techniques is trace playback, which consists in step-by-step reproduction of the trace. This allows reducing runtime overhead, which is relevant for runtime system-based tools. The experimental results demonstrate suitability of the technique for a range of applications.

Keywords: Automatic program construction · Fragmented programming technology · LuNA system · Trace playback

1 Introduction

Development of high performance scientific parallel programs for supercomputers is often complicated and hard due to the necessity to decompose data and computations, organize parallel data processing, provide non-functional properties of the programs. Such properties may include efficiency (execution time, memory consumption, network load, etc.), static or dynamic workload balancing, fault tolerance, checkpointing, etc. All this requires in-depth knowledge of hardware architecture, skill in parallel programming, familiarity with appropriate parallel programming methods and tools. This makes manual programming troublesome for an average supercomputer user, who is an expert in the subject domain, not in system parallel programming. Usage of parallel programming automation systems, languages and tools allows to significantly reduce complexity of parallel programming, improve quality of produced programs and reduce knowledge and skill requirements a programmer has to possess.

© Springer Nature Switzerland AG 2021
V. Malyshkin (Ed.): PaCT 2021, LNCS 12942, pp. 3–10, 2021.
https://doi.org/10.1007/978-3-030-86359-3_1

In general automatic construction of an efficient parallel program is algorithmically hard, which why no effective general approach is expected to exist, so a diversity of various approaches, heuristics, languages and programming systems are being constantly developed to support parallel programming automation in different particular cases and subject domains. One of the promising approaches of parallel programs automatic optimization is trace-based optimization. This approach assumes that a program is first run on some characteristic input data, and its performance is being recorded as a trace of events (computational, communicational, etc.). The trace is then analyzed to extract quantitative information and pass it to a programming system (compiler, interpreter, etc.) to produce more efficient code. This is similar to profile-based optimization, except that a profile contains statistical information, while a trace contains the full log of significant events. In particular, trace can be used to reproduce the computation process recorded ("trace playback"), which can be more efficient than the normal program execution if the latter involves dynamic decision-making or other overhead, which can be omitted with trace playback. This, however, is not always possible, because change in input data or the computing system state may cause inconsistent execution. This paper is devoted to implementation of this idea in LuNA system for distributed parallel programs construction [1].

The rest of the paper is organized as follows. Section 2 contains a brief necessary introduction into LuNA system computational model in comparison with other systems. Section 3 describes how trace gathering and playback are implemented in LuNA. Section 4 presents results of the experimental study.

2 Trace Playback in LuNA System

2.1 LuNA System

LuNA (Language for Numerical Algorithms) is a language and a system for automatic construction of numerical distributed parallel programs for distributed memory parallel computers (multicomputers). It is an academic project of the Institute of Computational Mathematics and Mathematical Geophysics of the Siberian Branch of Russian Academy of Sciences. The system is based on the theory of structured synthesis of parallel programs [2], and its purpose is to support the active knowledge technology [3]. LuNA program is a high-level coarse-grained explicitly-parallel description of a numerical algorithm, which is basically a description of a bipartite oriented graph of computational fragments (CFs) and data fragments (DFs). DF is an immutable piece of data, the result of data decomposition. Each CF is a conventional subroutine call, which takes a number of DFs as inputs to compute values of a number of other DFs (these production-consuming relations correspond to arcs in the graph). So, LuNA program defines a set of informationally dependent tasks (CFs), which have to be executed in an order, which satisfies the dependencies. To execute such program LuNA has to distribute CFs to computing nodes, perform DFs transfer from producers to consumers and execute CFs after all their input DFs are available at the node. Efficiency of such execution is conditioned by the CFs distribution and execution order, by DFs network transfer delays and by the runtime system overhead. As our previous works show [4–11] the performance of LuNA programs is 1–100 times less than that of manually developed programs, depending on the subject domain. We continue to improve LuNA system algorithms to provide better performance for practical application classes, and this work is one of such improvements. More details on LuNA system can be found in [1] and in its public repository[1].

2.2 Trace Playback in LuNA System

Since LuNA program execution consists eventually of CFs executions, the trace information includes CF execution start and end times and the node on which the CF was executed. This information is sufficient to completely reproduce the computation of LuNA program. Once the trace is recorded, its playback on each computing node may be organized as follows:

1. Pick the earliest unexecuted CF a from the trace (on the node).
2. For each input DF x of the CF a find in the trace the CF b, which produced it.
3. If CF b was executed on the same computing node where CF a was executed, then DF x is available on the node; otherwise receive DF x as a message from CF b's node.
4. Invoke the conventional subroutine, related to CF a with input DFs passed to it.
5. For each output DF x of the CF a find all CFs c, which take DF x as input. If CF c is located on the same node as CF a, then store DF x locally, otherwise send DF x as a message to CF c's node.

This is an essential scheme, although some more or less obvious tuning should be done in practical implementation. For example, if multiple CFs are located on the same computing node and take the same DF as input, then only one copy of the DF should be passed via network. Note that trace playback can be performed in multiple threads for each node (normal LuNA operation is also multi-threaded on each computing node).

This scheme misses the garbage collection, which takes place with normal LuNA operation. It can be straightforwardly implemented by recording to the trace the relative time point where DF deallocation took place. However, this appeared to be redundant, since all actual DFs consumptions are explicitly seen in the trace, thus the DF deallocation is performed as soon as last consumption on the node has occurred.

With trace playback the run-time overhead is reduced to the minimum. No decision making on CFs distribution, CFs execution ordering or DFs garbage collection is needed. In particular, LuNA dynamically balances workload by redistributing CFs to computing nodes, but only a final location where execution took place matters. All multi-hop DF transfers become single-hop transfers. Reduction of most kinds of overhead is the main source of performance improvement for trace playback as compared to normal LuNA operation.

Note, that LuNA programs execution is non-deterministic in sense of CFs distribution and execution ordering, and in sense of timings. Even minor factors (such as network delays or external CPU load) may influence the decisions LuNA system makes and implements. Dynamic load balancing is especially sensible to such factors. The trace, however, is much more deterministic, since most events are rigidly fixed.

---

[1] https://gitlab.ssd.sscc.ru/luna/luna.

3 Discussion and Related Works

3.1 Analysis and Discussion

The main drawback of the approach is that the set of CFs may depend on input data. As long as the task graph (the set of CFs) persists trace playback produces valid execution for any input data. But, for example, if the number of loop iterations depends on input data, then trace playback may be erroneous. This drawback can be partially compensated by two factors. First, there are a lot of applications where tasks graph does not depend on input data (e.g. dense linear algebra operations). Second, the fact that the task graph appeared to be different for given input can be detected automatically. In particular, in LuNA there are three operators, which can produce data-dependent task graphs: if, for and while. Each of the operators can be supplied with straightforward checks, which will ensure that each if condition was resolved to the same true/false value and that every for and while operator has the same iteration range. So, trace playback engine can inform the user on unsuccessful playback (rather than silently perform erroneous execution) and suggest normal program execution.

To some extent this drawback can be overcome further. E.g. some kind of induction techniques can be used to stack for or while loop iterations into a parametric range-independent form. For the if operator both *then* and *else* branches can be traced at first precedent, and after that the execution of both branches can be done via trace playback. Study of these possibilities is out of the scope of the paper.

Another drawback of the approach is that no decisions on CFs distribution and execution order are made – only the decisions made by LuNA system in the traced run are recorded and reproduced. These decisions may be not good for a number of reasons. E.g., hardware configuration or its external load may be different; CF execution time may depend on input data; absence of run-time system overhead may influence timings, etc. The decisions themselves, that LuNA system has made, can be not good, because LuNA system algorithms are not perfect. This brings us to the idea of trace optimization and tuning before doing the playback. Study of the idea is beyond the current work, but a brief overview of the problem can be given. Firstly, the trace can be analyzed for work imbalance or inefficient CFs execution order. Secondly, any CF can be reassigned to another node with no risk of bringing error to the execution. Also, CFs execution can be reordered unless informational dependencies are violated. Such trace transformations can either eliminate work imbalance or retarget the trace to another hardware configuration (computing nodes number, network topology, relative nodes performance, etc.).

Besides trace optimization, trace execution engine can be improved. For example, the above mentioned thread pool-based execution is one of such possible optimizations. More dynamic improvements can be made. For example, dynamic workload balancing may be employed to eliminate work imbalance that occurs during trace playback. Study of these possibilities is also out of the scope of the paper.

3.2 Related Works

Trace playback is practical in LuNA system because of the computational model it employs. In particular, the "computational" part of a LuNA program is separate from distributed management logic, which allows to replace the latter with a trace playback engine. There are other programming languages and systems, where trace playback may be of use. For manually developed conventional distributed parallel programs trace playback appears to be inapplicable, since it is impossible to distinguish the essential computational part from the rest of the code, which organizes parallel computations, communications and data storage.

Specialized computational models, such as the map-reduce model [12, 13], allow to distinguish the computational part and computations structure since it is explicitly formulated in the code. This, in turn, allows to trace execution and playback the trace. For example, such systems implement dynamic workload balancing, which causes some run-time overhead. It can be reduced by the trace playback technique. Of course, this makes sense for a series of computations where imbalance is known to be the same. The rest of overhead is usually negligible.

Task-based systems, such as Charm++ [14] or OpenTS [15] allow trace playback mostly the same way it is possible in LuNA system. In Charm++, however, it may be harder to implement, because *chares* (Charm++ decomposition units) may behave differently depending on the order in which they receive messages from other chares. To allow safe trace playback some additional constraints to chare codes may be required.

For systems with explicit program behavior control, such as PaRSEC [16] or Legion [17] trace playback seems to be as easily implemented as in LuNA, since the computational part is explicitly formulated in the computational model.

Some possibilities of trace playback exist in systems for automated serial code parallelization, such as DVM-H [18]. Here a serial code is annotated (either manually or automatically) with "parallelization pragmas", and a parallel program is generated automatically. In particular, a dynamic workload balancing mechanism may be included into the generated program. Code annotations allow identifying the computational part, and since the distributed code is generated automatically, it can be instrumented to trace events, necessary to perform the playback.

It can be concluded, that trace playback is a reasonable technique for programming languages, systems and tools, which employ run-time systems, or at least provide some dynamic properties (such as dynamic load balancing) at cost of some overhead.

4   Experiments

To playback a trace a series of actions to perform is generated for each computing node. Possible actions are invocation of a serial subroutine, DF transfer to another node and DF deletion. Such series is easily constructed from trace. Implementation of the series of actions on each node causes the trace playback.

The naïve way to implement trace playback is to generate the series of actions as a conventional (e.g. MPI-based) program. Such an approach possesses minimal possible overhead. In practice, however, such source code listing grows large and takes too much time to be compiled into binary (e.g. hours of compilation for a large program). To overcome this issue the series of events was encoded into a binary file (to reduce size), and a trivial interpreter was developed, which decodes the file and performs the actions using a worker thread pool on each node. This decoding does add some overhead, but it is usually negligible due to coarse granularity of CFs. A separate thread was dedicated to receiving messages from other nodes.

For experimental performance evaluation a Particle-In-Cell application for self-gravitating dust-cloud simulation [19] was used as an example of a rather complicated real supercomputing-targeted application. Tests were conducted on MVS-10p cluster of Joint Supercomputing Center of Russian Academy of Sciences[2]. The testing was conducted for various parameters (see Table 1) to investigate performance in different conditions.

Table 1. Experimental results

Parameters			Execution time (sec.)		
Mesh size	Particles	Cores	MPI	LuNA-TB	LuNA
100^3	10^6	64	5.287	13.69	355.5
150^3	10^6	64	18.896	31.088	732.833
150^3	10^7	64	23.594	111.194	2983
150^3	10^7	125	23.352	118.32	3086.411
150^3	10^6	343	33.697	39.651	1008.655

Two programs, developed by S. Kireev for [20], which implement the same algorithm, were used. The first program is a conventional C++ distributed parallel program, based on Message Passing Interface (MPI). The second one is a LuNA program. Execution time for these programs is shown in Table 1. There is also a column labeled LuNA-TB. This is the execution time of the same LuNA program, but using the trace-playback technique. The MPI program can be considered as a reference point, its efficiency is what one can expect as a result of manual development of an experienced applied programmer. The LuNA program is an automatically constructed program using a general approach, and its efficiency is expectedly much lower, than that of the MPI program. And the LuNA-TB is somewhere in the middle, an automatically generated program using the particular trace playback approach.

The main result of the testing is that LuNA-playback indeed significantly speeds up execution of LuNA programs. This confirms that the trace playback is a useful technique for optimizing efficiency of automatically constructed parallel programs. Its efficiency is still lower than that of the MPI program, but this is obviously a practically usable result, considering that it is obtained automatically.

It can also be seen from Table 1 that trace playback approach is more advantageous for programs with finer granularity, where fragments are of lesser size. The advantage is the bigger the more computing nodes are involved in computation. This is also expected, since dynamic decentralized algorithms employed in LuNA produce significant overhead, which is cut off with trace playback.

---

[2] http://www.jscc.ru.

5 Conclusion

The trace playback technique is investigated as a distributed programs optimization technique for parallel programming automation systems. Trace playback was implemented for LuNA system for automatic numerical parallel programs construction. The experiments showed a significant improvement of the efficiency of constructed programs. Possible improvements of the technique, aimed at overcoming its drawbacks are briefly discussed. It can be concluded that the trace playback technique is practical for high performance distributed parallel programs construction automation, which can be used automatically (along with other particular system algorithms and heuristics). In future we plan to further investigate the approach within LuNA system to widen the application class this technique is applicable to.

Acknowledgements. The work was supported by the budget project of the ICMMG SB RAS No. 0251-2021-0005.

References

1. Malyshkin, V.E., Perepelkin, V.A.: LuNA fragmented programming system, main functions and peculiarities of run-time subsystem. In: Malyshkin, V. (ed.) PaCT 2011. LNCS, vol. 6873, pp. 53–61. Springer, Heidelberg (2011). https://doi.org/10.1007/978-3-642-23178-0_5
2. Valkovsky, V.A., Malyshkin, V.E.: Synthesis of parallel programs and systems on the basis of computational models. Nauka, Novosibirsk (1988). (in Russian)
3. Malyshkin, V.: Active knowledge, LuNA and literacy for oncoming centuries. In: Bodei, C., Ferrari, G.-L., Priami, C. (eds.) Programming Languages with Applications to Biology and Security. LNCS, vol. 9465, pp. 292–303. Springer, Cham (2015). https://doi.org/10.1007/978-3-319-25527-9_19
4. Akhmed-Zaki, D., Lebedev, D., Malyshkin, V., Perepelkin, V.: Automated construction of high performance distributed programs in LuNA system. In: Malyshkin, V. (ed.) PaCT 2019. LNCS, vol. 11657, pp. 3–9. Springer, Cham (2019). https://doi.org/10.1007/978-3-030-25636-4_1
5. Akhmed-Zaki, D., Lebedev, D., Perepelkin, V.: Implementation of a 3D model heat equation using fragmented programming technology. J. Supercomput. **75**(12), 7827–7832 (2018). https://doi.org/10.1007/s11227-018-2710-1
6. Daribayev, B., Perepelkin, V., Lebedev, D., Akhmed-Zaki, D.: Implementation of the two-dimensional elliptic equation model in LuNA fragmented programming system. In: 2018 IEEE 12th International Conference on Application of Information and Communication Technologies (AICT), pp. 1–4 (2018)
7. Nikolay, B., Perepelkin, V.: Automated GPU support in LuNA fragmented programming system. In: Malyshkin, V. (ed.) PaCT 2017. LNCS, vol. 10421, pp. 272–277. Springer, Cham (2017). https://doi.org/10.1007/978-3-319-62932-2_26
8. Malyshkin, V., Perepelkin, V., Schukin, G.: Scalable distributed data allocation in LuNA fragmented programming system. J. Supercomput. **73**(2), 726–732 (2016). https://doi.org/10.1007/s11227-016-1781-0
9. Malyshkin, V.E., Perepelkin, V.A., Tkacheva, A.A.: Control flow usage to improve performance of fragmented programs execution. In: Malyshkin, V. (ed.) PaCT 2015. LNCS, vol. 9251, pp. 86–90. Springer, Cham (2015). https://doi.org/10.1007/978-3-319-21909-7_9

10. Malyshkin, V.E., Perepelkin, V.A., Schukin, G.A.: Distributed algorithm of data allocation in the fragmented programming system LuNA. In: Malyshkin, V. (ed.) PaCT 2015. LNCS, vol. 9251, pp. 80–85. Springer, Cham (2015). https://doi.org/10.1007/978-3-319-21909-7_8
11. Alias, N., Kireev, S.: Fragmentation of IADE method using LuNA system. In: Malyshkin, V. (ed.) PaCT 2017. LNCS, vol. 10421, pp. 85–93. Springer, Cham (2017). https://doi.org/10.1007/978-3-319-62932-2_7
12. Dean, J., Ghemawat, S.: MapReduce: simplified data processing on large clusters. In: Sixth Symposium on Operating System Design and Implementation, OSDI 2004, San Francisco, CA, pp. 137–150 (2004)
13. White, T.: Hadoop: The Definitive Guide: Storage and Analysis at Internet Scale, 4th edn, 756 p. O'Reilly Media, Sebastopol (2015). ISBN-13 978-1491901632
14. Kale, L.V., Bhatele, A.: Parallel Science and Engineering Applications: The Charm++ Approach. Taylor & Francis Group, CRC Press (2013). ISBN 9781466504127
15. Moskovsky, A., Roganov, V., Abramov, S.: Parallelism granules aggregation with the T-system. In: Malyshkin, V. (ed.) PaCT 2007. LNCS, vol. 4671, pp. 293–302. Springer, Heidelberg (2007). https://doi.org/10.1007/978-3-540-73940-1_30
16. Bosilca, G., Bouteiller, A., Danalis, A., Faverge, M., Hérault, T., Dongarra, J.: PaRSEC: a programming paradigm exploiting heterogeneity for enhancing scalability. Comput. Sci. Eng. 99(2013), 1 (2013)
17. Bauer, M., Treichler, S., Slaughter, E., Aiken, A.: Legion: expressing locality and independence with logical regions. In: Conference on High Performance Computing Networking, Storage and Analysis, SC 2012, Salt Lake City, UT, USA, 11–15 November 2012 (2012). https://doi.org/10.1109/SC.2012.71
18. Kataev, N.A., Kolganov, A.S.: The experience of using DVM and SAPFOR systems in semi automatic parallelization of an application for 3D modeling in geophysics. J. Supercomput. 75, 7833–7843 (2018)
19. Kireev, S.: A parallel 3D code for simulation of self-gravitating gas-dust systems. In: Malyshkin, V. (ed.) PaCT 2009. LNCS, vol. 5698, pp. 406–413. Springer, Heidelberg (2009). https://doi.org/10.1007/978-3-642-03275-2_40
20. Belyaev, N., Kireev, S.: LuNA-ICLU compiler for automated generation of iterative fragmented programs. In: Malyshkin, V. (ed.) PaCT 2019. LNCS, vol. 11657, pp. 10–17. Springer, Cham (2019). https://doi.org/10.1007/978-3-030-25636-4_2

A New Model-Based Approach to Performance Comparison of MPI Collective Algorithms

Emin Nuriyev⬛ and Alexey Lastovetsky$^{(\boxtimes)}$⬛

School of Computer Science, University College Dublin, Dublin, Ireland
emin.nuriyev@ucdconnect.ie, alexey.lastovetsky@ucd.ie

Abstract. The performance of collective operations has been a critical issue since the advent of Message Passing Interface (MPI). Many algorithms have been proposed for each MPI collective operation but none of them proved optimal in all situations. Different algorithms demonstrate superior performance depending on the platform, the message size, the number of processes, etc. MPI implementations perform the selection of the collective algorithm empirically, executing a simple runtime decision function. While efficient, this approach does not guarantee the optimal selection. As a more accurate but equally efficient alternative, the use of analytical performance models of collective algorithms for the selection process was proposed and studied. Unfortunately, the previous attempts in this direction have not been successful.

We revisit the analytical model-based approach and propose two innovations that significantly improve the selective accuracy of analytical models: (1) We derive analytical models from the code implementing the algorithms rather than from their high-level mathematical definitions. This results in more detailed models. (2) We estimate model parameters separately for each collective algorithm and include the execution of this algorithm in the corresponding communication experiment.

We experimentally demonstrate the accuracy and efficiency of our approach using Open MPI broadcast algorithms and two different Grid'5000 clusters.

Keywords: Message Passing · Collective communication algorithms · Communication performance modelling · MPI

1 Introduction

The message passing interface (MPI) [1] is the de-facto standard, which provides a reliable and portable environment for developing high-performance parallel applications on different platforms. The study [2] shows that collective operations consume more than eighty percent of the total communication time of a typical

This publication has emanated from research conducted with the financial support of Science Foundation Ireland (SFI) under Grant Number 14/IA/2474.

© Springer Nature Switzerland AG 2021
V. Malyshkin (Ed.): PaCT 2021, LNCS 12942, pp. 11–25, 2021.
https://doi.org/10.1007/978-3-030-86359-3_2

MPI application. Therefore, a significant amount of research has been invested into optimisation of MPI collectives. Those researches have resulted in a large number of algorithms, each of which comes up optimal for specific message sizes, platforms, numbers of processes, and so forth. Mainstream MPI libraries [3,4] provide multiple collective algorithms for each collective routine.

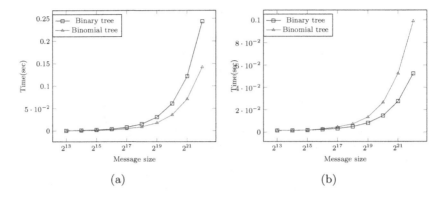

(a)   (b)

Fig. 1. Performance estimation of the binary and binomial tree broadcast algorithms by the traditional analytical models in comparison with experimental curves. The experiments involve ninety processes ($P = 90$). (a) Estimation by the existing analytical models. (b) Experimental performance curves.

There are two ways how this selection can be made in the MPI program. The first one, MPI_T interface [1], is provided by the MPI standard and allows the MPI programmer to select the collective algorithm explicitly from the list of available algorithms for each collective call at run-time. It does not solve the problem of optimal selection delegating its solution to the programmer. The second one is transparent to the MPI programmer and provided by MPI implementations. It uses a simple *decision function* in each collective routine, which is used to select the algorithm at runtime. The decision function is empirically derived from extensive testing on the dedicated system. For example, for each collective operation, both MPICH and Open MPI use a simple decision routine selecting the algorithm based on the message size and number of processes [5–7]. The main advantage of this solution is its efficiency. The algorithm selection is very fast and does not affect the performance of the program. The main disadvantage of the existing decision functions is that they do not guarantee the optimal selection in all situations.

As an alternative approach, the use of analytical performance models of collective algorithms for the selection process has been proposed and studied [8]. Unfortunately, the analytical performance models proposed in this work could not reach the level of accuracy sufficient for selection of the optimal algorithm (Fig. 1).

In this paper, we revisit the model-based approach and propose a number of innovations that significantly improve the selective accuracy of analytical models to the extent that allows them to be used for accurate selection of optimal collective algorithms.

The main contributions of this paper can be summarized as follows:

- We propose and implement a new analytical performance modelling approach for MPI collective algorithms, which derives the models from the code implementing the algorithms.
- We propose and implement a novel approach to estimation of the parameters of analytical performance models of MPI collective algorithms, which estimates the parameters separately for each algorithm and includes the modelled collective algorithm in the communication experiment, which is used to estimate the model parameters.
- We experimentally validate the proposed approach to selection of optimal collective algorithms on two different clusters of the Grid'5000 platform.

The rest of the paper is structured as follows. Section 2 reviews the existing approaches to performance modelling and algorithm selection problems. Section 3 describes our approach to construction of analytical performance models of MPI collective algorithms by deriving them from the MPI implementation. Section 4 presents our method to measure analytical model parameters. Section 5 presents experimental validation of the proposed approach. Section 6 concludes the paper with a discussion of the results and an outline of the future work.

2   Related Work

In order to select the optimal algorithm for a given collective operation, we have to be able to accurately compare the performance of the available algorithms. Analytical performance models are one of the efficient ways to express and compare the performance of collective algorithms. In this section, we overview the state-of-the-art in analytical performance modelling, measurement of model parameters and selection of the optimal collective algorithms.

2.1   Analytical Performance Models of MPI Collective Algorithms

Thakur et al. [5] propose analytical performance models of several collective algorithms using the Hockney model [9]. Chan et al. [10] build analytical performance models of *Minimum-spanning tree* algorithms and *Bucket* algorithms for MPI_Bcast, MPI_Reduce, MPI_Scatter, MPI_Gather, MPI_Allgather collectives and later extend this work for multidimensional mesh architecture in [11]. Neither of the studies listed above uses the build models for selecting the optimal collective algorithms. Pjevsivac-Grbovic et al. [8] study selection of optimal collective algorithms using analytical performance models for *barrier, broadcast, reduce* and *alltoall* collective operations. The models are built up with the traditional approach using high-level mathematical definitions of the collective algorithms.

In order to predict the cost of a collective algorithm by analytical formula, model parameters are measured using point-to-point communication experiments. After experimental validation of their modelling approach, the authors conclude that the proposed models are not accurate enough for selection of optimal algorithms.

2.2   Measurement of Model Parameters

Hockney [9] presents a measurement method to find the α and β parameters of the Hockney model. The set of communication experiments consists of point-to-point round-trips. Culler et al. [12] propose a method of measurement of parameters of the LogP model, namely, L, the upper bound on the latency, o_s, the overhead of processor involving sending a message, o_r, the overhead of processor involving receiving a message, and g, the gap between consecutive message transmission. Kielmann et al. [13] propose a method of measurement of parameters of the PLogP (Parametrized LogP) model. PLogP defines its model parameters, except for latency L, as functions of message size. All approaches listed above to measure model parameters are based on point-to-point communication experiments.

From this overview, we can conclude that the state-of-the-art analytical performance models are built using only high-level mathematical definition of the algorithms, and methods for measurement of parameters of communication performance models are all based on *point-to-point communication experiments*. The only exception from this rule is a method for measurement of parameters of the LMO heterogeneous communication model [14–16]. LMO is a communication model of heterogeneous clusters, and the total number of its parameters is significantly larger than the maximum number of independent point-to-point communication experiments that can be designed to derive a system of independent linear equations with the model parameters as unknowns. To address this problem and obtain the sufficient number of independent linear equations involving model parameters, the method additionally introduces simple collective communication experiments, each using three processors and consisting of a one-to-two communication operation (scatter) followed by a two-to-one communication operation (gather). This method however is not designed to improve the accuracy of predictive analytical models of communication algorithms.

In this work, we propose to use *collective communication experiments* in the measurement method in order to improve the predictive accuracy of analytical models of collective algorithms. A more detailed survey in analytical performance modelling and estimation of the model parameters can be found in [17].

2.3   Selection of Collective Algorithms Using Machine Learning Algorithms

Machine learning (ML) techniques have been also tried to solve the problem of selection of optimal MPI algorithms.

In [18], applicability of the quadtree encoding method to this problem is studied. The goal of this work is to select the best performing algorithm and segment size for a particular collective on a particular platform. The experimental results show that the decision function performs poorly on unseen data. Applicability of the C4.5 algorithm to the MPI collective selection problem is explored in [19]. The C4.5 algorithm [20] is a decision tree classifier, which is employed to generate a decision function, based on a detailed profiling data of MPI collectives. While the accuracy of the decision function built by the C4.5 classification algorithm is higher than that of the decision function built by quadtree encoding algorithm, still, the performance penalty is higher than 50%.

Most recently Hunold et al. [21] studied the applicability of six different ML algorithms including Random Forests, Neural Networks, Linear Regressions, XGBoost, K-nearest Neighbor, and generalized additive models (GAM) for selection of optimal MPI collective algorithms. First, it is very expansive and difficult to build a regression model even for a relatively small cluster. There is no clear guidance how to do it to achieve better results. Second, even the best regression models do not accurately predict the fastest collective algorithm in most of the reported cases. Moreover, in many cases the selected algorithm performs worse than the default algorithm, that is, the one selected by a simple native decision function.

To the best of the authors' knowledge, the works outlined in this subsection are the only research done in MPI collective algorithm selection using ML algorithms. The results show that the selection of the optimal algorithm without any information about the semantics of the algorithm yields inaccurate results. While the ML-based methods treat a collective algorithm as a black box, we derive its performance model from the implementation code and estimate the model parameters using statistical techniques. The limitations of the application of the statistical techniques (AI/ML) to collective performance modelling and selection problem can be found in a detailed survey [22].

3   Implementation-Derived Analytical Models of Collective Algorithms

As stated in Sect. 1, we propose a new approach to analytical performance modelling of collective algorithms. While the traditional approach only takes into account high-level mathematical definitions of the algorithms, we derive our models from their implementation. This way, our models take into account important details of their execution having a significant impact on their performance. Open MPI uses six tree-based broadcast algorithms to implement MPI_Bcast including Linear tree algorithm, Chain tree algorithm, Binary tree algorithm, Split binary tree algorithm, K-Chain tree algorithm and Binomial tree algorithm. Because of the limited space, we present our analytical modelling approach by applying it to only binomial tree broadcast algorithm implemented in Open MPI.

To model point-to-point communications, we use the Hockney model, which estimates the time $T_{p2p}(m)$ of sending a message of size m between two processes as $T_{p2p}(m) = \alpha + \beta \cdot m$, where α and β are the latency and the reciprocal bandwidth respectively. For segmented collective algorithms, we assume that $m = n_s \cdot m_s$, where n_s and m_s are the number of segments and the segment size respectively. We assume that each algorithm involves P processes ranked from 0 to $P - 1$.

3.1   Binomial Tree Broadcast Algorithm

Fig. 2. Balanced binomial tree

In Open MPI, the binomial tree broadcast algorithm is segmentation-based and implemented as a combination of linear tree broadcast algorithms using non-blocking *send* and *receive* operations. The height of the binomial tree is the order of the tree, $H = \lfloor \log_2 P \rfloor$ (Fig. 2).

Figure 3 shows the stages of execution of the binomial tree broadcast algorithm. Each stage consists of parallel execution of a number of linear broadcast algorithms using non-blocking communication. The linear broadcast algorithms running in parallel have a different number of children. Therefore, the execution time of each stage will be equal to the execution time of the linear broadcast algorithm with the maximum number of children. The execution time of the whole binomial broadcast algorithm will be equal to the sum of the execution times of these stages.

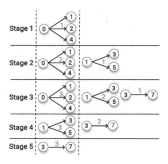

Fig. 3. Execution stages of the binomial tree broadcast algorithm, employing the non-blocking linear broadcast ($P = 8$, $n_s = 3$). Nodes are labelled by the process ranks. Each arrow represents transmission of a segment. The number over the arrow gives the index of the broadcast segment.

In the non-blocking linear broadcast algorithm, $P - 1$ non-blocking *send*s will run on the *root* concurrently. Therefore, the execution time of the linear broadcast algorithm using non-blocking point-to-point communications and buffered mode, $T_{linear\_bcast}^{nonblock}(P, m)$, can be bounded as follows:

$$T_{p2p}(m) \leq T_{linear\_bcast}^{nonblock}(P, m) \leq (P - 1) \cdot T_{p2p}(m). \tag{1}$$

We will approximate $T^{nonblock}_{linear\_bcast}(P, m)$ as

$$T^{nonblock}_{linear\_bcast}(P, m) = \gamma(P, m) \cdot (\alpha + m \cdot \beta), \qquad (2)$$

where

$$\gamma(P, m) = \frac{T^{nonblock}_{linear\_bcast}(P, m)}{T_{p2p}(m)}. \qquad (3)$$

In Open MPI, the binomial tree broadcast algorithm employs the balanced binomial tree virtual topology (Fig. 2). Therefore, the number of stages in the binomial broadcast algorithm can be calculated as

$$N_{steps} = \lfloor log_2 P \rfloor + n_s - 1. \qquad (4)$$

Thus, the time to complete the binomial tree broadcast algorithm can be estimated as follows:

$$T_{binomial\_bcast}(P, m, n_s) =$$

$$\sum_{i=1}^{\lfloor log_2 P \rfloor + n_s - 1} \max_{1 \leq j \leq \min(\lfloor log_2 P \rfloor, n_s)} T^{nonblock}_{linear\_bcast}(P_{ij}, \frac{m}{n_s}), \qquad (5)$$

where P_{ij} denotes the number of nodes in the j-th linear tree of the i-th stage. Using the property of the binomial tree and Formula 2, we have

$$T_{binomial\_bcast}(P, m, n_s) = (n_s \cdot \gamma(\lceil \log_2 P \rceil + 1)$$

$$+ \sum_{i=1}^{\lfloor \log_2 P \rfloor - 1} \gamma(\lceil \log_2 P \rceil - i + 1) - 1) \cdot (\alpha + \frac{m}{n_s} \cdot \beta). \qquad (6)$$

4   Estimation of Model Parameters

4.1   Estimation of $\gamma(P)$

The model parameter $\gamma(P)$ appears in the formula estimating the execution time of the linear tree broadcast algorithm with non-blocking communication, which is only used for broadcasting of a segment in the tree-based segmented broadcast algorithms. Thus, in the context of Open MPI, the linear tree broadcast algorithm with non-blocking communication will always broadcast a message of size m_s to a relatively small number of processes.

According to Formula 3,

$$\gamma(P) = \frac{T^{nonblock}_{linear\_bcast}(P, m_s)}{T_{p2p}(m_s)} = \frac{T^{nonblock}_{linear\_bcast}(P, m_s)}{T^{nonblock}_{linear\_bcast}(2, m_s)}.$$

Therefore, in order to estimate $\gamma(P)$ for a given range of the number of processes, $P \in \{2, ..., P_{max}\}$, we need a method for estimation of $T^{nonblock}_{linear\_bcast}(P, m_s)$. We use the following method:

– For each $2 \leq q \leq P_{max}$, we measure on the root the execution time $T_1(P, N)$ of N successive calls to the *linear tree with non-blocking communication* broadcast routine separated by barriers. The routine broadcasts a message of size m_s.

– We estimate $T_{linear\_bcast}^{nonblock}(P, m_s)$ as $T_2(P) = \frac{T_1(P,N)}{N}$.

The experimentally obtained discrete function $\frac{T_2(P)}{T_2(2)}$ is used as a platform-specific but algorithm-independent estimation of $\gamma(P)$.

From our experiments, we observed that the discrete estimation of $\gamma(P)$ is near linear. Therefore, as an alternative for platforms with very large numbers of processors, we can build by linear regression a linear approximation of the discrete function $\frac{T_2(P)}{T_2(2)}$, obtained for a representative subset of the full range of P, and use this linear approximation as an analytical estimation of $\gamma(P)$.

$$
\begin{cases}
(n_{s_1} \cdot \gamma(\lceil \log_2 P \rceil + 1) + \sum_{i=1}^{\lfloor \log_2 P \rfloor - 1} \gamma(\lceil \log_2 P \rceil - i + 1) - 1) \cdot (\alpha + \frac{m_1}{n_{s_1}} \cdot \beta) + (P-1) \cdot (\alpha + m_{g_1} \cdot \beta) = T_1 \\
(n_{s_2} \cdot \gamma(\lceil \log_2 P \rceil + 1) + \sum_{i=1}^{\lfloor \log_2 P \rfloor - 1} \gamma(\lceil \log_2 P \rceil - i + 1) - 1) \cdot (\alpha + \frac{m_2}{n_{s_2}} \cdot \beta) + (P-1) \cdot (\alpha + m_{g_2} \cdot \beta) = T_2 \\
\cdots \\
(n_{s_M} \cdot \gamma(\lceil \log_2 P \rceil + 1) + \sum_{i=1}^{\lfloor \log_2 P \rfloor - 1} \gamma(\lceil \log_2 P \rceil - i + 1) - 1) \cdot (\alpha + \frac{m_M}{n_{s_M}} \cdot \beta) + (P-1) \cdot (\alpha + m_{g_M} \cdot \beta) = T_M
\end{cases}
$$

$\Downarrow$

$$
\begin{cases}
\alpha + \beta \cdot \frac{(n_{s_1} \cdot \gamma(\lceil \log_2 P \rceil + 1) + \sum_{i=1}^{\lfloor \log_2 P \rfloor - 1} \gamma(\lceil \log_2 P \rceil - i + 1) - 1) \cdot \frac{m_1}{n_{s_1}} + (P-1) \cdot m_{g_1}}{(n_{s_1} \cdot \gamma(\lceil \log_2 P \rceil + 1) + \sum_{i=1}^{\lfloor \log_2 P \rfloor - 1} \gamma(\lceil \log_2 P \rceil - i + 1) - 1) + P - 1} = \frac{T_1}{\left((n_{s_1} \cdot \gamma(\lceil \log_2 P \rceil + 1) + \sum_{i=1}^{\lfloor \log_2 P \rfloor - 1} \gamma(\lceil \log_2 P \rceil - i + 1) - 1) + P - 1 \right)} \\
\alpha + \beta \cdot \frac{(n_{s_2} \cdot \gamma(\lceil \log_2 P \rceil + 1) + \sum_{i=1}^{\lfloor \log_2 P \rfloor - 1} \gamma(\lceil \log_2 P \rceil - i + 1) - 1) \cdot \frac{m_2}{n_{s_2}} + (P-1) \cdot m_{g_2}}{(n_{s_2} \cdot \gamma(\lceil \log_2 P \rceil + 1) + \sum_{i=1}^{\lfloor \log_2 P \rfloor - 1} \gamma(\lceil \log_2 P \rceil - i + 1) - 1) + P - 1} = \frac{T_2}{\left((n_{s_2} \cdot \gamma(\lceil \log_2 P \rceil + 1) + \sum_{i=1}^{\lfloor \log_2 P \rfloor - 1} \gamma(\lceil \log_2 P \rceil - i + 1) - 1) + P - 1 \right)} \\
\cdots \\
\alpha + \beta \cdot \frac{(n_{s_M} \cdot \gamma(\lceil \log_2 P \rceil + 1) + \sum_{i=1}^{\lfloor \log_2 P \rfloor - 1} \gamma(\lceil \log_2 P \rceil - i + 1) - 1) \cdot \frac{m_M}{n_{s_M}} + (P-1) \cdot m_{g_M}}{(n_{s_M} \cdot \gamma(\lceil \log_2 P \rceil + 1) + \sum_{i=1}^{\lfloor \log_2 P \rfloor - 1} \gamma(\lceil \log_2 P \rceil - i + 1) - 1) + P - 1} = \frac{T_M}{\left((n_{s_M} \cdot \gamma(\lceil \log_2 P \rceil + 1) + \sum_{i=1}^{\lfloor \log_2 P \rfloor - 1} \gamma(\lceil \log_2 P \rceil - i + 1) - 1) + P - 1 \right)}
\end{cases}
$$

Fig. 4. A system of M non-linear equations with α, β, $\gamma(\lceil \log_2 P \rceil + 1)$ and $\gamma(\lceil \log_2 P \rceil - i + 1)$ as unknowns, derived from M communication experiments, each consisting of the execution of the binomial tree broadcast algorithm, broadcasting a message of size m_i $(i = 1, ..., M)$ from the root to the remaining $P - 1$ processes, followed by the linear gather algorithm without synchronisation, gathering messages of size $m_{g_i} (m_{g_i} \neq m_s)$ on the root. The execution times, T_i, of these experiments are measured on the root. Given $\gamma(\lceil \log_2 P \rceil + 1)$ and $\gamma(\lceil \log_2 P \rceil - i + 1)$ are evaluated separately, the system becomes a system of M linear equations with α and β as unknowns.

4.2 Estimation of Algorithm Specific α and β

To estimate the model parameters α and β for a given collective algorithm, we design a communication experiment, which starts and finishes on the root (in order to accurately measure its execution time using the root clock), and involves the execution of the modelled collective algorithm so that the total time of the experiment would be dominated by the time of its execution.

For example, for all broadcast algorithms, the communication experiment consists of a broadcast of a message of size m (where m is a multiple of segment size m_s), using the modelled broadcast algorithm, followed by a *linear-without-synchronisation* gather algorithm, gathering messages of size $m_g (m_g \neq m_s)$ on the root. The execution time of this experiment on P nodes, $T_{bcast\_exp}(P, m)$, can be estimated as follows:

$$T_{bcast\_exp}(P, m) = T_{bcast\_alg}(P, m) + T_{linear\_gather}(P, m_g) \qquad (7)$$

The execution time of the linear-without-synchronisation gather algorithm, gathering a message size of m_g on the root from $\mathrm{P} - 1$ processes where $m_g \neq m_s$, is estimated as follows,

$$T_{linear\_gather}(P, m_g) = (P - 1) \cdot (\alpha + m_g \cdot \beta) \qquad (8)$$

Using Formula 6 and 8 for each combination of P and m this experiment will yield one linear equation with α and β as unknowns. By repeating this experiment with different p and m, we obtain a system of linear equations for α and β. Each equation in this system can be represented in the canonical form, $\alpha + \beta \times m_i = T_i$ $(i = 1, ..., M)$. Finally, we use the least-square regression to find α and β, giving us the best linear approximation $\alpha + \beta \times m$ of the discrete function $f(m_i) = T_i$ $(i = 1, ..., M)$.

Figure 4 shows a system of linear equations built for the binomial tree broadcast algorithm for our experimental platform. To build this system, we used the same P nodes in all experiments but varied the message size $m \in \{m_1, ..., m_M\}$ and $m_g \in \{m_{g_1}, ..., m_{g_M}\}$. With M different message sizes, we obtained a system of M equations. The number of nodes, P, was approximately equal to the half of the total number of nodes. We observed that the use of larger numbers of nodes in the experiments will not change the estimation of α and β.

5 Experimental Results and Analysis

This section presents experimental evaluation of the proposed approach to selection of optimal collective algorithms using Open MPI broadcast operation. In all experiments. We use the default Open MPI configuration (without any collective optimization tuning).

5.1 Experiment Setup

For experiments, we use Open MPI 3.1 running on a dedicated Grisou and Gros clusters of the Nancy site of the Grid'5000 infrastructure [23]. The Grisou cluster consists of 51 nodes each with 2 Intel Xeon E5-2630 v3 CPUs (8 cores/CPU), 128 GB RAM, 2x558 GB HDD, interconnected via 10 Gbps Ethernet. The Gros cluster consists of 124 nodes each with Intel Xeon Gold 5220 (18 cores/CPU), 96 GB RAM, 894 GB SSD, interconnected via 2×25 Gb Ethernet.

To make sure that the experimental results are reliable, we follow a detailed methodology: 1) We make sure that the cluster is fully reserved and dedicated to our experiments. 2) For each data point in the execution time of collective algorithms, the sample mean is used, which is calculated by executing the application repeatedly until the sample mean lies in the 95% confidence interval and a precision of 0.025 (2.5%) has been achieved. We also check that the individual observations are independent and their population follows the normal distribution. For this purpose, MPIBlib [24] is used.

In our communication experiments, MPI programs use the one-process-per-CPU configuration, and the maximal total number of processes is equal to 90 on Grisou and 124 on Gros clusters. The message segment size, m_s, for segmented broadcast algorithms is set to 8 KB and is the same in all experiments. This segment size is commonly used for segmented broadcast algorithms in Open MPI. Selection of optimal segment size is out of the scope of this paper.

5.2 Experimental Estimation of Model Parameters

First of all, we would like to stress again that we estimate model parameters for each cluster separately.

Estimation of parameter $\gamma(p)$ for our experimental platforms follows the method presented in Sect. 4.1. With the maximal number of processes equal to 90 (Grisou) and 124 (Gros), the maximal number of children in the linear tree broadcast algorithm with non-blocking communication, used in the segmented Open MPI broadcast algorithms, will be equal to seven. Therefore, the number of processes in our communication experiments ranges from 2 to 7 for both clusters. By definition, $\gamma(2) = 1$. The estimated values of $\gamma(p)$ for p from 3 to 7 are given in Table 1.

Table 1. Estimated values of $\gamma(P)$ on Grisou and Gros clusters.

P	$\gamma(P)$	
	Grisou	Gros
3	1.114	1.084
4	1.219	1.17
5	1.283	1.254
6	1.451	1.339
7	1.540	1.424

After estimation of $\gamma(p)$, we conduct communication experiments to estimate algorithm-specific values of parameters α and β for six broadcast algorithms following the method described in Sect. 4.2. In these experiments we use 40 processes on Grisou and 124 on Gros. The message size, m, varies in the range from 8 KB to 4 MB in the broadcast experiments. We use 10 different sizes for broadcast algorithms, $\{m_i\}_{i=1}^{10}$, separated by a constant step in the logarithmic scale, $\log m_{i-1} - \log m_i = const$. Thus, for each collective algorithm, we obtain a system of 10 linear equations with α and β as unknowns. We use the Huber regressor [25] to find their values from the system.

The values of parameters α and β obtained this way can be found in Table 2. We can see that the values of α and β do vary depending on the collective algorithm, and the difference is more significant between algorithms implementing different collective operations. The results support our original hypothesis that the average execution time of a point-to-point communication will very much depend on the context of the use of the point-to-point communications in the algorithm. Therefore, the estimated values of the α and β capture more than just sheer network characteristics. One interesting example is the Split-binary tree and Binary tree broadcast algorithms. They both use the same virtual topology, but the estimated time of a point-to-point communication, $\alpha + \beta \times m$, is smaller in the context of the Split-binary one. This can be explained by a higher level of parallelism of the Split-binary algorithm, where a significant part of point-to-point communications is performed in parallel by a large number of independent pairs of processes from the left and right subtrees.

Table 2. Estimated values of α and β for the Grisou and Gros clusters and Open MPI broadcast algorithms.

Collective algorithm	$\alpha(sec)$	$\beta\left(\frac{sec}{byte}\right)$	Collective algorithm	$\alpha(sec)$	$\beta\left(\frac{sec}{byte}\right)$
Broadcast			Broadcast		
Linear tree	2.2×10^{-12}	1.8×10^{-8}	Linear tree	1.4×10^{-12}	1.1×10^{-8}
K-Chain tree	5.7×10^{-13}	4.7×10^{-9}	K-Chain tree	5.4×10^{-13}	4.5×10^{-9}
Chain tree	6.1×10^{-13}	4.9×10^{-9}	Chain tree	4.7×10^{-12}	3.8×10^{-8}
Split-binary tree	3.7×10^{-13}	3.6×10^{-9}	Split-binary tree	5.5×10^{-13}	4.5×10^{-9}
Binary tree	5.8×10^{-13}	4.7×10^{-9}	Binary tree	5.8×10^{-13}	4.7×10^{-9}
Binomial tree	5.8×10^{-13}	4.8×10^{-9}	Binomial tree	1.2×10^{-13}	1.0×10^{-9}

5.3 Accuracy of Selection of Optimal Collective Algorithms Using the Constructed Analytical Performance Models

The constructed analytical performance models of the Open MPI broadcast collective algorithms are designed for the use in the MPI_Bcast routines for runtime selection of the optimal algorithm, depending on the number of processes and the message size. While the efficiency of the selection procedure is evident from the low complexity of the analytical formulas derived in Sect. 3, the experimental results on the accuracy are presented in this section.

Figure 5 shows the results of our experiments for MPI_Bcast. We present results of experiments with 50, 80 and 90 processes on Grisou, and 80, 100 and 124 on Gros. The message size, m, varies in the range from 8 KB to 4 MB in the broadcast experiments. We use 10 different sizes for broadcast algorithms, $\{m_i\}_{i=1}^{10}$, separated by a constant step in the logarithmic scale, $\log m_{i-1} - \log m_i = const$. The graphs show the execution time of the collective operation as a function of message size. Each data point on a blue line shows the performance of the algorithm selected by the Open MPI decision function for the given operation, number of processes and message size. Each point on a red line shows the performance of the algorithm selected by our decision function, which uses the constructed analytical models. Each point on a green line shows the performance of the best Open MPI algorithm for the given collective operation, number of processes and message size.

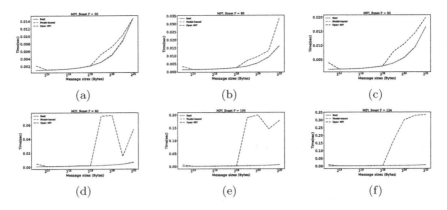

(a)   (b)   (c)

(d)   (e)   (f)

Fig. 5. Comparison of the selection accuracy of the Open MPI decision function and the proposed model-based method for MPI_Bcast. (a–c) and (d–f) present performance of collectives on Grisou and Gros clusters respectively.

Table 3. Comparison of the model-based and Open MPI selections with the best performing MPI_Bcast algorithm. For each selected algorithm, its performance degradation against the optimal one is given in braces.

<table>
<tr><th colspan="4">P=90, MPI_Bcast, Grisou</th><th colspan="4">P=100, MPI_Bcast, Gros</th></tr>
<tr><th>m (KB)</th><th>Best</th><th>Model-based (%)</th><th>Open MPI (%)</th><th>m (KB)</th><th>Best</th><th>Model-based (%)</th><th>Open MPI (%)</th></tr>
<tr><td>8</td><td>binomial</td><td>binary (3)</td><td>split_binary (160)</td><td>8</td><td>binary</td><td>binomial (3)</td><td>split_binary (549)</td></tr>
<tr><td>16</td><td>binary</td><td>binary (0)</td><td>split_binary (1)</td><td>16</td><td>binomial</td><td>binomial (0)</td><td>split_binary (32)</td></tr>
<tr><td>32</td><td>binary</td><td>binary (0)</td><td>split_binary (0)</td><td>32</td><td>binomial</td><td>binomial (0)</td><td>split_binary (3)</td></tr>
<tr><td>64</td><td>split_binary</td><td>binary (1)</td><td>split_binary (0)</td><td>64</td><td>split_binary</td><td>binomial (8)</td><td>split_binary (0)</td></tr>
<tr><td>128</td><td>binary</td><td>binary (0)</td><td>split_binary (1)</td><td>128</td><td>split_binary</td><td>binomial (8)</td><td>split_binary (0)</td></tr>
<tr><td>256</td><td>split_binary</td><td>binary (2)</td><td>split_binary (0)</td><td>256</td><td>binary</td><td>binary (0)</td><td>split_binary (6)</td></tr>
<tr><td>512</td><td>split_binary</td><td>binary (2)</td><td>chain (111)</td><td>512</td><td>binary</td><td>binary (0)</td><td>chain (7297)</td></tr>
<tr><td>1024</td><td>split_binary</td><td>binary (3)</td><td>chain (88)</td><td>1024</td><td>split_binary</td><td>binary (7)</td><td>chain (6094)</td></tr>
<tr><td>2048</td><td>split_binary</td><td>binary (2)</td><td>chain (55)</td><td>2048</td><td>split_binary</td><td>binary (4)</td><td>chain (3227)</td></tr>
<tr><td>4096</td><td>split_binary</td><td>binary (1)</td><td>chain (20)</td><td>4096</td><td>split_binary</td><td>binary (9)</td><td>chain (2568)</td></tr>
</table>

Table 3 presents selections made for MPI_Bcast using the proposed model-based runtime procedure and the Open MPI decision function. For each message size m, the best performing algorithm, the model-based selected algorithm, and the Open MPI selected algorithm are given. For the latter two, the performance degradation in percents in comparison with the best performing algorithm is also given. We can see that for the Grisou cluster, the model-based selection either pick the best performing algorithm, or the algorithm, the performance of which deviates from the best no more than 3%. Given the accuracy of measurements, this means that the model-based selection is practically always optimal as the performance of the selected algorithm is indistinguishable from the best performance. The Open MPI selection is near optimal in 50% cases and causes significant, up to 160%, degradation in the remaining cases. For the Gros cluster, the model-based selection picks either the best performing algorithm or the algorithm with near optimal performance, no worse than 10% in comparison with the best performing algorithm. At the same time, while near optimal in

40% cases, the algorithms selected by the Open MPI demonstrate catastrophic degradation (up to 7297%) in 50% cases.

The Open MPI decision functions select the algorithm depending on the message size and the number of processes. For example, the Open MPI broadcast decision function selects the chain broadcast algorithm for large message sizes. However, from Table 3 it is evident that chain broadcast algorithm is not the best performing algorithm for large message sizes on both clusters. From the same table, one can see that the model-based selection procedure accurately picks the best performing binomial tree broadcast algorithm for 16 KB and 32 KB message sizes on the Gros cluster, where Open MPI only selects the binomial tree algorithm for broadcasting messages smaller than 2 KB.

6   Conclusions

In this paper, we proposed a novel model-based approach to automatic selection of optimal algorithms for MPI collective operations, which proved to be both efficient and accurate. The novelty of the approach is two-fold. First, we proposed to derive analytical models of collective algorithms from the code of their implementation rather than from high-level mathematical definitions. Second, we proposed to estimate model parameters separately for each algorithm, using a communication experiment, where the execution of the algorithm itself dominates the execution time of the experiment.

We also developed this approach into a detailed method and applied it to Open MPI 3.1 and its MPI_Bcast. We experimentally validated this method on two different clusters and demonstrated its accuracy and efficiency. These results suggest that the proposed approach, based on analytical performance modelling of collective algorithms, can be successful in the solution of the problem of accurate and efficient runtime selection of optimal algorithms for MPI collective operations.

References

1. A Message-Passing Interface Standard. https://www.mpi-forum.org/. Accessed 8 Mar 2021
2. Rabenseifner, R.: Automatic profiling of MPI applications with hardware performance counters. In: Dongarra, J., Luque, E., Margalef, T. (eds.) EuroPVM/MPI 1999. LNCS, vol. 1697, pp. 35–42. Springer, Heidelberg (1999). https://doi.org/10.1007/3-540-48158-3_5
3. Open MPI: Open Source High Performance Computing. https://www.open-mpi.org/. Accessed 8 Mar 2021
4. MPICH - A Portable Implementation of MPI. http://www.mpich.org/. Accessed 8 Mar 2021
5. Thakur, R., Rabenseifner, R., Gropp, W.: Optimization of collective communication operations in MPICH. Int. J. High Perform. Comput. Appl. **19**(1), 49–66 (2005)

6. Gabriel, E., et al.: Open MPI: goals, concept, and design of a next generation MPI implementation. In: Kranzlmüller, D., Kacsuk, P., Dongarra, J. (eds.) EuroPVM/MPI 2004. LNCS, vol. 3241, pp. 97–104. Springer, Heidelberg (2004). https://doi.org/10.1007/978-3-540-30218-6_19
7. Fagg, G.E., Pjesivac-Grbovic, J., Bosilca, G., Angskun, T., Dongarra, J., Jeannot, E.: Flexible collective communication tuning architecture applied to Open MPI. In: Euro PVM/MPI (2006)
8. Pješivac-Grbović, J., Angskun, T., Bosilca, G., Fagg, G.E., Gabriel, E., Dongarra, J.J.: Performance analysis of MPI collective operations. Clust. Comput. **10**(2), 127–143 (2007)
9. Hockney, R.W.: The communication challenge for MPP: Intel Paragon and Meiko CS-2. Parallel Comput. **20**(3), 389–398 (1994)
10. Chan, E.W., Heimlich, M.F., Purkayastha, A., van de Geijn, R.A.: On optimizing collective communication. In: IEEE International Conference on Cluster Computing 2004, pp. 145–155 (2004)
11. Chan, E., Heimlich, M., Purkayastha, A., van de Geijn, R.: Collective communication: theory, practice, and experience: research articles. Concurr. Comput. Pract. Exper. **19**(13), 1749–1783 (2007)
12. Culler, D., Liu, L.T., Martin, R.P., Yoshikawa, C.: LogP performance assessment of fast network interfaces. IEEE Micro **16**(1), 35–43 (1996)
13. Kielmann, T., Bal, H.E., Verstoep, K.: Fast measurement of LogP parameters for message passing platforms. In: Rolim, J. (ed.) IPDPS 2000. LNCS, vol. 1800, pp. 1176–1183. Springer, Heidelberg (2000). https://doi.org/10.1007/3-540-45591-4_162
14. Lastovetsky, A., Rychkov, V.: Building the communication performance model of heterogeneous clusters based on a switched network. In: IEEE International Conference on Cluster Computing 2007, pp. 568–575 (2007)
15. Lastovetsky, A., Rychkov, V.: Accurate and efficient estimation of parameters of heterogeneous communication performance models. Int. J. High Perform. Comput. Appl. **23**(2), 123–139 (2009)
16. Lastovetsky, A., Rychkov, V., O'Flynn, M.: Accurate heterogeneous communication models and a software tool for their efficient estimation. Int. J. High Perform. Comput. Appl. **24**(1), 34–48 (2010)
17. Rico-Gallego, J.A., Díaz-Martín, J.C., Manumachu, R.R., Lastovetsky, A.L.: A survey of communication performance models for high-performance computing. ACM Comput. Surv. **51**(6), 1–36 (2019)
18. Pješivac–Grbović, J., Fagg, G.E., Angskun, T., Bosilca, G., Dongarra, J.J.: MPI collective algorithm selection and quadtree encoding. In: Mohr, B., Träff, J.L., Worringen, J., Dongarra, J. (eds.) EuroPVM/MPI 2006. LNCS, vol. 4192, pp. 40–48. Springer, Heidelberg (2006). https://doi.org/10.1007/11846802_14
19. Pješivac-Grbović, J., Bosilca, G., Fagg, G.E., Angskun, T., Dongarra, J.J.: Decision trees and MPI collective algorithm selection problem. In: Kermarrec, A.-M., Bougé, L., Priol, T. (eds.) Euro-Par 2007. LNCS, vol. 4641, pp. 107–117. Springer, Heidelberg (2007). https://doi.org/10.1007/978-3-540-74466-5_13
20. Quinlan, J.R.: C4.5: Programs for Machine Learning. Morgan Kaufmann Publishers Inc., Burlington (1993)
21. Hunold, S., Bhatele, A., Bosilca, G., Knees, P.: Predicting MPI collective communication performance using machine learning. In: IEEE International Conference on Cluster Computing 2020, pp. 259–269 (2020)
22. Wickramasinghe, U., Lumsdaine, A.: A survey of methods for collective communication optimization and tuning. arXiv preprint arXiv:1611.06334 (2016)

23. Grid5000. http://www.grid5000.fr. Accessed 8 Mar 2021
24. Lastovetsky, A., Rychkov, V., O'Flynn, M.: MPIBlib: benchmarking MPI communications for parallel computing on homogeneous and heterogeneous clusters. In: Lastovetsky, A., Kechadi, T., Dongarra, J. (eds.) EuroPVM/MPI 2008. LNCS, vol. 5205, pp. 227–238. Springer, Heidelberg (2008). https://doi.org/10.1007/978-3-540-87475-1_32
25. Huber, P.J.: Robust estimation of a location parameter. In: Kotz, S., Johnson, N.L. (eds.) Breakthroughs in Statistics. Springer Series in Statistics (Perspectives in Statistics), pp. 492–518. Springer, New York (1992). https://doi.org/10.1007/978-1-4612-4380-9_35

Deterministic OpenMP and the LBP Parallelizing Manycore Processor

Bernard Goossens[1,2](✉), Kenelm Louetsi[1,2], and David Parello[1,2]

[1] LIRMM, Univ Montpellier, Montpellier, France
{bernard.goossens,kenelm.louetsi,david.parello}@lirmm.fr
[2] DALI, Univ Perpignan, Perpignan, France
{bernard.goossens,kenelm.louetsi,david.parello}@univ-perp.fr
http://www.lirmm.fr,
http://www.univ-perp.fr/

Abstract. Multicore processors are becoming standard called as COTS (Commercial Off The Shelf) processors but can not be fully used in the context of critical real time systems. OpenMP is one of the most used programming models to build parallel programs able to exploit such multicore processors. A lot of work try to tackle the issue of the determinism of parallel programming models. The critical real time system face an *unpredictability* wall of parallel programs. This paper presents Deterministic OpenMP, a new runtime for OpenMP programs, and the Little Big Processor (LBP) manycore processor design. Their aim is to help to solve the non determinism problem at the programming level but also at the execution level. When run on LBP, a Deterministic OpenMP code produces cycle by cycle deterministic computations. LBP and Deterministic OpenMP are particularly suited to safely accelerate real time embedded applications through their parallel execution.

1 Introduction

Multicore processor is becoming the standard for embedded applications and also for real-time critical applications. OpenMP [1,2] is a reference and widely used to parallelize applications on such microarchitectures. OpenMP is based on well known operating system multithreading primitives such as Pthreads [3]. A lot of recent works have been done to control execution time on such microarchitectures [4,5]. Recently, great efforts aimed at reducing the overhead of such OS kernel threads with user light-weight threads [6,7].

Non-determinism makes parallel programs hard to debug as a bug may be non repeatable and the action of a debugger may alter the run in a way which eliminates the emergence of the bug.

Effort have been done to build a deterministic parallel programming model as in DOMP [8] based on a subset of OpenMP API. Even if these works mainly deal with the non determinism of the results at the programming model level, the runtime determinism is still an important issue for critical real-time systems.

© Springer Nature Switzerland AG 2021
V. Malyshkin (Ed.): PaCT 2021, LNCS 12942, pp. 26–40, 2021.
https://doi.org/10.1007/978-3-030-86359-3_3

One side effect is that parallelization can hardly benefit to real time critical applications [9] as a precise timing cannot be ensured.

In this paper, we introduce Deterministic OpenMP, Parallel Instruction Set Computer (PISC) and the Little Big Processor (LBP). They were designed to ensure at the same time, determinism at the programming model level but also at the execution level. From Deterministic OpenMP source we build self parallelizing binary programs and we run them on a bare-metal parallel multicore hardware, i.e. with no operating system to manage.

The rest of the paper is organized as follows: Sect. 2 presents Deterministic OpenMP, Sect. 3 introduces PISC instruction set extension, Sect. 4 shows LBP microarchitecture, Sect. 5 presents some experimental results and Sect. 6 shows the conclusions.

2 Deterministic OpenMP

```
#include <det_omp.h>
#define NUM_HART 8
void thread(/*...*/){
  /*... (1);*/
}
void main(){
  int t;
  omp_set_num_threads(NUM_HART);
  #pragma omp parallel for
  for (t=0; t<NUM_HART; t++)
    thread(/*...*/);
  /*... (2);*/
}
```

(a) Source program.

```
#define NUM_HART 8
#define HART_PER_CORE 4
unsigned omp_num_threads;
typedef struct type_s{int t; /*...*/} type_t;
type_t st;
void thread ( void *arg ){
  type_t *pt=(type_t *)arg;
  /*...(1);*/
}
static inline void
fork_on_current(void(*f)(void*), void *data){/*p_fc(data);*/}
static inline void
fork_on_next(void(*f)(void*), void *data){/*p_fn(data);*/}
void LBP_parallel_start(void(*f)(void*), void *data){
  type_t *pt=(type_t *)data;
  unsigned nt=omp_num_threads, h, t;
  for (t=0; t<nt-1; t++){
    h=t%HART_PER_CORE;
    pt->t=t;
    if (h<HART_PER_CORE-1) fork_on_current(f, data);
    else fork_on_next(f, data);
  }
  pt->t=nt-1;
  f(data);
}
void main(){
  omp_num_threads=NUM_HART;
  LBP_parallel_start( thread , (void *)&st);
  /*...(2);*/
}
```

(b) Transformed program.

Fig. 1. The Deterministic OpenMP program transformation.

A Deterministic OpenMP program is quite not distinguishable from a classic OpenMP one [1]. Figure 1a shows an example of a Deterministic OpenMP code to distribute and parallelize a *thread* function on a set of eight harts. A hart is a hardware thread as defined into the riscv specification [10]. The difference with a pure OpenMP version lies in the header file inclusion (*det_omp.h* instead of *omp.h*, in red on the figure).

In a classic OpenMP implementation, the *parallel for* pragma builds a team of OMP_NUM_THREADS threads which the OS maps on the available harts, optionally balancing their loads. In the Gnu implementation, this is done through the *GOMP_parallel* function (OMP API [2]). In Deterministic OpenMP, a team of harts -not threads- is created, each matching a unique and constant placement on the processor. One drawback is that on LBP, load balancing is the problem of the programmer. It is her responsability to evenly divide her job into parallel tasks. If properly done, the efficiency is improved compared to a dynamic load balancing handled by the OS because dynamic load balancing implies costly thread migrations.

The Deterministic OpenMP code in Fig. 1a is translated into the code in Fig. 1b. The text in black on the figure comes from the original OpenMP source code on Fig. 1a. The green text is added by the translator (i.e., the compiler).

The *LBP_parallel_start* function creates and starts the team of harts. It organizes the distribution of the copies of function *thread* on the harts. It calls *fork_on_current* which creates a new hart on the current core or *fork_on_next* which creates a new hart on the next core (LBP cores are ordered). The machine code for *fork_on_current* is given in Sect. 3. The *LBP_parallel_start* function fills the harts available in a core before expanding to the next one.

Functions *fork_on_current* and *fork_on_next* do not interact with the OS by calling a forking or cloning system call. Instead, they directly use the hardware capability to fork the fetch point, running the *thread* function locally and the continuation remotely (on the same core or on the next one).

The hardware fork mechanism has two advantages over the classic OS one:

– it concatenates the continuation thread to the creating one in the sequential referential order, on which the hardware is able to synchronize and connect producers and consumers,
– it places the continuation thread on a fixed hart, in the same or next core.

In a classic OpenMP run of the code on Fig. 1a, all the function *thread* copies would become non-ordered and independent threads. In contrast in Deterministic OpenMP, the created harts are ordered (in the iterations order) which simplifies communications: a creating hart sends continuation values to the created one through direct core-to-core links.

LBP offers the programmer the possibility to map her code and data on the computing resources according to the application structure. A producing function can be parallelized on the same set of cores and harts than the consuming one, eliminating any non local memory access. The OS is not able to do the same for OpenMP runs as it has no knowledge of which thread produces and which thread consumes. The OS can only act on load balancing. The task of good mapping in classic OpenMP is the programmer's duty. To do her job properly, she has to deal with her application, but also with the OS and the computing and memory resources (e.g. load balancing, pagination). This leads to difficult decisions, with a complexity proportional (if not worse) to the number of cores.

The next section describes the PISC ISA extension which is an extension of the RISCV ISA (Instruction Set Architecture) [10] needed to implement the Deterministic OpenMP library.

3   The PISC ISA

syntax	semantic
p_lwcv rd, offset	restore rd from local stack at offset
p_swcv rs1, rs2, offset	save rs2 on rs1 hart stack at offset (allocated hart)
p_lwre rd, offset	restore rd from local result buffer number offset
p_swre rs1, rs2, offset	save rs2 to rs1 hart (any prior hart) result buffer number offset
p_jal rd, rs1, offset	send pc+4 to rs1 hart (allocated hart) clear rd goto pc+offset

p_jalr zero, rs1, rs2
(p_ret)
	if rs1==0 && rs2==−1: exit
	if rs1==0 && rs2!=current hart: end current hart
	                   send ending hart signal to next hart
	if rs1==0 && rs2==current hart: keep current hart waiting
	                   send ending hart signal to next hart
	if rs1!=0: send rs1 to rs2 hart (join hart)

p_jalr rd, rs1, rs2	send pc+4 to rs2 hart (allocated hart) clear rd goto rs1
p_merge rd, rs1, rs2	rd=(rs1&0x7fff0000) | (rs2&0x0000ffff)
p_fc rd	allocate a free pc on current core (fork) rd=(4*c+allocated hart)
p_fn rd	allocate a free pc on next core (fork) rd=(4*(c+1)+allocated hart)
p_syncm	stop fetch until all decoded memory accesses in local hart are run
p_set rd, rs1	rd=(rs1&0x0000ffff) | ((4*core+hart)<<16) | 0x80000000

Fig. 2. The X_PAR RISCV PISC ISA extension.

The PISC ISA extension is a set of 12 new machine instructions. A RISCV extension named X_PAR has been defined. It is summarized on Fig. 2.

The *p_swcv* and *p_lwcv* instructions (*cv* stands for *continuation value*) serve to achieve a hardware synchronized communication between a producer and a consumer of the same team, for example to transmit an input argument from member to member (e.g. the iteration loop index). The consumer should be the hart next after the producer (same or next core).

The *p_swre* and *p_lwre* instructions (*re* stands for *result*) serve to achieve a hardware synchronized communication between a producer and a consumer of different teams, with the consumer physically preceding the producer (same or preceding core; the connection used to transmit the value is the intercore backward link). They allow for example a team to produce a reduction value and have its last member send it to the join hart.

The *p_jal* instruction is a parallelized call. Instead of pushing the return address on the stack, it sends it to an allocated continuation hart. The *p_jalr* instruction is the indirect variant of the *p_jal* one. It can also be used as a return from a parallelized hart (pseudo instruction *p_ret* standing for *p_jalr zero, rs1, rs2*).

The *p_fc* and *p_fn* instructions serve to allocate a new hart, either on the same core or on the next one.

The *p_merge* and *p_set* are instructions used to manipulate hart identities. They are used to prepare and propagate the first team member identity to allow the join from the last team member back to the first.

The *p_syncm* instruction serves to synchronize memory accesses within a hart. In a hart, loads and stores are unordered. The hardware provides no control on the out-of-order behaviour of loads and stores. For example, to ensure a load depending on a store is run after it, a *p_syncm* instruction should be inserted between them. The *p_syncm* acts by blocking the fetch (as soon as it is decoded) until all the in flight memory accesses of the hart are done.

More details on PISC can be found at URL [11].

```
main:  li      t0,-1        #t0 = exit code
       addi    sp,sp,-8     #allocate two words on local stack
       sw      ra,0(sp)     #save reg. ra on local stack, offset 0
       sw      t0,4(sp)     #save reg. t0 on local stack, offset 4
       p_set   t0           #t0 = 4*core+hart (current hart identity)
       li      a0,thread    #a0 = thread function pointer
       li      a1,data      #a1 = pointer on data structure
       jal     LBP_parallel_start
rp:    /*...(2)*/
       lw      ra,0(sp)     #restore ra from local stack, offset 0
       lw      t0,4(sp)     #restore t0 from local stack, offset 4
       addi    sp,sp,8      #free two words on local stack
       p_ret                #ra==0 && t0==-1 => exit
```

Fig. 3. The PISC RISCV code for the *main* function.

Figures 3, 4 and 5 show the machine instructions compiled for the Deterministic OpenMP code on Fig. 1b. The target processor is assumed to be bare-metal (no OS). The LBP processor implemented in the FPGA directly starts running the *main* function and stops when the *p_ret* instruction is met (with register *ra*=0 and register *t0*=-1, meaning exit).

Figure 3 shows the compiled PISC RISCV code for the *main* function. Registers *ra* and *t0* play a special role. Register *ra* has its normal usage: it holds the return address. When *LBP_parallel_start* is called, *ra* receives the future team join address (labeled *rp* on the figure). Register *t0* holds the current hart number set with the *p_set* instruction and propagated along the team members through register transmission (*t0* contains a value combining the hosting core and the current hart identity in the core). The *LBP_parallel_start* function creates the team of harts to run the parallelized loop. It returns to *rp* label when a *p_ret* instruction in the last created team member is reached. The ending hart sends *ra* to hart *t0* (i.e. core 0, hart 0 in the example), which resumes the run at *rp* label.

There are four types of team member endings (continuation after a *p_ret* instruction) according to the received *ra* and *t0*:

1. *ra* is null and *t0* is not the current hart: the hart ends,
2. *ra* is null and *t0* is the current hart: the hart waits for a join,
3. *ra* is null and *t0* is −1: the process exits,
4. *ra* is not null: the hart ends and sends *ra* to the *t0* hart which restarts fetch (the parallel section ends and is continued by a sequential one).

The *p_ret* instructions are committed in-order (in the sequential referential order) to implement a hardware separation barrier between a team of concurrent harts and the following sequential OpenMP section. The barrier is implemented as a hardware signal transmitted from hart to hart along the team members. Hence, the harts are released in order (each hart commits its *p_ret* only when it has received the *ending hart* signal from its predecessor; it sends its own *ending hart* signal to its successor after this commit).

Figure 4 shows how function *LBP_parallel_start* calls the last occurrence of function *thread* with a RISCV *jalr a0* indirect call instruction. This last call is run by the last created team member on the last allocated hart (i.e. no fork). The same hart runs the code after the return point at *rp2* label. The ending *p_ret* instruction joins with the following sequential ending part of *main* (team member ending type 4 with *ra* being not null).

```
         addi    sp,sp,-8     #allocate two words on local stack
         sw      ra,0(sp)     #save reg. ra on stack, offset 0
         sw      t0,4(sp)     #save reg. t0 on stack, offset 4
         p_set   t0           #t0 = 4*core+hart (current hart identity)
         jalr    a0           #a0 is the pointer on function thread
rp2:     lw      ra,0(sp)     #restore reg. ra from stack, offset 0
         lw      t0,4(sp)     #restore reg. t0 from stack, offset 4
         addi    sp,sp,8      #free two words on local stack
         p_ret                #ra!=0 => end and send ra to t0 hart
```

Fig. 4. The PISC RISCV code for the end of the *LBP_parallel_start* function.

The *fork_on_current* function called in *LBP_parallel_start* (Fig. 1b) is a fork protocol composed of the machine instructions presented on Fig. 5. The code allocates a new hart on the current core (*p_fc* X_PAR machine instruction; the allocated hart identity is saved to the destination register *t6*), sends registers to the allocated hart (three *p_swcv* X_PAR instructions; *a1* holds a pointer on the *data* argument), starts the new hart (*p_jalr* X_PAR instruction; *a0* holds the *thread* address) which receives the transmitted registers (three matching *p_lwcv* X_PAR instructions; the join address is restored from stack to *ra*, the join core/hart to *t0* and the data pointer to *a1*).

```
p_fc        t6          #t6 = allocated hart number (4*core+hart)
p_swcv      ra,t6,0     #save ra on t6 hart stack, off. 0
p_swcv      t0,t6,4     #save t0 on t6 hart stack, off. 4
p_swcv      a1,t6,8     #save a1 on t6 hart stack, off. 8 (data)
p_merge     t0,t0,t6    #merge reg. t0 and t6 into t0
p_syncm                 #block fetch until mem. accesses are done
p_jalr      ra,t0,a0    #call thread locally, start pc+4 remotely
p_lwcv      ra,0        #restore ra from stack, off. 0
p_lwcv      t0,4        #restore t0 from stack, off. 4
p_lwcv      a1,8        #restore a1 from stack, off. 8 (data)
```

Fig. 5. The PISC RISCV forking protocol. (Color figure online)

The seven first instructions (in red; down to the *p_jalr*) are run by the forking hart and the three last ones (in blue) are run by the forked hart. The *p_jalr* instruction calls the *thread* function on the local hart and sends the return address[1] to the allocated hart, which starts fetching at the next cycle. After the *p_jalr* instruction has been issued, the function called and the code after return are run in parallel by two different harts.

The *p_syncm* X_PAR instruction synchronizes the send/receive transmission protocol (*p_swcv* and *p_lwcv* pairs). The sending hart is blocked until all memory writes are done (*ra*, *t0* and *a1* registers saved on the allocated hart stack). The allocated hart starts only when its arguments have been copied on its stack by the allocating hart.

This new PISC ISA extension is strongly connected to its microarchitecture implementation named LBP to ensure the execution time and cycle by cycle determinism. The next section presents the LBP processor implementing X_PAR.

4   The LBP Parallelizing Processor

4.1   The Cores

Figure 6 shows the general structure of the 64-core LBP processor as it is implemented on the FPGA[2]. Cores are represented by blue squares labeled *c*. Links between cores are represented by magenta and blue arrows. Dashed lines represent optional extensions, either to have a larger manycore or to connect multiple LBP chips. There are 64 ordered cores. The first core in order (core 0) is surrounded by a red circle (top of the figure). Its successor is just aside, on the left. The last core (core 63) is surrounded by a green circle (at the right of core 0). The line of cores has a serpentine shape. The last core is not connected to the first one. Hence, teams may only expand along successive cores until the last one, no further.

Each core is directly connected to its successor (blue arrow). Each core is indirectly connected to any predecessor through a unidirectional line (magenta arrows). The direct connections (blue arrows) are used to allocate harts (fork

---

[1] *pc+4* points on the *p_lwcv* instruction just following the *p_jalr* one.

[2] The FPGA implementation uses a small FPGA which limits the processor to 8 cores.

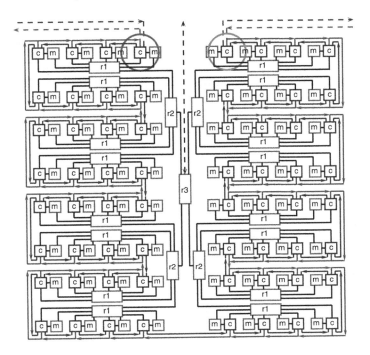

Fig. 6. The LBP processor.

with p_fc or p_fn), send continuation values (p_swcv) and propagate *ending hart* signals (p_ret). The backward line (magenta arrows) is used to send join addresses (p_ret), function results and reduction values (send a result with p_swre).

4.2   The Pipeline

Figure 7 shows the LBP core pipeline. It has a classic five stages out-of-order organization. Each stage selects one active hart at every cycle as shown on Figs. 8 and 9. In one cycle, a core fetches one instruction for the selected fetching hart, renames one instruction of the selected renaming hart, issues one instruction of the selected issuing hart, writes back one result to the register file of the selected writing hart and commits one instruction for the selected committing hart. The five selections are independent from each other.

A hart may be selected to fetch if its pc is full (a thread is running), if it has not been suspended and if the fetched instruction may be saved in the decode stage instruction buffer (labeled *ib* on Fig. 8) (the buffer remains full until the hart is selected for decode/rename).

In particular, a hart is suspended after fetch until the next pc is known, at best after the decoding which produces $nextPC$ as shown on Fig. 7 ($pc+1$ or the target of an unconditional direct branch). During the suspension, other active harts on the core are selected. LBP hides branch latency through multithreading

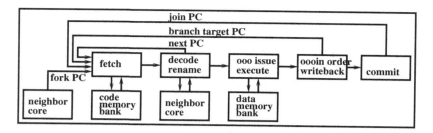

Fig. 7. The core pipeline.

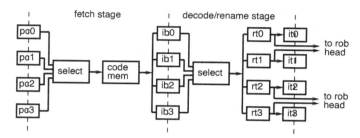

Fig. 8. Fetch and decode/rename stages.

instead of eliminating it through prediction (because every hart is suspended after fetch, at least two full harts are necessary to fill the pipeline).

A hart may be selected for renaming if its instruction buffer is full with a fetched instruction, if there are available resources to do the renaming (renaming table labeled rt on Fig. 8, decode/rename stage) and there is a free entry in the hart reorder buffer (labeled rob in the commit stage on Fig. 9). Once renamed an instruction is saved in the hart instruction table (labeled it in the issue stage) and in the hart reorder buffer.

A hart may be selected for issue if it has at least one ready instruction in its instruction table (renamed instructions wait in the hart instruction table until the sources are ready; there is one table per hart) and if the result buffer of the hart in the write back stage is empty (labeled rb in the write back stage; hence, a multicycle computation blocks the hart for issue until the result has been written back, releasing the result buffer). Once issued, the renamed instruction reads its renamed sources in the renaming register file (labeled rrf), crosses a single or multiple cycle functional unit (labeled fu) and saves the result in the hart result buffer.

A hart may be selected for write back if its result buffer is full and the commit buffer of the hart is empty (labeled cb). The selected result is written to the renaming register file of the hart. The written back instruction is notified as terminated in the hart reorder buffer.

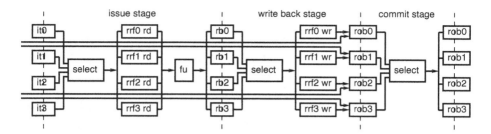

Fig. 9. Issue, write back and commit stages.

A hart may be selected for commit if its reorder buffer tail entry is terminated. If the instruction is a hart ending *p_ret*, the *ending hart* signal must have been received from the preceding hart.

The pipeline has the minimum hardware to make the out-of-order engine work (to keep each core as simple as possible, which allows either to maximize the number of cores on the die for a high performance manycore or to build a very low-cost parallelizing microcontroller with one core and 4 harts).

A consequence is that a hart may have to wait in multiple situations: to fetch because the *pc* is unknown (after a branch; this is frequent), to decode because there is no renaming register available to rename the destination (this is rare) or to issue because the result buffer is occupied (waiting for a computation in progress or waiting to be selected for write back; this is frequent in programs with a lot of memory accesses and/or a lot of multiplications/divisions). However, our experiments have shown that when the 4 harts are active, the core pipeline achieves a rate close to the peak of one instruction per cycle.

Even though harts are interleaved in the pipeline on a cycle by cycle basis, this interleaving keeps deterministic as it only involves harts belonging to the same application.

4.3   The Memory

Figure 6 shows the LBP manycore with its memory organization. Each core is associated to a set of memory banks (red square labeled *m*). There are three banks per core. One bank holds the code, another holds local data (a stack) and the last one is used as a shared global memory.

The shared banks have two access ports. One port is used for a local access and the other port is used for distant accesses through a hierarchy of routers which interconnect the banks[3].

---

[3] The routers are not yet implemented on the FPGA but simulated for the reported experiment.

Each core has a bidirectional access to a level one router (green rectangle labeled $r1$ and shared by four cores). Each $r1$ router is connected to a level two $r2$ router (shared by four $r1$). Eventually, $r2$ routers are connected through a level three router $r3$. The pattern is extensible (for example to extend the shared memory out of the LBP chip or for future extensions of an LBP manycore).

Each $r1$ router is able to handle one access per link per cycle (i.e. 8 transactions with the connected cores plus 4 transactions with the connected memory banks; the router has the necessary internal buffers to pipeline the transactions from core to memory and back to core). Every cycle, each $r2$ router is able to receive 4 incoming requests from the 4 connected $r1$, send 4 outgoing request results to the 4 $r1$, propagate one request to $r3$ and receive one request result from $r3$. Eventually, every cycle the $r3$ router is able to propagate 4 requests and 4 results to/from the 4 connected $r2$.

5   A Matrix Multiplication Program Example Experiment

Figure 10 shows a Deterministic OpenMP program to multiply integer matrices. Except for the *det_omp.h* reference in red, the remaining of the text is standard OpenMP code and can be compiled with *gcc -fopenmp*.

This program (the *base*) has been run on three sizes of a *vivado_HLS* simulation (Xilinx High Level Synthesis tool, version 2019.2) of the LBP processor (4, 16 and 64 cores). Four other versions have also been implemented and run on the simulated LBP: *copy*, *distributed*, *d+c* and *tiled*. The different codes are shown at URL [11].

The aim of the experience is to show that the LBP design is able to fill the harts pipelines with instructions all along the run, thanks to the high level of distant ILP exhibited by the Deterministic OpenMP parallelization, despite the multiple latencies each hart has to wait for. A second goal is to verify that the shared memory interconnection is dimensioned proportionally to the number of harts. As the number of cores is increased in LBP, the distant memory access requests are more frequent and have a longer latency. The experience should check that the hardware is able to sustain a high proportion of distant accesses without stalling the harts, i.e. keeping the IPC as close as possible to its peak.

Each run multiplies a matrix X with h lines and $h/2$ columns and a matrix Y with $h/2$ line and h columns, where h is the number of harts (i.e. 16, resp. 64 and resp. 256 for a 4, resp. 16 and resp. 64 core LBP processor).

The *copy* code copies a line of matrix X in the local stack to avoid its multiple accesses in the shared memory. The *distributed* code distributes and interleaves the three matrices evenly on the memory banks (four lines of X, two lines of Y and four lines of Z in each bank), to avoid the concentration of memory accesses on the same banks (which happens if matrix Y is not distributed). The *c+d* version copies and distributes. The *tiled* version is the classic five nested loops tiled matrix multiplication algorithm. Each tile has $h/2$ elements for matrices X and Y ($\sqrt{h} * \sqrt{h}/2$) and h for the result matrix Z ($\sqrt{h} * \sqrt{h}$).

```
#include <stdio.h>
#include <det_omp.h>
#define LINE_X        16
#define COLUMN_X  8
#define LINE_Y        COLUMN_X
#define COLUMN_Y  16
#define LINE_Z        LINE_X
#define COLUMN_Z  COLUMN_Y
#define NUM_HART  16
int X[LINE_X*COLUMN_X]={[0...LINE_X*COLUMN_X−1]=1};
int Y[LINE_Y*COLUMN_Y]={[0...LINE_Y*COLUMN_Y−1]=1};
int Z[LINE_Z*COLUMN_Z];
void thread(int t){
  int i, j, k, l, tmp;
  for (l=0, i=t*LINE_Z/NUM_HART; l<LINE_Z/NUM_HART; l++, i++)
    for (j=0; j<COLUMN_Z; j++)
      tmp=0;
      for (k=0; k<COLUMN_X; k++)
        tmp+=*(X+(i*COLUMN_X+k)) * *(Y+(k*COLUMN_Y+j));
      *(Z+(i*COLUMN_Z+j))=tmp;
    }
}
void main(){
  int t;
  omp_set_num_threads(NUM_HART);
#pragma omp parallel for
  for (t=0; t<NUM_HART; t++) thread(t);
}
```

Fig. 10. A Deterministic OpenMP matrix multiplication program.

Figures 11, 12 and 13 show nine histograms (number of cycles, IPC and number of retired instructions) for the five codes on the three sizes of LBP. These values are reproducible thanks to cycle determinism. The three bottom histograms also include the best measures done on a Xeon Phi2 for the tiled version (MCDRAM configured in flat mode and all-to-all cluster mode; OMP_NUM_THREADS = 256, OMP_PLACES = threads, OMP_PROC_BIND = close). The measures are the minimum ones after 1000 runs. They were obtained with a PAPI instrumentation of the original tiled version.

What matters is the number of cycles, i.e. the duration of the run. The IPC is an indication whether the parallelization is effective. However, a high IPC does not mean that useful work is done. The number of retired instructions is important to see the overcost of parallelization.

On a 4-core LBP (Fig. 11), even though the tiled version has the highest IPC (3.67 for a peak at 4), the base version is better as it is twice faster. The innermost loop has seven instructions (two loads, one multiplication, one addition, two address incrementations and a conditional branch), which are repeated $h^3/2$ times, i.e. 14336 instructions when $h = 16$. The base version has 16722 retired instructions, which leaves 2386 instructions for the two outer loops, the parallelization and its control (creation of 16 threads and their join).

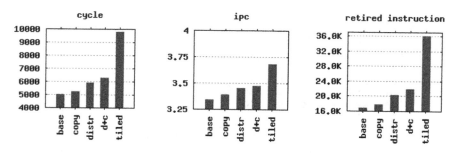

Fig. 11. Number of cycles, IPC and retired instructions for the matrix multiplication five versions on a 4-core LBP (16 harts).

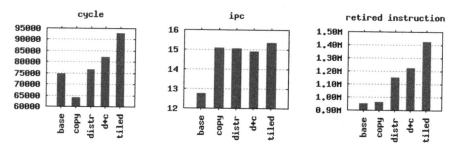

Fig. 12. Number of cycles, IPC and retired instructions for the matrix multiplication five versions on a 16-core LBP (64 harts).

On a 16-core LBP (Fig. 12), the fastest is the copy version. The base version achieves a poor 12.7 IPC when the copy version IPC is over 15 (for a peak of 16), saving more than 10000 cycles (16% faster). The overhead is moderate (14500 instructions, i.e. 1.5%).

On the 64-core LBP (Fig. 13), the tiled version is the best because it saves many long distance communications and because it distributes the remaining ones more evenly over time and space. It is twice faster than the distributed version and four times faster than the base version (1.18M cycles vs 2.08M and 4.14M). The IPC is 61.7 (for a peak of 64), showing that the LBP interconnect is strong enough to handle the high demand. The tiling overhead is not negligible (73M instructions versus 59M for the base version, i.e. +23%).

The 64-core LBP is not as fast as the Xeon Phi2 (1.18M cycles vs 391K, 3 times more). Firstly, there is no vector unit in LBP, which explains that the Xeon runs 32M instructions and LBP runs 73M, i.e. 2.28 times more. Secondly, LBP peak performance is 1 IPC per core when the Xeon peak is 6 (2 int, 2 mem and 2 vector ops per cycle). Hence, LBP reaches 0.96 IPC per core (96% of 1 IPC peak) and the Xeon reaches 1.28 IPC per core (81.86/64 ; 21% of 6 IPC peak). LBP is aiming embedded applications and should keep low-power and energy efficient, which the Xeon Phi2 is not.

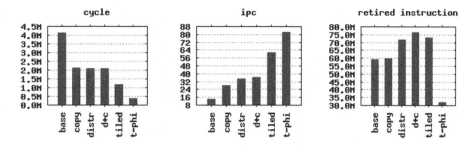

Fig. 13. Number of cycles, IPC and retired instructions for the matrix multiplication five versions on a 64-core LBP (256 harts).

6   Conclusion and Perspectives

Safety critical real time applications can benefit from parallel manycore processors, if a high level of determinism is ensured to guarantee repeatable timings, as on the LBP processor. Moreover, the reported experiment shows that a low-power manycore processor can be built for the embedded high performance computations. The design of the LBP processor is suited to either offer parallelism to microcontrollers or to safely accelerate computations through their parallelization and capture the distant ILP by hundreds of distributed harts.

Deterministic OpenMP is subset of the standard OpenMP with a new runtime. For the programmer, the difference resides in the new *det_omp.h* header file and the hardware placement of code and data according to the program structure. The main difference between OpenMP classic runtime and Deterministic OpenMP new one comes from the ordering of harts in a parallel team. This ordering is optional in standard OpenMP but mandatory in Deterministic OpenMP because the hardware synchronization which ensures safety relies on the referential sequential order. As an example, a producing hart has to precede a consuming one in the referential sequential order to exhibit the read-after-write dependency linking the producer to the consumer. In Deterministic OpenMP, a later hart cannot send anything to a prior one (a data cannot go back in time).

In a future work, we will extend the actual 8-core FPGA implementation of LBP to fit a 16 core and two levels of routers on the Xilinx ZCU106 development board. We will also complete the Deterministic OpenMP translator to automatize the translation of standard OpenMP codes into our LBP specific machine code.

References

1. Dagum, L., Menon, R.: OpenMP: an industry standard API for shared-memory programming. IEEE Comput. Sci. Eng. **5**, 46–55 (1998)
2. OpenMP Architecture Review Board. OpenMP Application Program Interface Version 5.0 (2018). https://www.openmp.org/wp-content/uploads/OpenMP-API-Specification-5.0.pdf

3. Padua, D.: Encyclopedia of Parallel Computing, pp. 1592–1593. Springer, Boston (2011). ISBN 978-0-387-09766-4. https://doi.org/10.1007/978-0-387-09766-4_447
4. Lee, E.: The problem with threads. Computer **39**(5), 33–42 (2006). ISSN 0018–9162
5. Pinho, L.M., et al.: High-Performance and Time-Predictable Embedded Computing. River Publishers, Wharton (2018).ISBN 8793609698
6. Shiina, S., Iwasaki, S., Taura, K., Balaji, P.: Lightweight preemptive user-level threads. In: Proceedings of the 26th ACM SIGPLAN Symposium on Principles and Practice of Parallel Programming, PPoPP 2021, Virtual Event Republic of Korea, 17 February 2021, pp. 374–388. ACM (2021). ISBN 978-1-4503-8294-6. https://doi.org/10.1145/3437801.3441610. https://dl.acm.doi.org/10.1145/3437801.3441610
7. Iwasaki, S., Amer, A., Taura, K., Balaji, P.: Analyzing the performance trade-off in implementing user-level threads. IEEE Trans. Parallel Distrib. Syst. **31**, 1859–1877 (2020). https://doi.org/10.1109/TPDS.2020.2976057
8. Aviram, A., Ford, B.: Deterministic OpenMP for race-free parallelism. In: Proceedings of the 3rd USENIX Conference on Hot Topic in Parallelism, HotPar 2011, p. 4. USENIX Association, Berkeley (2011)
9. Lee, E.A.: What is real time computing? A personal view. IEEE Design Test **35**, 64–72 (2018). https://doi.org/10.1109/MDAT.2017.2766560
10. Waterman, A., Asanović, K. (eds.): The RISC-V Instruction Set Manual, Volume I: User-Level ISA. Document version20191213 (2019)
11. LBP Project Review Board. LBP Project (2020). https://gite.lirmm.fr/lbp-group/lbp-projects/-/wikis/Little-Big-Processor-project

Additional Parallelization of Existing MPI Programs Using SAPFOR

Nikita Kataev$^{(\boxtimes)}$ ⓘ and Alexander Kolganov ⓘ

Keldysh Institute of Applied Mathematics RAS, Moscow, Russia
dvm@keldysh.ru
http://dvm-system.org/en/

Abstract. The SAPFOR and DVM systems were primary designed to simplify the development of parallel programs of scientific-technical calculations. SAPFOR is a software development suite that aims to produce a parallel version of a sequential program in a semi-automatic way. Fully automatic parallelization is also possible if the program is well-formed and satisfies certain requirements. SAPFOR uses the DVMH directive-based programming model to expose parallelism in the code. The DVMH model introduces CDVMH and Fortran-DVMH (FDVMH) programming languages which extend standard C and Fortran languages by parallelism specifications. We present MPI-aware extension of the SAPFOR system that exploits opportunities provided by the new features of the DVMH model to extend existing MPI programs with intra-node parallelism. In that way, our approach reduces the cost of parallel program maintainability and allows the MPI program to utilize accelerators and multi-core processors. SAPFOR extension has been implemented for both Fortran and C programming languages. In this paper, we use the NAS Parallel Benchmarks to evaluate the performance of generated programs.

Keywords: SAPFOR · Automation of parallelization · Additional parallelization · MPI programs · Heterogeneous computational clusters · GPUs · DVMH

1 Introduction

Since the first release of the Message Passing Interface (MPI) [1] standard in 1994, a lot of parallel MPI applications have been written to utilize available high-performance clusters. Currently, MPI is one of the most common programming models used to develop compute-intensive applications on distributed-memory systems. However, every single node in a cluster tends to provide intra-node parallelism. Current trends in architecture make heterogeneous systems mainstream. To write portable code and to exploit multi-level program parallelism today's developers are challenged to use a variety of parallel programming models.

Moreover, a choice of the best-suited programming model becomes very important. Low-level models, such as POSIX Threads, CUDA, OpenCL give

© Springer Nature Switzerland AG 2021
V. Malyshkin (Ed.): PaCT 2021, LNCS 12942, pp. 41–52, 2021.
https://doi.org/10.1007/978-3-030-86359-3_4

programmers fine-grained control over the program execution and allow them to gain the best performance. At the same time, these models make the code less portable. Thus, the migration between different architectures, for example, NVIDIA and AMD GPUs, requires additional effort from software developers.

The possible solution is the use of higher-level programming approaches, such as directive-based programming models, DSL compilers [2–4], or general-purpose libraries [5,6]. While DSLs are restricted to a given domain and demand to study narrowly specialized language constructs, directive-based models and high-performance libraries preserve the sequential code written in general-purpose programming languages.

A great effort has been made to extend the OpenMP [7] standard to support heterogeneous architectures. With new features, added to the standard, it becomes possible to use accelerators as well as cores of the central processor unit. Unfortunately, many compilers have only limited implementation of these features, and there is no full support of the latest OpenMP 5.1 specification in any compiler [8]. OpenACC [9], which emerged in 2011, is another well-known directive-based programming model. It defines an abstract model for accelerated computing to ensure support for currently available devices and for future ones as well.

However, even though these models simplify parallel programming and increase software maintainability, they still demand developers to be very aware of GPU programming. To gain parallel program performance the user has to understand how certain high-level specifications affect program execution. Implementation details may differ between the compilers. Thus, diving into a particular compiler implementation is desirable to understand optimization reports and find out appropriate optimization options and necessary program transformations. Moreover, the absence of convenient tools to debug and tune OpenMP and OpenACC programs executed on accelerators makes the developers use lower-level tools, such as Nvidia Visual Profiler.

Parallel programming automation tools are very much desirable in this case. Unfortunately, automatic compilers often suffer from the inability to reveal parallelism in arbitrary applications.

We advocate the use of a blended approach that comprises three layers: a directive-based programming model, automation tools, and user participation. The foundation is the DVMH [10,11] parallel programming model that provides high-level directives to expose parallelism in C and Fortran code. The DVMH-based programming preserves the sequential C code and allows experienced users to write parallel programs manually or tune existing ones if necessary. The advanced runtime system manages the program execution and adapts it to all available resources. Since it is aware of program execution as well as higher-level parallelism specifications, the DVM system provides tools for program debugging and performance analysis.

The next layer is the System FOR Automated Parallelization (SAPFOR) [12, 13] which is a software development suite that aims to produce a parallel version of a sequential program in a semi-automatic way. The SAPFOR core is an automatic parallelizing compiler [14] that relies on static [15] and dynamic [16] analysis techniques to reveal available parallelism in a source code. However, unlike conventional automatic parallelizing compilers, which may suffer from a lack of user participation, SAPFOR relies on an implicitly parallel programming model and involves the user in the parallelization process.

If necessary, the developer uses the interactive subsystem of SAPFOR [17] or manually inserts directives in a sequential source code to assert high-level program properties. Although some assertions break the sequential semantics, they do not require programmers to understand parallel programming under the hood. The system also provides the user with a set of automatically performed source-to-source transformations (inline expansion, dead code elimination, expression propagation and other) that he can apply to the original sequential program.

SAPFOR produces a parallel version according to DVMH programming model. Hence, the developer can pay attention to the decisions SAPFOR has been made. Moreover, DVMH-based parallelizations allow SAPFOR to transfer some decision-making (for example, low-level data transfer optimization) to the DVMH system runtime library.

Initially, if the DVMH programming model is used, it utilizes all available computational resources. It takes control over inter-node data transfer and distributed computations. Utilizing multi-core processors and accelerators, it also manages parallelism inside a compute node. The presence of a large number of MPI applications convinces us to extend the DVMH model, as well as SAPFOR, and to enable DVMH-based parallelization of MPI programs.

In summary, our main contributions are:

- the blended approach (a directive-based programming model, automation tools, and user participation) to the exploitation of intra-node parallelism in existing MPI programs (C and Fortran),
- MPI-aware extension of the SAPFOR system that exploits opportunities provided by the new features of the DVMH model to extend existing MPI programs with intra-node parallelism,
- experimental evaluation of our blended approach on some programs from the NAS Parallel Benchmark.

The rest of the paper is organized as follows. Section 2 outlines implementation details of MPI-aware program parallelization in SAPFOR. It also briefly describes some useful program transformations on the example of programs from the NAS Parallel Benchmarks [18]. Section 3 presents the performance of the NAS Parallel Benchmarks execution on heterogeneous computational clusters. We examine the DVMH programs obtained in a semi-automatic way from MPI versions of this benchmark. Fortran and C sources are considered. Section 4 discusses the related work and finally, Sect. 5 concludes this paper.

2 MPI-Aware Parallelization in SAPFOR

Initially, DVMH-based program parallelization requires the developer to distribute array elements between the processors and then to map iterations of parallel loops on the elements of these distributed arrays. If the MPI program is parallelized, the DVMH model does not imply any distributed data, so it is not necessary to set the mapping on distributed arrays. However, to enable compile-time and runtime optimizations a relation between loop iterations and array elements has to be specified. The new DVMH specification **tie** was introduced [19] to imply this relation. In this context, it is possible not to specify any distributed arrays in terms of the DVMH model and at the same time use the capabilities of the DVM system:

- to use parallelism on shared memory with using CPU cores (OpenMP threads) or to use graphics accelerators;
- to perform the automatic data transformation on GPUs, and to use simplified management of data movements between CPU and GPUs memories;
- to select optimization parameters of DVMH runtime system;
- to use tools for debugging and performance analyzing of parallel programs.

To place the **tie** specification in the code in an automatic way we add the corresponding analysis techniques to SAPFOR.

In the first step, SAPFOR searches for the outermost perfect loop nests to execute in parallel. It examines whether some properties of the loop nest prevent its parallel execution (safety of control flow and memory accesses, canonical loop form, etc.) [14]. It determines the direction of data usage to place the corresponding specification (**in**, **out**, **local**) because the DVMH compilers do not implement interprocedural analysis and they have to make conservative assumptions.

In the next step, SAPFOR explores memory accesses inside the loop nest body. It relies on expression propagation and scalar evolution techniques to determine the presence of the loop induction variable in subscript expressions. As we do not use DVMH to expose data distribution in MPI programs, we do not determine memory access patterns properly. This allows us to also process non-affine expressions.

After parallel loop nests are identified, the MPI program will be divided into computational regions, which can be executed on different computational devices (multi-core CPU and accelerators), and into sequential code fragments. SAPFOR places neighboring parallel nests at the same level of a loop tree in a single computational region to decrease initialization overhead at the region entry point.

To ensure the memory consistency between CPU and GPUs SAPFOR needs to determine points to put the actualization specifications. The deferred semantic of actualization directives allows DVMH runtime system to avoid redundant data transfer. For example, if the user inserts two identical consecutive directives into his program, the copying will be performed only once. Moreover, there is no difference whether all regions are executed on GPU or a part of them is targeted to the CPU-only execution. Actualization directives affect only transfer between regions and sequential code fragments outside them. The DVMH runtime system will manage the necessary data transfer in an automatic way.

This approach helps SAPFOR to automatically place these directives and to decrease communication overhead. The goal of SAPFOR is to determine data to be updated. It uses the previously collected information that identifies the direction of data usage. It also applies alias analysis to determine indirect memory accesses in C programs. The only variables, which are accessed inside the region body, need to be updated. According to this information, SAPFOR places the **actual** directive, which specifies variables that have the newest values in CPU memory, after each assignment to the variable outside computational regions. The **get_actual** directive, which specifies variables to copy from GPU to CPU memory, is placed before each statement which uses the variable changed in any DVMH region.

Industrial programs usually comprise a large number of procedures (functions or subroutines). Initially, SAPFOR divided all procedures into three groups: built-in routines (intrinsic Fortran procedures and functions from the C standard library), user-defined procedures, and external procedures.

As the body of the external procedure is not available, SAPFOR unable to determine the variables to be used in a procedure call. In this case, the system inserts the actualization specifications before and after this procedure call for all variables accessed in any computational region. To avoid redundant data transfer the user may manually specify the memory the procedure accesses. For example, it is possible to insert SAPFOR assertion to mark the absence of side effects.

SAPFOR treated MPI procedures as external procedures, so we add the fourth group of procedures to SAPFOR (MPI procedures) and specify the data usage direction for procedures in this group.

If the reduction variable participates in MPI communications, SAPFOR was not able to analyze it because the corresponding argument in a data transfer function has a pointer type. We use SAPFOR intermediate representation of the program to make an implicit copy of the scalar variables that cause data dependencies. In that way, we break the explicit relation between reduction variables and MPI functions to force the reduction variable analysis in SAPFOR. However, we do not modify the original source code and perform necessary transformations implicitly. These transformations extend the support of reduction computations in SAPFOR.

One of the most important transformations that allows us to analyze large programs is function inlining [14]. We also use SAPFOR intermediate representation to implement function inlining implicitly. We determine function calls that degrade analysis and automatically schedule them to inline before SAPFOR analysis passes are executed. At the same time, we do not affect the source code.

Unfortunately, manual source-to-source transformations are still necessary to parallelize some sources. The accelerator memory has a limited size and a large number of threads in conjunction with large privitizable arrays prevents offloading computations to GPU. An example is the main computational loop in the EP application from the NAS Parallel Benchmarks [18] (Listing 1.1).

Listing 1.1. Parallel loop with a large privitizable array in the EP benchmark

```
double x[2*NK];  // NK is 65536, NP depends on an input
#pragma dvm parallel([k]) private(x) ...
for (k = 1; k <= NP; k++) {
  ...
  for (i = 0; i < NK; i++) {
    ...
    x[i] = r46 * (*x4);
  }
  for (i = 0; i < NK; i++) {
    x1 = 2.0 * x[2 * i] - 1.0;
    x2 = 2.0 * x[2 * i + 1] - 1.0;
    t1 = x1 * x1 + x2 * x2;
    ...
  }
  ...
}
```

In order to eliminate the array **x**, we manually fused two adjacent loops into a single loop (the first loop initializes this array and the second loop accesses the calculated values) and added a re-calculation of the required elements (two neighboring array elements) at each iteration of the new loop. As a result, the array was replaced with two scalar variables (Listing 1.2).

Listing 1.2. The result of fusion of i-loops in Listing 1.1

```
#pragma dvm parallel([k]) ...
for (k = 1; k <= NP; k++) {
  ...
  for (i = 0; i < NK; i++) {
    double x_2i, x_2i1;
    { ...
      x_2i = r46 * (*x4);
    }
    { ...
      x_2i1 = r46 * (*x4);
    }
    x1 = 2.0 * x_2i - 1.0;
    x2 = 2.0 * x_2i1 - 1.0;
    ...
  }
  ...
}
```

3   Results

In this section, we use the NAS Parallel Benchmarks [18] to demonstrate the capabilities of SAPFOR to embed DVMH specifications in MPI programs. The BT, CG, and EP applications are considered. The performance of the resulting MPI programs with DVMH extensions and parallel programs, using only MPI, was evaluated on the K60 supercomputer [20] which is equipped with Intel Xeon Gold 6142v4 CPUs and NVIDIA V100 GPUs (Volta architecture).

Each node of K60 (GPU partition) has two 16-cores processors and four GPUs V100. Two processors are linked by a shared memory (NUMA architecture). A single node has about 60 TFPLOS single-precision performance and about 30 TFLOPS double-precision performance. Hence, we can use a small number of nodes to achieve the high performance of our parallel programs. For all experiments, we use the total power of two nodes: for MPI programs we use up to 64 cores, for MPI programs with DVMH extensions we use up to 64 cores and 8 GPUs.

Table 1 and Table 2 show the execution time of different MPI programs written in Fortran and C languages, respectively.

Table 1. Times in seconds of Fortran programs, NPB 3.3 class D.

	MPI programs			Transformed MPI programs			MPI programs + **FDVMH**		
	BT	CG	EP	BT	CG	EP	BT	CG	EP
1 node	665.1	397.5	93.68	785.29	376.8	83.34	63.3	80.99	0.62
2 nodes	361.6	209.6	46.53	428.07	229.61	42.06	50.3	42.6	0.38

The first group of columns called MPI programs represents the original version written by the developers of the NAS Parallel Benchmark. Actually, in the suite, there are no built-in C versions of analyzed programs, so we translated Fortran to C manually. The second group called Transformed MPI programs represents versions that were obtained as a result of automated transformations (functions inlining, loop fusion, loop distribution, and etc.) using the SAPFOR system. Also, some transformations we did manually since at the moment they have not been implemented in SAPFOR yet. And finally, the last group called MPI programs + DVMH represents the results of automatic parallelization of Transformed MPI programs using the SAPFOR system.

Table 2. Times in seconds of C programs, NPB 3.3 class D.

	MPI programs			Transformed MPI programs			MPI programs + **CDVMH**		
	BT	CG	EP	BT	CG	EP	BT	CG	EP
1 node	694.6	326.16	98.41	768.5	328.8	99.37	97.7	186.12	0.67
2 nodes	386	218.9	49.29	421.3	214.3	50.05	75.7	96.75	0.38

For each node, we use the maximum number of CPU cores to execute original and transformed versions. To execute DVMH versions we use all GPUs available in the node. We can see that the transformations do not lead to severe performance degradation. And in doing so, SAPFOR parallelizes the transformed versions in an automatic way and offloads computations to GPU.

The lack of some useful analyzes in the CDVMH compiler explains the difference in the speedup of Fortran and C applications. The analysis capability of the Fortran DVMH compiler has been expanded. The compiler analyses imperfectly nested loops inside a parallel loop nest. If it is able to determine the presence of reduction computations, the compiler exploits additional parallelism inside the parallel loop nest. A lot of indirect accesses in the CG application make this approach especially effective because it allows the compiler to parallelize imperfect loop nests.

The maximum speedup of BT application, if GPUs are used, is 10.5 times compared to 32 MPI processes and is 7.18 times compared to 64 cores. The maximum speedup of CG application is 4.9 times in both configurations. And the maximum speedup of EP application is 151 times in both configurations.

The difference in the complexity of these applications leads to different speedup of parallel programs. EP executes a large number of independent operations without accessing the device memory. While CG implements sparse matrix-vector multiplication which produces indirect accesses to the device memory.

And finally, BT is a compute-intensive application with a lot of memory accesses, and loops with regular dependencies. As a result, a large number of communications affect the program performance. Figure 1 shows the ratio of computation time to communication time. It shows that communication time between MPI processes (total communications) does not depend on the total number of cores. As MPI API is separated from DVMH runtime, SAPFOR cannot affect this time, so the CPU-to-GPU data transfer remains the same. However, the computation time decreases in proportion to the number of GPUs.

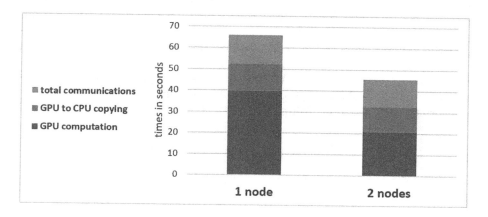

Fig. 1. The ratio of computation time to communication time in the BT application

4 Related Works

There are two different approaches to program parallelization. The first one, adopted by SAPFOR, is a stepwise transformation of a source code in the direction of a parallel program. There are a lot of typical techniques like loop fusion, loop reversal, loop distribution, function inlining, expression propagation and etc. [21]. Some of them have been already implemented in SAPFOR and could be chosen automatically to apply or the user may guide SAPFOR through a desirable transformation sequence. Model-based parallelization is another approach that uses a mathematical model to represent program fragments to be parallelized. The main advantage of this approach is the ability to represent the entire transformation sequence as a single transformation and discover it in an automatic way using mathematical optimization methods.

The latter approach is more suitable for automatic compilers than for automation tools that depend on active user participation. In the context of model-based approach, it may be difficult to understand the decisions made by the compiler. The following compilers adopt this approach [22,23,25,26,28].

Pluto [22] focuses on loop transformations to optimize data locality and to exploit OpenMP level parallelism as well as to vectorize loops. PPCG [23] was designed to offload data-parallel computation to a GPU and uses CUDA or OpenCL to expose parallelism in a source code.

Both these tools use only C code as input and do not support Fortran language. In spite of these tools implement source-to-source program transformation and use high-level language to expose available parallelism, they still suffer from readability issues. For example, PPCG relies on CUDA and OpenCL low-level models that require the user to well understand GPU programming.

Moreover, to apply the polyhedral model the well-structured code fragments (SCoPs) should be revealed. The user has to insert corresponding specifications (for example, a **scop** pragmas) into a source code. The way these specifications are inserted may affect the performance of the resulting parallel code. For example, PPCG handles data transfer separately for each static control loop nest. Hence, if adjacent loop nests are placed in different SCoPs, the data transfer happens at the beginning and at the end of each loop nest. As a result, this leads to drastic performance degradation. Pluto and PPCG also suffer from an inability to reveal reduction computations and to support them in a parallel code.

The LLVM-based [24] compilers Polly [25] and Polly-ACC [26] solve some of the mentioned problems. A reduction-enabled scheduling approach has been implemented in Polly [27]. These compilers also implement automatic detection of SCoPs. Actually, Polly-ACC is an extension of Polly enabling accelerator support in LLVM-based compilers. Therefore, it maintains the parallelization for different languages if the corresponding front-end for LLVM-project exists.

On the other hand, it operates with low-level LLVM IR. That makes it impossible for the user to explore the parallel program. Even though Polly automatically detects SCoPs in the source code it may require the user to change the original program if the code cannot be analyzed. In this context, the user has to explore the hierarchy of SCoPs and be aware of LLVM IR (Intermediate Representation).

Unlike all mentioned tools, Apollo [28] optimizer, which also relies on the polyhedral model, applies speculative optimizations at run time. Thus, it overcomes static analysis issues, but it does not offload computations to accelerators.

If we summarize all mentioned issues, all these tools are not suitable to parallelize the NAS Parallel Benchmarks. The direct application of PPCG and Polly has not given any performance impact. It increases program execution time instead.

5   Conclusion

The paper examines the approach to additional parallelization of existing MPI programs which was implemented in SAPFOR and DVM systems. We present performance evaluation results of the built programs on the example of some applications from the NAS Parallel Benchmarks.

SAPFOR relies on the new features of the DVM system that allows us to offload computations in MPI programs to a GPU in a semi-automatic way. We advocate the use of a blended approach to parallel programming that comprises three layers: a directive-based programming model, automation tools, and user participation. In this context, SAPFOR and DVM complement each other and bring advantages over other parallel programming models, like MPI+OpenMP or MPI+OpenACC.

Firstly, the SAPFOR system implements an automatic parallelizing compiler that is suitable to parallelize well-formed programs. If the complexity of the original program hinders its analysis the user may assert program properties or guide SAPFOR through the sequence of source-to-source transformations. It is important, that available assertions do not require programmers to understand parallel programming in detail.

To gain parallel program performance the SAPFOR system can rely on various optimizations implemented in the DVMH compiler and runtime system: data transformation at runtime to choose the right memory access pattern, dynamic CUDA handler compilation during the program runtime, parallel execution of loops with regular loop carried dependences on GPU and other. These optimizations are hidden from the user, and to enable them SAPFOR uses higher-level parallelism specifications that do not affect source code readability and maintainability.

Moreover, the DVM system provides performance analysis tools that operate in terms understandable to a user. These tools accumulate the characteristics of parallel program performance and associate them with DVMH constructs. SAPFOR can use these performance characteristics to perform further optimizations.

Thus, the SAPFOR and DVM systems can significantly reduce the effort required to embed intra-node parallelism into the existing MPI programs and to utilize available architectures such as multi-core CPUs or GPUs. We believe that they can also help to develop and to optimize scalable algorithms for supercomputers.

References

1. MPI Documents. http://www.mpi-forum.org/docs/. Accessed 8 Apr 2021
2. Ragan-Kelley, J., Barnes, C., Adams, A., Paris, S., Durand, F., Amarasinghe, S.P.: Halide: a language and compiler for optimizing parallelism, locality, and recomputation in image processing pipelines. In: Proceedings of the 34th ACM SIGPLAN Conference on Programming Language Design and Implementation, PLDI 2013, pp. 519–530 (2013)
3. Beaugnon, U., Kravets, A., van Haastregt, S., Baghdadi, R., Tweed, D., Absar, J., Lokhmotov, A.: VOBLA: a vehicle for optimized basic linear algebra. In: Proceedings of the 2014 SIGPLAN/SIGBED Conference on Languages, Compilers and Tools for Embedded Systems, LCTES 2014, New York, NY, USA, pp. 115–124 (2014)
4. Zhang, Y., Yang, M., Baghdadi, R., Kamil, S., Shun, J., Amarasinghe, S.: GraphIt: a high-performance graph DSL. Proc. ACM Program. Lang. **2**(OOPSLA), 121:1-121:30 (2018)
5. An, P., et al.: STAPL: an adaptive, generic parallel C++ library. In: Dietz, H.G. (ed.) LCPC 2001. LNCS, vol. 2624, pp. 193–208. Springer, Heidelberg (2003). https://doi.org/10.1007/3-540-35767-X_13
6. Bell, N., Hoberock, J.: Thrust: a productivity-oriented library for CUDA. In: Hwu, W.-M.W. (ed.) GPU Computing Gems, Jade Edition, pp. 359–371 (2012). https://doi.org/10.1016/B978-0-12-385963-1.00026-5
7. OpenMP Application Programming Interface. Version 5.1, November 2020. https://www.openmp.org/wp-content/uploads/OpenMP-API-Specification-5-1.pdf. Accessed 8 Apr 2021
8. OpenMP Compilers & Tools. https://www.openmp.org/resources/openmp-compilers-tools/. Accessed 8 Apr 2021
9. The OpenACC Application Programming Interface. Version 3.1. November 2020. https://www.openacc.org/sites/default/files/inline-images/Specification/OpenACC-3.1-final.pdf. Accessed 8 Apr 2021
10. Konovalov, N.A., Krukov, V.A., Mikhajlov, S.N., Pogrebtsov, A.A.: Fortan DVM: a language for portable parallel program development. Program. Comput. Softw. **21**(1), 35–38 (1995)
11. Bakhtin, V.A., Klinov, M.S., Krukov, V.A., Podderugina, N.V., Pritula, M.N., Sazanov, Yu.L.: Extension of the DVM-model of parallel programming for clusters with heterogeneous nodes (in Russian). Bull. South Ural State Univ. Seri.: Math. Model. Program. Comput. Softw. **18**(277), 82–92(2012). Issue 12
12. Klinov, M.S., Krukov, V.A.: Automatic parallelization of fortran programs. Mapping to cluster (in Russian). Vest. Lobachevsky Univ. Nizhni Novgorod **2**, 128–134 (2009)
13. Bakhtin, V.A., et al.: Interaction with the programmer in the system for automation parallelization SAPFOR (in Russian). Vest. Lobachevsky State Univ. Nizhni Novgorod **5**(2), 242–245 (2012)
14. Kataev, N.: LLVM based parallelization of C programs for GPU. In: Voevodin, V., Sobolev, S. (eds.) RuSCDays 2020. CCIS, vol. 1331, pp. 436–448. Springer, Cham (2020). https://doi.org/10.1007/978-3-030-64616-5_38
15. Kataev, N.: Application of the LLVM compiler infrastructure to the program analysis in SAPFOR. In: Voevodin, V., Sobolev, S. (eds.) RuSCDays 2018. CCIS, vol. 965, pp. 487–499. Springer, Cham (2019). https://doi.org/10.1007/978-3-030-05807-4_41

16. Kataev, N., Smirnov, A., Zhukov, A.: Dynamic data-dependence analysis in SAP-FOR. In: CEUR Workshop Proceedings, vol. 2543, pp. 199–208 (2020)
17. Kataev, N.: Interactive parallelization of C programs in SAPFOR. In: Scientific Services & Internet 2020. CEUR Workshop Proceedings, vol. 2784, pp. 139–148 (2020)
18. NAS Parallel Benchmarks. https://www.nas.nasa.gov/publications/npb.html. Accessed 8 Apr 2021
19. Bakhtin, V., et al.: New features of DVM-System for additional parallelization of MPI programs. In: Scientific Services & Internet 2020. CEUR Workshop Proceedings, vol. 2784, pp. 23–38 (2020)
20. Heterogeneous cluster K60. https://www.kiam.ru/MVS/resourses/k60.html. Accessed 8 Apr 2021
21. Wolfe, M.: High Performance Compilers for Parallel Computing. Addison-Wesley (1995)
22. Bondhugula, U., Hartono, A., Ramanujam, J., Sadayappan, P.: A practical automatic polyhedral parallelizer and locality optimizer. SIGPLAN Not. **43**(6), 101–113 (2008)
23. Verdoolaege, S., Juega, J.C., Cohen, A., Gomez, J.I., Tenllado, C., Catthoor, F.: Polyhedral parallel code generation for CUDA. ACM Trans. Archit. Code Optim. **9**(4), 1–23 (2013)
24. Lattner, C., Adve, V.: LLVM: a compilation framework for lifelong program analysis & transformation. In: Proceedings of the 2004 International Symposium on Code Generation and Optimization (CGO 2004). Palo Alto, California (2004)
25. Grosser, T., Groesslinger, A., Lengauer, C.: Polly—performing polyhedral optimizations on a low-level intermediate representation. Parallel Process. Lett. **22**(04), 1250010 (2012)
26. Grosser, T., Hoefler, T.: Polly-ACC transparent compilation to heterogeneous hardware. In: ICS 2016: Proceedings of the 2016 International Conference on Supercomputing, June 2016, pp. 1–13 (2016) https://doi.org/10.1145/2925426.2926286
27. Doerfert, J., Streit, K., Hack, S., Benaissa, Z.: Polly's polyhedral scheduling in the presence of reductions. In: 5th International Workshop on Polyhedral Compilation Techniques (IMPACT) (2015)
28. Caamano, J.M.M., Sukumaran-Rajam, A., Baloian, A., Selva, M., Clauss, P.: APOLLO: automatic speculative POLyhedral loop optimizer. In: 7th International Workshop on Polyhedral Compilation Techniques (IMPACT), January 2017, Stockholm, Sweden (2017)

Sparse System Solution Methods for Complex Problems

Igor Konshin[1,2,3,4(✉)] and Kirill Terekhov[1,2]

[1] Marchuk Institute of Numerical Mathematics, RAS, Moscow 119333, Russia
`igor.konshin@gmail.com, terekhov@inm.ras.ru`
[2] Moscow Institute of Physics and Technology, Moscow 141701, Russia
[3] Dorodnicyn Computing Centre of FRC CSC, RAS, Moscow 119333, Russia
[4] Sechenov University, Moscow 119991, Russia

Abstract. Sparse system solution methods (S^3M) is a collection of interoperable linear solvers and preconditioners organized into a C++ header-only library. The current set of methods in the collection span both rather traditional Krylov space acceleration methods and smoothers as well as advanced incomplete factorization methods and rescaling and reordering methods. The methods can be integrated into algebraic multigrid and multi-stage fashion to construct solution strategies for complex linear systems that originate from coupled multi-physics problems. Several examples are considered in this work, that includes Constrained Pressure Residual (CPR) multi-stage strategy for oil & gas problem and Schur complement method for the system obtained with mimetic finite difference discretization for anisotropic diffusion problem.

Keywords: Sparse linear system · Numerical modeling · Constrained pressure residual · Mimetic finite difference · Parallel efficiency

1 Introduction

Modern industrial applications require numerical analysis of large complex coupled multi-physics problems. Strong coupling of a variety of physical processes of different nature and properties result in inability or inefficiency in solving arising linear systems using usual methods such as algebraic multigrid or incomplete factorization. To gain traction in solving such systems the problem is subdivided into systems with known properties and then solved in a multi-stage fashion with each part addressed with the best possible preconditioner. A weak preconditioner for the whole system or block Gauss–Seidel method is used to resolve the coupling. The examples are two-stage constrained pressure residual (CPR) method [1] for multiphase oil recovery problems, its three-stage extension when thermal effects are considered [2], Uzawa method for the Navier–Stokes systems, the fixed-stress method for coupled fluid-structure interaction [3]. The Sparse System Solution Methods (S^3M) package is designed to tackle such linear systems and solution strategies.

© Springer Nature Switzerland AG 2021
V. Malyshkin (Ed.): PaCT 2021, LNCS 12942, pp. 53–73, 2021.
https://doi.org/10.1007/978-3-030-86359-3_5

There is a number of various open-source packages directed at solving complex problems in parallel, such as PETSc [4], Hypre [5], Trilinos [6], AMGCL [7], INMOST [8–10]. All of them are tailored to address either close to elliptic diagonally-dominant systems (PETSc GAMG, Hypre BoomerAMG, Trilinos ML, AMGCL, and others) or for raw utilization of parallel computing power. Among the commercially available solvers, the notable example is SAMG [11], the successor of the first commercially-successful AMG1R5 multigrid solver [12]. SAMG implements a number of strategies to address various industrial problems [13,14]. On the other side the distributed direct solvers such as open-source MUMPS [15] and SuperLU [16] or commercial PARDISO [17] are available but are limited by linear system size even in distributive memory implementations.

INMOST software platform was developed earlier by the authors (see [9,10]). It is mainly based on distributed and partially shared memory parallel implementations. The built-in methods are ubiquitously based on the additive Schwartz method that, with a large number of cores, may either drastically reduce the performance or require much memory to locally store the overlapped matrix. The flagship preconditioner is the multi-level second-order inverse-based incomplete factorization method [18]. The robustness of this built-in preconditioner is provided based on adaptive tuning of dropping tolerances [19] and efficiency of Schur complement approximation [20] that becomes memory hungry and nontractable for large complex systems. To some extent, the S^3M package serves as a test bed for the linear solver subsystem in INMOST, especially once MPI-based distributed-memory parallelization is implemented.

Section 2 describes the structure of the S3M package. Section 3 clarifies the mathematical aspects of the implemented algorithms. In Sect. 4, the main properties of the S3M package are formulated. Section 5 contains the results of numerical experiments. The final section summarizes the findings.

2 S^3M Package Structure

The basic idea of S^3M structure organization is the use of C++ templates. Each template class is placed in its own header file along with the implementation of all the methods. In this way, the S^3M package is organized into a C++ header-only library and a separate set of ready-to-use programs for the linear system solution. In Fig. 1 the general structure of S^3M library is presented.

The respective header files allows one to support:

- *Data storage.* These data structures are used at all levels, from operations on vectors to operations on sparse matrices.
  - *CSR matrix.* Matrix data is stored in the most widely used Compressed Sparse Row (CSR) format. It allows to load and save matrix in Matrix-Market and internal binary formats, provide functionality to multiply a vector by direct and transpose matrix as wells as transposition, multiplication, addition and subtraction of sparse matrices and some other helpful accessory methods.

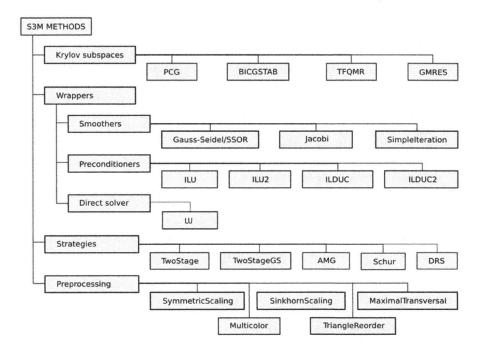

Fig. 1. The S^{3}M library structure of methods and class.

  ∘ class `CSRMatrix<KeyType>`, where `KeyType` is the type of matrix values.
- *CSR matrix graph.* Matrix graph has a similar structure expect it does not store values.
  ∘ class `CSRGraph`.
- *Sparse row accumulator.* This class is used for SAXPY operations on sparse vectors. For a survey on the implementation of row accumulators, see [21].
  ∘ class `RowAccumulator<KeyType>`, where `KeyType` is the type of sparse row values.
- *Vector* operations. At the present, this file contains the set of operations over `std::vector<double>` vectors. The functionality covers loading and saving MatrixMarket and binary file formats and computing dot product and vector norm.
– *Iterative schemes.* A set of iterative schemes is implemented.
 - *Preconditioned Conjugate Gradient* method. This iterative scheme should only be used for linear systems with symmetric positive definite matrices. Class `PCG` contains the coefficient matrix and the preconditioner.
  ∘ class `PCG< Preconditioner >`, here and further `Preconditioner` stands for the class defining preconditioning method.

- *Biconjugate gradient stabilized method.* The most popular iterative scheme for nonsymmetric systems of linear equations.
  - class **BICGSTAB< Preconditioner >**.
- *Generalized minimal residual method.* The method approximates the solution by the vector in a Krylov subspace with minimal residual. The Arnoldi iteration is used to find this vector. The working memory required by this method depends on the number of restarting iterations (the default value is **restart** = 25).
  - class **GMRES< Preconditioner >**.
- *Transpose-Free Quasi-Minimal Residual method.*
  - class **TFQMR< Preconditioner >**.
- *Relaxation methods.* These methods mainly serve as smoothers for algebraic multigrid method.
  - *Conjugate Gradient.* The Conjugate Gradient method contains no preconditioner. This iterative scheme should be used for linear systems with symmetric positive definite matrices.
    - class **ConjugateGradient**.
  - *Jacobi* method. The Jacobi method with relaxation parameter ω.
    - class **Jacobi**.
  - *Gauss–Seidel* method. The Gauss–Seidel method with alternate backward and forward substitutions as well as relaxation parameter ω.
    - class **GaussSeidel**.
  - *Multicolor Gauss–Seidel* method. The multicolor Gauss–Seidel method includes the multicoloring reordering based on parallel maximal independent set algorithm to uncover the parallelism. It is equivalent to symmetric Gauss–Seidel method if the number of iterations is even, or symmetric successive over-relaxation (SSOR) if relaxation parameter $\omega > 1$.
    - class **MulticolorGaussSeidel**.
  - *Chebyshev polynomial* method. The Chebyshev polynomial method can be used as a standalone iterative process or as an AMG smoother with the eigenvalues estimated using Gershgorin's disks.
    - class **Chebyshev**.
  - *Simple-iteration* method. The simple-iteration method can also be used as an AMG smoother.
    - class **SimpleIteration**.
  - *Dummy* method. This is the do-nothing method.
    - class **DummySolver**.
- Triangular factorizations. Both complete and incomplete triangular factorizations are implemented. It can be used as a standalone preconditioner or as a smoother to algebraic multigrid method.
  - *LU factorization* method. This method uses full pivoting and is applied to a dense matrix.
    - class **LU**.
  - *The second-order ILU factorization* method. The second-order rowwise ILU factorization [22] or a two thresholds ILU(τ_1, τ_2) factorization

provides a robust preconditioner for sparse matrices. A setting $\tau_2 = \tau_1 = \tau$ coarsens it to a regular 1-st order ILU(τ) factorization.

  ○ class ILU2.

- *Second-order Crout incomplete LDU factorization.* The Crout version of the second-order ILU factorization using the estimation on the norms of the inverse triangular factors described in [18]. It is the most powerful method to construct an accurate preconditioner for sparse ill-conditioned matrices. The method also has diagonal perturbation parameter τ_D. Increasing this parameter reduces factorization accuracy and complexity. It proves to be helpful for large systems.

  ○ class ILDUC2.

- *Crout incomplete LDU factorization.* This is a simplified version without a second-order threshold and equivalent to the one in [23].

  ○ class ILDUC.

- *Preprocessing.* The preprocessing stage includes matrix scaling and matrix reordering algorithms. Preprocessing is an important part of the linear system solution. It may provide relevant properties for the matrix [24, 25].

  - *Symmetric scaling* method. The scaling algorithm is implemented following [24]. Note, that description of the method in Sect. 4.3 (see [24, p. 353]) and Algorithm 4.1 (see [24, p. 354]) contain square roots. The current implementation uses the power of quarter instead.

    ○ class SymmetricScaling< Solver >, here and further Solver is the class of the subsequent solution method.

  - *Sinkhorn scaling* method. The row-column alternating scaling algorithm to doubly-stochastic form is implemented following [26].

    ○ class SinkhornScaling< Solver >.

  - *Multicoloring* algorithm. Multicoloring algorithm is implemented using either sequential RCM-like approach or using parallel maximum independent set algorithm for matrix reordering.

    ○ class Multicolor< Solver >.

  - *Triangle reordering* algorithm. This method tries to reorder the system as much as possible to lower-triangular matrix.

    ○ class TriangleReorder< Solver >.

  - *Maximal transversal* algorithm. This algorithm constructs both scaling and reordering to provide the use of maximal pivots during factorization. Maximal transversal algorithm is implemented following [25, 27]. It is the most powerful instrument for preprocessing the ill-conditioned linear systems.

    ○ class MaximalTransversal< Solver >.

  - *Dummy preprocessing* method. This is the do-nothing algorithm.

    ○ class DummyPreprocessor< Solver >.

- *A set of wrappers.* The wrappers provides parameter-based choice of particular method.

  ○ class KrylovWrapper< Preconditioner >;
  ○ class PreprocessorWrapper< Solver >;

○ class `PreconditionerWrapper`;
○ class `SmootherWrapper`.

– *Two-stage and AMG solution strategies.*
  • *Two-stage* method. This is the method that applies two different preconditioners to the linear system. The first preconditioner is applied to a (usually pressure) sub-block of the system. The second preconditioner is applied to the full system with the account of the solution of the first system in the right-hand side. The method requires definition of sub-block via parameters. For details see Subsect. 3.4.
  ○ class `TwoStage< BlockSolver, SystemSolver >`.
  • *Two-stage Gauss–Seidel* method. The symmetric block Gauss–Seidel method for two blocks. The method requires definition of sub-block via parameters. For details see Subsect. 3.5.
  ○ class `TwoStageGaussSeidel< Block1Solver, Block2Solver >`.
  • *Algebraic multigrid* method. The AMG method is implemented following [28]. It accepts matrix preprocessing on each level, similar to the approach in multi-level methods.
  ○ class `AMG< Smoother, Preprocessor, CoarsestSolver >`.
  • *Ruge–Stüben algebraic multigrid* method. The classical Ruge–Stüben AMG class is implemented following [29]. For the details of the AMG implementation see Sect. 3.
  ○ class `AMGRugeStuben< Smoother, CoarsestSolver>`.
– *Schur complement.* Some methods based on Schur complement are implemented.
  • *Schur complement* method. This method is used for solve the linear systems appeared after mixed hybrid finite elements or mimetic finite difference discretizations. This method uses the fact that the leading block of the matrix is diagonal.
  ○ class `SchurMFD< SchurSolver >`.
  • *Schur complement series.* This method uses power series to solve block system but avoid constructing Schur complement.
  ○ class `SchurSeries< BSolver, CSolver >`.
– *Constrained Pressure Residual* method. The method involves preliminary matrix scaling and a two-stage approach. This method is implemented following [30,31]. For details see Subsect. 3.3.
  ○ class `CPR< PSolver, SSolver, TwoStageMethod >`.
– *Dynamic Row Scaling* method. A smarter matrix scaling method that preservers diagonal-dominance of pressure system. The implementation follows by [13,32,33].
  ○ class `DRS< PSolver, SSolver, TwoStageMethod >`.
– *Parallelization* method. An OpenMP based parallelization strategy based on block-Jacobi method for a serial solver.
  ○ class `BlockJacobi< Solver >`.

It should be noted, that partial parallelization is currently only performed at the OpenMP level. OpenMP directives are incorporated in the code wherever it is required. In addition, there is a special structure for thread-safe construction of a sparse matrix using OpenMP.

As a result of the organization of C++ classes, the CPR method based on block Gauss–Seidel approach with the pressure system solved by multigrid method with the smoother based on Crout incomplete factorization method acting on symmetrically pre-scaled matrix and saturation system solved by parallel Gauss–Seidel method can be expressed as an object of the class `CPR< AMG RugeStuben< SymmetricScaling< ILDUC >, LU >, MulticolorGaussSeidel, TwoStageGaussSeidel>`. Here the direct `LU` method is used at the coarsest level. It translates into `CPR` method with:

- `PSolver` equals to `AMGRugeStuben` with
  - `Smoother` equals to `SymmetricScaling` with `Solver` equal to `ILDUC`;
  - `CoarsestSolver` equals to `LU`;
- `SSolver` equals to `MulticolorGaussSeidel`;
- `TwoStageMethod` equals to `TwoStageGaussSeidel`.

3   Mathematical Aspects

In this section, we describe in detail some of the algorithms involved in S^3M package and used in the above numerical experiments. The methods described below are used as a preconditioner to one of Krylov subspace methods.

3.1   AMG Method

AMG algorithm is implemented following the classic description from [28]. It prepares the coarse space and operators of prolongation, restriction, and relaxation. In the following we denote the matrix on level m by $A^m = \left\{a_{ij}^m\right\}, i, j \in \overline{1, N^m}$. Currently, the preparation of the coarse space is sequential, the parallel alternatives are PMIS, HMIS, CLJP and others [34] that we intend to implement later as alternative choice. In the original Ruge & Stuben algorithm [29] an extension of coarse space is performed. This step is optionally available in our implementation, but we do not exercise this option in numerical tests and thus omit its presentation.

3.2   Smoothers

Further in numerical tests we use multi-color Symmetric Gauss–Seidel method, polynomial Chebyshev method and Crout incomplete factorization as relaxation methods in AMG. In numerical experiments we compare the results of AMG methods to the single-level second-order variant of the Crout incomplete factorization with the dropping tolerances tuned according to the inverse-based condition estimation [19]. The description and algorithms corresponding to the method can be found in [18].

3.3 Constrained Pressure Residual Algorithm

The CPR method is specifically tailored to address the linear systems arising in fully-implicit reservoir simulators. The method mimics IMPES solution strategy, that stands for "implicit pressure explicit saturation", by uncoupling the pressure equations from saturations. Let the linear system $Ax = b$ be split into "pressure" and "saturation" parts

$$\begin{bmatrix} A_{pp} & A_{ps} \\ A_{sp} & A_{ss} \end{bmatrix} \cdot \begin{bmatrix} p \\ s \end{bmatrix} = \begin{bmatrix} b_p \\ b_s \end{bmatrix}. \tag{1}$$

After left scaling by $S = \begin{bmatrix} I & -D_{ps}D_{ss}^{-1} \\ 0 & I \end{bmatrix}$ the system is transferred to

$$\begin{bmatrix} B_{pp} & Z_{ps} \\ A_{sp} & A_{ss} \end{bmatrix} \cdot \begin{bmatrix} p \\ s \end{bmatrix} = \begin{bmatrix} b_p - D_{ps}D_{ss}^{-1}b_s \\ b_s \end{bmatrix}, \tag{2}$$

where $B_{pp} \equiv A_{pp} - D_{ps}D_{ss}^{-1}A_{ps}$ and $Z_{ps} \equiv A_{ps} - D_{ps}D_{ss}^{-1}A_{ss} \approx 0$ is assumed.
 The solution is similar to the inexact Uzawa iteration:

1. Solve pressure system

$$B_{pp}\tilde{p} = b_p - D_{ps}D_{ss}^{-1}b_s.$$

2. Solve full system

$$x = A^{-1}\left(b - A \cdot \begin{bmatrix} \tilde{p} \\ 0 \end{bmatrix}\right) + \begin{bmatrix} \tilde{p} \\ 0 \end{bmatrix}.$$

The two variants of CPR method may be considered based on choice of D_{ps} and D_{ss}:

- "true-IMPES": $D_{ps} = \text{colsum}(A_{ps})$, $D_{ss} = \text{colsum}(A_{ss})$;
- "quasi-IMPES": $D_{ps} = \text{diag}(A_{ps})$, $D_{ss} = \text{diag}(A_{ss})$.

Here, $\text{diag}(\cdot)$ is the diagonal of the matrix, while $\text{colsum}(\cdot)$ is the diagonal matrix with the respective column sums. Besides, since D_{ps} may not be square, it is also assumed that the first order upstream method is used for the mobility advection and the diagonal terms are negative. This is true for oil-water system of equations. As a result the method automatically detects the diagonal of the block, corresponding to saturations by negative value.

3.4 Two-Stage Method

Let the preconditioned linear system has the form

$$(AM^{-1})(Mx) = b,$$

where a general multi-stage preconditioner M^{-1} is formally written as [2]:

$$M^{-1} = M_1^{-1} + \sum_{i=2}^{n_{st}} M_i^{-1} \prod_{j=1}^{i-1}(I - AM_j^{-1}),$$

with n_{st} the number of stages and M_j^{-1} the jth preconditioner for A. Then the CPR method represents a special case of two-stage algorithm, where the preconditioner matrices are defined as follows:

$$M_1^{-1} = \begin{bmatrix} 0 & 0 \\ 0 & M_{pp}^{-1} \end{bmatrix}, \quad M_{pp}^{-1} \approx B_{pp}^{-1}, \quad M_2^{-1} \approx (SA)^{-1}.$$

3.5   Two-Stage Symmetric Gauss–Seidel Algorithm

The alternative to the two-stage algorithm considered in [1] is a block Gauss–Seidel algorithm. Here we consider its symmetric variant. Let the linear system (1) after left-scaling be presented in two-by-two block form (2).

Let $M_1^{-1} \approx B_{pp}^{-1}$ and $M_2 \approx A_{ss}^{-1}$ be some preconditioners for the matrix diagonal blocks in (1). Then the solution process can be presented as follows:

$$\begin{bmatrix} \tilde{x}_p \\ \tilde{x}_s \end{bmatrix} = \begin{bmatrix} B_{pp} & \\ A_{sp} & A_{ss} \end{bmatrix}^{-1} \left(\begin{bmatrix} b_p \\ b_s \end{bmatrix} - \begin{bmatrix} 0 & Z_{ps} \\ & 0 \end{bmatrix} \cdot \begin{bmatrix} x_p \\ x_s \end{bmatrix} \right),$$

$$\begin{bmatrix} x_p \\ x_s \end{bmatrix} = \begin{bmatrix} B_{pp} & Z_{ps} \\ & A_{ss} \end{bmatrix}^{-1} \left(\begin{bmatrix} b_p \\ b_s \end{bmatrix} - \begin{bmatrix} 0 & \\ A_{sp} & 0 \end{bmatrix} \cdot \begin{bmatrix} \tilde{x}_p \\ \tilde{x}_s \end{bmatrix} \right).$$

In this way the next approximation to the solution x can be computed by

$$\tilde{x}_p = M_1^{-1}(b_p - Z_{ps}x_s), \quad x_s = M_2^{-1}(b_s - A_{sp}\tilde{x}_p), \quad x_p = M_1^{-1}(b_p - Z_{ps}x_s).$$

4   S³M Properties

The S³M package properties can be summarized as follows:

(a) *Cross-platform software.* The S³M package is written in pure C++ and does not use any platform dependant components. The S³M library was successfully tested on Linux, UNIX-based, Windows, and Mac OS systems. Compilation with Visual Studio posed a challenged due to old OpenMP standard. We defined specific signed integer type for parallel loops and replaced OpenMP task constructs.

(b) *Ease to install.* To install the S³M library "cmake" is used. It is sufficient to download S³M package which consists of a set of headers (include files) in the main directory and a set of ready-to-use C++ examples located in "utils" directory. All solvers can be invoked from the command line with the parameters "matrix.mtx [rhs.mtx] [sol.mtx]".

(c) *Simple to change and construct linear solvers.* To change the linear solver, the modification of the solver type is required (see Sect. 2).

(d) *Ease to change the linear solvers' parameters.* An easy way to change the solver parameters is to run the executable file without parameters. The file "params_default.txt" with default parameters will appear in the current directory (see Fig. 2). If necessary, it can be changed with any text editor and copied to "params.txt" for further reading by an executable file. For example, the main default parameters of "AMGRugeStuben" solver are:

```
Method:
    name = BICGSTAB
    dtol = 1e+50
    rtol = 1e-06
    tol = 1e-08
    maxiters = 5000
    true_residual = 0
    verbosity = 1
    Preconditioner:
        name = AMGRugeStuben
        check = 1
        cycle = V
        level = *
        operator_type = 2
        order = 0
        phi = 0.25
        refine_splitting = 0
        verbosity = 1
        write_matrix = 0
        CoarsestSolver:
            name = LU
            verbosity = 0
        /
        Smoother:
            name = Chebyshev
            maxiters = 2
            tol = 0
            verbosity = 1
        /
    /
/
```

Fig. 2. The default parameters set for the AMG solver.

- Method parameters (the name of the iterative method name = BICGSTAB, drop tolerance tol = 10^{-8}, relative tolerance rtol = 10^{-6}, maximal number of outer iterations maxiters = 5000);
- Preconditioner solver parameters (the name of the preconditioner solver name = AMG, the type of the inner loop cycle = V, the maximal number of levels level = *, where "*" means "unlimited");
- The coarsest solver parameters (the name of the coarsest solver name = LU);
- Smoother parameters (the name of the smoother name = Chebyshev, maximal number of iterations maxiters = 2).

Most of the solver stages contain a "verbosity" parameter to output the solution trace.

(e) *Traditional sparse storage format (CSR)*. This simplifies the development of new linear solver components and also makes loading data from external files more standard.

(f) *Support for external data format*, MatrixMarket's format (MTX). This makes it possible to use for testing collections of sparse matrices, which primarily support this particular format of storing sparse matrices. The use of data recording in text form makes the MTX format a universal and machine-independent tool.

(g) *Availability of internal binary data format*. The binary data format allows using up to 3 times less storage space and also speeds up the matrix reading by up to 100 times.

(h) *Parallel implementation.* For now, most parts of S^{3}M are shared memory parallel using OpenMP. We have used the standard approach of parallelization by OpenMP directives for loops with independent data, such as all of the vector operations, sparse matrix-vector and sparse matrix-matrix multiplications. So far it is sufficient as the package is still a research one.But we are working on distributed memory parallelization using MPI and discover the modern ability of OpenMP to offload task to the GPU.

(i) *Open access.* The simplified version of S^{3}M is now uploaded to "github" as the basis for practical training for students [35]. When MPI parallelization is complete and the package has passed extensive testing, it will be released for public use.

5   Numerical Experiments

5.1   Symmetric Matrices

The variety of combinations of methods available within S^{3}M is very large. For brevity we test the following combinations:

- "MT-ILUC2": Biconjugate gradient stabilized method with inverse-based second-order Crout incomplete factorization as preconditioner with the matrix pre-ordered to maximize diagonal product and rescaled into I-dominant matrix. This solver is identical to the INNER_MPTILUC method from the INMOST package.
  ○ class `BICGSTAB< MaximalTransversal< ILDUC2 > >`.
- "AMG-GS": Biconjugate gradient stabilizedgradient method with algebraic-multigrid method utilizing multicolor Gauss–Seidel smoother and direct LU factorization on the coarsest level.
  ○ class `PCG< AMGRugeStuben< MulticolorGaussSeidel,LU > >`
- "AMG-CHEB": Preconditioned conjugate gradient method with algebraic-multigrid method utilizing polynomial Chebyshev smoother and direct LU factorization on the coarsest level.
  ○ class `PCG< AMGRugeStuben< Chebyshev,LU> >`
- "AMG-ILU": Biconjugate gradient stabilized method with algebraic-multigrid method utilizing Crout incomplete factorization as smoother and direct LU factorization on the coarsest level.
  ○ class `BICGSTAB< AMGRugeStuben< ILDUC,LU> >`
- "AMG-MT-GS": Biconjugate gradient stabilized method with algebraic-multigrid method utilizing multicolor Gauss–Seidel smoother and direct LU factorization on the coarsest method.
  ○ class `BICGSTAB< AMGRugeStuben< MaximalTransversal< Multicolor GaussSeidel >, LU> >`

Although it is possible to construct stronger smoothers within the package, we limit our considerations to simpler smoothers.

Fig. 3. The solution to a problem with two wells by three different schemes: tpfa (left), finite-volume for saddle-point formulation (middle), mfd (right).

First, we consider a set of symmetric matrices of various nature:

- "poisson(N)": A discretization of Poisson problem $(-\nabla \cdot \nabla p = 0)$ with Dirichlet boundary conditions on a 3D structured grid in unit cube with N^3 elements.
- "two-wells(scheme)": An anisotropic diffusion problem $(-\nabla \cdot \mathbb{K}\nabla p = 0)$ with Dirichlet boundary conditions on a unit cube with $11 \times 11 \times 11$ elements. Two elements are extracted as on Fig. 3 and the pressure is defined to represent wells. The problem is usually used to test the monotonicity properties of discretization schemes.
- "Norne(scheme)": An anisotropic diffusion problem $(-\nabla \cdot \mathbb{K}\nabla p = 0)$ with Neumann boundary conditions and two wells on realistic data of Norne oil field [36]. The solution to the problem with two different schemes is demonstrated in Fig. 4.

The scheme choices are:

- "tpfa": A conventional two-points flux approximation method. The method produces a symmetric positive definite system, but does not correctly honor problem anisotropy and results in a monotone but the incorrect solution. Due to the good properties of the system, the method remains the industry standard.
- "mfd": A mimetic finite difference method [37,38], also known as the mixed hybrid finite element method. The method requires additional degrees of freedom at interfaces that help enforce continuity of the flux. The resulting system is symmetric and has a block structure with a diagonal leading block.
- "saddle": A finite volume method applied to the mixed form of the Darcy equation [39]. Velocity and pressure unknowns are introduced for each cell. The method results in a symmetric quasi-definite system, correctly captures the anisotropy, and produces a monotone solution.

The solvers are run on a single node of the INM RAS cluster with 24 cores and 128 GB memory [40]. Each problem is solved at least 50 times or 1 s, whatever happens earlier, and the average time for the iterations (Tit), solver setup (Ts), and total time (T) are reported. In addition to that, we report the number of iterations in the Krylov solver (Nit), number of levels in multigrid (Lvl),

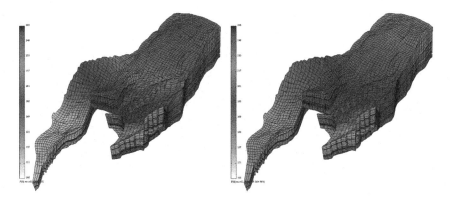

Fig. 4. The solution to a problem with two wells by two different schemes: tpfa (left), finite-volume for saddle-point formulation (right).

and memory usage during solve phase (Mem). The peak memory usage during preconditioner construction may be higher. The time required for memory allocation is hidden due to averaging. The numerical results are presented in Table 1. Most of the tests were successfully solved up to 10^7 times reduction of the initial residual norm. The AMG-CHEB method showed the fastest performance on simple matrices but the weakest reliability, while all the others demonstrate a small number of iterations and an acceptable solution time. MT-ILUC2 appears to be quite a robust choice but usually requires substantially more memory than simpler AMG-GS and AMG-CHEB methods. AMG-MT-GS shows the smallest iteration count and the best performance on linear systems of the Norne oil field. The current implementation of rescaling and reordering requires reassembling the matrix which is not memory efficient. A special class should be introduced to represent such matrices.

The need to solve problems of large dimensions requires the use of parallelization to obtain a solution in a reasonable time. To analyze the scalability of S^3M parallelization by means of OpenMP, the solution to the poisson(100) problem was chosen with the AMG-CHEB solver. The corresponding speedup values when solving from 1 to 24 OpenMP threads for the total solution time and time for iterations are given in Table 2. The insufficiently high efficiency of calculations is due not so much to the properties of S^3M code parallelization as to the hardware features of the computational nodes of the INM RAS cluster.

5.2 Schur Complement Method for Mimetic Finite Difference Scheme

With mimetic finite difference method it is possible to assemble the system in such a way that it has symmetric structure

$$A \begin{bmatrix} p_c \\ p_f \end{bmatrix} = \begin{bmatrix} B & E \\ E^T & C \end{bmatrix} \begin{bmatrix} p_c \\ p_f \end{bmatrix} = \begin{bmatrix} q \\ 0 \end{bmatrix}, \tag{3}$$

Table 1. Several test problems solved by S^3M. †The system not solved to prescribed tolerances.

		MT-ILUC2	AMG-GS	AMG-CHEB	AMG-ILUC	AMG-MT-GS
poisson(10)	T	0.0056	0.015	0.0078	0.011	0.018
size 1000	Ts	0.0047	0.0068	0.0027	0.0084	0.0072
nnz 6400	Tit	0.00088	0.0086	0.0051	0.0021	0.011
	Nit	8	9	15	7	6
	Lvl	—	5	5	5	5
	Mem	0.38 MB	0.23 MB	0.24 MB	0.59 MB	0.47 MB
poisson(100)	T	10	8.1	4.1	24.1	9.1
size 1 000 000	Ts	6.0	5.3	2.6	8.8	5.5
nnz 6 940 000	Tit	4.1	2.8	1.5	15.3	3.5
	Nit	57	25	36	63	16
	Lvl	—	10	10	10	10
	Mem	578 MB	272 MB	277 MB	761 MB	563 MB
two-wells(tpfa)	T	0.3	0.26	0.11	0.75	0.30
size 35 883	Ts	0.21	0.16	0.068	0.32	0.18
nnz 244 539	Tit	0.089	0.10	0.04	0.43	0.11
	Nit	35	17	21	43	11
	Lvl	—	8	8	8	8
	Mem	19.1 MB	9.6 MB	9.9 MB	33.7 MB	20.7 MB
two-wells(mfd)	T	7.9	3.5	1.5	236.7	5.4
size 146 745	Ts	2.7	0.74	0.29	1.1	0.83
nnz 1 651 909	Tit	5.2	2.8	1.2	235.5	4.6
	Nit	160	160	227	5001†	117
	Lvl	—	9	9	9	9
	Mem	132.5 MB	31.6 MB	32.3 MB	155.5 MB	78.0 MB
two-wells(saddle)	T	3.5	2.1	56.3	11.3	2.6
size 143 532	Ts	2.9	0.68	0.29	2.1	0.97
nnz 1 206 738	Tit	0.59	1.4	56.0	9.2	1.6
	Nit	24	59	5001†	214	30
	Lvl	—	10	10	10	10
	Mem	127.1 MB	45.1 MB	46.1 MB	106.2 MB	98.9 MB
Norne(tpfa)	T	0.60	0.35	11.5	2.3	0.34
size 44 915	Ts	0.22	0.19	0.067	0.29	0.22
nnz 316 867	Tit	0.38	0.15	11.4	2.0	0.11
	Nit	130	22	5001†	182	9
	Lvl	—	9	9	9	9
	Mem	17.9 MB	10.0 MB	10.3 MB	26.6 MB	22.4 MB
Norne(saddle)	T	3.1	34.2	78.7	130.2	25.1
size 179 660	Ts	1.6	1.7	0.50	1.6	1.9
nnz 5 069 872	Tit	1.4	32.4	78.2	128.6	23.1
	Nit	114	686	5001†	2368	234
	Lvl	—	11	11	11	11
	Mem	103.2 MB	55.6 MB	56.8 MB	110.3 MB	164.5 MB

Table 2. The speedup S and Sit for solution of problem poisson(100) by AMG-CHEB solver for total solution time and iterations stage, respectively.

p	2	4	8	12	16	24
S	1.41	1.57	1.87	2.34	2.30	2.64
Sit	1.72	2.92	4.37	4.89	5.76	5.3

where B is a diagonal matrix corresponding to the cell-centered pressure unknowns p_c. We compare the methods directly applied to system (3) with the methods applied to the Schur complement $S = C - E^T B^{-1} E$.

The methods we compare are:

- "MT-ILUC2": Similar to the method from Subsect. 5.1.
  ○ class `BICGSTAB< MaximalTransversal<ILDUC2> >`
- "S-MT-ILUC2": The same method but applied to the Schur complement.
  ○ class `SchurMFD< BICGSTAB< MaximalTransversal<ILDUC2> > >`
- "AMG-GS": Similar to the algebraic multigrid method with Gauss–Seidel smoother described in Subsect. 5.1.
  ○ class `PCG< AMGRugeStuben< MulticolorGaussSeidel, LU> > >`
- "S-AMG-GS": The same method but applied to the Schur complement.
  ○ class `SchurMFD< PCG< AMGRugeStuben< MulticolorGaussSeidel, LU > > >`

First, we consider a problem with two wells, illustrated in Fig. 3 on a coarse $11 \times 11 \times 11$, a medium $33 \times 33 \times 33$ and the finest $99 \times 99 \times 99$ grids. Then we consider a large mesh, generated from the SPE10 dataset [41] that is characterized by a high anisotropy ratio. The mesh is vertically distorted and the permeability tensor is rotated following the distortion resulting in a full 3×3 permeability tensor. The original dataset has $60 \times 220 \times 85$ dimensions resulting in 1122000 cells. We additionally consider refinement of each cell by $3 \times 3 \times 1$ resulting in 10098000 cells. All boundary conditions are Neumann, thus to avoid system singularity we pick five random cells and place a source in it with well index (connection strength) equal to 1000 and random bottom hole pressure equal to 100 ± 50 bar.

The results with the tests are presented in Table 3 and the mesh and solution are demonstrated in Fig. 5. From the results, one may find that the reduced system has less nonzeroes and is solved slightly faster. However, it requires storing the reduced system, and typically the number of iterations and multigrid levels do not change between full and reduced system. For MT-ILUC2 and S-MT-ILUC2 methods applied to "spe10(mfd)" matrices we have to tune diagonal perturbation parameter to rather large value of $\tau_D = 5 \cdot 10^{-2}$. Otherwise, the method gets either stuck in the matrix factorization step (with too low τ_D) or iterations diverge (with moderate τ_D). Nevertheless, MT-ILUC2 variants are not able to solve the system to prescribed tolerance. In the best case, they are able to reduce the initial tolerance by only $\sim 10^3$. During factorization of the 10 M test case, the methods MT-ILUC2 and S-MT-ILUC2 used 47 GB and 38 GB of memory, respectively.

Fig. 5. The porosity of SPE10 dataset (left) and a solution to the anisotropic diffusion problem (right).

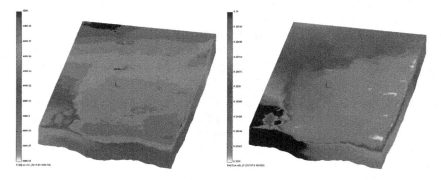

Fig. 6. Pressure (left) and water saturation (right) after 1000 days of simulation in a middle cutaway of the grid.

5.3 CPR for Two-Phase Oil Recovery Problem

Finally, we consider a two-phase oil recovery problem using a two-point flux approximation scheme on the same meshes originating from the spe10 dataset. In addition, we consider the $3 \times 3 \times 3$ refinement of each block resulting in 30 294 000 cells. We prescribe the wells in the corners of the mesh, i.e. to the cells with the first index and the last index, and assign them well indices 10 and 10 and bottom hole pressures 4500 psi and 3900 psi, respectively. Initial water saturation is $S_w = 0.25$ and pressure $P = 4000$ psi. The detailed oil and water properties are omitted for brevity, see [42] for details. The gravity and capillarity are taken into account.

We consider the first nonlinear iteration after 10 days of simulation and output the matrix. The comparison of the following solution methods is performed on this matrix:

Table 3. Comparison of solvers for linear systems arising from mimetic finite difference discretization. [†]The system not solved to prescribed tolerances.

		MT-ILUC2	S-MT-ILUC2	AMG-GS	S-AMG-GS
two-wells(mfd)	T	0.11	0.1	0.58	0.49
size 5 673	Ts	0.08	0.07	0.04	0.05
nnz 61 333	Tit	0.03	0.02	0.54	0.44
S size 4 344	Nit	34	28	130	104
S nnz 44 094	Lvl	—	—	6	6
	Mem	4.4 MB	5 MB	1 MB	1.8 MB
two-wells(mfd)	T	7	5	3.5	2.6
size 146 745	Ts	2.9	2.8	0.74	0.65
nnz 1 651 909	Tit	4.1	2.2	2.8	1.95
S size 110 862	Nit	138	116	160	139
S nnz 1 186 272	Lvl	—	—	9	8
	Mem	133 MB	147 MB	32 MB	47 MB
two-wells(mfd)	T	819	317	121	80
size 3 904 281	Ts	90	79	26.4	22.7
nnz 44 577 664	Tit	730	238	94.8	57.6
S size 2 935 440	Nit	767	431	189	201
S nnz 31 990 950	Lvl	—	—	12	11
	Mem	4 GB	4.4 GB	927 MB	1.3 GB
spe10(mfd)	T	2211	1656	279	211
size 4 525 000	Ts	73	83	35	33
nnz 51 649 000	Tit	2138	1573	245	178
S size 3 403 000	Nit	5001[†]	5001[†]	368	357
S nnz 37 063 000	Lvl	—	—	12	12
	Mem	1.9 GB	2.3 GB	1.4 GB	2 GB
spe10(mfd)	T	26062	21429	32723	10724
size 40 582 200	Ts	767	852	310	297
nnz 464 698 200	Tit	25295	20577	32413	10427
S size 30 484 200	Nit	5001[†]	5001[†]	4733	2029
S nnz 333 424 200	Lvl	—	—	14	13
	Mem	22 GB	20 GB	12 GB	18 GB

- "MT-ILUC2": Equivalent to the method from Subsect. 5.1.
    ○ class BICGSTAB< MaximalTransversal< ILDUC2 > >
- "CPR-MT-ILUC2": The method MT-ILUC2 is applied to the pre-scaled matrix according to the Quasi-IMPES method. The two-stage strategy is not used in this case.
    ○ class BICGSTAB< CPRScaling< MaximalTransversal< ILDUC2 > > >

Table 4. Comparison of solvers for the linear system after 10 days of simulation of two-phase oil recovery problem.

		MT-ILUC2	CPR-MT-ILUC2	CPR-TS	CPR-TSGS
spe10 (tpfa)	T	283	187	92	46
size 2 244 000	Ts	99	63	71	16
nnz 31 120 000	Tit	184	123.7	20	29
	Nit	405	356	38	76
	Lvl	—	—	10	10
	Mem	2.5 GB	2.2 GB	2.6 GB	1.4 GB
spe10 (tpfa)	T	4332	2940	1067	975
size 20 196 000	Ts	687	522	597	150
nnz 281 222 400	Tit	3645	2417	470	825
	Nit	1065	799	93	225
	Lvl	—	—	12	12
	Mem	21 GB	19 GB	22 GB	13 GB
spe10 (tpfa)	T	20857	20276	3758	3976
size 60 588 000	Ts	1693	1440	1564	466
nnz 845 568 000	Tit	19164	18836	2194	2510
	Nit	2156	2241	150	321
	Lvl	—	—	12	12
	Mem	54 GB	52 GB	63 GB	39 GB

- "CPR-TS" Quasi-IMPES variant of CPR with two-stage strategy. The MT-ILUC2 method is used as the full system preconditioner.
  o class `CPR< AMGRugeStuben< MulticolorGaussSeidel, LU >, Maximal Transversal< ILDUC2 >, TwoStage >`
- "CPR-TSGS": Quasi-IMPES variant of CPR with two-stage block symmetric Gauss-Seidel strategy. The MT-ILUC2 method is used as the saturation system preconditioner.
  o class `CPR< AMGRugeStuben< MulticolorGaussSeidel, LU >, Maximal Transversal< ILDUC2 >, TwoStageGaussSeidel >`

From Table 4 it follows that complex strategies significantly outperform the plain MT-ILUC2 method both in terms of speed and memory. Although MT-ILUC2 serves as a full system preconditioner for CPR-TS, its setup phase is faster for the pre-scaled matrix as also seen from the results with CPR-MT-ILUC2. Apparently, the application of the algebraic multigrid method to the pressure block significantly improves the convergence. CPR-TSGS shows to outperform CPR-TS due to an even smaller setup time but slightly larger iterations time. In 10 M case the iterations time becomes large enough to make methods on par, and in 30 M case CPR-TS is slightly faster. However, the block Gauss–Seidel variant consumes far less memory. For all system sizes we used diagonal perturbation parameter $\tau_D = 10^{-7}$ in incomplete factorization method.

Note that the peak memory, consumed during factorization step of MT-ILUC2 in spe10 example with 20 M unknowns is 58 GB for the unscaled system in MT-ILUC2 and 46 GB for the scaled system in CPR-MT-ILUC2 and CPR-TS. When only the saturations system is factored in CPR-TSGS, only 9 GB is required. In spe10 example with 60 M unknowns these numbers are 127.5 GB for the unscaled system, 115 GB for the scaled system, and only 31 GB for saturations system. Most of this memory is discarded after the factorization due to second-order fill-in, but this may severely impact the ability to perform large-scale simulations.

The simulation of 1000 days of oil recovery with the maximal time step of 50 days using CPR-TSGS on a mesh with 1 M cells is illustrated in Fig. 6. The method used from the S^3M package has solved all the linear systems arising at each nonlinear iteration, despite the high condition number for matrices of these linear systems and changes in their properties during simulation. It indicates that the method is a robust choice for the simulator.

Conclusions

The S^3M library was designed and implemented as a collection of interoperable linear solvers and preconditioners organized into a C++ header-only library. S^3M provides the ability to construct Krylov-type iterative solution methods with multilevel and algebraic multigrid preconditioners using rescaling and reordering. A description of the S^3M library structure is presented, as well as a detailed review of the literature on the methods used in the library.

The main feature of the S^3M library is its flexibility that allows finding the most optimal solution method. Numerical experiments both with a set of test linear systems and in solving real-life problems allow us to conclude about the superiority of the considered complex solution strategies.

Acknowledgements. This work has been supported by Russian Science Foundation grant 21-71-20024.

References

1. Lacroix, S., Vassilevski, Y., Wheeler, J., Wheeler, M.: Iterative solution methods for modeling multiphase flow in porous media fully implicitly. SIAM J. Sci. Comput. **25**(3), 905–926 (2003)
2. Cremon, M.A., Castelletto, N., White, J.A.: Multi-stage preconditioners for thermal-compositional-reactive flow in porous media. J. Comput. Phys. **418**, 109607 (2020)
3. Castelletto, N., White, J.A., Tchelepi, H.A.: Accuracy and convergence properties of the fixed-stress iterative solution of two-way coupled poromechanics. Int. J. Numer. Anal. Methods Geomech. **39**(14), 1593–1618 (2015)

4. PETSc - Portable, Extensible Toolkit for Scientific Computation. https://www.mcs.anl.gov/petsc. Accessed 25 March 2021
5. HYPRE: Scalable Linear Solvers and Multigrid Methods. https://computing.llnl.gov/projects/hypre. Accessed 25 March 2021
6. Trilinos - platform for the solution of large-scale, complex multi-physics engineering and scientific problems. http://trilinos.org/. Accessed 25 March 2021
7. AMGCL - a header-only C++ library for solving with AMG method. https://amgcl.readthedocs.io/en/latest/index.html. Accessed 25 March 2021
8. INMOST: a toolkit for distributed mathematical modelling. http://www.inmost.org. Accessed 25 March 2021
9. Vassilevski, Y.V., Konshin, I.N., Kopytov, G.V., Terekhov, K.M.: INMOST - Programming Platform and Graphical Environment for Development of Parallel Numerical Models on General Grids. Moscow University Press, Moscow (2013). (in Russian)
10. Vassilevski, Y., Terekhov, K., Nikitin, K., Kapyrin, I.: Parallel Finite Volume Computation on General Meshes. Springer, Cham (2020). https://doi.org/10.1007/978-3-030-47232-0
11. SAMG (Algebraic Multigrid Methods for Systems) - Efficiently solving large linear systems of equations. https://www.scai.fraunhofer.de/en/business-research-areas/fast-solvers/products/samg.html. Accessed 25 March 2021
12. Stüben, K., Ruge, J.W., Clees, T., Gries, S.: Algebraic Multigrid: from Academia to Industry, pp. 83–119. Scientific Computing and Algorithms in Industrial Simulations. Springer, Cham (2017). https://doi.org/10.1007/978-3-319-62458-7_5
13. Gries, S., Stüben, K., Brown, G.L., Chen, D., Collins, D.A.: Preconditioning for efficiently applying algebraic multigrid in fully implicit reservoir simulations. In: Proceedings of SPE Reservoir Simulation Symposium 2013, Vol. 1. The Woodlands, Texas, USA, pp. 18–20 (2013)
14. Gries, S., Metsch, B., Terekhov, K.M., Tomin, P.: System-AMG for fully coupled reservoir simulation with geomechanics. In: SPE Reservoir Simulation Conference. Society of Petroleum Engineers (2019)
15. MUMPS - MUltifrontal Massively Parallel sparse direct Solver. http://mumps.enseeiht.fr/. Accessed 25 March 2021
16. SuperLU - Supernodal LU solver for large, sparse, nonsymmetric linear systems. https://portal.nersc.gov/project/sparse/superlu/. Accessed 25 March 2021
17. PARDISO - PARallel DIrect SOlver. https://www.pardiso-project.org/. Accessed 25 March 2021
18. Terekhov, K.: Parallel multilevel linear solver within INMOST platform. In: Voevodin, V., Sobolev, S. (eds.) RuSCDays 2020. CCIS, vol. 1331, pp. 297–309. Springer, Cham (2020). https://doi.org/10.1007/978-3-030-64616-5_26
19. Bollhöfer, M.: A robust ILU with pivoting based on monitoring the growth of the inverse factors. Linear Algebra Appl. **338**(1–3), 201–218 (2001)
20. Bollhöfer, M., Saad, Y.: Multilevel preconditioners constructed from inverse-based ILUs. SIAM J. Sci. Comput. **27**(5), 1627–1650 (2006)
21. Deveci, M., Trott, C., Rajamanickam, S.: Multithreaded sparse matrix-matrix multiplication for many-core and GPU architectures. Parallel Comput. **78**, 33–46 (2018)
22. Kaporin, I.E.: High quality preconditioning of a general symmetric positive definite matrix based on its $U^T U + U^T R + R^T U$-decomposition. Numer. Lin. Alg. Applic. **5**(6), 483–509 (1998)
23. Li, N., Saad, Y., Chow, E.: Crout versions of ILU for general sparse matrices. SIAM J. Sci. Comput. **25**(2), 716–728 (2003)

24. Kaporin, I.E.: Scaling, reordering, and diagonal pivoting in ILU preconditionings. Russ. J. Numer. Anal. Math. Model. **22**(4), 341–375 (2007)
25. Olschowka, M., Neumaier, A.: A new pivoting strategy for Gaussian elimination. Linear Algebra its Appl. **240**, 131–151 (1996)
26. Sinkhorn, R.: Diagonal equivalence to matrices with prescribed row and column sums. II. Proc. Amer. Math. Soc. **45**, 195–198 (1974)
27. Duff, I.S., Kaya, K, Ucar, B.: Design, implementation and analysis of maximum transversal algorithms. Technical Report TR/PA/10/76 (2010)
28. Stüben, K.: A review of algebraic multigrid. In: Numerical Analysis: Historical Developments in the 20th Century. Elsevier, pp. 331–359 (2001)
29. Ruge, J.W., Stüben, K.: Algebraic multigrid. In: Multigrid Methods. Society for Industrial and Applied Mathematics, pp. 73–130 (1987)
30. Cusini, M., Lukyanov, A., Natvig, J.R., Hajibeygi, H.: A constrained pressure residual multiscale (CPR-MS) compositional solver. In: Proceedings of ECMOR XIV-14th European Conference on the Mathematics of Oil Recovery. Catania, Sicily, Italy (2014)
31. Lacroix, S., Vassilevski, Y.V., Wheeler, M.F.: Decoupling preconditioners in the implicit parallel accurate reservoir simulator (IPARS). Numer. Lin. Alg. with Appl. **8**(8), 537–549 (2001)
32. Gries, S.: System-AMG approaches for industrial fully and adaptive implicit oil reservoir simulations. Ph.D Thesis. Der Universität zu Köln, Köln (2016)
33. Kayum, S., Cancellierei, M., Rogowski, M., Al-Zawawi, A.A.: Application of algebraic multigrid in fully implicit massive reservoir simulations. In: Proceedings of SPE Europec featured at 81st EAGE Conference and Exhibition. SPE-195472-MS (2019)
34. De Sterck, H., Yang, U.M., Heys, J.J.: Reducing complexity in parallel algebraic multigrid preconditioners. SIAM J. Matrix Anal. Appl. **27**(4), 1019–1039 (2006)
35. Supplimentary material for MIPT course Practical methods for system solutions. https://github.com/kirill-terekhov/mipt-solvers.git. Accessed 25 March 2021
36. Open datasets from open porous media initiative. https://opm-project.org/?page_id=559. Accessed 25 March 2021
37. Brezzi, F., Lipnikov, K., Shashkov, M.: Convergence of the mimetic finite difference method for diffusion problems on polyhedral meshes. SIAM J. Numer. Anal. **43**(5), 1872–1896 (2005)
38. Abushaikha, A.S., Terekhov, K.M.: A fully implicit mimetic finite difference scheme for general purpose subsurface reservoir simulation with full tensor permeability. J. Comput. Phys. **406**, 109194 (2020)
39. Terekhov, K.M., Vassilevski, Y.V.: Finite volume method for coupled subsurface flow problems, i: Darcy problem. J. Comput. Phys. **395**, 298–306 (2019)
40. INM RAS cluster. http://cluster2.inm.ras.ru/en. Accessed 25 March 2021
41. SPE10 dataset. https://www.spe.org/web/csp/datasets/set01.htm
42. Nikitin, K., Terekhov, K., Vassilevski, Y.: A monotone nonlinear finite volume method for diffusion equations and multiphase flows. Comput. Geosci. **18**(3–4), 311–324 (2014)

Resource-Independent Description of Information Graphs with Associative Operations in Set@l Programming Language

Ilya I. Levin[1] ⓘ, Alexey I. Dordopulo[2] ⓘ, Ivan Pisarenko[2(✉)] ⓘ, Denis Mikhailov[1], and Andrey Melnikov[3]

[1] Academy for Engineering and Technology, Institute of Computer Technologies and Information Security, Southern Federal University, Taganrog, Russia
iilevin@sfedu.ru
[2] Supercomputers and Neurocomputers Research Center, Taganrog, Russia
{dordopulo,pisarenko}@superevm.ru
[3] "InformInvestGroup" CJSC, Moscow, Russia
ak@iigroup.ru

Abstract. Usually, an information graph with associative operations has a sequential ("head/tail") or parallel ("half-splitting") topology with invariable quantity of operational vertices. If computational resource is insufficient for the implementation of all vertices, the reduction transformations of graphs with basic structures do not allow for the creation of an efficient resource-independent program for reconfigurable computer systems. In this paper, we propose to transform the topology of a graph with associative operations into a combined variant with sequential and parallel fragments of calculations. The resultant combined topology depends on computational resource of a reconfigurable computer system, and such transformation provides the improvement of specific performance for the reduced computing structure. We develop an algorithm for the conversion of the initial sequential graph to various combined topologies or to the limiting case of the "half-splitting" topology with regard to available hardware resource. This technique is described using the Set@l programming language.

Keywords: Associative operations · Resource-independent programming · Reconfigurable computer systems · Performance reduction · Set@l · "Head/tail" and "half-splitting" attributes

1 Introduction

Associativity is a fundamental property of binary operations, which determines the independence of the calculation result from the order of actions [1, 2]. There are two basic variants for the topology of such graphs with the same number of vertices: sequential (a linear or a "head/tail" structure) and parallel (a pyramid or a "half-splitting" structure) [3–5]. In fact, besides two aforementioned cases, multiple combined topologies exist, and they are composed of alternating parallel and sequential fragments of calculations

© Springer Nature Switzerland AG 2021
V. Malyshkin (Ed.): PaCT 2021, LNCS 12942, pp. 74–87, 2021.
https://doi.org/10.1007/978-3-030-86359-3_6

[6]. Usually, combined topologies are more efficient in terms of time required for the problem solution when there is a lack of hardware resources in the case of processing large amounts of data.

To scale parallel calculations for the solution of applied problems on computer systems with reconfigurable architecture [7–10], performance reduction methods are used [11]. Reduction transformations of associative information graphs with sequential and parallel topologies do not provide the creation of an efficient resource-independent parallel program. Therefore, for the synthesis of computational structure for graphs with sequential and parallel topologies, it is reasonable to use the combined structure, which contains isomorphic subgraphs with a maximal degree of operation parallelism. In paper [12], it is shown that for a given computational resource it is possible to synthesize an information equivalent graph containing set of hardware implemented isomorphic subgraphs and a single block for processing intermediate results with a sequential topology. In contrast to traditional parallel programming languages, the Set@1 (Set Aspect-Oriented Language) [13–15] allows to describe many variants of topologies for different configurations of a computer system in the form of an entire aspect-oriented program and shift between different implementations by converting types and partitions of sets without the source code changing.

Paper [12] provides the example of resource-independent topology description for information graph with associative operations in the Set@1 programming language. However, it considers the idealized case of operational vertices with a unit latency. In practice, the latency of a vertex performing an associative operation (e.g. the addition or multiplication of fixed-point numbers) typically exceeds one cycle. So, the transformation technique proposed in [12] does not provide the obtainment of efficient pipeline implementation of calculations: if the delay of feedback circuit is increased to the latency of an operational vertex, the addition of all partial sums appeared at the output of the pyramid structure is not ensured. To take into account the non-unit latency of operational vertices and to form the correct sequence of partial sums' adding, we propose to modify the topology of the information graph with associative operations. By combining sequential and parallel fragments of calculations, it is possible to synthesize the topology in accordance with the available amount of computing resource and latency of the operating vertex and reduce the total time to solve the problem by providing a dense data flow at the input of the computing structure. The description of the information graph in the aspect-oriented Set@1 programming language as a set with parameterized partition allows to transform the topology when the source code remains unchanged.

2   Topological Modification of Graphs with Associative Operations

If the available computational resource R of a reconfigurable computer system is not sufficient to realize the full information graph with associative operations and the latency L of one vertex exceeds one cycle, the conversion technique for groups of associative vertices proposed in paper [12] does not ensure an efficient pipeline implementation of the problem, since an intermediate data tuple will be formed at the output of the computational structure, but not the required final result of operations. For the further synthesis of an efficient computational structure, additional transformations of the original topology are generally needed.

Let us divide all operational vertices of the original information graph with the "head/tail" topology into l groups $Gr_1, Gr_2, \ldots, Gr_l$ of $Q = \lfloor (N - l + 1)/l \rfloor$ vertices in each, where N is the dimension of the processed data array ($N \gg l$), $\lfloor \ \rfloor$ is the floor function. Between neighboring groups, leave one vertex for processing intermediate data, as shown in Fig. 1-a. If the last group includes less than Q elements, add several operational vertices in order to reach the required dimension. Next, convert the vertices in each selected group according to the algorithm discussed in paper [12]. As a result of this transformation, we obtain the topology given in Fig. 1-b. The i-th group Gr_i contains:

- m isomorphic subgraphs $psG_{(i-1) \cdot m+1}, psG_{(i-1) \cdot m+2}, \ldots, psG_{i \cdot m}$ with the "half-splitting" topology and k vertices in each one;
- one block for processing intermediate results hsG_i with the "head/tail" topology and $(m - 1)$ operational vertices.

The parameters of the information graph partition m, k and l depend on the amount of available computational resource R and on the latency of operational vertex L. They are determined further during the computational structure synthesis.

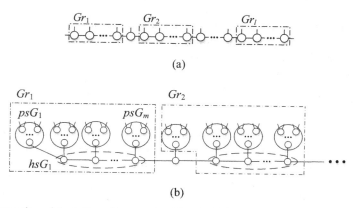

(a)

(b)

Fig. 1. Conversion of the information graph with associative operations: the selection of groups of operational vertices in the original "head/tail" graph (a); vertex reorganization in each group according to the algorithm developed in [12] (b)

In the next stage of topological modification, we apply the transformation, the principle of which is represented in Fig. 2. Consider adjacent "head/tail" blocks hsG linked by the single operational vertex (Fig. 2-a). If the pairs "single vertex + group of vertices" are converted according to the principle discussed in paper [12], we obtain the fragment of topology shown in Fig. 2-b. After performing this transformation, the set of single vertices is rearranged so that each of them has a common arc with at least one more vertex. Then, the transformation [12] is carried out again. As a result, the graph fragments are brought to pyramid view, as demonstrated in Fig. 2-c.

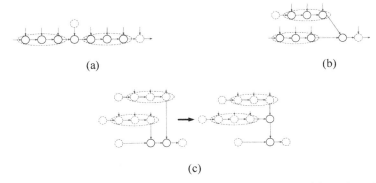

Fig. 2. Conversion of sequential fragments in the information graph: initial topology (a); "single vertex + vertex group" transformation (b); bringing the group of single vertices to pyramidal form (c)

After the conversion of the topology shown in Fig. 1-b according to the principle proposed in Fig. 2, we obtain the modified information graph with associative operations G (see Fig. 3-a). It contains l isomorphic subgraphs $Gr_1, Gr_2, \ldots, Gr_l$ connected by means of the "half-splitting" principle through the pyramid subgraph pG. Each subgraph Gr_i is formed using the combined principle "half-splitting + head/tail": it includes m isomorphic and pyramid subgraphs $psG_{(i-1)\cdot m+1}, psG_{(i-1)\cdot m+2}, \ldots, psG_{i\cdot m}$ and one sequential unit hsG_i that calculates intermediate results and incorporates $(m-1)$ operational vertices. In turn, each pyramid subgraph psG_j processes k elements of the original data array. It is worth noting that the final topology and the original "head/tail" structure have the same number of operational and input vertices, but the proposed topological transformation allows for the synthesis of more efficient computing structure for arbitrary hardware resource and latency of operational vertex. If $l = 1$, the subgraph pG does not contain operational vertices and the topology corresponds to the limit case described in paper [12].

3 Synthesis of Computing Structure

Treat the conversion of the information graph topology shown in Fig. 3 into a computing structure for the structural and procedural implementation of a problem on a computer system with reconfigurable architecture. At the first stage, the information-independent subgraphs $psG_{(i-1)\cdot m+1}, psG_{(i-1)\cdot m+2}, \ldots, psG_{i\cdot m}$ in each subgraph Gr_i are transformed into a *subcadr* C_i [16] (see Fig. 3-b). In the *cadr*, tuples of data elements are supplied to the inputs of C_i. The block hsG_i of information-dependent operational vertices form an additional vertex v_i with the feedback circuit delay of l cycles. However, the computing structure obtained at this stage is inefficient, since it is impossible to supply the data element to the inputs of its fragments $mGr_1, mGr_2, \ldots, mGr_l$ as a dense stream. In order to process all intermediate results correctly, it is necessary to cut the input data stream so that its duty cycle (data supply interval) becomes equal to l. Obviously, the l-fold increase in duty cycle leads to the same growth of time required for the problem solution.

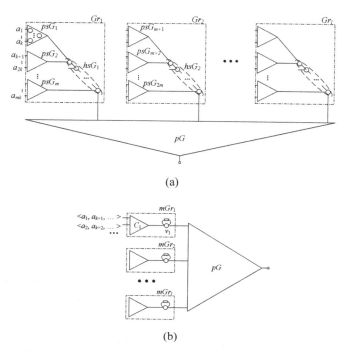

Fig. 3. The modified topology of the information graph with associative operations (a); the first step of forming the computing structure corresponding to the modified topology of the information graph with associative operations (b)

The computing structure given in Fig. 3-b processes intermediate results using the pG fragment with the "half-splitting" topology. In this case, l values should be supplied to the pG inputs simultaneously, and such mode leads to increase in duty cycle of data flows at the inputs of the blocks mGr_1, mGr_2, ..., mGr_l. To reduce the time of the problem solution, it is reasonable to perform the optimization of the first-stage computing structure. At the second stage of the transformation, the pG fragment is replaced with the advanced pG^* structure, which performs associative operation on l elements per l cycles: the correct result of processing of l operands appears at the output during the last l-th cycle. For this purpose, every data element is delayed for the corresponding number of cycles from 0 to $(l - 1)$ as shown in Fig. 4. At the same time, the blocks mGr_1, mGr_2, ..., mGr_l are replaced with the single fragment MG, to the inputs of which flows or tuples of input data with the duty cycle of 1 are supplied. In this case, the data streams at each input are represented as nested tuples of elements of the original set.

Consider the pG^* fragment of the computing structure given in Fig. 4. Operational vertices located in this part of the structure perform $(l - 1)$ actions in l cycles. So, at every time step only one operation is performed, and this feature allows to reduce the number of vertices in pG^*. For the convenience of further transformations, rearrange the pG^* delays as shown in Fig. 5 to highlight isomorphic blocks W that contain a vertex with a single delay at one of the inputs.

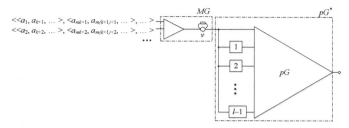

Fig. 4. The second stage of forming the computing structure that corresponds to the modified topology of the information graph with associative operations

As mentioned above, fragments W (see Fig. 5-a) calculate intermediate results sequentially. Therefore, according to the "embedded pipeline" principle [16], it is reasonable to replace them by the structure represented in Fig. 5-b. The serial connection of the multiplexer (MX in Fig. 5-b) and demultiplexer (DMX) with the delays can be reduced by leaving the single W fragment. After the analogous transformation of each iteration in the pyramid P (see Fig. 5-a), we obtain the accumulating pipeline structure with operational vertices and delays that is shown in Fig. 6.

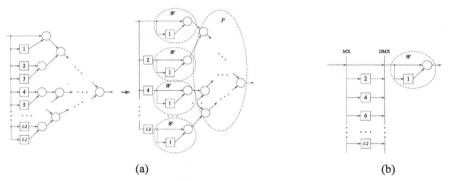

(a)   (b)

Fig. 5. The transformation of the pG^* fragment of the computing structure (a); the further conversion of multiple W blocks with delays (b)

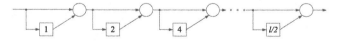

Fig. 6. The accumulating pipeline structure with associative operations

At the output of the block v with the feedback (see Fig. 4), we obtain l intermediate results $y_1 = f(x_1, x_{l+1})$, $y_2 = f(x_2, x_{l+2})$, ..., $y_l = f(x_l, x_{2l})$, and there is a need to perform additional operations f on these data elements. In order to calculate $f(y_1, y_2), f(y_2, y_3), ..., f(y_{l-1}, y_l)$, the first vertex of the accumulating pipeline structure (see Fig. 6) is supplied with the same stream of operands delayed for one cycle. In 2nd, 4th, 8th etc. cycles, at the output of the first vertex, we get the result of executing operation f on data elements y_1 and y_2, y_3 and y_4, y_5 and y_6 etc. At this stage, the size of the intermediate data sequence is $l/2$. Therefore, for the second vertex, data operands are delayed by two cycles, and its output gives $l/4$ results f (y_1, y_2, y_3, y_4), f (y_5, y_6, y_7, y_8) etc. Similarly, in the third and fourth vertices, the delays are 4 and 8 cycles, respectively. At the output of the last vertex with the number $\log_2(l) + 1$, after passing the entire array of input data, we obtain the result of performing associative operation f on two operands that is equal to the result of processing all elements included in the initial data array.

It is possible to organize the described computing structure only if the value of l is equal to the integer degree of two. Otherwise, the data flow is extended by neutral elements, and the feedback circuit is supplemented with additional delay elements up to the following value:

$$l = 2^{\lceil \log_2 L \rceil}, \tag{1}$$

where $\lceil\ \rceil$ is the ceil function; L is the latency of the operational vertex. Only in this case the correct result is achieved, because each vertex of the structure in Fig. 6 perform operation f on exactly half of the operands received at the inputs of the previous block. The final form of the computing structure is shown in Fig. 7. Using the value of l calculated by formula (1), it is possible to estimate other parameters of the information graph partition according to the following equations:

$$k = \lfloor R/R_0 \rfloor - \log_2 l; \tag{2}$$

$$m = \left\lceil \frac{N}{l \cdot (\mathbf{floor}(R/R_0) - \log_2 l)} \right\rceil, \tag{3}$$

where R_0 is the hardware resource required for the implementation of single operational vertex. It is worth noting that if $l = 1$, the computing structure discussed in paper [12] is obtained.

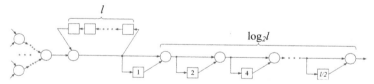

Fig. 7. The resultant computing structure that corresponds to the modified topology of the information graph with associative operations shown in Fig. 3-a

4 Description of Basic Topologies for Information Graphs with Associative Operations in Set@l Programming Language

Traditional parallel programming methods tend to operate with information graphs with fixed structures. Therefore, their application for the description of topological transformations in accordance with the proposed algorithm (see Fig. 1, 2 and 3) is quite cumbersome and inefficient. In terms of classical programming languages, the code of resource-independent program that implements the aforementioned graph conversions consists of multiple subprograms connected by conditional operators. Each subprogram specifies only one variant of topology. In contrast to the multiprocedural paradigm, the capabilities of the Set@1 architecture-independent programming language allows to describe the principles of graph constructing in the form of special processing method attributes assigned to the basic set of data A. In this case, the program source code describes not individual implementations, but the whole set of possible graph topologies for given dimension of the computational problem. Aspects select specific topology with regard to the values of computer system's configuration parameters. To modify the structure of the information graph, it is enough to edit the type and partition of the basic set A, while the source code of the program remains unchanged in general.

Let us consider the description of the "head/tail" and "half-splitting" principles for the standard topologies of information graphs with associative operations (see Fig. 8) in the Set@1 programming language.

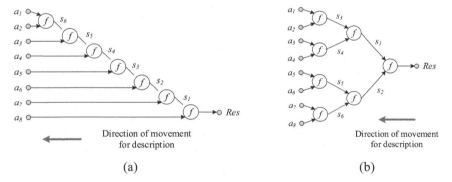

Fig. 8. Topologies of graphs with two-place associative operations f based on the "head/tail" (a) and "half-splitting" (b) principles

Using the `attribute` syntax construct [17], introduce additional features with the code given in Fig. 9. Attribute of the basic associative operation f is specified in generalized form: the certain operation type `Op` is declared in another module of parallel program and can take different values (e.g. "+" or "*"). The `operand` directive describes the types of objects to which an attribute can be assigned.

Figure 10 demonstrates the description of the "head/tail" principle (`H/T`, see graph in Fig. 8-a) in the Set@1 programming language. The attribute of serial operations `Lf` is declared recursively (`Rec`) using the previously mentioned basic binary operation `f` (see code in Fig. 9) and defines the relation between the set of processed data `A` and the result of calculations `Res`. The graph is built in the direction from the output vertex to the inputs. At each iteration, the sequential set of operations `Lf` on the elements of collection `A` can be represented as the combination of sequential operations `Lf` on the "tail" of set `A` and separate vertex `f`. The "head" of set `A` and intermediate result s of `Tail(A)` sequential processing are the inputs of the selected vertex, and its output `Res`

`attribute [f(a,b,c)|type(f)=Op]:`
　`operand(element(a,b,c),attribute(Op));`
　`c=f(a,b);`
`end(Op2);`
	`attribute Head(A):`
	　`operand(set(A),element(Head(A)));`
	　`Head(A)=A(1);`
	`end(Head);`

`attribute Tail(A):`
　`operand(set(A),set(Tail(A)));`
　`Tail(A)=dif(A,A(1));`
`end(Tail);`
	`attribute d2(A,A1,A2):`
	　`operand(set(A,A1,A2));`
	　`n=card(A);`
	　`A1=(A(k)|k<=n/2);`
	　`A2=dif(A,A1);`
	`end(d2);`

Fig. 9. Attributes of abstract binary operation (`f`), of allocation of "head" (`Head`) and "tail" (`Tail`) in a set, and of the partition of a collection into two subsets with the same cardinality (`d2`) in the Set@1 programming language

is the final or intermediate result of calculations (see line 5 in Fig. 10). Syntax construct `break[<condition>:<operation>]` (line 4 in Fig. 10) highlights the termination condition of recursion and describes the operation that completes formation of the graph structure. If the condition is met, the last operator `Lf` is converted into a special vertex, the inputs of which are supplied with two remaining elements of set `A`.

```
(1)    attribute [Lf(A,Res)|Lf=Rec(f),type(A)='H/T']:
(2)        operand(set(A),element(Res));
(3)        element(s);
(4)        Lf(A,Res)=break[card(Tail(A))=1: f(Head(Tail(A)),Head(A),Res)],
(5)                    union[Lf(Tail(A),s),f(s,Head(A),Res)];
(6)    end(Lf);
```

Fig. 10. The code of `Lf` attribute in the Set@1 programming language; it implements the "head/tail" principle (`H/T`) for constructing the information graph with associative operations `f`

Using the same method of recursive description, it is possible to define the parallel "half-splitting" principle of graph constructing (`DIV2`, graph in Fig. 8-b) in the Set@1 programming language. Figure 11 demonstrates the code where information graph is described in the direction from the output vertex to the input. At each iteration, attribute `d2` divides original set `A` into two subsets `A1` and `A2` with the same number of elements (line 3 in Fig. 11). In this case, the pyramid of operations `Pf` on elements in `A` can be represented as the combination of parallel operations `Pf` on elements in subset `A1`, parallel operations `Pf` on elements in subset `A2` and separate vertex `f` (line 6 in Fig. 11). Intermediate results `s1`, `s2` of performing pyramid operations `Pf` on subsets of the collection `A` are the inputs of this vertex `f`, and its output `Res` is the final or intermediate result of calculations. Recursion completes if the condition shown in line 5 of Fig. 11 is met. Parallelization of calculations is achieved by doubling the number of recursion branches at each step of the transformation.

Thus, the architecture-independent Set@1 programming language allows to describe the basic principles for the construction of graphs with single-output associative operations in the form of special attributes of processing method `H/T` and `DIV2`, which are assigned to the set of input data `A`. In contrast to previously proposed parallelism

```
(1)  attribute [Pf(A,Res)|Pf=Rec(f),type(A)='DIV2')]:
(2)     operand(set(A),element(Res));
(3)     d2(A,A1,A2);
(4)     element(s1,s2);
(5)     Pf(A,Res)=break[card(A1)=1 and card(A2)=1: f(Head(A1),Head(A2),Res)],
(6)               union[Pf(A1,s1),Pf(A2,s2),f(s1,s2,Res)];
(7)  end(Pf);
```

Fig. 11. The code of Pf attribute in the Set@1 programming language; it implements the "half-splitting" principle (DIV2) for the description of parallel information graph with associative operations f

types par, seq, pipe, conc and imp [13, 14] that specifies methods of calculations' parallelizing, these attributes determine the general structure of an information graph and modify it according to the architecture and configuration of a parallel computer system. If the "head/tail" and "half-splitting" principles are described once, it is possible to obtain various topologies without the change in program's source code. Some examples of code fragments using H/T and DIV2 processing types to synthesize different information graphs are given in Fig. 12.

Line of adders:	Pyramid of adders:
G=Gf(A,Res);	G=Gf(A,Res);
Gf=(Rec(f),type(A)='H/T');	Gf=(Rec(f),type(A)='DIV2');
type(f)='+';	type(f)='+';
Line of multipliers:	Pyramid for maximum search:
G=Gf(A,Res);	G=Gf(A,Res);
Gf=(Rec(f),type(A)='H/T');	Gf=(Rec(f),type(A)='DIV2');
type(f)='*';	type(f)='max';

Fig. 12. Code fragments that employs the "head/tail" (H/T) and "half-splitting" (DIV2) attributes of set processing to describe different information graphs based on associative single-output operations

The structure of the information graph G (see program code in Fig. 12) is determined by the relation Gf between the processed set A and the result of calculations Res. For the user, it is enough to change only the type of collection A to obtain an information graph with a completely different interconnection structure, while the generalized descriptions of the "head/tail" and "half-splitting" attributes remain unchanged. The type of basic associative operation f defines the functionality of operational vertices in the synthesized information graph.

5 Development of Resource-Independent Program in Set@1 Language

Utilizing the processing method attributes H/T and DIV2 (see code in Fig. 10 and 11), it is possible to describe the topological transformation of the information graph with associative operations according to the amount of available computing resource R and latency of operational vertex L as a change in the typing and partitioning of processed data set A. In general, collection A should have the following form, which ensures the further conversion into the efficient computing structure:

$$A = \mathbf{DIV2}\,[AsG_1,\, AsG_2,\, \ldots,\, AsG_l]; \tag{4}$$

$$AsG_i = \mathbf{H/T}\{Ap_{(i-1)\cdot m+1},\, Ap_{(i-1)\cdot m+2},\, \ldots,\, Ap_{i\cdot m}\}; \tag{5}$$

$$Ap_j = \mathbf{DIV2}\,[a_{(j-1)\cdot k+1},\, a_{(j-1)\cdot k+2},\, \ldots,\, a_{j\cdot k}], \tag{6}$$

where AsG_i is the set of input vertices of the i-th subgraph Gr_i with the "half-splitting + head/tail" topology (see Fig. 3-a); Ap_j is the set of input vertices of the j-th subgraph psG_j with the pyramid structure; a_z is the z-th element of the processed set A. In the source code of resource-independent program in the Set@l language, set A that defines the information graph topology is declared generally in accordance with Eqs. (4)–(6) (see Fig. 13). Within the source code, parameters m, k, l and, consequently, the specific partition of the collection A and the topology of the information graph are not defined.

```
(1)   A=DIV2(AsG(i)|i in (1...1));
(2)   AsG(i)=H/T(Ap(j)|j in ((i-1)*m+1...i*m));
(3)   Ap(j)=DIV2(a(z)|z in ((j-1)*k+1...j*k));
```

Fig. 13. General description of the topology of the information graph with associative operations (see Fig. 3-a) in the source code of the resource-independent program in Set@l

After the translation of the source code shown in Fig. 13, the following sets with imposed types of parallelism are formed:

$$G = \vec{[}\{Gr_1,\, Gr_2,\, \ldots,\, Gr_l\},\, p\vec{G}]; \tag{7}$$

$$Gr_i = \vec{[}\{psG_{(i-1)\cdot m+1},\, psG_{(i-1)\cdot m+2},\, \ldots,\, psG_{i\cdot m}\},\, h\vec{sG_i}]; \tag{8}$$

$$hsG_i = \vec{\{}v_{i,1},\, v_{i,2},\, \ldots,\, v_{i,m-1}\vec{\}}, \tag{9}$$

where G is the full information graph; Gr_i, psG_j, hsG_i are the subgraphs allocated in Fig. 3-a; $v_{i,q}$ is the vertex of the sequential unit hsG_i; $\vec{\{}\ \vec{\}}$, $\{\}$ are the parallel-dependent and parallel-independent types of processing. During the formation of the computing structure shown in Fig. 3-b, each Gr_i set is converted as follows:

$$mGr_i = \,<psG_{(i-1)\cdot m+1},\, \vec{\{}psG_{(i-1)\cdot m+2}, v_{i,1}\},\, \ldots,\, \vec{\{}psG_{i\cdot m}, v_{i,m-1}\}>, \tag{10}$$

where $< \,>$ is the pipeline processing type. The code of aspect in Set@l that performs this transformation is given in Fig. 14.

After the next stage of the transformation, when we obtain the computing structure shown in Fig. 4, set G takes the following form:

$$mG_1 = \vec{\{}\,<mGr_1, mGr_2, \ldots, mGr_l>,\, pG^*\vec{\}}; \tag{11}$$

```
(1)    Gr(i) -> mGr(i);
(2)    mGr(i)=pipe(psG((i-1)*m+1),dG);
(3)    dG=pipe( z(j)|j in (1...m-1) and z(j)=conc(psG((i-1)*m+1+j),v(i,j))));
```

Fig. 14. Code of the aspect in the Set@l programming language, which forms the computing structure, shown in Fig. 3-b

$$pG^* = \vec{\{}ds, pG\vec{\}};\tag{12}$$

$$ds = \{del(0), del(1), \ldots, del(l-1)\},\tag{13}$$

where $del(i)$ is the set element that describes the circuit unit providing delay for i cycles. The code of the aspect in the Set@l programming language, which defines the transition from the original collection G to the modified set mG_1, is represented in Fig. 15-a. At the final transformation stage (when we form the computing structure shown in Fig. 7), the pyramid fragment pG^* is replaced with the updated structure mpG with multiple delay blocks (see Fig. 6):

$$mG_2 = \vec{\{} < mGr_1, mGr_2, \ldots, mGr_l >, pG^*\vec{\}};\tag{14}$$

$$mpG = \vec{\{}\{del(0), del(1)\}, f_1, \{del(0), del(2)\}, f_2, \ldots, \{del(0), del(l/2)\}, f_{\log_2 l}\vec{\}},\tag{15}$$

where f_i is the i-th operational vertex in the accumulating fragment of the computing structure. The program code in the Set@l language that corresponds to the considered conversion of the original set mG_1 into a set mG_2 is shown in Fig. 15-b.

The considered technique for resource-independent description of information graphs with associative operations allows to synthesize the topology and corresponding computing structure optimized for the specified hardware resource R of reconfigurable computer system and basic operation latency L. Herewith, a dense flow of input data is provided, and the time of the problem solution is reduced by approximately L times compared to the non-optimized implementation.

From the user's point of view, the topological transformation and creation of an efficient computing structure are described by the following one-line code fragment in the Set@l programming language

```
G=AsG(R,L,f),
```

:where AsG is the attribute that forms the information graph with associative operations; f is the type of basic operation (addition, multiplication, search for a maximum or minimum etc.).

```
(1)  G -> mG1;
(2)  mG1=conc(Gpipe,pG*);
(3)  Gpipe=pipe(mGr(b)|b in (1...1));
(4)  pG*=conc(ds,pG);
(5)  ds=par(del(c)|c in (0...1-1));
```

(a)

```
(1)  mG1 -> mG2;
(2)  mG2=conc(Gpipe,mpG);
(3)  mpG=conc(p(y)|y in (1...log2(1)) and
                  p(y)=conc(par(del(0),del(2^(y-1))),f(y)));
```

(b)

Fig. 15. Fragments of the Set@l program code describing the formation of the computing structures shown in Fig. 4 and Fig. 7.

6 Conclusions

Thus, in this paper, we propose the method that rearrange the vertices of information graph with associative operations and perform further optimization of computing structure in order to reduce the time of problem solution by the number of times corresponding to the latency of operational vertex. The designed general graph topology combines sequential and parallel fragments of calculations and provides the formation of dense data flow at available hardware resource. The developed method extends the technique considered in our previous paper [12] to multiple cases when the latency of associative vertex exceeds one cycle. The architecture-independent Set@l programming language allows to describe the transformations in compact resource-independent form. In comparison to traditional parallel programming languages, in which the change of information graph topology requires the modification of the program source code, Set@l specifies many implementation variants in one program. The synthesis of particular computing structure is performed automatically according to the configuration parameters specified by the user (the amount of available computational resource and latency of basic operation).

Acknowledgments. The reported study was funded by the Russian Foundation for Basic Research, project number 20-07-00545.

References

1. Knuth, D.E.: The Art of Computer Programming. Vol. 4A: Combinatorial Algorithms. Addison-Wesley Professional, Boston (2011)
2. Novikov, F.: Discrete Mathematics. 3rd edn. Piter, Saint Petersburg (2019). (in Russian)
3. Karepova, E.D.: The Fundamentals of Multithread and Parallel Programming. Siberian Federal University Publishing, Krasnoyarsk (2016). (in Russian)
4. The problem of array element summation (in Russian). https://parallel.ru/fpga/Summ2
5. Starchenko, A.V., Bercun, V.N.: Methods of Parallel Computing. Tomsk University Publishing, Tomsk (2013). (in Russian)

6. Efimov, S.S.: Review of parallelizing methods for algorithms aimed at solution of certain problems of computational discrete mathematics. Math. Struct. Model. **17**, 72–93 (2007). (in Russian)

7. Kalyaev, I.A., Levin, I.I., Semernikov, E.A., SHmojlov, V.I.: Evolution domestic of multichip reconfigurable computer systems: from air to liquid cooling. SPIIRAS Proc. (1), 5–31 (2017). (in Russian). https://doi.org/10.15622/sp.50.1

8. Mittal, S., Vetter, J.: A survey of CPU-GPU heterogeneous computing techniques. ACM Comput. Surv. **47**(4), art. no. 69 (2015). https://doi.org/10.1145/2788396

9. Waidyasooriya, H.M., Hariyama, M., Uchiyama, K.: Design of FPGA-Based Computing Systems with OpenCL. Springer, Cham (2018). https://doi.org/10.1007/978-3-319-68161-0

10. Tessier, R., Pocek, K., DeHon, A.: Reconfigurable computing architectures. Proc. IEEE **103**(3), 332–354 (2015)

11. Levin, I.I., Dordopulo, A.I.: On the problem of automatic development of parallel applications for reconfigurable computer systems. Comput. Technol. **25**(1), 66–81 (2020). (in Russian)

12. Levin, I.I., Dordopulo, A.I., Pisarenko, I.V., Mihaylov, D.V.: Description of graphs with associative operations in Set@l programming language. Izv. SFedU. Eng. Sci. **3**, 98–111 (2020). (in Russian)

13. Pisarenko, I.V., Alekseev, K.N., Mel'nikov, A.K.: Resource-independent representation of sorting networks in Set@l programming language. Herald Comput. Inform. Technol. **11**, 53–60 (2019). (in Russian)

14. Levin, I.I., Dordopulo, A.I., Pisarenko, I.V., Melnikov, A.K.: Aspect-oriented Set@l language for architecture-independent programming of high-performance computer systems. In: Voevodin, V., Sobolev, S. (eds.) RuSCDays 2019. CCIS, vol. 1129, pp. 517–528. Springer, Cham (2019). https://doi.org/10.1007/978-3-030-36592-9_42

15. Levin, I.I., Dordopulo, A.I., Pisarenko, I.V., Mel'nikov, A.K.: Architecture-independent Set@l programming language for computer systems. Herald Comput. Inform. Technol. **3**, 48–56 (2019). (in Russian)

16. Kalyaev, A.V., Levin, I.I.: Modular-Expandable Multiprocessor Systems with Structural and Procedural Organization of Calculations. YAnus-K, Moscow (2003)

17. Levin, I.I., Dordopulo, A.I., Pisarenko, I.V., Melnikov, A.K.: Objects of alternative set theory in Set@l programming language. In: Malyshkin, V. (ed.) PaCT 2019. LNCS, vol. 11657, pp. 18–31. Springer, Cham (2019). https://doi.org/10.1007/978-3-030-25636-4_3

High-Level Synthesis of Scalable Solutions from C-Programs for Reconfigurable Computer Systems

Alexey I. Dordopulo[1]([⊠]), Ilya I. Levin[1,2], V. A. Gudkov[1,2], and A. A. Gulenok[1]

[1] Scientific Research Center of Supercomputers and Neurocomputers, Taganrog, Russia
{dordopulo,levin,gudkov,gulenok}@superevm.ru
[2] Southern Federal University, Taganrog, Russia

Abstract. In the paper we review high-level synthesis software tools for special-purpose hardware circuit configurations for reconfigurable computer systems that consist of a numerous FPGA chips interconnected by a spatial communication system. The distinctive feature of the software tools is mapping of the source C-program into the completely parallel form (an information graph) which is transformed into the resource-independent parallel pipeline form and automatically scaled. As a result, a reasonable solution for an available hardware resource is generated. The information graph consists of tasks with data dependencies and different rates of data flows. The parallel-pipeline form is scaled by the methods of performance reduction with the same reduction coefficient for all subgraphs. Owing to this, the different fragments of the problem have the same data processing rate. The result of the transformations is balanced and reasonable computing structure of the whole problem with the same rate of data flows among its fragments. Besides, we review the results of the suggested methods applied to several implemented problems.

Keywords: High-level synthesis · Translation of programs · C-language · Performance reduction · Reconfigurable computer system · Programming of multiprocessor computer systems

1 Introduction

The main aim of high-performance computations is decreasing of the problem solution time due to speedup of calculations. We can achieve this by increasing the performance of the computational nodes of a computer system, or by the maximal parallelization of computing operations [1]. At present, it is almost impossible to increase the performance of computational nodes, because technological capabilities of design improvement have practically achieved their limits. Therefore, increasing attention is being paid to both promising computing architectures based on GPU [2], FPGA [3] and hybrid computers, as well as new methods for parallelization and speedup of calculations.

In the paper, we consider transformation of a sequential C-program into an information graph, and its automatic scaling for the available hardware resource of the selected

© Springer Nature Switzerland AG 2021
V. Malyshkin (Ed.): PaCT 2021, LNCS 12942, pp. 88–102, 2021.
https://doi.org/10.1007/978-3-030-86359-3_7

reconfigurable computer system with multiple FPGAs [4]. The distinctive feature of this approach is performance and hardware costs reduction for scaling of calculations. It considerably decreases the number of variants of parallel calculations to be analyzed when a reconfigurable computer system computing structure is synthesized. The synthesized target design has the performance not less than 50% in comparison with the solution, designed by a circuit engineer. The number of steps, required to get this balanced solution is considerably less in comparison with parallelizing compilers, so the overall time to get good performance is significantly reduced. Along with functional similarities with Xilinx Vivado HLS and Xilinx Vitis, our software tools support scaling of the target design for a numerous of interconnected FPGA chips and automatic synchronization of data and control signals.

In this paper, we consider the automated calculation scaling methods for high level synthesis of a target solution for multichip reconfigurable computer systems from C-programs. This topic is extensive enough for a detailed consideration in one article, so detailed descriptions of some algorithms of scalable solution synthesis are given in [5, 6]. For example, [5] describes the methods and algorithms for analysis of information dependencies (Sect. 4.1 and 4.2 of the paper) and the methods for synthesis of a scalable parallel-pipeline form. The paper [6] is devoted to the methods of performance and hardware costs reduction (Sect. 4.4 and 4.5 of the paper), which are the mathematical basis for calculation scaling. This article is focused on calculation scaling (Sect. 3) for reconfigurable computer systems, and on calculation of the parallelism parameters to ensure balanced processing of the subtasks with different data flow rates (Sect. 4.4 and 4.5).

Here is the structure of the paper. In the second part we make a review of existing high-level synthesis (HLS) tools for automatic translation of sequential programs. In the third part we review methods of calculation scaling in RCSs. In the fourth part we consider principles of development of scalable solutions for our complex of high-level synthesis. In the fifth part we present the main steps and the methodology of sequential C-program transformation into reconfigurable computer system scalable solutions. The sixth part contains descriptions of the results obtained during translation of test applications. In conclusion we generalize the obtained results.

2 A Review of Existing High-Level Synthesis Tools

The translators that transform sequential C-programs into a circuit configuration of special-purpose hardware tools in HDL-languages [5, 7–14], are called high-level synthesis tools or HLS-compilers. According to the type of the input programming language, we can divide HLS-tools into two main types (see Fig. 1) [7] such as translators of domain-specific languages and translators of general-purpose languages. Domain Specific Languages are the versions of the C-language, adapted to some problem domain. General Purpose Languages are the dialects of the C-language with some special features and limitations. At present, both academic (DWARV [8], BAMBU [9] и LEGUP [10]) and commercial (CatapultC, Intel HLS Compiler [11], Cadence Stratus[12], Vivado HLS [13], Vivado Vitis [14]) complexes are being actively developed and used for design of high-performance and power efficient solutions.

DWARV is an academic HLS-compiler, developed at Delft University of Technology on basis of the commercial compiler CoSy[8] with modular infrastructure, which allows extension of functionality due to optimization modules that can be easily added. DWARV contains 255 kinds of transformations and optimizations, available as self-units such as transformations of conditional statements, pipelining of operations, memory extension, simple analysis of capacity (for standard integer data types), and loops unrolling (for performance optimization).

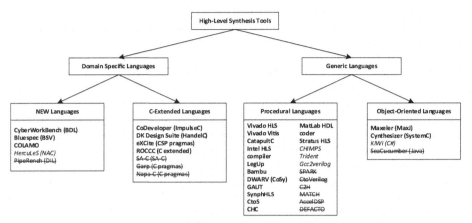

Fig. 1. Classification of high-level synthesis tools based on the type of the input language and the problem domain (**In Use**, *Not Available*, ~~Abandoned~~)

BAMBU [9] is an academic HLS-tool developed at Politecnico di Milano and first released in 2012. BAMBU is based on the GCC compiler and multiple optimizations developed specially for it. Owing to this, it supports complex structures of an input program (for example, function calls, pointers, multidimensional arrays and structures). BAMBU generates microarchitectures (IP-cores) with the optimal proportion between the solution time and the required resources, including support of floating point operations.

LEGUP [10] was developed in 2011, at Toronto University, on basis of the LLVM-compiler specially for Altera FPGAs of various families. LEGUP supports the Pthreads and OpenMP technologies with automatic synthesis of dataflow-based parallel-operating hardware devices and automatic variation of the capacity of processing data, and register optimization for multicycle paths. At present, only Microchip PolarFire FPGAs are supported. The detailed review and comparison analysis of these high-level synthesis tools are given in [6].

Concerning commercial HLS-compilers, we should mention Intel HLS Compiler [11], Cadence Stratus HLS [12], Xilinx Vivado HLS [13] and a new tool Xilinx Vitis [14] which provide powerful optimization of designed solutions. Xilinx Vivado HLS and Vitis support the most part of the complex structures of the C-language and use the comprehensive library of optimization transformations of computing structures from the earlier development environments Xilinx ISE and Vivado.

The most of existing HLS-compilers analyze the computationally extensive fragment of a C-program and transform it into a special-purpose calculator (an IP-core), based of a finite automata or a processor paradigm. Despite of a considerable computing speed gain in comparison with a processor [7], they generate an IP-core which takes relatively small hardware resource. All data flows required for this IP-core, and scaling (even within an FPGA chip) should be organized by the user. Therefore, in spite of automatic translation of the C-program, the user is completely responsible to obtain the solution of the whole problem on the available hardware resource. The problem becomes much more complex when we use several FPGAs of multichip reconfigurable computer systems interconnected by a spatial communication system [2] because calculations scaling becomes obligatory.

3 Calculation Scaling for Reconfigurable Computer System

3.1 Calculation Scaling Based on Inductive Programs Technology

When a problem is implemented in an multichip reconfigurable computer system, it is represented as an information graph [15], i.e. an oriented graph with vertices distributed into layers and iterations (see Fig. 2) so that its number of vertices is equal to the dimension of the problem (the number of data operations).

For example, Fig. 2 shows the information graph of forward elimination of the Gaussian method, corresponding to the source C-program (Fig. 5, Sect. 4.2). Here, each subgraph $g_{j,k}^i$ describes the calculation of the element m[j][k] from the internal cycle of forward elimination of the Gaussian method (Fig. 5). The layers contain vertices with no data dependences among the vertices. The iterations describe data dependences among the vertices of the different layers and have only forward connections from every current iteration to the next iterations. The layers and iterations of the information graph contain complex objects such as subgraphs $g_{j,k}^i$ that consist of several interconnected operations.

For structural procedural organization of calculations [15], a functionally complete subgraph (or a fragment) of the information graph is chosen. Such subgraph is called a basic subgraph g ($g_{1,0}^0, \ldots g_{N,N}^0, \ldots g_{2,1}^1, \ldots g_{j,k}^i, \ldots, g_{N,N}^{N-1}$). The basic subgraph of the information graph is selected according to three requirements such as:

- the subgraph g must be isomorphic within the computational problem structure;
- it is possible to create the general mapping functions for layers/iterations with basic subgraph;
- hardware implementation of the basic subgraph g as a cadr, i.e. as a computing structure and the input and output data flows, within the available hardware resource.

The basic subgraph g is always selected manually for each problem and depends not only on the problem's structure and data dependences, but also on the developer's experience. The basic subgraph g was always implemented by a performance engineer as a pipeline computing structure with the minimal possible interval of data processing to minimize the overall problem solution time. To scale problems with the implemented basic subgraph g the technology of inductive programs is used. If the reconfigurable computer system's hardware resource is available, then the number of hardwarily (as a

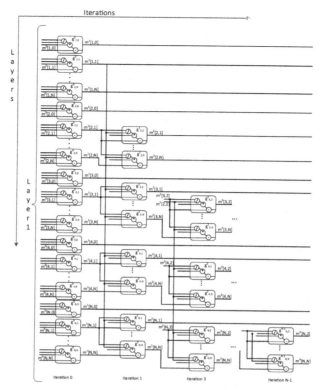

Fig. 2. The information graph of a problem with regular structure

pipeline) implemented basic subgraphs g is inductively increasing. Since the hardware implementation of the basic subgraph g keeps all data dependences then its scaled version keeps all dependences among the layers and iterations in the structure of the problem too. Figure 3 shows the way of scaling of structural procedural calculations.

The layers L and iterations It move from the start point with the coordinates $(1,1)$ which corresponds to the minimal cadr structure g. The vertical dashed line corresponds to the number of distributed memory channels (DMC) for the implementation of several data independent basic subgraphs g. The horizontal dashed line corresponds to the generalized hardware resource of FPGAs A_{RCS} (including LookUp Tables (LUT and MLUT), Flip-Flops (FF), Block RAM (BRAM) and other FPGA resources). Inductive increasing of the number of implemented basic subgraphs g was performed, first of all, by layers due to small available FPGA resource [2] which in the early 2000s was barely enough for hardware implementation of even one basic subgraph g. With increasing of FPGA hardware resource, data independent replication of cadr structures g are limited not by resource, but by the number of distributed memory channels. Therefore, to scale the problems with an iteration structure (the solution of SLAEs by the method of Gaussian elimination, the Jacob method, the Gauss-Seidel method, etc.), we use parallelizing by iterations [15], i.e. increasing of data dependent cadr structures g – moving along It-axis

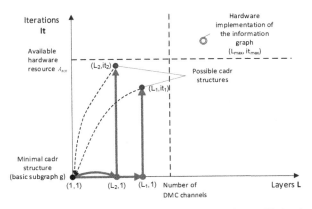

Fig. 3. Scaling of parallel calculations based on the technology of inductive programs

from $(L_1, 1)$ to (L_1, It_1) and from $(L_2, 1)$ to (L_2, It_2) in Fig. 3). The extreme variant of problem scaling for unlimited (endless) hardware resource is a cadr structure which corresponds to the hardware implementation of the information graph (L_{max}, It_{max}) in Fig. 3. The main problem for the technology of inductive programs is manual selection and obligatory high performance pipeline implementation of the basic subgraph g performed by a circuit engineer.

3.2 Calculation Scaling Based on Performance and Hardware Costs Reduction

In contrast to well-known calculation parallelizing methods that are used in parallelizing compilers, scaling of a problem with the help of methods of performance and hardware costs reduction [6] is based on an operation which is inverse of parallelizing. Let us call it de-parallelizing or sequencing. Parallelizing distributes calculations of a sequential program into multiprocessor computer system nodes according to the data dependences of the problem in order to minimize the solution time. De-parallelizing works with the completely parallel form of the problem, which requires a huge hardware resource, and which is scaled into a less parallel (or parallel-pipeline) structure implementable on the available reconfigurable computer system hardware resource. To illustrate the differences between the principles of the inductive scaling and the principles of the developed complex of high-level synthesis tools, let us represent problem scaling as a trajectory in a 3D-space. The space has three axes such as "Number of layers-Data width", "Number of iterations-Commands", and "Time-Interval" (see Fig. 4). The conditional "origin of coordinates" (like that in Fig. 3) is the structural pipeline implementation of the basic subgraph g [15]. The first and the seventh octants contain the points that describe 2 boundary (opposite) variants of organization of calculations such as the completely parallel hardware implementation of the information graph (all operations of the problem) and the completely sequential bitwise implementation of every operation of the problem. For the technology of inductive programs, scaling means tracing in the plane of layers and iterations from the basic subgraph g to the completely parallel hardware implementation of the information graph (the dashed line in Fig. 4).

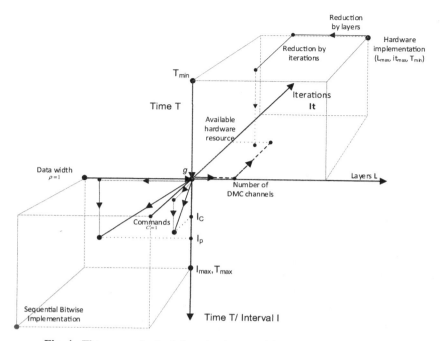

Fig. 4. The space of calculations implemented for various cadr structures

In the case of reduction of performance and hardware costs [6], we move from the point (L_{max}, It_{max}, T_{min}) in the upper octant to the area of the available hardware resource (the solid line) first by decreasing layers, then by decreasing iterations. So, at first we decrease the number of data independent basic subgraphs g and then the number of data dependent subgraphs g. In the lower octant, for the problems whose basic subgraph g exceeds the available hardware resource, the number of devices, the data width (capacity) and the data processing interval are decreasing. In this case the performance reduction always increases the problem solution time and the data processing interval, but in several cases leads to decreasing of the resource for hardware implementation. Such approach to scaling does not require manual selection of the basic subgraph g and provides solution when the available hardware resource is insufficient to implement even the minimal cadr structure g. Here, the calculations are reduced to special computational structures – micro-cadrs (or m-cadrs) [6] that sequentially perform the operations of the basic subgraph g on a lower hardware resource. Both m-cadrs and the minimal cadr structure g can be inductively scaled in the plane of layers and iterations to get a reasonable solution.

For this approach, well-known scaling technologies that required participation of a circuit engineer, are united into the single automatic methodology within the developed complex. The methodology is oriented to inductive scaling of computing structures, and reduction in case of lack of hardware resource for structural implementation of calculations.

4   The Methodology of High-Level Synthesis for Scalable Solutions of C-Programs

The methodology of high-level synthesis is based on the principles of reasonable computing structure search in the space of calculations (see Fig. 4). The source program written in the C-language (ISO/IEC 9899:1999) is translated into a program written in the dataflow programming language COLAMO. The transformations of translation and high-level synthesis are performed by the software tools [5]:

- the *Angel* translator which transforms the C-program into the completely parallel COLAMO-form (the information graph);
- the *Mermaid* processor which transforms the completely parallel COLAMO-program into the resource independent parallel pipeline form;
- the *Procrustes* processor which scales the resource independent program for the reconfigurable computer system architecture, estimating and modifying the parameters of the COLAMO-program;
- the *Nutcracker* processor which performs the performance reduction if the hardware resource is insufficient.

The translation of the generated COLAMO-program into FPGA VHDL files of multichip reconfigurable computer systems is performed by the COLAMO-translator and the synthesizer Fire!Constructor. The synthesis of bitstream files (*.bit) is performed by the synthesizer of the Xilinx Vivado CAD-system for every separate FPGA. Let us consider transformations of the input program according the suggested methodology.

4.1   Creation of the Problem Information Graph

At this stage, the input C-program is transformed into the completely parallel form – the information graph. The *Angel* translator transforms the sequential program with memory random access into the parallel program working with data flows when all arrays of the source program are transformed into the arrays with parallel (vector) access [15] of the COLAMO-program. The detailed review of this transformation is given in [5].

4.2   Analysis of the Structure of the Problem Information Graph

At this stage, the Angel translator selects tasks, defines the number of layers and iterations for every fragment, analyzes data dependences of every task and among all of them, splitting of the scalar variables, and distributes the arrays into the iterations to avoid violations of the single assignment and the single substitution rules. On the base of the structure of the source program (loops, functions, and procedures), the information graph is represented as a set of tasks. In every task, one or several functional subgraphs (FS) are selected. In other words, the functional subgraphs are the loops of the source program, or the fragments of calculations with the specified functions of scaling by layers and iterations. The number of layers and the number of iterations of every fragment is defined on the base of the loop analysis [16–18]. Besides, the data dependencies of

the variable and arrays of the loop body are taken into account. So, for the fragment
of the sequential program of the SLAE (system of linear algebraic equations) problem
solved by the Gaussian method for the specified matrix (see Fig. 5), the *Angel* translator
selects (and marks with #FuncGraph) three functional subgraphs such as forward elim-
ination(#FuncGraph_0), calculation of the last unknown variable (#FuncGraph_1), and
backward substitution (#FuncGraph_2).

```
#FuncGraph_0
for (i = 0; i < N; i++) {
   for (j = i+1 ; j < N ; j++) {
      d = (m[j][i]/m[i][i]);
      for (k = i; k < N+1 ; k++)
         m[j][k] = m[j][k] - (m[i][k]*d);
   }
}
#EndFuncGraph
#FuncGraph_1
otvet[N-1] = m[N-1][N-1]/m[N-1][N];
#EndFuncGraph
#FuncGraph_2
for (t=N-2;t >= 0;t --) {
      d=0;
      for (r=0;r<N-t-1;r ++)
         d = d - (otvet[N-r-1]*m[t][N-r-1]);
       otvet[t]=(m[t][N]+d)/m[t][t];
}
#EndFuncGraph
```

Fig. 5. The functional subgraphs in the source C-program for the SLAE problem

Forward elimination is represented as the triple loop, calculation of the last unknown
variable – as the single assignment expression, and backward substitution – as the loop
over the rows of the matrix The number of iterations of the loops by every variable corresponds to the number of layers and iterations. The functional subgraph that calculates the
last unknown variable, consists of one expression. The functional subgraph of backward
substitution contains three subgraphs because each of them has its own number of layers
and iterations (and has its own loop). For the selected functional subgraphs, the *Angel*
translator performs all transformations and generates a well-formed COLAMO-program
in the completely parallel form similar to the "Initial description" and "Initial loop" (see
Fig. 6).

4.3 Transformation of the Information Graph into the Scalable Parallel Pipeline Form

Owing to the scalable parallel pipeline form, generated by the *Mermaid* processor, it
is possible to automatically recalculate the number of implemented subgraphs and the

size of their data flows for the dimension of every array (data flow) with the help of the only one constant (the degree of parallelism). For this purpose, we describe every dimension of every array of the program with two interrelating parameters such as the vector parameter with parallel access, and the stream parameter with sequential access. That is why on this stage, the ***Mermaid*** processor adds the stream parameter to all vector dimensions of the arrays and splits all loops of the COLAMO-program as follows (see Fig. 7). It is clear, that the product of the vector and stream dimensions is equal to the size of the initial array in the sequence program. If we set the values of the parallelism parameters (dp1, dp2, etc.) equal to their initial values, we obtain the initial parallel program generated by the ***Angel*** translator.

Initial description	Transformed description

M: **Array** [N: **Vector**,
        N+1: **Vector**]
	M: **Array** [dp1: **Vector**, (N+dp1−
	        1)/dp1: **Stream**;
	        dp2: **Vector**,
	        (N+dp2−1)/dp2: **Stream**]

Initial loop	Transformed loop

For i:= 0 **To** N−1−1
  Step 1 **Do**
  For j:= i+1 **To** N−1
    Step 1 Do
    d:= m[j,i] / m[i,i];
    For k:=i To N+1−1
      Step 1 Do
      m[j,k]:=m[j,k]−
      m[i,k]*d;
	For vi:= 0 **to** dp1−1 **Step** 1 **Do**
	  For si:=0 **to** (N+dp1−1)/dp1−1
	    Step 1 **Do**
	    For vj:=0 **to** dp2−1 **Step** 1 **Do**
	      For sj:=0 **to** (N+dp2−1)/dp2−1
	        Step 1 **Do**
	        d:= m[vj,sj,vi,si] /
	          m[vi,si,vi,si]
	      For vk:= 0 **to** dp2−1 **Step** 1 **Do**
	        For sk:=0 **to** (N+dp2−1)/dp2−1
	          Step 1 **Do**
	      m[vj,sj,vk,sk]:=m[vj,sj,vk,sk]−
	      m[vi,si,vk,sk]*d

Fig. 6. The transformed loops with access to the variables in the scalable parallel pipeline form

As a result of this transformation, the completely parallel COLAMO-form turns into the parallel-pipeline one, and we can control parallelism with the help of parameters (dp1, dp2, etc.) with automatic recalculation of the stream parameter according to the syntax requirements.

4.4 Calculation of the Problem Parallelism Parameters for the Available Hardware Resource

At this stage, the ***Procrustes*** processor calculates the possible values of the scaling parameters of every functional subgraph according to the available hardware resource and the number of memory channels. Starting from the point (L_{max}, It_{max}, T_{min}) and tracing along layers, iterations and instructions of every functional subgraph (see Fig. 4), the *Procrustes* processor searches the reasonable computing structure of the whole problem. This transformation is the most complicated part of the considered methodology because we have to find a solution with balanced rates of data flows for various fragments of the problem, taking into account the data dependences among the fragments, and the reasonable implementation of the solution on the available hardware resource. For every functional subgraph, it is necessary to choose the most reasonable form of organization of calculations according to the computing structures of other subgraphs and the whole problem.

The scaling strategy is based on the idea that the total problem solution time depends on the most computationally expensive fragment most of all. Therefore, it is necessary to implement such fragment in the most efficient way; ideally, as a (multi) pipeline structure with the minimal data processing interval. The functional subgraph, which requires the largest part of the hardware resource (the product of the resource, taken by the loop body, and the number of the loop iterations), is the start point for scaling and matching of the parameters of all subgraphs of the problem. Let us call such functional subgraph a "flagship". The problem structure can contain several "flagships" that take a similar hardware resource. In this case, we start scaling from the largest fragment (or from any fragment if they are equal). Let us call all other functional subgraphs "boats", i.e. rather small functional subgraphs of a task or a problem that require considerably smaller resource in comparison with the "flagship". We do not claim that our classification of problems is comprehensive; however, it is possible to emphasize the most wide-spread variants of their data dependences (see Fig. 7).For example, for the SLAE problem solved by the Gaussian method (see Fig. 5) forward elimination (#FuncGraph_0) is a "flagship". Calculation of the last unknown variable (#FuncGraph_1), and backward substitution (#FuncGraph_2) both are "boats" because of less number of loop iterations comparing with forward elimination.

During scaling, the "boats" take the available hardware resource after implementation of the "flagship". The reduction of the "flagship" is based on hardware implementation of the iteration loop with all iteration steps (i.e. with no reduction) because it decreases the problem solution time, does not change (does not increase) the number of memory channels, and occupies the hardware resource that is quite enough in modern FPGAs. If it is impossible then several iteration steps, interconnected by a feedback, are implemented. The data processing interval increases due to a feedback, but it can be decreased with the help of the optimization transformations to prevent increasing of the interval of the whole problem.

The complex of high-level synthesis is based on a set of rules that decrease the number of analyzed variants and the total problem solution time, and provide reasonable scaling. Within a functional subgraph, if it contains several expressions (or operators), we use the same approach: the largest expression, which requires the largest resource, is scaled (or reduced) first. A "boat", especially if its hardware costs are small, can be implemented hardwarily with no reduction methods. Since such functional subgraphs do not considerably increase the hardware resource of the problem, their reduction can increase hardware costs due to the complex communication system of the reduced structure. To achieve the specified performance reduction coefficient, it is better to increase the data processing interval for small functional subgraphs. If there is a data dependence among the layers (i.e. a graph is functionally irregular), it is necessary to reduce the functional subgraph and implement it as a procedure.

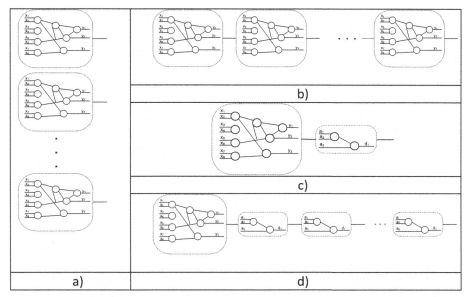

Fig. 7. The structures of the information graphs for some problems: a) a data independent "flagships"; b) a data dependent "flagships"; c) a "flagship" – a "boat"; d) a "flagship" – several "boats"

To scale the problem, i.e. to proportionally increase/decrease its tasks, it is necessary to provide one and the same reduction coefficient for all tasks, and proportional variation of the rate of data flows among the tasks. In the context of reduction, it means that it is reasonable to use the same types of reduction with the same coefficients if we want (and if it is possible) to keep the rate of data flows. If we use different types of reduction in different tasks, then, in the general case, it is necessary to match the rates of data flows. As a rule, this leads to additional hardware costs and increasing of the problem solution time, because hardware implementation of match units requires delay elements with multiplexers/demultiplexers, buffers, internal dual-port RAM (BRAM).

4.5   Data Processing Interval Optimization for the Generated Problem Solution

To match the rates of data flows after scaling of all functional subgraphs of the problem, the *Procrustes* processor applies the optimization methods that decrease the data processing interval, increased during reduction procedure, to its minimal value (ideally to 1) balanced for all subgraphs. To decrease the data processing interval [19], it is necessary to transform the problem into a pipeline of pipelines [15] or into a macro pipeline [15].

It is necessary to decrease the data processing interval when the number of implemented iteration steps is less than it is required for implementation of the whole iteration.

In this case we have an inevitable feedback, and the data processing interval increases. In the case of a feedback, the data processing interval is equal to the ratio of the latency of the implemented iteration steps to the number of registers of the feedback. So, if we cannot increase the number of implemented iteration steps, it is necessary to increase the number of registers in the feedback up to the latency. We decrease the data processing interval and get the contiguous data flow. This is the sense of transformation to a pipeline of pipelines (or a nested pipeline, a pipeline in a pipeline). Here, the data processing interval can be decreased to its minimal value of a unity. It is utterly important for scaling the "flagship" due to its influence on the total problem solution time. The transformation into a macro pipeline is also aimed at decrease of the data processing interval, but is applied for the fragments implemented as procedures. The sense of this transformation is to increase the number of procedural fragments up to the number of clock cycles of one procedural device. As a result, we get at least one unit free for the next part of a contiguous data flow, processed with the data processing interval equal to unity.

Owing to the described system of limitations and optimizations, it is possible to synthesize a reasonable solution such as a pipeline in a pipeline or a macro pipeline for the most computationally expensive "flagship". As a result, the total problem solution time is reduced.

5   Results of Experimental Research for the Developed Complex

With the help of the prototype version of our HLS-complex, which contains the **Angel** translator, and the processors **Mermaid** and **Procrustes**, we have successfully implemented several problems of linear algebra on various reconfigurable computer systems [4, 20]. Table 1 shows the solution time and speedup results for the SLAE problem, solved by the Gaussian forward elimination for matrices of 8000×8001 elements on the reconfigurable computer system Tertius [4] with a clock rate of 250 MHz, and on the personal computer based on Intel i5-7300 with a clock rate of 2500 MHz. Target design was obtained with the help of the method of parallelizing by iterations, i.e. reconfigurable computer system hardware implementation of the steps of the algorithm interconnected due to data dependences. These results were obtained automatically, without any pragmas like in the Xilinx Vivado HLS or other HLS-compilers. Instead, we use the methodology of high-level synthesis for scalable solutions, discussed in Sec.4. The first three rows (1, 10 and 100 steps of the target design) correspond to one FPGA, and 830 steps correspond to 4 FPGAs of the reconfigurable computer system Tertius. It is clear, that the personal computer based on Intel i5 is 10 times faster than Tertius, but even in this case one FPGA provides 10-fold speedup. If we increase the number of computational FPGAs (the last row), the performance ramps.

The time of translation of sequential C-programs into COLAMO-programs with the help of the prototype version of our HLS-complex does not exceed 10 min for 830 steps for the 4 FPGAs of the Tertius. The time of synthesis of FPGA VHDL-files for this design was about 4,5–5 h with average utilization 94% for each FPGA. The specific performance of the generated solutions is not less than 85% in comparison with those designed by software developers in the programming language COLAMO, and is not less than 70% in comparison with the solutions designed by circuit engineers.

Table 1. The time results for the SLAE problem, solved by the Gaussian elimination on the Tertius reconfigurable computer system

Number of iteration steps	Personal computer, sec	Tertius reconfigurable computer system, sec	Speedup
1 step	0.023	0.029	0.8
10 steps	0.91	0.28	3.25
100 steps	12.8	1.21	10.6
830 steps	103.07	1.67	61

6 Conclusion

The developed complex for high-level synthesis of scalable solutions for reconfigurable computer systems is based on the original methodology of transformation of sequential C-programs into FPGA configuration files. In contrast to well-known HLS-compilers, the complex provides scaling of the problem, which consists of fragments with different computational complexity, and generates the computing structure with the balanced rates of data flows. Each fragment and the whole problem are scaled by the *Procrustes* processor with the help of the methods of performance and hardware costs reduction without any pragmas. In contrast to the most part of HLS-compilers, the complex provides automatic synchronization of data and control signals for multichip solutions.

Owing to the experimental results of the translation of several tests, it is possible to conclude that the developed complex is applicable to automatic high-level synthesis of scalable solutions of linear algebra computational problems with complex structure.

References

1. Voevodin, V.V., Voevodin, Vl.V.: Parallel Computing. BHV-Petersburg (2002)
2. Mittal, S., Vetter, J.S.: A survey of CPU-GPU heterogeneous computing techniques. ACM Comput. Surv. **47**(4), 1–35 (2015)
3. Trimberger, S.M.: Three ages of FPGAs: a retrospective on the first thirty years of FPGA technology. Proc. IEEE **103**(3), 318–331 (2015)
4. Multiprocessor computer systems with reconfigurable architecture. http://superevm.ru/index.php?page=hardware-2. Accessed 27 May 2021
5. Levin, I., et al.: Software development tools for FPGA-based reconfigurable systems programming. In: Voevodin, V., Sobolev, S. (eds.) RuSCDays 2019. CCIS, vol. 1129, pp. 625–640. Springer, Cham (2019). https://doi.org/10.1007/978-3-030-36592-9_51
6. Dordopulo, A.I., Levin, I.I.: Performance reduction for automatic development of parallel applications for reconfigurable computer systems. Supercomput. Front. Innov. **7**(2), 4–23 (2020)
7. Nane, R., et al.: A survey and evaluation of FPGA high-level synthesis tools. IEEE Trans. Comput. Aided Des. Integr. Circuits Syst. **35**(10), 1591–1604 (2016)
8. Nane, R., Sima, V.-M., Olivier, B., Meeuws, R., Yankova, Y., Bertels, K.: DWARV 2.0: a cosy-based C-to-VHDL hardware compiler. In: FPL, pp. 619–622 (2012)

9. Pilato, C., Ferrandi, F.: Bambu: a modular framework for the high level synthesis of memory-intensive applications. In: FPL, pp. 1–4 (2013)
10. Canis, A., et al.: LegUp: high-level synthesis for FPGA-based processor/accelerator systems. In: ACM FPGA, pp. 33–36 (2011)
11. Intel HLS Compiler. https://www.intel.com/content/www/us/en/software/programmable/quartus-prime/hls-compiler.html?wapkw=Intel%20HLS%20compiler. Accessed 27 May 2021
12. Cadence Stratus HLS. https://www.cadence.com/en_US/home/tools/digital-design-and-signoff/synthesis/stratus-high-level-synthesis.html. Accessed 27 May 2021
13. Make Slow Software Run Fast with Vivado HLS. https://www.xilinx.com/publications/xcellonline/run-fast-with-Vivado-HLS.pdf. Accessed 27 May 2021
14. Vitis Unified Software Platform Documentation. Application Acceleration Development. https://www.xilinx.com/support/documentation/sw_manuals/xilinx2019_2/ug1393-vitis-application-acceleration.pdf. Accessed 27 May 2021
15. Kalyaev, I.A., Levin, I.I., Semernikov, E.A., Shmoilov, V.I.: Reconfigurable Multipipeline Computing Structures. Nova Science Publishers, New York (2012)
16. Bielecki, W., Palkowski, M.: Perfectly nested loop tiling transformations based on the transitive closure of the program dependence graph. In: Wiliński, A., El Fray, I., Pejaś, J. (eds.) Soft Computing in Computer and Information Science. AISC, vol. 342, pp. 309–320. Springer, Cham (2015). https://doi.org/10.1007/978-3-319-15147-2_26
17. Devan, P.S., Kamat, R.K.: A review – LOOP dependence analysis for parallelizing compiler. Int. J. Comput. Sci. Inf. Technol. 5(3) (2014). https://www.ijcsit.com/docs/Volume%205/vol5issue03/ijcsit20140503305.pdf. Accessed 27 May 2021
18. Giorgi, R., Khalili, F., Procaccini, M.: Translating timing into an architecture: the synergy of COTSon and HLS (domain expertise—designing a computer architecture via HLS). Int. J. Reconfig. Comput **2019** (2019). https://downloads.hindawi.com/journals/ijrc/2019/2624938.pdf. Accessed 27 May 2021. https://doi.org/10.1155/2019/2624938
19. Licht, J., Besta, M., Meierhans, S., Hoefler, T.: Transformations of high-level synthesis codes for high-performance computing, 23 November 2020. https://doi.org/10.1109/TPDS.2020.3039409
20. Levin, I., Dordopulo, A., Fedorov, A., Kalyaev, I.: Reconfigurable computer systems: from the first FPGAs towards liquid cooling systems. Supercomput. Front. Innov. 3(1), 22–40 (2016)

Precompiler for the ACELAN-COMPOS Package Solvers

Aleksandr Vasilenko, Vadim Veselovskiy, Elena Metelitsa, Nikita Zhivykh,
Boris Steinberg, and Oleg Steinberg$^{(\boxtimes)}$

Southern Federal University, Rostov-on-Don, Russia
{avas,veselovsky,metelica,zhivyh,byshtyaynberg,obshtyaynberg}@sfedu.ru

Abstract. This article proposes precompilers to accelerate the solvers
of the application software package. The substantiations of this approach
are given. The precompiler can be most valuable for programs that are
intended to be ported to different computing architectures. There are
given some examples of speeding up programs that implement iterative
numerical methods using a precompiler in this paper. In particular, skew
tiling and parallelization of loops with a linear recurrent dependence
are implemented in the presented precompilers. The results of numerical
experiments demonstrate the acceleration of programs by the precompiler by tens of percents and sometimes tens of times.

Keywords: High-performance computing · Optimizing compilers ·
Loop nest · Loop tiling · Program transformations · Parallel computing

Introduction

A precompiler is a preliminary source-to-source compiler. It is a compiler that
transforms high-level source program code into the same language code, which
will execute faster than original program (after compilation by the standard
compiler).

Program performance became more and more dependent on efficient using
of memory hierarchy and parallel computing devices [1]. Consequently, other
optimizations became more essential. These transformations are not processor
specific. Such as the block allocation of matrices in RAM [2], parallelization
of loops with a linear recurrent dependence [3,4], optimization of work with
structures [5], etc. Many of such transformations (for example, aimed at data
localization) can be useful for various modern computation systems with different
command systems. The transformations can be added to the compiler only if it
has open source code, for example, GCC or LLVM compilers. But popular and,
in many cases, most effective (see Table 1) compilers, such as Intel or PGI, are
proprietary. In many libraries, solvers are created not for a single problem, but

The article was supported by the Russian Federation Government grant No. 075-15-
2019-1928.

V. Malyshkin (Ed.): PaCT 2021, LNCS 12942, pp. 103–116, 2021.
https://doi.org/10.1007/978-3-030-86359-3_8

a set of similar problems that differ in parameter values. At the same time, different optimizing transformations can be optimal for different values of the parameters. For example, with some kinds of data dependence, rectangular tiling can be used, and with others, skew tiling; loop parallelizing with linear recurrent dependency can speed the code up in some cases and slowdown in others.

Optimized program code may be larger and more complex than source program code. For different values of the parameters, the code differs not only in the values of the parameters, but also in the text: for example, loop unrolling with 5 iterations contains 5 copies of the original loop body, and with 10 iterations - 10 ones; inlining in one case can be done for 10 function calls, and in the other - for 1000 [6]. The parallelization of the Floyd-Warshall algorithm is correct if the graph weights matrix contain only non-negative numbers. Notice, the compiler is not aware of the signs of the graph weights matrix elements. Thus, a compiler intended for a wide range of users (such as GCC, LLVM, ICC) cannot parallelize such an algorithm automatically. But within the specific package framework of applied programs with appropriate documentation, such the transformation can be implemented [7].

Compilers of the LLVM family can transform loop nests by using Polly [7]. In particular, the possibility of tiling was stated. To do this, compilers must receive a lot of instructions, since Polly does not determine the optimal tile sizes, etc. The skewed tiling is not implemented in Polly. Intel C++ compiler provides the ability to apply loop tiling (loop blocking) [9,10] for loop nests (perfect loop nests) by using compiler's pragma directive `#pragma block_loop factor (int) level (int)`, where `factor(int)` is the block size, `level(int)` are the numbers of the loops to which tiling will be applied. If the compiler directive is used without parameters, tiling is applied to all loops, and the tile sizes are determined automatically based on the processor type and memory access patterns. The loop-carried dependencies are ignored during the processing of `block_loop` pragmas, it means that user is responsible for the transformation's correctness.

Even classic program transformations are not always used effectively. There are dozens of program transformations. For example, the list of LLVM transformations [11]. Some transformations conflict with each other. For example, the expression $ab + ac + bc$ can be replaced with $a(b + c) + bc$ or $ab + (a + b)c$, but two substitutions cannot exist at the same time. There are transformations with parameters. For example, loop unrolling decreases the number of loop iterations by a factor of k, but at the same time increases the size of the body of the loop by a factor of k. Choosing a desirable combination of transformations can turn out to be an even more complex intellectual task [12], which is not solved by modern compilers. For a specific program, the developer can create his own combinations of transformations within the precompiler.

The article [13] acknowledges that optimizing linear algebra software can be a tedious and time consuming process. The programmer must understand the architecture, how the memory hierarchy can be used to provide data in an optimum fashion. Care must be taken to optimize the operations to account for many parameters such as blocking factors, loop unrolling depths, loop ordering, register allocations. However, compiler technology is far from mature enough to perform these optimizations automatically. Compilers for less widely marketed machines are almost certain not to be developed. Many cases in which optimizing compilers poorly optimize are described in [14].

The article [15] describes the Source-to-Source outliner developed by the authors, based on an open infrastructure for creating compilers and related utilities ROSE (ROSE Compiler).

The results presented in this article confirm the conclusions [16] that modern optimizing compilers do not optimize programs well enough.

This article presents new transformations that speed up programs significantly and can be done both manually and automatically: parallelization of recurrent loops, skewed tiling, and their parallel execution of skewed tiles. The last transformation gives an acceleration of more than 10 times. Numerical experiments have shown that the code transformed with skewed tiling can be further accelerated by classical compiler transformations. It is shown that the compilers GCC, MS-VS, ICC do not perform the classic optimizing transformations available to them in combination with the new transformations presented in the article. To create fast and portable programs, the article suggests creating preliminary compilers. The cost of creating a precompiler can be justified for the solvers of applied software packages.

Precompilers can be created for the specific program that receives parameters. Such a precompiler can be useful in case if this program is intended to be ported to another architectures while maintaining high efficiency.

The precompilers are presented in this paper are oriented to the solvers of the ACELAN-COMPOS application software package [17,18]. The article presents a solver for SLE with a block-band matrix. A special representation of the block-band matrix ensures high performance of the iterative algorithm. A new special parallel algorithm for solving SLE with a tridiagonal matrix has been developed for this solver. The article presents a Dirichlet problem solver that combines skewed tiling and OpenMP parallelization. The acceleration of this solver is more than 20 times compared to the sequential algorithm.

A precompiler is useful if the following conditions are met: 1) the usual optimizing compiler speeds up the program poorly; 2) you know the transformations that can speed up the program; 3) more than one variant of the solver is required.

1   Precompiler for Accelerating Solvers for SLE with Block-Band Matrix

A SLE (System of Linear Equations) with a block-band matrix arises in the mathematical modeling when using such numerical methods as finite difference method or finite element method.

The solver is designed for a set of similar types of certain problems. It receives some initial data of the problem, for example, size of the block and other as a parameters.

1.1   Iterative Algorithms for Solving the Target SLEs

Consider a system of linear equations

$Ax = b_0.$

and the iterative algorithm for solving it

$$x^{k+1} = B \cdot x^k + b$$
$$B = I - t \cdot C^{-1} \cdot A, \ b = t \cdot C^{-1} \cdot b_0$$

where k is an iteration number, b_0 is an initial approximation, C is a non-singular matrix; t is a positive numerical parameter. For the iterative process convergence, the spectral radius of matrix B must be less than 1.

It is assumed that C is a block-tridiagonal matrix.

After LU-decomposition of matrix C (or $C = L \cdot D \cdot L^*$ for a symmetric matrix), $C^{-1} \cdot x$ can be quickly calculated at each iteration of the algorithm. In SLEs with saddle point singularity [19] the matrix C can be symmetric.

Programs for solving SLE with block-band matrix have their own features. By understanding these features, the performance of the algorithm can be improved.

After LU decomposition (shown in Fig. 1), the blocks of matrices L and U located on the main diagonals have a triangular shape

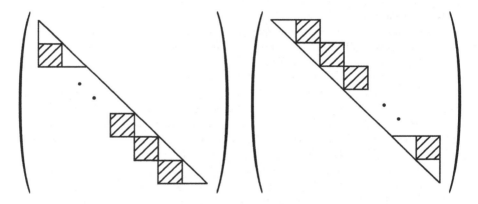

Fig. 1. LU-decomposition of the block-tridiagonal matrix. The blocks on the main diagonal have a triangular form. The decomposition of the symmetric matrix can be obtained by the Cholesky method [20]

The program for solving the system of linear equations with block matrices contains double loops with square and triangular iteration space (these double loops scan square and triangular blocks of data, respectively). The number of iterations in these loops is small, which means that full loop unrolling could be applied. The unrolling of such loops can give a noticeable acceleration, as confirmed by numerical experiments.

Matrix blocks are stored in memory in one-dimensional arrays. It improves data localization but complicates array index computing. The index expressions of arrays depend on loop counters as affine functions. Optimization of calculation of such index expressions can be achieved by transformations: "Common Subexpression Elimination"; "Loop-invariant code motion" [21–24].

The authors of the article have developed the first version of solver for the SLEs with block-band matrices with saddle point singularity in C language.

1.2   Precompiler for the Solver for SLE

The precompiler for the solver is based on the OPS (Optimizing paralleling system), which is the technology developed and distributed by the authors of this article [25, 26]. OPS is a parallelizing system owned by Southern Federal University designed to transform programs written in C. It is a system that consists of a high-level internal representation, program dependency analyzers, program transformation library, and helper functions. The next program transformations were used: carried invariants out of loops, linearization, loop unrolling, and loop canonization.

Example 1.

A chain of transformations for the next perfect loop nest is considered:

```
for (i = 0; i < 4; i++)
    for (j = 0; j < 4; j++)
        B_C[b * 4 + 4 + j]  -= A_X[b * BLK_SZ_2 + i * 4 + j]
                            * B_C[b * 4 + i];
```

The following optimizations are performed: loop-invariant code motion, linearization, full loop unrolling. The following is the fragment of the transformed code for $i = 0$ and $i = 1$ (for values $i = 2$ and $i = 3$, code similar to consider).

```
__a = b * BLK_SZ_2;
// i = 0
__b = b * 4;
__c = B_C[__b];
B_C[4 + __b] = B_C[4 + __b] - A_X[__a] * __c;
B_C[5 + __b] = B_C[5 + __b] - A_X[1 + __a] * __c;
B_C[6 + __b] = B_C[6 + __b] - A_X[2 + __a] * __c;
B_C[7 + __b] = B_C[7 + __b] - A_X[3 + __a] * __c;
// i = 1
__b = 4 + b * 4;
__c = B_C[1 + __b];
B_C[4 + __b] = B_C[4 + __b] - A_X[4 + __a] * __c;
B_C[5 + __b] = B_C[5 + __b] - A_X[5 + __a] * __c;
B_C[6 + __b] = B_C[6 + __b] - A_X[6 + __a] * __c;
B_C[7 + __b] = B_C[7 + __b] - A_X[7 + __a] * __c;


...
```

Note that the optimized code is much larger than the source code. If the block-size would be 12 instead 4, then the size of the converted code would be 9 times larger than result code in this Example.

1.3 Numerical Experiments with the Solver for SLEs with Block-Band Matrices

The solver was tested on a computer with the following characteristics:

- Processor: Intel i7-9700; 3.00 GHz; Core - 8; L1 - 256 KB (for each core); L2 - 2 MB (for each core); L3 - 12 MB
- System bus frequency 8 GT/s; Max. 41.6 GB/s Bandwidth
- RAM: DDR4 16 GB; Min. frequency 1600 MHz; Max. frequency 2666 MHz

The iterative algorithm stops when the norm of the difference between the results of the iterations k and $(k+1)$ is less than some positive number. Numerical experiments were performed for the algorithm with the following characteristics.

Parameters of the problem in numerical experiments: number of blocks - 999999, block size - 4, scalar matrix size - 3999996, number of non-zero elements in one triangular block - 6, number of iterations in one step - 50, error - 0.001; Matrix A has a saddle point, the number of positive eigenvalues is twice more than negative; matrix C is assumed to be symmetric represented in the form $C = L \cdot L^*$ (i.e. in the expansion $C = L \cdot D \cdot L^*$ diagonal matrix is the identity).

Table 1. Results of numerical experiments for the transformed program for different compilers.

Compiler and key	GCC, O2	GCC, O3	ICC, O2	ICC, O3	MSVS O2
Original program	5.035	5.313	7.014	6.763	10.157
Transformed program	4.919	4.609	4.259	4.208	9.270

The table of numerical experiments (shown in Table 1) shows that the optimizing precompiler speeds up the solver by 30% (when the fastest code is produced by the ICC proprietary compiler). This acceleration is obtained without parallelization, which is considered in the next section.

1.4 Parallelization of a Loop with Linear Recurrent Dependence

The solution of a SLE with matrix C can sometimes be accelerated by parallelization using OpenMP. This problem is reduced to the parallelization of loops with a linear recurrent dependence, which is effective for some parameter sets of the algorithm and some computational architectures [3,4].

For example, if the block sizes are equal to 1 and the Intel processor is used, the algorithm [4] can speed up the solution of the SLE with the matrix C.

The loop below solves a SLE with a two-diagonal matrix.

```
for (i = 0; i < N; i++)
    x[i] -= a[i] * x[i-1] + b[i];
```

After LU decomposition of the matrix C, the algorithm [4] can be applied both for solving SLEs with lower triangular and upper triangular matrices. The program is implemented for the case when the block size is 1 (that is, there are only three non-zero diagonals in the matrix).

It should be noted that in the case of multiple iterations of calculations (in an iterative algorithm), an unchanging matrix, but the changing right side, some auxiliary calculations of the algorithm can be done only once and then reused. In addition, since the column on the right side of the SLE does not need to be saved, the resulting vector can be written in its place. This also leads to an acceleration of the parallel version of the program. The results of numerical experiments are presented in the Table 2.

Table 2. Results of numerical experiments for parallelizing a loop with a linear recurrent dependence.

Compiler and key	GCC, O2	GCC, O3	ICC, O2	ICC, O3	MSVS O2
Original program	2.189	1.940	2.000	1.895	2.197
Transformed program	1.314	1.216	1.231	1.236	1.221

It can be assumed that for processors with addressable local memory [27–29] or GPU, algorithms [3] or [4] will give the acceleration even in the case of block sizes greater than 1, since intermediate data can be left in the processor without writing to the RAM (Intel processor cache does not have this feature).

2   Precompiler for Accelerating the Gauss-Seidel Algorithm for Solving the Dirichlet Problem

2.1   Automatic Tiling for the Gauss-Seidel Algorithm for Solving the Dirichlet Problem

The loop nest of the original program:

```
for (int k = 0; k < K; ++k)
    for (int i = 1; i < N - 1; ++i)
        for (int j = 1; j < M - 1; ++j)
            u[i][j] = (u[i - 1][j] + u[i + 1][j]
                    + u[i][j - 1] + u[i][j + 1]) / 4.0;
```

The code of the transformed program for solving the Dirichlet problem with the Gauss-Seidel algorithm:

```
int i, j, k;
for(int jjjj = 0; jjjj < ((((((M - 1) - 1) + K - 1) - 1)
/ d3 + 1) + (((((N - 1) - 1) + K - 1) - 1) / d2 + 1) - 1)
+ K / d1 - 1; jjjj += 1) {
    int j14, k18, k19;
    if (0 > (jjjj - ((((((M - 1) - 1) + K - 1) - 1) / d3 + 1)
    + (((((N - 1) - 1) + K - 1) - 1) / d2 + 1) - 1)) + 1)
        k18 = 0;
    else
        k18 = (jjjj - ((((((M - 1) - 1) + K - 1) - 1) / d3
        + 1) + (((((N - 1) - 1) + K - 1) - 1) / d2 + 1)
        - 1)) + 1;
    if (K / d1 < jjjj + 1)
        k19 = K / d1;
    else
        k19 = jjjj + 1;
    for (int kk = k18; kk < k19; kk += 1) {
        int i15, i16;
        j14 = jjjj - kk;
        if (0 > (j14 - ((((((M - 1) - 1) + K - 1) - 1) / d3
        + 1)) + 1)
            i15 = 0;
        else
            i15 = (j14 - ((((((M - 1) - 1) + K - 1) - 1)
            / d3 + 1)) + 1;
        if (((((N - 1) - 1) + K - 1) - 1) / d2 + 1 < j14
        + 1)
            i16 = ((((N - 1) - 1) + K - 1) - 1) / d2 + 1;
        else
            i16 = j14 + 1;
        for (int iii = 15i; iii < 16i; iii += 1) {
            jjj = j14 - iii;
            for (int k = kk * d1; k < (kk + 1) * d1; k += 1) {
                int i10, i11;
                if (k > iii * d2)
                    i10 = k;
                else
                    i10 = iii * d2;
                if (((N - 1) - 1) + k < (iii + 1) * d2)
                    i11 = ((N - 1) - 1) + k;
                else
                    i11 = (iii + 1) * d2;
                for (int ii = i10; ii < i11; ii += 1) {
                    int j12, j13;
                    if (k > jjj * d3)
```

```
            j12 = k;
        else
            j12 = jjj * d3;
        if (((M - 1) - 1) + k < (jjj + 1) * d3)
            j13 = ((M - 1) - 1) + k;
        else
            j13 = (jjj + 1) * d3;
        for (int jj = j12; jj < j13; jj += 1)
        {
            i = ii - k;
            j = jj - k;
            u[(1 + i)][(1 + j)] = (((u[((1
            + i) - 1)][(1 + j)] + u[((1 + i)
            + 1)][(1 + j)]) + u[(1 + i)][((1
            + j) - 1)]) + u[(1 + i)][((1 + j)
            + 1)]) / 4.;
        }
      }
    }
  }
 }
}
```

The code of transformed program was obtained by the precompiler for Gauss-Seidel problem, variable names were changed for readability.

Note that the optimized code is not only larger but also much more complex. For a three-dimensional problem, the code would be even more complicated. The code has parameters - the size of the tile. For different processors, these optimal sizes will be different.

Table 3. Results of numerical experiments of the Gauss-Seidel algorithm for solving the Dirichlet problem.

Compiler and key	GCC, O2	GCC, O3	ICC, O2	ICC, O3	MSVS O2
Original program	16.431	11.634	18.888	18.611	20.565
Transform program	1.247	1.003	1.085	0.875	2.235

The table of numerical experiments (shown in Table 3) shows that the optimizing precompiler speeds up the solver on all compilers.

The manually obtained code (considering the peculiarities of the issue) has a much higher speedup, which means that the precompiler for this program has potential for development.

The paper [30] presents a method for combining parallelization on distributed memory and skewed tiling with OpenMP parallelization. So, the Gauss-Seidel algorithm presented in this section for the Dirichlet problem can be used in a parallel algorithm for solving this problem on a supercomputer with nodes based on Intel Core i7 CPUs. The acceleration of such an algorithm can be equal to the product of the number of processors and the acceleration of this algorithm on one processor (20 times).

2.2 Linearization of Expressions in the OPS

OPS is a tool for developing C2C (C-language to C-language) accelerating converters. The OPS consists of an IR (internal representation), a library of program transformations, a set of program analyzers and other functions.

The transformation "linearization of expressions" [31], implemented in Optimising Parallelising System (OPS), is used to convert expressions to a standard form before constructing a data dependence graph and for other purposes. Linearization is achieved by constant propagation [32] and simplifying arithmetic expressions. Linearization was applied only to expressions in assignment operators in the previous version of OPS. The application of this transformation is extended in this paper.

Linearization in the new version is redefined for comparison operations of the form a @ b, where a and b are integer expressions, and @ is the comparison operator. This operation is replaced by $(a - b)$@0, and then an attempt is made to calculate the difference to reduce the given expression to some constant value that can be compared to zero. Such comparison operations, which are identical to zero or one, can occur in the program after automatic code transformations.

Extension of linearization for a composite operator is a sequential application of optimization for each operator of this block. For a conditional operator of the form

```
IF condition THEN thenBlock ELSE elseBlock
```

linearization consists of two steps: the first step is optimization of a logical expression in condition statement. If after linearization the expression degenerates into a constant, then, depending on its value, the conditional operator is replaced by a composite operator *thenBlock* or *elseBlock*. The second step is the linearization of composite operators.

For a loop operator of the form

```
for (I = Expr1; I op Expr2; I = I + Step)
    LoopBody
```

explicit linearization of the I op $Expr2$ operation can lead to violation of the canonical form of the loop, which is required for further transformations.

We need to linearize the expression $Expr1$ op $Expr2$ and check if it is identically false.

If the expression $Expr1$ op $Expr2$ is not identically false, the Step expression and the *LoopBody* will be linearized. Otherwise, the loop operator is replaced by the assignment operator $I = Expr1$.

2.3 Gauss-Seidel Algorithm Accelerating by Expression Linearization

Example 2. We consider a code fragment in the "transformed program of Dirichlet problem solution by Gauss-Seidel algorithm" (Sect. 2.1), highlighted in bold. This code, obtained by applying the "automatic tiling" transformation, contains expressions to which linearization can be applied. The transformed code of the selected program fragment:

```
for ( int jjjj = 0; jjjj < (-4 + K + M) / d3
+ (-4 + K + N) / d2 + K / d1 ; jjjj += 1 ) {
    int j14, k18, k19;
    if (0 < jjjj - ((-4 + K + M) / d3
    + (-4 + K + N) / d2 + K / d1))
        k18 = 0;
    else
        k18 = (-4 + K + M) / d3 + (-4 + K + N) / d2 + K / d1;
}
```

3 On the Problem of High-Efficiency Software Portability

On the one hand, CS (computing systems) architectures are getting more complicated every year. On the other hand, their diversity is expanding. The popular X86 and ARM processor architectures are evolving, changing, in particular, the characteristics of the cache memory, the possibility of parallelization and vectorization, and other elements that affect performance. The usage of Nvidia graphics cards as accelerators is expanding. It is impossible not to notice the success of the line of Russian processors "Elbrus" with the VLIW architecture. Recently, manycore processors with addressable local memory have appeared: Tile64 processor (2007, 64 asynchronous cores) [27]; Epiphany processor - 1024 cores [28]; developed in Russia manycore system on a chip "1879VM8Ya" from "Module" [29].

The increasing diversity and complexity of architectures lead to the complication of high-performance software development. Moreover, there is a well-known problem of software portability from one CS to another. But for portable high-performance software, an additional problem of maintaining high efficiency arises. Here, high efficiency is understood as the fraction of the peak CS performance to which the software is ported. Indeed, there are many types of high-performance software for which the consumer would like even better performance. Such are, for example, solvers of application software package for problems of mathematical modeling, bioinformatics, weather forecasts, artificial intelligence, nanotechnology, etc. Such packages are always desired to be ported to new CS in order to obtain higher performance.

The OpenCL language standard has been developed for porting parallel programs. But a simple porting of such programs can lead to a significant loss of efficiency since for high performance it is necessary to take into account not only parallelism and memory stability [33].

Fast programs are developed in high-level languages. The compilers included in the processor system software optimize the mapping of instructions to the microarchitecture of that processor. Optimizing program transformations are performed in low-level (register) internal representations of popular open-source compilers GCC, LLVM and in closed-source compilers MS-compiler, ICC, LCC ("Elbrus").

Table 1 shows the advantages of the Intel compiler [34] for generating code for the Intel processor. If we want to transform the program with our own transformation using the precompiler, and then get better performance on the Intel processor, then the Intel compiler should be used after the precompiler. We cannot develop a precompiler in IR ICC (Intel Compiler Intermediate Representation), since we can't obtain the source code of the compiler. This means that the output of the precompiler must be a language that is input to the Intel: C or Fortran. Similarly, if after the precompiler optimizations we need to generate the GPU code with the closed-source PGI compiler, then the output of the precompiler should also contain a program in C/C++ or Fortran languages.

So we come to the conclusion that the portability of a high-performance solver while maintaining high performance implies that the solver must be written in a high-level portable language. Pre-optimization of the program (solver) should return code in this language.

Many memory optimizations can be performed independently of the processor microarchitecture. Such transformations include tiling, alignment, block allocation of matrices in RAM, linearization of multidimensional arrays, etc. Thus, transformations that minimize memory accesses can be common for different CS. You can perform high-level transformations in the precompiler, and low-level transformations in the optimizing compiler. The precompiler can carry out the portable part of the optimizations that don't depend on the microarchitecture of the processor, and the compiler can create machine code.

Transformations such as tiling or block allocation of matrices use parameters (block sizes). These parameters can be selected taking into account the characteristics of the algorithm and memory modules.

Complex array index expressions may appear after high-level transformations such as tiling or block matrix placements. This is solved by additional optimization with several traditional transformations. The resulting code after such automatic conversions can become so complex (see Fig. 1) that well-known compilers with the necessary transformations will not be able to optimize it well - this can be seen from Table 3.

It is reasonable to develop precompilers in program transformation systems such as Rose source to source compiler [35], OPS [25, 26] and others.

Conclusion

The article substantiates the need for precompilers for high-performance programs. Precompilers based on OPS for two solvers written for Intel processors are presented in this article. The results of numerical experiments demonstrating acceleration are presented. This document may be useful for developers of application software packages or application libraries, especially those who are focused on software portability.

References

1. Graham, S., et al.: Getting Up to Speed: The Future of Supercomputing, p. 289 (2005). https://doi.org/10.17226/11148
2. Yurushkin, M.: Block Data Layout Automation in C Language Compiler. Program. Ingeneria 16–18 (2014). ISSN 2220-3397
3. Shteinberg, O.B.: The parallelizing of recurent loops with using of non-regular superpositions computations. Bull. High. Educ. Inst. North Caucasus Region Nat. Sci. **2**, 16–18 (2009). ISSN 1026-2237
4. Shteinberg, O.B.: Parallelization of recurrent loops due to the preliminary computation of superpositions. Vest. Yuzhno-Ural'skogo Univ. Ser. Mat. Model. Program. **13**(3), 59–67 (2020). https://doi.org/10.14529/mmp200305
5. Structure Splitting for Elbrus processor compiler. Proceedings of the 14th Central & Eastern European Software Engineering Conference Russia (CEE-SECR 2019), St. Petersburg, Russia (2019). https://2019.secrus.org/program/submitted-presentations/structure-splitting-for-elbrus-processor-compiler/
6. Aggressive Inlining for VLIW, vol. 27, pp. 189–198 (2015). https://doi.org/10.15514/ISPRAS-2015-27(6)-13
7. Morylev, R., Shapovalov, V., Steinberg, B.Y.: Symbolic analisis in dialog-based parallelization of programs. Inform. Technol. **2**, 33–36 (2013). ISSN 1684-6400
8. Polly. https://www.sites.google.com/site/parallelizationforllvm/why-not-polly. Accessed 1 July 2020
9. Loop Optimizations Where Blocks are Required. https://software.intel.com/content/www/us/en/develop/articles/loop-optimizations-where-blocks-are-required.html. Accessed 1 July 2020
10. Developer Guide and Reference. block_loop=noblock_loop. https://software.intel.com/en-us/cpp-compiler-developer-guide-and-reference-block-loop-noblockloop. Accessed 1 July 2020
11. LLVM analyzers and converters. https://llvm.org/docs/Passes.html. Accessed 1 July 2020
12. Makoshenko, D.V.: Analiticheskoye predskazaniye vremeni ispolneniya programm i osnovannyye na nem metody optimizatsii, p. 122. Novosibirsk (2011)
13. Whaley, R.C., Dongarra, J.J.: Automatically tuned linear algebra software. In: SC 1998: Proceedings of the 1998 ACM/IEEE Conference on Supercomputing, p. 38. IEEE (1998)
14. Steinberg, B.Y., Steinberg, O.B.: Program transformations are the fundamental basis for creating optimizing parallelizing compilers. Softw. Syst.: Theory Appl. **12**(1), 21–113 (2021)
15. Liao, C., Quinlan, D.J., Vuduc, R., Panas, T.: Effective source-to-source outlining to support whole program empirical optimization. In: Gao, G.R., Pollock, L.L., Cavazos, J., Li, X. (eds.) LCPC 2009. LNCS, vol. 5898, pp. 308–322. Springer, Heidelberg (2010). https://doi.org/10.1007/978-3-642-13374-9_21
16. Gong, Z., et al.: An empirical study of the effect of source-level loop transformations on compiler stability. Proc. ACM Program. Lang. **2**(OOPSLA), 1–29 (2018)

17. Kudimova, A., et al.: Finite element homogenization models of bulk mixed piezo-composites with granular elastic inclusions in ACELAN package. Mater. Phys. Mech. **37**(1), 25–33 (2018)
18. Kurbatova, N.V., Nadolin, D.K., Nasedkin, A.V., Oganesyan, P.A., Soloviev, A.N.: Finite element approach for composite magneto-piezoelectric materials modeling in ACELAN-COMPOS package. In: Altenbach, H., Carrera, E., Kulikov, G. (eds.) Analysis and Modelling of Advanced Structures and Smart Systems. ASM, vol. 81, pp. 69–88. Springer, Singapore (2018). https://doi.org/10.1007/978-981-10-6895-9_5
19. Chen, F., Li, T.-Y., Lu, K.-Y.: Updated preconditioned Hermitian and skew- Hermitian splitting-type iteration methods for solving saddle-point problems. Comput. Appl. Math. **39**, 1–4 (2020)
20. Gill, P.E., Saunders, M.A., Shinnerl, J.R.: On the stability of Cholesky factorization for symmetric quasidefinite systems. SIAM J. Matrix Anal. Appl. **17**(1), 35–46 (1996)
21. Allen, R., Kennedy, K.: Optimizing Compilers for Mordern Architetures, p. 790. Morgan Kaufmann Publisher, Academic Press, USA (2002)
22. Evstigneev, V., Kasyanov, V.: Optimizing transformations in parallelizing compilers. Programirovanie **6**, 12–26 (1996). ISSN 0132-3474
23. Muchnick, S., et al.: Advanced Compiler Design Implementation. Morgan Kaufmann (1997)
24. Aho, A., et al.: Compilers Principles, Techniques, And Tools, 2nd edn., p. 1184 (2006)
25. Optimizing parallelization system (2013). www.ops.rsu.ru. Accessed 1 July 2020
26. Gervich, L.R., et al.: How OPS (optimizing parallelizing system) may be useful for clang. In: Proceedings of the 13th Central & Eastern European Software Engineering Conference in Russia (CEE-SECR 2017). Association for Computing Machinery, St. Petersburg (2017). ISBN 9781450363969. https://doi.org/10.1145/3166094.3166116
27. Tilera company. http://wiki-org.ru/wiki/Tilera. Accessed 1 July 2020
28. Epyphany processor. https://www.parallella.org/2016/10/05/epiphany-v-a-1024-core-64-bitrisc-processor/. Accessed 1 July 2020
29. The processor of NTC Modul. http://www.cnews.ru/news/top/2019-03-06_svet_uvidel_moshchnejshij_rossijskij_nejroprotsessor. Accessed 1 July 2020
30. Ammaev, S.G., Gervich, L.R., Steinberg, B.Y.: Combining parallelization with overlaps and optimization of cache memory usage. In: Malyshkin, V. (ed.) PaCT 2017. LNCS, vol. 10421, pp. 257–264. Springer, Cham (2017). https://doi.org/10.1007/978-3-319-62932-2_24
31. Cherdantsev, D.N.: Expressions linearization in optimization or parallelization compilers. Bull. High. Educ. Inst. North Caucasus Region Nat. Sci. **2**, 13–16 (2009). ISSN 1026-2237
32. Olmos, K., Visser, E.: Strategies for source-to-source constant propagation. Electron. Notes Theor. Comput. Sci. **70**(6), 156–175 (2002)
33. Abu-Khalil, Z., Guda, S., Steinberg, B.: Porting parallel programs without losing efficiency. Open Syst. J. **39**(4), 18–19 (2015). ISSN 1028-7493
34. Intel C++ Compiler. https://software.intel.com/content/www/us/en/develop/tools/oneapi/components/dpc-compiler.html. Accessed 1 July 2020
35. Rose source to source compiler. http://rosecompiler.org/. Accessed 1 July 2020

Execution of NVRAM Programs
with Persistent Stack

Vitaly Aksenov[1]([✉]), Ohad Ben-Baruch[2], Danny Hendler[2], Ilya Kokorin[1], and Matan Rusanovsky[2]

[1] ITMO University, 49, Kronverksky ave, 197101 Saint-Petersburg, Russia
[2] Ben-Gurion University, P.O.B. 653, 8410501 Beer-Sheva, Israel
`hendlerd@bgu.ac.il, matanru@post.bgu.ac.il`

Abstract. Non-Volatile Random Access Memory (NVRAM) is a novel type of hardware that combines the benefits of traditional persistent memory (persistency of data over hardware failures) and DRAM (fast random access). In this work, we describe an algorithm that can be used to execute NVRAM programs and recover the system after a hardware failure while taking the architecture of real-world NVRAM systems into account. Moreover, the algorithm can be used to execute NVRAM-destined programs on commodity persistent hardware, such as hard drives. That allows us to test NVRAM algorithms using only cheap hardware, without having access to the NVRAM. We report the usage of our algorithm to implement and test NVRAM CAS algorithm.

Keywords: Concurrency · Shared memory · Persistency · NVRAM

1 Introduction

For a long time the industry assumed the existence of two distinct types of the memory. The first one is a persistent memory that preserves its content even in the presence of hardware (e.g., power) failures. This type of memory was assumed to support mainly sequential block access with the poor performance of random access. Due to its ability to persist data this kind of memory is widely used to recover the system after a hardware failure: one can load the data from the persistent memory and restore the state of the application before the crash. The second type of the memory is DRAM that supports fast random byte-addressable access but loses its content on hardware failures. Due to its speed, this kind of memory is widely used in high-performance computations.

Nowadays, we can get benefits from both of these worlds due to the invention of Non-Volatile Random Access Memory (NVRAM)—a novel type of hardware that combines both the persistency and the fast random access. This allows us to implement low-latency persistent data structures that require random access to the memory, e.g., binary search trees, linked lists, and etc. A lot of work has been done to come up with data structures, hand-tuned for the NVRAM [10,11,13].

© Springer Nature Switzerland AG 2021
V. Malyshkin (Ed.): PaCT 2021, LNCS 12942, pp. 117–131, 2021.
https://doi.org/10.1007/978-3-030-86359-3_9

Some authors propose techniques, that can be used to transform DRAM-resident data structures into the ones suitable for the NVRAM [9,12].

Despite the speed of the NVRAM is compatible with the speed of the DRAM, the NVRAM is not expected to replace volatile memory totally since processor registers and the NVRAM cache are expected to remain volatile. Thus, even on NVRAM systems, a system failure leads to: 1) the loss of the results of recent computations since x86 computations are performed using volatile processor registers, and 2) the loss of data that was written to the NVRAM cache and has not been flushed to the NVRAM.

To make sure that the written data becomes persistent, we should flush one or several cache lines to the NVRAM. Flush of a single cache line is an atomic action: if a crash occurs during cache line flushing, the whole cache line is either persisted or not. However, if we want to flush multiple cache lines at a time, a crash event can occur between flushes—in such a case, only a part of the data becomes persistent while the rest is lost.

This yields one of the major challenges of NVRAM. If a system failure happens during a complex update, when some updated values have been flushed to the NVRAM from the cache while others still reside in the cache, non-flushed memory is lost and after the restart the NVRAM appears to be in an inconsistent state.

Due to the difficulty of ensuring storage consistency in the presence of the volatile NVRAM cache, a lot of works assume the absence of such cache [3–5,7]. However, in this work we consider real-life systems, thus we take the volatility of the NVRAM cache into account.

Another problem with the NVRAM is defining which executions are considered "correct" in the presence of hardware failures, that can lead to the loss of data. Despite a lot of correctness conditions were defined in the previous years [1,3,5,14,17], only *Nesting Safe Recoverable Linearizability* [3] describes the work with nested functions. Thus, maintaining persistent call stack is a crucial part of systems based on this concept. However, while methods of maintaining NVRAM heap are well-studied [6,8], methods of maintaining the persistent program stack are not studied at all: other works just assume the existence of a persistent call stack [3,4,12].

Moreover, the persistent stack allows us to design and implement novel complex system recovery algorithms, which can be faster than traditional log-based system recovery methods. Previously, such complex algorithms were considered impractical for traditional persistent memory systems due to the high latency of random access of traditional persistent memory, following directly from its mechanical nature, but on NVRAM-based systems such complex algorithms may be found useful.

In this work, we describe an algorithm, based on the implementation of the persistent call stack, that can be used to execute NVRAM programs and recover the system after a hardware failure while taking the architecture of real-world NVRAM systems into account. Moreover, the algorithm can be used to execute NVRAM-destined programs on commodity persistent hardware, such as hard drives. That allows us to test NVRAM algorithms using only cheap persistent hardware, such as HDD, SSD, etc., without having access to the NVRAM. We report the usage of our algorithm to implement and test correct and incorrect versions of the NVRAM CAS algorithm [3]. Also, we describe a method, that can be used to verify executions of NVRAM CAS algorithm for serializability.

The rest of the work is organized as follows. In Sect. 2, we discuss the system model, various failure models, operation execution model and talk about different correctness conditions, suitable for the NVRAM. In Sect. 3, we discuss the concept of the persistent program stack and its implementation. In Sect. 4 we present the solutions for the challenges we faced during the implementation of our algorithm. Also, we show there the architecture of the system along with the system recovery algorithm. In Sect. 5, we discuss the usage of our algorithm to implement and verify the NVRAM CAS algorithm, along with the method of checking executions of the NVRAM CAS algorithm for serializability. In Sect. 6, we discuss the directions of the future research. We conclude our work with Sect. 7.

2   Model

2.1   System Model

Our system model is based on the model described in [3].

There are N processes $\{p_i\}_{i=1}^N$ executing operations concurrently. Also, there are M objects $\{O_j\}_{j=1}^M$ located in the shared non-volatile memory. Processes communicate with each other by executing operations on shared objects (see Fig. 1a), that can support read, write or read-modify-write [15] operations.

In our model, all shared memory is considered non-volatile, i.e., it does not lose its content even after a crash event. However, we assume the existence of a volatile memory in the system. Each object LO, located in the volatile memory, is considered local to some process p. In other words, only process p can access object LO. Thus, besides being able to execute operations on shared objects, each process can access its local objects. Such objects support only read and write operations (see Fig. 1b).

However, our model still does not reflect some properties of the real-world hardware: for example, it does not take into account the existence of the volatile NVRAM cache and the existence of shared volatile memory.

2.2   Failure Model

There exist two general failure models:

- *Individual* crash-recovery model [3]. In such a model, each process can face a crash event independently of all other processes. When a process faces a crash event, it stops working until it is restarted. All data, stored in the volatile memory of the failed process, is lost. However, all data persisted to the NVRAM is not lost and remains available to the failed process after its restart.

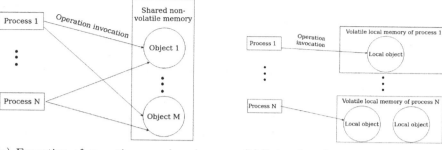

(a) Execution of operations on shared objects

(b) Execution of operations on local objects

Fig. 1. System execution model

- *System* crash-recovery model. In such a model, a crash event happens in the whole system instead of an individual process. The whole systems stops working until it is restarted. After the system restarts, the contents of all the volatile memory is lost. As in the previous model, the data, persisted in the NVRAM, is not lost and remains available to all processes after the system restarts.

Note, that the *system* crash-recovery model is a special case of the *individual* crash-recovery model, since a crash of the whole system can be represented as a set of N simultaneous crash events of individual processes—one crash event per each process. Despite the fact that *individual* crash-recovery model is a more general model, in this work we focus mainly on *system* crash-recovery model. In real-world shared memory systems multiple computational units are placed in a single server and thus a failure of a single computational unit is impossible without a failure of the entire system. That is why, in our opinion, *system* crash-recovery model describes more accurately the real-life crash event—for example, power loss.

2.3   Operation Execution

We say that function F is being executed by process p if execution of F has been started by p but has not been finished yet. As described in [3], we work with the nested invocation of functions: at any moment, multiple functions can be executed by any process. It happens when function F invokes function G. Thus, executed functions in each process form a nested sequence. In the above example execution of G is nested into the execution of F.

To allow the recovery of the system, we provide each function F with a dual function F.Recover, which receives the same arguments as F. F.Recover is called after the system restart to perform system recovery and it should either finish the execution of F or roll back F.

To perform the system recovery, for each process p we should call F.Recover for each function F being executed by p at the crash moment. Moreover, recovery functions should be called in the certain order: if the execution of G is nested into the execution of F, G.Recover should be called before F.Recover. Thus, each process should perform the recovery in the LIFO (stack) order.

Also we should consider the possibility of *repeated failures*—failures which happen during the recovery procedure. Consider the system failure after F was invoked. After the restart, we should call F.Recover to complete the recovery. Suppose another system failure happens before F.Recover is finished. After the second restart, we should again continue the recovery at executing F.Recover. It means that there is no difference between the system failure happening during the execution of F or during the execution of F.Recover: in both cases, we should call F.Recover to complete the recovery. Thus, F.Recover should be designed so that it can complete the operation (or roll it back) no matter whether the crash occurred when executing F or F.Recover.

2.4   Correctness

Multiple correctness conditions for NVRAM exist. Here, we outline three most important (from the strongest to the weakest):

1. *Nesting Safe Recoverable Linearizability* [3]. It requires each invoked function F to be completed even if a crash event occurs while executing F. Thus, under that correctness condition, F.Recover should finish the execution of F either by completing it successfully or by rolling it back.
2. *Durable Linearizability* [17]. It requires that each function F, execution of which has finished before a crash, should be completed. If a crash event occurs while executing function F, such function may be either completed or not.
3. *Buffered Durable Linearizability* [17]. It is a weaker form of *Durable Linearizability*. Its difference is in that it allows function F not to be completed even if its execution finished before a crash. However, that correctness condition requires each object to provide sync operation—all functions, finished before a call to sync must be completed, even if a crash event occurs.

In this work, we propose an algorithm that can be used to run NVRAM-destined programs under *Nesting Safe Recoverable Linearizability*—the strongest correctness condition.

3   Persistent Stack

3.1   Program Stack Concept

In order to execute programs for NVRAM, for each thread[1] t we maintain an information about functions, which were executed by t when a crash occurred.

---

[1] When talking about practical aspects of concurrent programming, we use the word "thread" in the same context, as the word "process" in the theory of concurrent programming.

Also, to invoke recover functions in the correct order we maintain the order in which these functions were invoked.

We maintain that order by using the notion of program stack: each thread t has its own NVRAM-located stack, and each function executed by t corresponds to a single frame of the stack. When a function is invoked, the corresponding frame is added to the top of the stack. After the end of the execution, the frame is removed from the top. Therefore, when a crash occurs, the stack of thread t contains frames, that correspond to functions that were executed by t at the crash moment. Moreover, such frames are located in the correct order: if execution of G was nested into the execution of F, a frame of G is located closer to the top of the stack, than a frame of F.

3.2 Issues of Existing Implementations

The functionality of the program stack is already implemented by standard execution systems: for example, x86 program stack. However, we cannot use them as-is, even if we transfer it from the DRAM to the NVRAM.

Here we remind the implementation of the function call via the x86 stack. Suppose function F calls function G using x86 command CALL G. To perform such an invocation, we should store a return address on the stack—the address of the instruction in function F that follows the instruction CALL G. After the execution of G is finished, we continue execution of F from that instruction. This is exactly how x86 instruction RET works—it simply reads the return address from the stack and performs JMP to that address, allowing it to the continue the execution from the desired point.

Note that such a program stack implementation has a number of drawbacks, that makes it impossible for us to use such implementation as a persistent stack:

- After the system restart due to the crash, the code segment may be relocated, i.e., have a different offset in the virtual address space. That will make us incapable of identifying which functions were executed at the crash moment— we simply won't be able to match return address from the stack with an address of some instruction after the code segment relocation.
- We cannot guarantee an atomicity of adding a new frame to the stack or removing a frame from the top of the stack – if a crash occurs during adding or removing stack frame, after the system restarts the stack might be in an inconsistent state.

Thus, instead of using existing program stack implementations, we present our persisted stack structure that overcomes the above drawbacks.

3.3 Persistent Stack Structure

Each thread has an access to its own persistent stack. For simplicity, in this section we assume that the persistent stack is allocated in the NVRAM as a continuous memory region of constant size. However, we explain how to make a stack of unbounded size in Appendix A of the full version of the paper, available at [2].

Persistent stack consists of consequent persistent stack frames—one frame per function that accesses NVRAM.[2] Each frame ends with a one-byte end marker: it is 0x1 (*stack end marker*) if the frame is the last frame of the stack; otherwise, it is 0x0 (*frame end marker*). Any data located after the *stack end marker* is considered invalid—it should never be read or interpreted in any way (see Fig. 2).

Fig. 2. Persistent stack structure

To finish the description of the data layout, each persistent stack frame consists of: 1) a unique identifier of the invoked function that allows us to call the appropriate recover function during the system recovery; 2) arguments of the function, serialized into a byte array—during the system recovery we pass them to the recover function; 3) a one-byte end marker (either 0x0 or 0x1).

3.4 Update of the Persistent Stack

The persistent stack should be updated: 1) when the function is invoked—a new frame should be added to the top of the stack, 2) when the function execution is finished—the top frame of the stack should be removed.

Adding the New Frame to the Top of the Stack. Suppose the stack at the beginning of the operation has two frames in it (see Fig. 3a):

To add a new frame to the top of the stack, we perform the following actions:

1. After the *stack end marker*, we write a new frame with the *stack end marker* set. Note that the new frame (frame 3) is located after the *stack end marker* of the previous frame (frame 2). Therefore, the new frame is not considered as a stack frame, while the previous frame (frame 2) is still the last stack frame (see Fig. 3b).
2. Change the end marker of the current last stack frame (frame 2) from 0x1 to 0x0. Thus, the last stack frame (frame 2) becomes the penultimate stack frame and the new frame (frame 3) becomes the last stack frame (see Fig. 3c). We name that one-byte end marker changing operation as *moving the stack end forward*.

---

[2] Each such function must have a recover version, as described above.

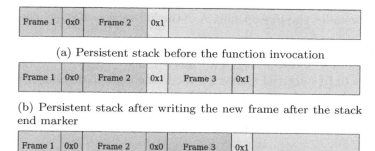

(a) Persistent stack before the function invocation

(b) Persistent stack after writing the new frame after the stack end marker

(c) Persistent stack after adding the new frame to the top of the stack

Fig. 3. Adding new frame to the top of the stack

Removing the Top Frame from the Stack. Suppose the stack at the beginning of the operation has three frames in it (see Fig. 4a):

To remove the top frame from the stack, we simply change the end marker of the penultimate stack frame (frame 2) from 0x0 to 0x1, thus making the penultimate stack frame the last stack frame (see Fig. 4b). We name that one-byte end marker changing operation as *moving the stack end backward*. Note, that frame 3 becomes the part of the invalid data and, therefore, it will not be considered as a stack frame anymore.

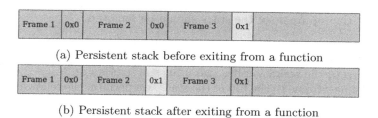

(a) Persistent stack before exiting from a function

(b) Persistent stack after exiting from a function

Fig. 4. Removing the top frame from the stack

Dummy Frame. Note that both frame removal and frame addition procedures assume the existence of at least one frame in the stack, besides the one that is being removed or added. Particularly, this assumption implies that the bottom stack frame cannot be removed from the stack. We can simply satisfy that assumption by introducing a dummy frame—the stack frame, located at the bottom of the stack (i.e. the first frame, added to the stack). That frame is added to the stack at the initialization of the stack and is never removed. By that, we ensure that there is always at least one frame, thus making it possible for us to use the stack update procedures, described above.

Flushing Long Frames. Note that sometimes a new stack frame does not fit into a single cache line—for example, that can happen when some function receives arguments list with length greater than the cache line size. In such case, we will not be able to add such frame to the stack atomically (since only single cache line can be persisted atomically). Therefore, we can face a crash event that will force us to write the new frame partially (see Fig. 5).

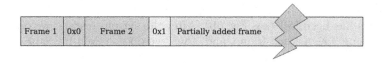

Fig. 5. Persistent stack with partially flushed frame

In our algorithm, we, at first, add a new frame to the stack, and only after the new frame has been written successfully we *move the stack end forward*. Thus, even if the crash event happens, the stack will remain consistent: partially written frame will be located after the *stack end marker* and will not be considered as a stack frame. Therefore, this scenario does not brake *Nesting-safe Recoverable Linearizability*, since the last function invocation was not linearized before the crash event. We can simply think that the crash happened before the function invocation and the function was never invoked.

The Atomicity of the Stack Update. We can say that a function invocation linearizes only when we *move the stack end forward*. This requires only the flushing of a single byte to the NVRAM. Since a single byte always resides in a single cache line, this flush always happens atomically.

The same observation can be made for *moving the stack end backward*: an execution of the function is finished when we change the end marker of the penultimate stack frame from 0x0 to 0x1. As was described above, such action happens atomically.

Persistent Stack and the NVRAM. The procedure of adding and removing a stack frame requires only the ability to flush a single byte atomically and not the entire cache line—this makes us capable of implementing the stack maintenance algorithm on a hardware that does not support atomic flushing of an entire cache line. Thus, the algorithm described above, can be easily emulated without having access to an expensive NVRAM hardware, using almost any existing persistent hardware such as HDD, SSD, etc.

For the above reasoning to remain correct, we should maintain two following invariants:

1. We should flush the new stack frame before *moving the stack end forward*. Suppose we violate that rule. Consider a crash event that happens at some time after the *moving the stack end forward*. Suppose also, that new stack frame (frame 3) has been written to the volatile NVRAM cache and was lost during the crash. After the system restart we will not be able to call the recover function for frame 3, because we have lost that frame (see Fig. 6a).

2. When changing the end marker of some frame (either from 0x0 to 0x1 or vice versa) we should immediately flush it before staring the execution of the invoked function or continuing the execution of the caller function.

   Suppose we violate that rule. Consider a crash event, happening while executing function F, corresponding to frame 3. Also consider that the *frame end marker*, written to frame 2, has been written to the volatile NVRAM cache and thus has been lost (see Fig. 6b). After the system restart, we do not consider frame 3 as a stack frame, and, thus, we do not even invoke F.Recover.

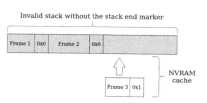

(a) New stack frame has been lost due to volatility of the NVRAM cache

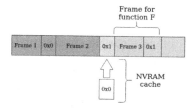

(b) End marker has been lost due to volatility of the NVRAM cache

Fig. 6. Results of violating flushing invariants

4   System Implementation

4.1   Pointers to the Memory in NVRAM

When working with pointers to the NVRAM we face the problems similar to those we faced when working with function addresses (Sect. 3.2). Suppose we have acquired pointer `ptr` pointing to the NVRAM. We store `ptr` in the NVRAM (for example, in some persistent stack frame, as an argument of some function F). After that, we face a crash event. And when we restart the system, the mapping of the NVRAM into the virtual address space can change, thus, making pointer `ptr` invalid, since it does not point to the NVRAM anymore.

The same problem happens when we emulate NVRAM using HDDs, mapped to the virtual address space using `mmap` syscall: on each system restart, HDD is mapped to a different location in the virtual address space.

This problem has a very simple solution: instead of using direct pointers to the NVRAM, we shall use offsets from the beginning of the NVRAM mapping into the virtual address space. Suppose the mapping of the NVRAM begins at address MAP_ADDR. Then, instead of storing ptr we store ptr - MAP_ADDR—an offset of the desired memory location. Note that such an offset does not depend on an exact location of the mapping, thus making it safe for us to store it in the NVRAM and use after the system restart.

4.2 Handling Return Values

Traditionally, on x86 architecture, functions return value using the volatile memory—either in x86 register EAX, if the return value is an integer, or in FPU register ST0, if the return value is floating-point. For example, cdecl, one of the most popular x86 calling conventions, implies the above rules for return value.

However, in our case we cannot use volatile processor registers to store return value. Consider a crash event occurring after the callee function G has saved the return value to the EAX and finished its execution by *moving the stack end backward*. At that time, the caller function F has not persisted the return value from EAX to the NVRAM. After the system restart, we will not invoke G.Recover, but start from F.Recover instead. However, we cannot execute F.Recover properly, because we have lost the result of G.

That is why functions should store their results directly in the NVRAM. We could come up with two approaches where to store them:

1. on the persistent stack. For example, we can use an especially-allocated place in a persistent stack frame for that purpose.
2. in the NVRAM heap. In such a case, the caller can preallocate a memory location for the answer before invoking the callee, and pass the pointer to that memory location in callee's arguments (note, that as was mentioned in Sect. 4.1, we should use offsets instead of pointers to the NVRAM). After that, callee can store its answer in that memory location.

In both cases, the callee should flush the answer to the NVRAM before *moving the stack end backward*. Our implementation supports returning of small values (up to 8 bytes) on the persistent stack, while big values are returned in the NVRAM heap.

4.3 Architecture of the System

The system consists of a single main thread and N worker threads.

Main thread can run in either a standard mode or a recovery mode.

When running in the standard mode, the main thread performs the following steps.

1. Initialize the NVRAM heap. This may include the initialization of the memory allocator, the mapping of the NVRAM to the virtual address space and the consequent initialization of the variable MAP_ADDR, mentioned in Sect. 4.1
2. Initialize N new persistent stacks.

3. Start N worker threads, giving each worker thread pointer to the beginning of its persistent stack.
4. Receive task that should be executed by the system and add them to the producer-consumer queue.

When running in the standard mode, worker threads receive tasks from the producer-consumer queue and execute them.

In case of a crash, the main thread starts in the recovery mode and performs the following steps:

1. Initialize the NVRAM heap.
2. Start N recovery threads, giving each recovery thread the pointer to a persistent stack of some worker thread.
3. Wait for all recovery threads to finish.
4. Restart the system in normal mode.

Each recovery thread executes the following algorithm:

1. Traverse its persistent stack from the top to the bottom.
2. Execute the corresponding recover operation for each stack frame.
3. After the recovering of an operation on the top of the stack is finished, pop the top frame.
4. After all the frames (except for the dummy one) are removed from the stack, finish the execution.

System recovery happens in parallel, which allows for a faster recovery than an ordinary single-threaded recovery.

We note that our algorithm deals well with *repeated failures*. If such a failure happens during the recovery, the new recovery continues not from the beginning, but from where the previous recovery was interrupted. More formally, consider a frame, corresponding to a function F. If during the recovery we have completed execution of F.Recover and removed that frame from the stack, even after the *repeated failure* we will not run a recover function for that frame once more. Thus, we achieve the progress even in the presence of *repeated failures*.

5   Verification

The described algorithm of the persistent stack can be used to implement and verify CAS algorithm for NVRAM, described in [3]. That paper assumes the absence of the volatile NVRAM cache, i.e., all writes are performed right into the memory. To emulate this, we should flush each written cache line to the NVRAM immediately after the corresponding write. Also, we should implement the algorithm so that each written value never crosses the border of a cache line to allow atomic flush of each written value.

Multiple correctness conditions for concurrent algorithms exist: the most popular are linearizability [16], sequential consistency and serializability [18]. We want to perform the verification against some of these correctness conditions.

From now on we take CAS algorithm for the NVRAM as the running example. Consider the following execution. Multiple threads run a set of CAS operations on a single register Reg: $\{CAS(Reg, old_i, new_i)\}_{i=1}^N$, and the initial value of Reg is init. And for each operation we know whether it was finished successfully or not.

We present an algorithm, that can be used to check such an execution for serializability in a polynomial time.

5.1 Serializability

To verify the execution for serializability in polynomial time, we build a graph $\langle V, E \rangle$, $G = \{old_i\}_{i=1}^N \cup \{new_i\}_{i=1}^N \cup \{init\}$ and construct the set of edges E the following way: $a \to b \in E$ if and only if there exists a successful $CAS(Reg, a, b)$ in the execution. Also, we read the final value of the register. We can read it after all the CAS operations are finished.

Since each edge of G corresponds to a successful CAS, each successful CAS was executed exactly once, and each successful $CAS(Reg, a, b)$ changed value of Reg from a to b, to verify the execution for serializability we should find some Eulerian circuit that starts in the initial value of the register and ends in the final value of the register—such a circuit corresponds to the sequential execution. Thus, the execution is serializable if and only if such a circuit can be found[3].

5.2 Running Examples

We have implemented the algorithm, described above, using HDD-based memory-mapped files to emulate the NVRAM. We used UNIX utility kill to interrupt the system at random moments by that emulating system crashes.

We have generated random executions of the algorithm in the following way:

1. Generate an initial integer value of the register;
2. Generate $\{new_i\}_{i=1}^N$ and $\{old_i\}_{i=1}^N$ as integer values, uniformly sampled from some range: either wide range $[-10^5, 10^5]$), or narrow range ($[-10, 10]$);
3. Start the system in the normal mode, add descriptors of $\{CAS(Reg, old_i, new_i)\}_{i=1}^N$ operations to the producer-consumer queue in the random order;
4. Run 4 working threads that execute these CAS operations;
5. At random moment, emulate system failure using the kill utility;
6. Restart the system in the recovery mode waiting for all CAS operations, that were executing at the crash moment, to complete;
7. Restart the system in the normal mode, add all remaining descriptors to the queue;
8. Run steps 4–7 until all operations are completed;
9. Get answers of all CAS operations, get the final value of the register, and, finally, verify the execution for serializability.

---

[3] Please, note that we can simply serialize unsuccessful operation at the times when the register holds a value different from old_i.

We have verified a lot of random executions along with emulated system failures at random moments. All executions of the CAS algorithm presented in [3] were found to be serializable. We also verified the executions of incorrect CAS algorithm with especially-added bugs: we have removed the matrix R from the CAS algorithm. The executions of such a wrong implementation were reported to be non-serializable.

The implementation is publicly available at https://github.com/KokorinIlya/ NVRAM_runner.

6   Future Work

We find three interesting directions for the future work: 1) implement and test other NVRAM algorithms; 2) find the polynomial algorithm that verifies executions of CAS algorithm for linearizability and sequential consistency, or prove that the problem of such a verification is NP-complete; 3) develop a plugin for one of the modern C++ compilers that can be used to reduce the boilerplate code: e.g., automatically create a new stack frame on each function call, remove the top frame when a function execution finishes, and etc.

7   Conclusion

In this paper we presented an algorithm that can be used to run NVRAM programs. The described algorithm takes into consideration different aspects of real-world NVRAM systems. Moreover, the algorithm can be used to run NVRAM-destined programs on commodity persistent hardware, which can be useful for implementing and testing novel NVRAM algorithms without having an access to an expensive NVRAM hardware. The algorithm was successfully used to implement and verify the CAS algorithm for NVRAM.

References

1. Aguilera, M.K., Frølund, S.: Strict linearizability and the power of aborting. Technical Report HPL-2003-241 (2003)
2. Aksenov, V., Ben-Baruch, O., Hendler, D., Kokorin, I., Rusanovsky, M.: Execution of nvram programs with persistent stack, full version (2021). https://arxiv.org/abs/2105.11932
3. Attiya, H., Ben-Baruch, O., Hendler, D.: Nesting-safe recoverable linearizability: modular constructions for non-volatile memory. In: Proceedings of the 2018 ACM Symposium on Principles of Distributed Computing, pp. 7–16 (2018)
4. Ben-David, N., Blelloch, G.E., Friedman, M., Wei, Y.: Delay-free concurrency on faulty persistent memory. In: The 31st ACM Symposium on Parallelism in Algorithms and Architectures, pp. 253–264 (2019)
5. Berryhill, R., Golab, W., Tripunitara, M.: Robust shared objects for non-volatile main memory. In: 19th International Conference on Principles of Distributed Systems (OPODIS 2015), Schloss Dagstuhl-Leibniz-Zentrum fuer Informatik (2016)

6. Bhandari, K., Chakrabarti, D.R., Boehm, H.J.: Makalu: fast recoverable allocation of non-volatile memory. ACM SIGPLAN Not. **51**(10), 677–694 (2016)
7. Blelloch, G.E., Gibbons, P.B., Gu, Y., McGuffey, C., Shun, J.: The parallel persistent memory model. In: Proceedings of the 30th on Symposium on Parallelism in Algorithms and Architectures, pp. 247–258 (2018)
8. Cai, W., Wen, H., Beadle, H.A., Hedayati, M., Scott, M.L.: Understanding and optimizing persistent memory allocation. In: Proceedings of the 25th ACM SIG-PLAN Symposium on Principles and Practice of Parallel Programming, pp. 421–422 (2020)
9. Chauhan, H., Calciu, I., Chidambaram, V., Schkufza, E., Mutlu, O., Subrahmanyam, P.: {NVMOVE}: helping programmers move to byte-based persistence. In: 4th Workshop on Interactions of NVM/Flash with Operating Systems and Workloads ({INFLOW} 16) (2016)
10. Chen, S., Jin, Q.: Persistent B+-trees in non-volatile main memory. Proc. VLDB Endow. **8**(7), 786–797 (2015)
11. David, T., Dragojevic, A., Guerraoui, R., Zablotchi, I.: Log-free concurrent data structures. In: 2018 {USENIX} Annual Technical Conference ({USENIX}{ATC} 18), pp. 373–386 (2018)
12. Friedman, M., Ben-David, N., Wei, Y., Blelloch, G.E., Petrank, E.: NVTraverse: in NVRAM data structures, the destination is more important than the journey. In: Proceedings of the 41st ACM SIGPLAN Conference on Programming Language Design and Implementation, pp. 377–392 (2020)
13. Friedman, M., Herlihy, M., Marathe, V., Petrank, E.: A persistent lock-free queue for non-volatile memory. ACM SIGPLAN Not. **53**(1), 28–40 (2018)
14. Guerraoui, R., Levy, R.R.: Robust emulations of shared memory in a crash-recovery model. In: 2004 Proceedings of 24th International Conference on Distributed Computing Systems, pp. 400–407. IEEE (2004)
15. Herlihy, M.: Wait-free synchronization. ACM Trans. Program. Lang. Syst. (TOPLAS) **13**(1), 124–149 (1991)
16. Herlihy, M.P., Wing, J.M.: Linearizability: a correctness condition for concurrent objects. ACM Trans. Program. Lang. Syst. (TOPLAS) **12**(3), 463–492 (1990)
17. Izraelevitz, J., Mendes, H., Scott, M.L.: Linearizability of persistent memory objects under a full-system-crash failure model. In: Gavoille, C., Ilcinkas, D. (eds.) DISC 2016. LNCS, vol. 9888, pp. 313–327. Springer, Heidelberg (2016). https://doi.org/10.1007/978-3-662-53426-7_23
18. Papadimitriou, C.H.: The serializability of concurrent database updates. J. ACM (JACM) **26**(4), 631–653 (1979)

A Study on the Influence of Monitoring System Noise on MPI Collective Operations

A. A. Khudoleeva[(⊠)] and K. S. Stefanov

Lomonosov Moscow State University, GSP-1, Leninskie Gory, Moscow 119991, Russian Federation
cstef@parallel.ru

Abstract. Parallel application can be studied with the means of the supercomputer performance monitoring system. The monitoring system agent activates during the parallel program run to collect data from system sensors. It occupies program's resources and causes perturbation in the monitored program workflow. The influence of the performance monitoring system on parallel applications is poorly studied. We propose to measure the monitoring system noise using a benchmark based on MPI collective operations All-to-All and Barrier—the noise detector. We present measurements of the detector execution time under influence of a real monitoring system and without it. The results demonstrate that monitoring system agent has negligible influence on the detector when the agent and the detector have different CPU core affinity. On the other hand, monitoring system agent impact is statistically significant on the execution of All-to-All and Barrier when bound to the same logical core as the collective operations.

Keywords: Supercomputer · Performance monitoring · Monitoring system noise · Parallel job slowdown

1 Introduction

Modern supercomputers have very complex and sophisticated architecture, they contain many hardware and software components. Development of correct parallel applications designed for scientific problem modeling on HPC systems is a difficult task itself. Optimization of the parallel program for the target system is a challenge as well. Supercomputer resources provide a wide selection of instruments and ample opportunities for parallel application design. The structure of the user applications is often complicated, making its optimization difficult. Full utilization of allocated resources is desirable and reaching the maximum possible performance of parallel application and decreasing its execution time is vital. Researchers may obtain information from the monitoring system of the supercomputer to configure the program optimally for the HPC system.

© Springer Nature Switzerland AG 2021
V. Malyshkin (Ed.): PaCT 2021, LNCS 12942, pp. 132–142, 2021.
https://doi.org/10.1007/978-3-030-86359-3_10

Parallel application can be studied by the means of performance monitoring systems. Monitoring system agents collect data from hardware and software sensors during the execution of parallel programs. Collected data contain different performance metrics such as cache misses, FLOPS, CPU load etc. Monitoring system agent is launched on the same supercomputer node as the monitored application. Agent aggregates statistics from the system sensors at fixed intervals and sends the data for further processing. Monitoring system agent shares resources with the monitored program. It interrupts application workflow, interferes with the program. In this paper we call this phenomenon monitoring system noise. Developers often say that monitoring system noise is insignificant and has no effect on parallel application performance, but provide no data supporting this statement. Generally, the influence of the monitoring system agent on a parallel application is poorly studied. We use a benchmark based on MPI collective operations to study performance monitoring system noise. We call the benchmark a monitoring system noise detector. We present the results using the detector. The detector is launched in presence of DiMMon monitoring system [11] and without it on the Lomonosov-2 supercomputer [13]. We consider several cases of the mutual detector and monitoring system agent allocation on a supercomputer node. The results show that the monitoring system agent under standard conditions has a negligible influence on the parallel application. We also show that the monitoring system noise has a statistically significant impact on the detector, which allocates all logical cores of a node. We study influence of the monitoring system with a higher frequency to complete the research and compare it with the standard frequency 1 Hz.

The main contribution of this paper is a method for estimating the noise produced by a performance monitoring system of a supercomputer. We apply this method to a real performance monitoring system and show the cases when the noise of the monitoring system is negligible and when it is not.

2   Background and Related Work

There is a wide selection of performance monitoring systems. The list includes systems like Supermon [10], NWPerf [7], HPCToolkit [1], LIKWID [9,12], Performance Co-Pilot [4], LDMS [2]. Authors of the HPCToolkit [1], Performance Co-Pilot [4] and LIKWID [9,12] tools state that the listed systems are lightweight i.e. noise generated by them does not influence the workflow of parallel applications. However, there are neither reported results that can prove such finding nor descriptions of conducted experiments. Only papers [7] and [2] contain surveys on monitoring system noise.

NWPerf monitoring system agent is known to be a Linux kernel module. NWPerf reads systems sensors once per minute in the standard configuration. NWPerf influence on performance of collective operations All-to-All and All-Reduce was reported in [7]. A loop of 10 000 collective operations was launched on 128 nodes, 256 cores with and without NWPerf. All-to-All execution was 27% longer and All-Reduce execution was 9.46% longer, when the monitoring rate was ten times higher than the standard (0,17 Hz). However, OS noise concealed NWPerf perturbation, when the monitoring rate was standard. On the other hand, LDMS [2] monitoring system was tested on two HPC clusters and its influence on performance of MPI collective operations was shown to be small. LDMS with the frequency 1 Hz has almost no effect on the time of MPI parallel application execution.

The results of the two studies are hard to generalize. NWPerf with high monitoring rate has a significant influence on MPI collective operations, while the noise of the second system is negligible. The contradictory results can be caused by the difference in target platforms, where the experiments were conducted. Moreover, NWPerf system is 10 years older than LDMS. Therefore the result described in the LDMS paper is more reliable and relevant but still is very specific. To summarize, there is still no clear understanding about the influence of monitoring systems on MPI collective operations.

There are more surveys dedicated to the problem of OS noise and its influence on supercomputer applications. Authors of the article [3] report that unsynchronised OS noise can cause a notable slowdown of Barrier and All-Reduce. Survey [8] describes an optimization of SAGE application on ASCI Q supercomputer. All-Reduce performance was shown to be significantly influenced by OS noise. The paper authors concluded that well synchronized parallel applications may be influenced by frequent but weak OS noise.

Noise generated by monitoring systems is expected to be weaker than the OS noise. However, we assume that MPI collective operations might "sense" the low impact of the monitoring system agent. We showed the decrease in MPI collective operations performance under influence of a lightweight noise generator in our previous work [6]. We investigate whether execution of DiMMon performance monitoring system causes notable desynchronization of a highly parallel MPI code in this paper.

3   The Proposed Method

3.1   Monitoring System Detector

We study an influence of the monitoring system on parallel applications that use MPI collective operations. We use a benchmark, which we call a monitoring system noise detector, to model MPI programs. The noise detector is outlined in Algorithm 1. We can vary the number of collective operations, set collective operation type and choose message size to configure the detector. In our previous study [6] we showed that MPI Barrier and MPI All-to-All with message size 2 KB are sensitive to an interference of a small artificial noise. We focus on the listed collective operations and find out whether their performance is influenced by injection of a real monitoring system. The following notation is used to indicate the detector configuration: *[Operation type] N[number of nodes in use] [number of iterations in the loop]it. [message length for All-to-All only]*. For example, All-to-All N4 1000 it. 2 KB.

Algorithm 1. Monitoring System Detector

start *timer*
for iteration count **do**
    MPI Collective Operation
end for
stop *timer*

3.2 Statistical Criteria

Hence we need to identify whether monitoring system influences the performance of the detector, we use statistical methods described in [5] to compare execution time of the detector in presence of the performance monitoring system and without it.

T_{no-mon} is the detector time execution sample without the monitoring system.

$T_{x\ Hz}$ is the detector time execution sample with the monitoring system of x Hz frequency.

Criterion 1 is based on the use of 99% confidence intervals. The criterion is applied for samples of a relatively small size, 10 elements in our case. We claim that the collective operations are slowed down by the monitoring system, if

$$boundary\ gap = lower\ bound\ 99\%CI(T_{xHz}) - upper\ bound\ 99\%CI(T_{no-mon}) \tag{1}$$

has a positive value, where $99\%CI(S)$ is 99% confidence interval of sample S. Therefore,

$$mean\ gap = mean(T_{x\ Hz}) - mean(T_{no-mon}) \tag{2}$$

would be positive. CIs mutual positions are shown in Fig. 1.

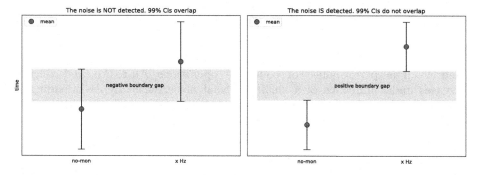

Fig. 1. Criterion 1. Left figure represents CIs position when noise is insignificant. Right figure represents CIs position when noise is significant.

Criterion 2 is based on the use of T_{no-mon} inlier range. The criterion is applied for large samples that contain hundreds of elements. We claim that the performance of collective operations statistically decreases under the influence of the monitoring system if elements in $T_{x\,Hz}$ can be classified as outliers comparing to elements in T_{no-mon}. The range for inlier values of T_{no-mon} is expressed via formula

$$inlier\ range = [Q_{25\%} - 1.5 \times IQR, Q_{75\%} + 1.5 \times IQR] \qquad (3)$$

where $Q_{25\%}$ and $Q_{75\%}$ are the lower and upper quartiles, IQR—interquartile range.

3.3 Experimental Setup

We conduct the experiments on the Lomonosov-2 supercomputer [13]. Each node of the supercomputer has 14 cores Intel Haswell-EP E5-2697v3 processor. We launch the detector in the test queue of the supercomputer, which has a 15 min time limit for users' programs runtime. It allows us to avoid waiting in a busy main queue of the supercomputer. We set the number of collective operations based on the time limit of the test queue of the supercomputer. DiMMon monitoring system [11] is installed on the supercomputer. Its frequency is set 1 Hz by default—it reads system sensors once per second. The frequency 1 Hz is typical for many monitoring tools nowadays and allows to collect enough data about running applications [11]. One DiMMon agent per node is launched to collect data about running application.

3.4 Allocation of the Detector

We distinguish the following three cases in the experiments:

    I. The detector is launched on all logical cores of a node (28 cores).
   II. The detector is launched on the number of logical cores equal to the number of physical cores of a node (14 cores).
 III. Case II, but affinity for the detector and the monitoring system is set manually.

Case I. The performance of collective operations All-to-All and Barrier decreases under influence of a small artificial noise, when hyper-threading is utilized [6]. We study the influence of the monitoring system on a parallel application that uses all available logical cores.

Case II. A regular way to submit a parallel job to the supercomputer is to request number of processes per node equal to the number of physical cores. The detector is launched on the number of cores equal to the number of physical cores.
    Criterion 1 is used for both **Cases I** and **II**.

Case III. We conduct additional experiments to draw a conclusion whether monitoring system with standard monitoring rate influences collective operations All-to-All and Barrier. When the detector is launched on half of the logical cores per node, free resources are left to OS for computations planning and workflow optimization. The detector and the monitoring system can be set on the same logical cores by the OS, as well as they can be set apart. This uncertainty in experiments causes a lower reproducibility of the detector execution, which leads to appearance of outliers in time measurements. Core affinity is set manually for the monitoring system and for the detector to remove the uncertainty in the experiments. We highlight the following cases:

A—NO monitoring system.
B—logical cores for the detector and the monitoring system have different numbers.
C—the monitoring system agent is bound to the same logical core as the detector.

We split the loop (See Algorithm 1) into chunks of collective operations to collect bigger samples and eliminate chunks, which were influenced by external non-periodic noise. We choose chunk size (number of loop iterations) to have an execution time between 10 s and 20 s for each type of collective operation. We use **Criterion 2** in **Case III** to identify performance monitoring influence on execution of MPI collective operations.

4   Experimental Results

4.1   Using All Logical Cores

Execution time of the detector with the monitoring system is higher than without it as seen in Fig. 2.

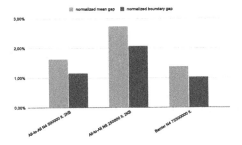

Fig. 2. Case I. 28 cores per node. The bins are scaled by the *mean* T_{no-mon} for every detector configuration. The experiments are conducted on 4 and 8 supercomputer nodes.

Collective operations pure execution time statistically differs from execution of MPI operations with monitoring system 1 Hz rate. The *mean time* overhead for every configuration is between 1,38% and 2,73% or 8s–16 s for 10 min runs of the detector without the monitoring system. The overhead becomes the higher the more nodes are used. The interference of the delays caused by the effects of the monitoring system and by the overhead on communication between the nodes may be the reason for this effect.

4.2   Using Only Physical Cores

The results of **Case II** experiments are shown in Fig. 3.

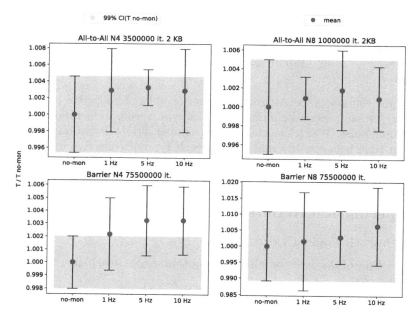

Fig. 3. Case II. 14 cores per node. Y-axis is time scaled by the $mean(T_{no-mon})$ for every detector configuration. The whiskers are $99\%CIs$ *median*. Gray— $99\%CI$ *median*(T_{no-mon}). $CI(T_{no-mon})$ and $CI(T_{x\ Hz})$ overlap. *boundary gap* (1) is negative for every configuration.

At first we considered monitoring system of 1 Hz frequency. The confidence intervals for T_{no-mon} and $T_{1\ Hz}$ overlap, MPI collective operations don't differ statistically in two cases. Therefore, we increased the frequency of monitoring system 5 Hz 10 Hz to see whether the detector is sensitive to a more intensive noise. As it is seen in Fig. 3, all confidence intervals overlap i.e. monitoring system noise is not recognizable for all considered configurations and monitoring system rates. The collected data samples are inconsistent—lengths of confidence intervals varies from 5 s to 10 s for 10 min runs.

4.3   Setting Core Affinity

Results for All-to-All and Barrier collective operations with set affinity are shown in Fig. 4 and Fig. 5 respectively.

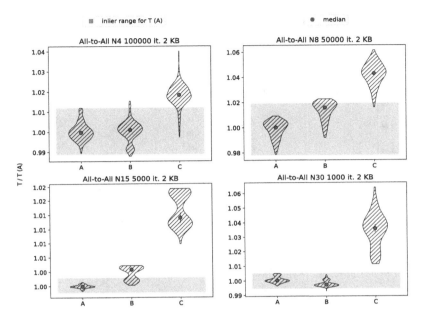

Fig. 4. Case III. All-to-All. 14 cores per node. Affinity is set. 4, 8, 15 and 30 supercomputer nodes are used. Y-axis is time sample scaled by the $median(T_A)$. Gray—*inlier range* (3) *for* T_A.

The results for All-to-All and Barrier can be generalized. Collective operations execution time in case C is classified as an outlier to execution time in case A. The collective operations chunk slowdown in case C, All-to-All and Barrier on 4, 8 and 15 nodes, is about 1–4%. For Barrier on 30 nodes the slowdown is 12%. The more nodes are used, the more influence monitoring system has on collective operations in case C. To summarize, the monitoring system noise has statistically significant impact on collective operations, when monitoring system agents are bound to the same logical cores as the parallel application.

On the other hand, the influence of monitoring system noise is not distinguishable, when monitoring system and the detector are bound to different logical cores—the majority time elements in case B fall in *inlier range* for T_A. The only exception is *All-to-All N15 5000 it. 2 KB*. However, the overhead in case B for this configuration (See Fig. 4) is less than 1% and can be considered insignificant.

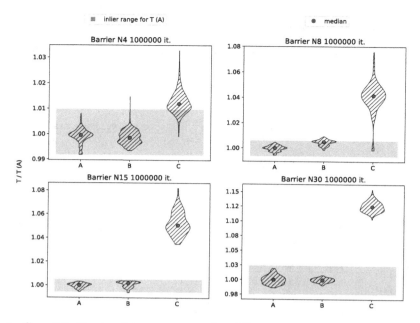

Fig. 5. Case III. Barrier. 14 cores per node. Affinity is set. 4, 8, 15 and 30 supercomputer nodes are used. Y-axis is time sample scaled by the $median(T_A)$. Gray— *inlier range* (3) *for* T_A

4.4   Increased Monitoring Frequency

We showed 1 Hz monitoring system does not influence the performance of collective operations All-to-All and Barrier, whether the monitoring system and the detector are being bound to different logical cores. In this section we study the limit for monitoring system rate. We try to answer the question: what is a borderline frequency for monitoring system to have no significant effect? The results of the experiments are shown in Fig. 6.

The monitoring system frequency is gradually increased until the noise has visible effects on the performance of MPI collective operation. The decrease in performance of Barrier grows gradually and becomes significant 20 Hz monitoring rate. The performance of All-to-All is influenced only by 100 Hz noise. The other studied monitoring rates have equal impact on the execution time of All-to-All operation. It is shown that monitoring system noise of a higher rate can have no significant impact on performance of All-to-All, when the agents are launched on separate logical cores.

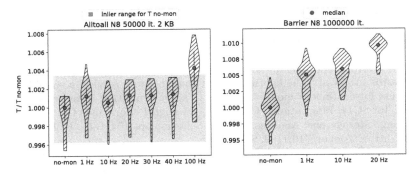

Fig. 6. Increased monitoring frequency. Affinity to logical cores is set as in case B. 8 supercomputer nodes are used. Y-axis is time sample scaled by the $median(T_{no-mon})$. Gray—*inlier range* (3) *for* T_{no-mon}

5   Conclusion

In summary, we have demonstrated a method for estimating the noise produced by the performance monitoring system of a supercomputer. The method is based on using MPI collective operations All-to-All and Barrier. We have considered several cases of the mutual monitoring system and collective operations allocation on a supercomputer node. We have shown that All-to-All and Barrier execution time are statistically significantly influenced by the noise of a real performance monitoring system when the monitoring system agent and a parallel program are bound to the same logical cores. On the other hand, monitoring system 1 Hz rate has insignificant impact on the performance of the collective operations when affinity is set to different logical cores for the monitoring system agent and the parallel program. We have also shown that the monitoring system with increased up 40 Hz frequency has the same influence on All-to-All operation as the monitoring system with standard rate 1 Hz. The difference in the performance of Barrier has been observed 20 Hz frequency. In conclusion, we have demonstrated that the performance monitoring system under standard configuration does not influence the performance of a parallel program with MPI collective operations, which is launched on the supercomputer a regular way.

The proposed method allows us to compare influence of the monitoring systems configured to collect data from different set of system sensors. We also plan to study the impact of the performance monitoring system noise on compute and memory bound parallel applications in our future work.

Acknowledgements. The reported study was funded by RFBR according to the research project № 19-07-00940. The research was carried out using the equipment of the shared research facilities of HPC computing resources at Lomonosov Moscow State University.

References

1. Adhianto, L., et al.: HPCTOOLKIT: tools for performance analysis of optimized parallel programs. Concurr. Comput.: Pract. Exp. **22**(6), 685–701 (2009). https://doi.org/10.1002/cpe.1553
2. Agelastos, A., et al.: The lightweight distributed metric service: a scalable infrastructure for continuous monitoring of large scale computing systems and applications. In: SC14: International Conference for High Performance Computing, Networking, Storage and Analysis, vol. 2015-Janua, pp. 154–165. IEEE, November 2014. https://doi.org/10.1109/SC.2014.18
3. Beckman, P., Iskra, K., Yoshii, K., Coghlan, S., Nataraj, A.: Benchmarking the effects of operating system interference on extreme-scale parallel machines. Clust. Comput. **11**(1), 3–16 (2008). https://doi.org/10.1007/s10586-007-0047-2
4. Gerstmayr, A., McDonell, K., Berk, L., Goodwin, M., Myllynen, M., Scott, N.: Performance co-pilot. https://pcp.io/. Accessed 06 Apr 2021
5. Hoefler, T., Belli, R.: Scientific benchmarking of parallel computing systems: twelve ways to tell the masses when reporting performance results. In: International Conference for High Performance Computing, Networking, Storage and Analysis, SC, vol. 15–20-Nove. IEEE Computer Society, November 2015. https://doi.org/10.1145/2807591.2807644
6. Khudoleeva, A., Stefanov, K.: Modeling influence of monitoring system on performance of MPI collective operations. Bull. South Ural State Univ. Ser. "Comput. Math. Softw. Engi." **10**(1), 62–74 (feb 2021). https://doi.org/10.14529/cmse210105. (in Russian)
7. Mooney, R., Schmidt, K., Studham, R.: NWPerf: a system wide performance monitoring tool for large Linux clusters. In: 2004 IEEE International Conference on Cluster Computing (IEEE Cat. No. 04EX935), pp. 379–389. IEEE (2004). https://doi.org/10.1109/CLUSTR.2004.1392637
8. Petrini, F., Kerbyson, D.J., Pakin, S.: The case of the missing supercomputer performance. In: Proceedings of the 2003 ACM/IEEE Conference on Supercomputing - SC 2003, p. 55. ACM Press, New York (2003). https://doi.org/10.1145/1048935.1050204
9. Rohl, T., Eitzinger, J., Hager, G., Wellein, G.: LIKWID monitoring stack: a flexible framework enabling job specific performance monitoring for the masses. In: 2017 IEEE International Conference on Cluster Computing (CLUSTER), vol. 2017-Septe, pp. 781–784. IEEE, September 2017. https://doi.org/10.1109/CLUSTER.2017.115
10. Sottile, M., Minnich, R.: Supermon: a high-speed cluster monitoring system. In: Proceedings of the IEEE International Conference on Cluster Computing, vol. 2002-Janua, pp. 39–46. IEEE Computer Society (2002). https://doi.org/10.1109/CLUSTR.2002.1137727
11. Stefanov, K., Voevodin, V., Zhumatiy, S., Voevodin, V.: Dynamically reconfigurable distributed modular monitoring system for supercomputers (DiMMon). Procedia Comput. Sci. **66**, 625–634 (2015). https://doi.org/10.1016/j.procs.2015.11.071
12. Treibig, J., Hager, G., Wellein, G.: LIKWID: a lightweight performance-oriented tool suite for x86 multicore environments. In: 2010 39th International Conference on Parallel Processing Workshops, pp. 207–216. IEEE, September 2010. https://doi.org/10.1109/ICPPW.2010.38
13. Voevodin, V.V., et al.: Supercomputer Lomonosov-2: large scale, deep monitoring and fine analytics for the user community. Supercomput. Front. Innov. **6**(2), 4–11 (2019). https://doi.org/10.14529/jsfi190201

High-Efficiency Specialized Support for Dense Linear Algebra Arithmetic in LuNA System

Nikolay Belyaev[1] and Vladislav Perepelkin[1,2(✉)]

[1] Institute of Computational Mathematics and Mathematical Geophysics SB RAS,
Novosibirsk, Russia
perepelkin@ssd.sscc.ru
[2] Novosibirsk State University, Novosibirsk, Russia

Abstract. Automatic synthesis of efficient scientific parallel programs for super-computers is in general a complex problem of system parallel programming. Therefore various specialized synthesis algorithms and heuristics are of use. LuNA system for automatic construction of distributed parallel programs provides a basis for accumulation of such algorithms to provide high-quality parallel programs generation in particular subject domains. If no specialized support is available in LuNA for given input, then the general synthesis algorithm is used, which does construct the required program, but its efficiency may be unsatisfactory. In the paper a specialized run-time system for LuNA is presented, which provides run-time support for dense linear algebra operations implementation on distributed memory multicomputers. Experimental results demonstrate, that automatically generated parallel programs of the class outperform corresponding ScaLAPACK library subroutines, which makes LuNA system practically applicable for generating high performance distributed parallel programs for supercomputers in the dense linear algebra application class.

Keywords: Parallel programming automation · Fragmented programming technology · LuNA system · Distributed dense linear algebra subroutines

1 Introduction

This paper is devoted to the problem of efficient parallel program construction automation in the field of high performance scientific computations on supercomputers. Efficiency is a mandatory requirement for such programs. Otherwise costly high performance computing resources are wasted. Provision of efficiency of a parallel program is a hard problem (NP-hard in general case), which makes such program construction automation challenging. The complexity of efficiency provision arises from the necessity to decompose data and computations and organize parallel data processing in such a way that as much of hardware resources as possible are loaded fully and evenly with useful computations. Manual development of efficient parallel programs requires knowledge of distributed hardware architecture, familiarity with methods and tools for distributed parallel programming, skills in system parallel programming. Such expertise is different

© Springer Nature Switzerland AG 2021
V. Malyshkin (Ed.): PaCT 2021, LNCS 12942, pp. 143–150, 2021.
https://doi.org/10.1007/978-3-030-86359-3_11

from the expertise in the subject domains, to which computations are related. Manual parallel programs development compels users to possess expertise in both domains. This conditions the importance of program construction automation tools, which allow one to describe computations with a higher level programming language (or an API), and expect an efficient parallel program to be constructed and executed automatically. Such an approach allows encapsulating much of the expertise a parallel programmer needs to possess into a programming system and automatically apply the encapsulated knowledge for program construction. Since no general solution exists, of practical interest are particular and heuristic solutions, capable of providing satisfactory efficiency for certain application classes. Also of practical interest are approaches, aimed at accumulation and automatic application of various particular solutions.

Nowadays the need in parallel programming automation means tends to increase, since supercomputers' hardware and software grow more complex. Heterogeneity of hardware increases, number of nodes and cores per node increases, network and memory subsystems become more lagging behind cores and therefore more critical, co-processors usage becomes essential to maximize performance, etc. Taking all this into account is both necessary and hard, so research in the field of parallel programming automation is more and more demanding.

Many programming systems, languages and tools exist and evolve to assist or replace programmers [1, 2].

Charm++ [3, 4] is an open-source parallel system which consists of distributed runtime system which is able to execute a distributed computational tasks (chares) graph on a supercomputer. Each task is able to communicate with others by sending and receiving messages. An applied programmer has to program communications between tasks by hand using low-level C++ interface. The task-based computational model, employed in Charm++ allows using particular system algorithms to support various classes of applications, but in general the peculiarities of the model make Charm++ programs partially opaque to the system because of low-level message passing means employed. That impedes Charm++'s capability to accumulate particular system algorithms.

PaRSEC [5] is a parallel programming system, designed specifically for automated generation of efficient parallel programs, which implement linear algebra operations. An applied programmer describes a tasks graph using the built-in high level language. This simplifies the process of development of high performance parallel programs. PaRSEC is able to generate programs only for the restricted class of linear algebra algorithms.

Legion [6], Regent [7–9] and LuNA [10] systems are also able to execute an algorithm described as a task graph on a supercomputer. These systems use general system algorithms to distribute tasks to computing nodes and execute the graph. The systems also provide powerful means to provide specialized support of program construction and execution, because execution control algorithms are excluded from the algorithm description, thus program construction and execution can be varied freely to support efficient execution of applied algorithms in particular subject domains. The systems are therefore suitable for accumulating various system algorithms for different subject domains.

It can be seen that a great effort is being put into automating programming. It is also clear that the efficiency problem is far from being solved for many subject domains.

In the presented work we employ LuNA as the system capable for particular system algorithms accumulation. LuNA is a system for automatic construction of scientific parallel programs for multicomputers. It is an academic project of the Institute of Computational Mathematics and Mathematical Geophysics of the Siberian Branch of Russian Academy of Sciences. This system is aimed at automatic construction of high performance distributed parallel programs for conducting numerical computations on supercomputers. It focuses on providing to a user an ability to describe computations, that need to be conducted, in a high-level platform-independent form. Also it provides some high level means (called recommendations) to express a programmer idea on how to organize efficient parallel execution on a supercomputer. This approach is based on the structured synthesis theory [11] and conforms to the active knowledge technology [12]. It allows to significantly reduce the complexity of efficient parallel program generation problem without the need for the programmer to do low-level parallel programming. Source code of LuNA system can be found in its public repository[1].

In this paper we investigate how satisfactory efficiency can be achieved in LuNA by making a specific system support for a particular subject domain, namely, dense linear algebra operations. This support is implemented as a particular run-time system, which is capable of execution of LuNA programs (or subprograms) of particular form, common for many dense linear algebra operations. The run-time system takes into account peculiarities of the operations to achieve high efficiency, comparable with that of ScaLA-PACK, which is a widely used library for such operations. This demonstrates that LuNA system can be a useful tool for practical construction of high performance scientific programs for subject domains, reasonably supported by specialized system algorithms.

The rest of the paper is organized as follows. Section 2 describes the proposed approach to support dense linear algebra operations in LuNA. Section 3 presents the experimental results, where LuNA performance is compared to that of ScaLAPACK on some operations. Conclusion ends the paper.

2   Particular Execution Algorithms Approach

2.1   Main Idea

This section describes the overall idea of the proposed solution. For the class of numerical algorithms particular distributed run-time system algorithms are developed and integrated to LuNA system. LuNA system analyzes the input algorithm description written in LuNA language and determines the class to which the input algorithm description belongs. Then LuNA compiler selects particular system algorithms which are used to automatically generate a parallel program by the input algorithm description. The result of compilation is a C++ code, which can be compiled by a conventional C++ compiler and linked against a library, which implements the run-time system. Then it is able to be executed on a supercomputer.

In this paper only a single class of numerical algorithms is considered to demonstrate the approach. This class contains widely used matrix algorithms such as LU, LL^T, LDL^T and similar matrix factorization algorithms. One of the advantages of the approach is

---

[1] https://gitlab.ssd.sscc.ru/luna/luna.

that it is possible to identify whether input algorithm belongs to the class or not. No sophisticated information dependencies analysis is required for that.

2.2 Main Definitions and Class of Algorithms Description

For further discussion the model of algorithm is described as it is one of the most important things when developing parallel programming systems. Firstly let some formal definitions be given. Secondly, the main idea of the model is given. Then the class of algorithms is described formally.

Definition 1. A **data fragment** (DF) is a the following tuple: $\langle N, V \rangle$, where N is a name (a regular string), V is an arbitrary value.

Definition 2. **DFs array** is the following set: $\{x | x = \langle h_1, ..., h_N \rangle, df_{\langle h_1, ..., h_N \rangle}, \forall i \in \{1, ..., N\} : 0 \le h_i < M_i, h_i \in \mathbb{N}_0\}$, where $df_{\langle h_1, ..., h_N \rangle}$ is a DF, M_i is the size of i-th dimension of the array, $\langle h_1, ..., h_N \rangle, N \in \mathbb{N}, \forall i \in \{1, ..., N\} : 0 \le h_i < M_i$ is a tuple of array element indices.

Definition 3. Let concept of **task argument** now be defined as follows:

1. Every DF is a task argument
2. The following tuple is a task argument: $\langle A, \langle h_1, ..., h_N \rangle \rangle$, where A is an N-dimensional array of DFs. This kind of argument is also called **array-argument**.

Definition 4. A **task** is the following tuple: $\langle n, I, O \rangle$, where n − name (regular string), $I = \{a_1, ..., a_M\}, M \in \mathbb{N}_0$ - set of task arguments called **input arguments**, $O = \{b_1, ..., b_K\}, K \in \mathbb{N}_0$ - set of task arguments called **output arguments**.

Definition 5. Algorithm is a tuple $\langle A, D, T \rangle$, where A – is a finite set of DFs arrays, D – is a finite set of DFs, T – is a set of tasks.

Let the main idea of the algorithm model now be explained. An applied programmer describes the data processed by a numerical algorithm with a set of DFs and DFs arrays (the description is the input for the system). Each DF is associated with a value which may store arbitrary data. For example, the value of some DF may store a dense matrix block, a vector part or a single value of some type. For each DFs array the applied programmer provides a mapping function. The mapping function maps DFs array elements to some memory location depending on array element indices and the computing node to which an array element is distributed. Many DFs array elements may be mapped to the same memory location. In this case, the applied programmer is responsible for avoiding collisions, i.e. when different elements, mapped to the same location, are in use at the same time span. Then the applied programmer describes a set of tasks. Each task transforms the values of its input DFs to the values of its output DFs by calling associated external routine. Such routine is implemented by the applied programmer with some conventional language, such as C, C++ or Fortran. For example, there may be implemented an external routine that multiplies two dense matrix blocks, represented as DFs. Multiple tasks may be associated with the same external routine. At run-time this external routine with the values of the input DFs forms a task that can be executed by LuNA system when all values of the input DFs are computed. After the task is executed the values of its output DFs become computed, so some other tasks may become executable.

Consider now a class of algorithms that is handled by the developed particular system algorithms. The class of algorithms consists of algorithms that meet the following requirements:

1. Every task within the algorithm has either only one output array-argument or all array element indices of all output array-arguments of the task are pairwise equal.
2. Dimension of all DFs arrays is the same, and the sizes of each dimension are pairwise equal.

For example, Cholesky (LL^T) factorization algorithm mentioned above meets the requirements. Also this class contains many other matrix algorithms such as LU factorization, LDL^T factorization and others.

2.3 Compiler

One of the important components of the developed LuNA system extension is a compiler that checks if the input algorithm description belongs to the supported class of algorithms. If the input numerical algorithm meets the above requirements the compiler generates a parallel program according the following principle. For each task described within the input algorithm description compiler generates a C++ lambda function (it is called run-time task). The body of the lambda-function consists of a C++ call statement of the routine associated with the task. Then compiler generates a call to the run-time library that implements distributed execution of the input algorithm (the execution algorithm is described in Sect. 2.4). This call submits the task to the executor.

At run-time a set of tasks with their arguments forms a bipartite directed acyclic tasks graph (DAG) which is submitted to a distributed executor implemented in the run-time library. The executor distributes DFs and DFs arrays to nodes and asynchronously executes the tasks graph on the multicomputer. Figure 1 shows the overall structure of generated program.

2.4 Run-Time Library and Task Graph Execution

Consider now the distributed tasks graph executor that is implemented in the run-time library. At first, consider the data distribution algorithm. The value of each DF (not an DFs array element) is stored in the memory of all computing nodes. Each DFs array is distributed according to the block-cyclic [13] principle. The parameters of the block-cyclic distribution may be set by the applied programmer. The dimension of the block-cyclic distribution is equal to the dimension of the DFs arrays declared in the input algorithm description.

Consider now the principle of tasks mapping to computing nodes and execution of the tasks graph. At run-time each task is mapped to the computing node to which

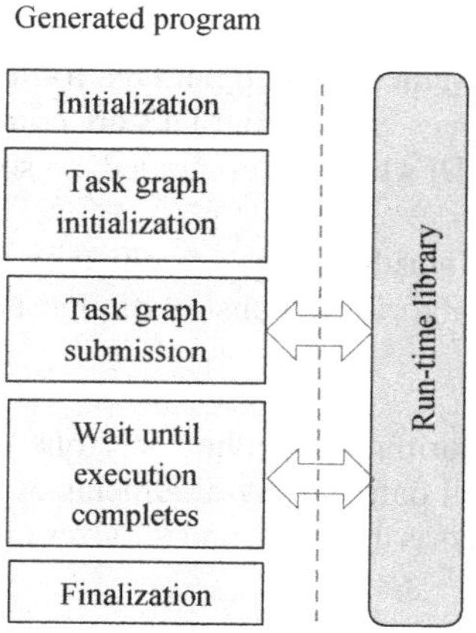

Generated program

Fig. 1. Structure of generated program.

its output DFs arrays elements is mapped (according the requirements indices of all output array-arguments of a task are pairwise equal and thus all corresponding DFs array elements are mapped to the same computing node). If an input argument of a task is mapped to a different node, an asynchronous message is sent after the producer task execution. In addition, each computing node runs a receiver loop in a dedicated thread. When some task argument value is received, corresponding consumer tasks are found. When the values of all input arguments of a consumer task are obtained, it is executed. The process continues until all tasks are executed.

3   Performance Evaluation

To measure the performance of the implemented extension of LuNA system, a test implementation of the Cholesky factorization of a dense matrix was developed with LuNA language. Such factorization is an example of an algorithm with complex structure and information dependencies. For performance evaluation the same test was implemented using a ScaLAPACK [14] implementation of Cholesky factorization. ScaLAPACK is a widely used library, where Cholesky factorization is implemented. Execution times of both implementations were compared. Both implementations used two-dimensional block cyclic distribution of the input matrix into square matrix of square blocks, and the block size was a parameter. OpenBLAS library (version 0.3.15) [15] implementation of BLAS and LAPACK subroutines was used for both tests. Both implementations used right-looking blocked Cholesky factorization algorithm [16].

Two square dense double-precision matrices of sizes 32768 and 65536 were used as input data. For each of the matrices a number of experiments were conducted using different matrix block sizes ranging from 256 to 2048. Execution times of both tests were measured.

Testing was conducted on MVS-10P cluster of the Joint Supercomputing Centre of Russian Academy of Sciences[2] on a two-dimensional grid of 2×2 computing nodes.

---

[2] http://www.jscc.ru.

Each node contains 32 cores and 16 GB of memory. All 32 cores of each CPU were used in all tests.

Figure 2 shows execution times comparison of the ScaLAPACK and the LuNA implementations for the input matrix of sizes 32768 (left) and 65536 (right).

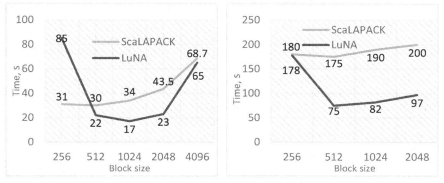

Fig. 2. Performance evaluation result for a square matrix of 32768 (left) and 65536 (right) elements.

Here in both cases the LuNA implementation outperforms the ScaLAPACK implementation of Cholesky factorization (by 2.4 times for matrix of 32768 elements and block size of 512 and by 2.1 for matrix of 65536 elements and block size of 1024).

The above results demonstrate that LuNA is able to generate an efficient parallel program from an algorithm description with complex information dependencies. The performance of the generated parallel program is approximately 2 times better than that of library developed by experts (for the studied test).

4   Conclusion

Automatic construction of efficient parallel programs generally requires different construction algorithms for different subject domains. LuNA system is capable of accumulating such algorithms. This ability was demonstrated by adding specialized support for dense linear algebra operations class. The achieved performance is comparable with that of a widely used library ScaLAPACK. This makes LuNA a practical tool for automatic construction of high performance distributed parallel programs for the applications class. Other classes of applications can also be particularly supported in LuNA in order to improve performance of automatically constructed programs if programs, constructed by general LuNA algorithms are not efficient enough.

Acknowledgments. The work was supported by the budget project of the ICMMG SB RAS No. 0251-2021-0005.

References

1. Sterling, T., Anderson, M., Brodowicz, M.: A survey: runtime software systems for high performance computing. Supercomput. Front. Innov. **4**(1), 48–68 (2017). https://doi.org/10.14529/jsfi170103
2. Thoman, P., et al.: A taxonomy of task-based parallel programming technologies for high-performance computing. J. Supercomput. **74**(4), 1422–1434 (2018). https://doi.org/10.1007/s11227-018-2238-4
3. Kale, L.V., Krishnan, S.: Charm++ a portable concurrent object oriented system based on C++. In: Proceedings of the Eighth Annual Conference on Object-Oriented Programming Systems, Languages, and Applications, pp. 91–108, October 1993
4. Acun, B., et al.: Parallel programming with migratable objects: Charm++ in practice. In: Proceedings of the International Conference for High Performance Computing, Networking, Storage and Analysis, SC 2014, pp. 647–658. IEEE, November 2014
5. Bosilca, G., Bouteiller, A., Danalis, A., Faverge, M., Hérault, T., Dongarra, J.J.: PaRSEC: exploiting heterogeneity to enhance scalability. Comput. Sci. Eng. **15**(6), 36–45 (2013)
6. Bauer, M., Treichler, S., Slaughter, E., Aiken, A.: Legion: expressing locality and independence with logical regions. In: Proceedings of the International Conference on High Performance Computing, Networking, Storage and Analysis, SC 2012, pp. 1–11. IEEE, November 2012.
7. Slaughter, E., Lee, W., Treichler, S., Bauer, M., Aiken, A.: Regent: a high-productivity programming language for HPC with logical regions. In: Proceedings of the International Conference for High Performance Computing, Networking, Storage and Analysis, pp. 1–12, November 2015
8. Slaughter, E.: Regent: a high-productivity programming language for implicit parallelism with logical regions. Doctoral dissertation, Stanford University (2017)
9. Torres, H., Papadakis, M., Jofre Cruanyes, L.: Soleil-X: turbulence, particles, and radiation in the Regent programming language. In: Proceedings of the International Conference for High Performance Computing, Networking, Storage and Analysis, SC 2019, pp. 1–4 (2019)
10. Malyshkin, V.E., Perepelkin, V.A.: LuNA fragmented programming system, main functions and peculiarities of run-time subsystem. In: Malyshkin, V. (ed.) PaCT 2011. LNCS, vol. 6873, pp. 53–61. Springer, Heidelberg (2011). https://doi.org/10.1007/978-3-642-23178-0_5
11. Valkovsky, V.A., Malyshkin, V.E.: Synthesis of Parallel Programs and Systems on the Basis of Computational Models. Nauka, Novosibirsk (1988). (in Russian)
12. Malyshkin, V.: Active knowledge, LuNA and literacy for oncoming centuries. In: Bodei, C., Ferrari, G.-L., Priami, C. (eds.) Programming Languages with Applications to Biology and Security. LNCS, vol. 9465, pp. 292–303. Springer, Cham (2015). https://doi.org/10.1007/978-3-319-25527-9_19
13. Hiranandani, S., Kennedy, K., Mellor-Crummey, J., Sethi, A.: Compilation techniques for block-cyclic distributions. In: Proceedings of the 8th International Conference on Supercomputing, pp. 392–403, July 1994
14. Choi, J., Dongarra, J.J., Pozo, R., Walker, D.W.: ScaLAPACK: a scalable linear algebra library for distributed memory concurrent computers. In: The Fourth Symposium on the Frontiers of Massively Parallel Computation, pp. 120–121. IEEE Computer Society, January 1992
15. Goto, K., Van De Geijn, R.: High-performance implementation of the level-3 BLAS. ACM Trans. Math. Softw. **35**(1), 1–14 (2008). Article 4. https://doi.org/10.1145/1377603.1377607
16. Kurzak, J., Ltaief, H., Dongarra, J., Badia, R.M.: Scheduling dense linear algebra operations on multicore processors. Concurr. Comput. Practice Exp. **22**(1), 15–44 (2010)

Applications

Efficient Cluster Parallelization Technology for Aerothermodynamics Problems

Oleg Bessonov[✉]

Ishlinsky Institute for Problems in Mechanics RAS,
101, Vernadsky ave., 119526 Moscow, Russia
bess@ipmnet.ru

Abstract. HPC modeling of gas dynamic and aerodynamic problems is very important for the development of aircrafts, missiles and space vehicles and requires a lot of processor time. For this reason, the numerical codes for such simulations must be efficiently parallelized. This paper presents a technological approach that greatly simplifies the parallelization of problems with unstructured grids. The paper introduces the principle of a unified mathematical address space of the problem for all used cluster nodes. This technology also simplifies grid partitioning. Parallelization of the code is carried out with minimal effort, without changing the main parts of the program. As a result, a single computational code is produced for all regimes – sequential, multi-threaded, and cluster. Performance measurements confirm the good scalability of the method.

Keywords: Computational aerodynamics · Navier-Stokes equation · Unstructured grids · Cuthill-McKee algorithm · Grid partitioning · Parallelization · OpenMP · MPI

1 Introduction

The numerical study of gas dynamics and aerodynamics of high-speed aircrafts and hypersonic vehicles requires very large computing resources. The main reason for this is the need for large grids (tens or hundreds of million of grid cells) to resolve thin boundary layers and shock waves at high speeds. Modeling complex multiphysics, chemical and radiation processes complicates the structure of the numerical code and further increases the requirements for computer resources [1,2]. With the proliferation of high-performance clusters and supercomputers, it becomes possible to dramatically accelerate modeling by massively parallelizing computational codes [3,4].

Massive parallelization in CFD is possible only if certain requirements are met. First of all, the problem must possess natural parallelism (usually geometric). This determines the choice of a numerical method and a time integration scheme – as a rule, this is a finite volume method [5] and an explicit integration scheme [6].

© Springer Nature Switzerland AG 2021
V. Malyshkin (Ed.): PaCT 2021, LNCS 12942, pp. 153–165, 2021.
https://doi.org/10.1007/978-3-030-86359-3_12

The use of structured or unstructured grids defines different approaches to partitioning the computational domain. Structured grids allow to implement a simple partition of the domain in several spatial directions [7]. In turn, unstructured grids that provide discretization of regions of arbitrary shape [8] require the use of complex algorithms for optimal partitioning [9, 10].

The requirement to minimize exchanges between subdomains necessitates optimization of cell numbering. The graph of connections of grid cells corresponds to a sparse matrix, therefore optimization turns into reducing the bandwidth of this matrix by renumbering the matrix entries [11].

Finally, software environment is an important element of massive parallelization. For multicore processors and multiprocessor systems, the most natural and economical approach is to use shared memory parallelization environment (usually OpenMP [12]). For systems with distributed memory (supercomputer clusters), a software tool is required to exchange data between nodes (MPI [13]). Accordingly, for clusters built of nodes with multi-core processors, it is natural to use the hybrid OpenMP + MPI approach [6, 14]. At the same time, additional attention should be paid to the peculiarities of optimal programming for systems with non-uniform memory access (NUMA) [15].

However, despite the accumulated practice, parallelization still remains a very difficult task and requires a lot of effort and experience, as well as training and coordination among programmers when creating new codes or adapting old ones. In this regard, it becomes more and more important to develop technological principles that could simplify and accelerate the process of parallelizing existing computational codes and creating new ones. Within these principles, it is desirable to ensure that the existing sequential code remains as unchanged as possible and that the grid partitioning scheme does not affect the structure of the code. At the same time, the efficiency of parallelization of new code should not be sacrificed much over a more complex (and less technological) approach.

The previous paper of the author in this area [16] discussed a hybrid parallelization approach for OpenMP and MPI environments based on a unified address space for an MPI-parallel program. The paper presents an example of such parallelization for the CFD problem in the domain of regular geometry. The current paper extends this approach and formulates the principle of a single mathematical address space for problems with unstructured grids. This principle fulfills the above requirements for both code structure and grid partitioning.

The rest of the paper is organized as follows. Section 2 briefly introduces the mathematical model. Section 3 presents the general organization of the code and illustrates the optimization of the grid. Section 4 describes OpenMP parallelization with non-uniform memory (NUMA) specific features. Section 5 gives all the details of MPI parallelization technology and grid partitioning. Finally, Sect. 6 presents and briefly analyzes the performance results for massive cluster parallelization.

2   Mathematical Model

The numerical code used in this work was developed for the analysis of aerother-modynamics of high-speed aircrafts and space vehicles and is applicable over a wide range of Mach numbers and altitudes. The code is based on solving the three-dimensional Navier-Stokes equation:

$$\frac{\partial w}{\partial t} + \frac{\partial F^x(w)}{\partial x} + \frac{\partial F^y(w)}{\partial y} + \frac{\partial F^z(w)}{\partial z} = \frac{\partial G^x(w)}{\partial x} + \frac{\partial G^y(w)}{\partial y} + \frac{\partial G^z(w)}{\partial z}$$

with $w = \rho, \rho u, \rho v, \rho w, \rho E$, where w represents the conservative variables, and F and G are convective and diffusive flux vectors, respectively.

The system of equations is integrated numerically using the HLLE method [17]. The method is an approximate solution to the Riemann problem. This app-roach can be classified as a method that extends the concept of calculating flows using the Riemann solver proposed by S. Godunov [18]. The time integration scheme is explicit, with a local time step.

The finite volume discretization of the computational domain is carried out using unstructured grids with two types of cells – tetrahedrons and prisms. To compute the boundary conditions, dummy cells are used next to the inner cells.

The computational code is written in Fortran. Figure 1 presents two examples of numerical simulations performed by this code.

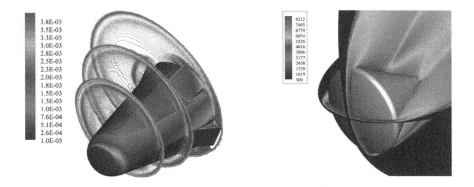

Fig. 1. Some results of numerical simulations: pressure distribution around Expert space vehicle [19] (left); temperature distribution near Apollo lander [20] (right)

3   General Organization of the Computational Code

The organization and structure of the computational code is determined by the use of an unstructured grid to discretize the computational domain.

Calculations are made in accordance with order of cells (control volumes). The main data structures for storing physical quantities and auxiliary variables have the dimension (N) or (M, N), where N is either the total number of cells including fictive cells outside the computational domain (N = Ntot), or the number of internal cells in the domain (N = Nsum), and M is the number of cell faces (4 or 5). Arrays of size (N) are used for values located within a cell, and arrays of size (M, N) – for values calculated on cell faces (for example, numerical fluxes) or belonging to adjacent cells. All fictive elements in arrays of dimension (Ntot) are placed at the end of the arrays, after the internal elements. Computations are organized in the form of DO-loops from 1 to Nsum.

The links between cells are defined in the array NB(M, N) which contains the addresses of the adjacent cells for each internal cell of the domain. Mathematically, this array is a sparse matrix, each row of which contains 4 or 5 nonzero elements. The positions of these elements in the rows of the matrix correspond to the adjacent cells' numbers. When carrying out calculations involving adjacent cells, the elements of different arrays are referred to by these numbers (indices). Algebraically, the processing of such arrays can be represented as $y = A \otimes x$, where A is the matrix representation of the array NB, x is the array indexed by this matrix, y is the result of calculations, and $\otimes$ is the operator performing calculations on the indexed elements of the array x. This representation is similar to the operator of multiplication of a sparse matrix by a vector.

Depending on how the cells are numbered in the computational domain, nonzero elements in the matrix A will be placed in a more or less regular way. If such elements are located as close as possible to the diagonal of the matrix, then the indexed elements of the arrays will be grouped into small subsets that can fit in the cache memory. Subsequent fetches of elements of this subset come from the cache memory, which leads to a decrease of the computation time.

In practice, grid generation programs enumerate nodes and cells in a non-optimal way. In this regard, the cells are renumbered using a simplified variant of the Cuthill-McKee method [11]. Due to such renumbering, the computation time can be reduced by 15–25% (depending on the quality of the original grid). Another goal of grid renumbering is to provide efficient partitioning for cluster parallelization.

Figure 2 shows examples of portraits of matrix A for the original and renumbered grids (left and center, respectively).

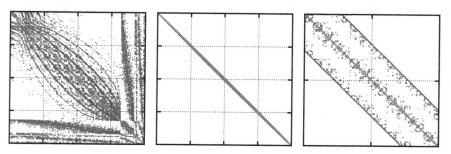

Fig. 2. Grid renumbering: portraits of the original (left) and renumbered (center) grid matrices; middle part of the renumbered matrix, enlarged 25 times (right)

It can be seen that the matrix A after renumbering has a band structure, and its bandwidth determines the spread of addresses at which memory accesses occur. With sufficiently optimal renumbering, the maximum bandwidth for a three-dimensional domain is approximately $O(N^{\frac{2}{3}})$ (this corresponds to natural numbering in a cubic or spherical domain).

The quality of the renumbered matrix may depend on the choice of the cell from which the new numbering begins. The program uses an adaptive algorithm with checking three options for the initial cell and choosing the option with the minimum bandwidth.

4 Shared Memory Parallelization in the OpenMP Environment

Parallelization in the OpenMP model is done in a standard way, using Fortran "!$OMP DO" operators. As a result of applying this operator, DO-loops of dimension N are automatically split into approximately equal parts, which are executed in parallel threads. With such a partitioning, most of the accesses to indexed array elements occur due to grid optimization within a small subset allocated in the cache.

Parallelization for shared memory computers with two or more processors requires special optimization due to the fact that such systems have a non-uniform memory organization (NUMA). Processors in NUMA systems are connected by high-speed channels, and each processor has its own memory block. When a program thread is executing, accessing data in the memory of its own processor occurs at full speed, while accessing data in another processor can become significantly slower. To ensure the correct data placement, special initialization of the main arrays is required in parallel loops, which are identical to the main computational loops of the algorithm. As a result, all physical pages with data are allocated in the local memory of the corresponding processor.

It should be noted that for some configurations of the problem it is beneficial to run one MPI process per processor rather than per node. In this case, the overhead of explicit MPI exchanges can become less than the loss due to slower access to remote memory in threads near interprocessor boundaries.

Running jobs on a shared-memory system must be performed with binding threads to processor cores. This is especially important for NUMA systems (e.g., dual-processor cluster nodes), since thread migration can lead to the fact that threads and their data fall into different processors, and the computation speed will noticeably decrease. On Linux, thread-bound jobs can be launched using the "taskset" or "numactl" command, and on Windows using the "start" command.

5 Cluster Parallelization Approach

Parallelization for distributed memory computers (clusters) requires splitting the computational domain into subdomains and organizing the work in such a way that computations within subdomains alternate with exchanges of near-boundary data.

Data exchanges between the cluster nodes are performed using calls to the MPI library routines. Typically, for such parallelization, the special computational code is created with the organization of independent data structures in each cluster node and with explicit calls for exchanges with neighboring nodes in accordance with the structure of the partition of the computational domain [21]. In this case, splitting the computational domain means dividing each original data array of dimension (N) into subarrays with independent numbering of elements, as well as preparing special data structures that describe the boundaries between subdomains and data exchange strategies.

5.1 Grid Optimization and Partitioning

In this work, a technological approach is used while maintaining a unified end-to-end numbering of cells and corresponding data arrays [16]. The splitting of the computational domain is performed algebraically by dividing each original array of dimension (N) into subarrays of equal size. In this case, the unified subarray index space is used (a single mathematical address space of the problem).

Algebraic partitioning of data arrays means that in main calculations, which have the form $y = A \otimes x$, there will be overlap at the boundaries between the subarrays. Figure 3 illustrates data access patterns when computing $A \otimes x \rightarrow y$ within a subarray (left) and near the boundary (right).

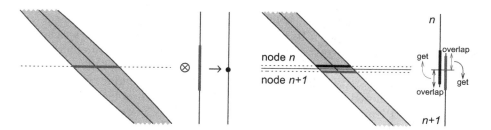

Fig. 3. Data access patterns: without overlap (left); with overlap and exchange of data between subdomains (right)

It can be seen that for each array that is accessed as a vector x, additional overlap areas need to be added when allocating subarrays. The size of each area of overlap is about half of the bandwidth of the matrix A at that position. After each computation of the form $y = A \otimes x$, the overlapping areas must be transferred between the corresponding cluster nodes. Simultaneous exchanges of data in opposite directions should be performed to save time.

In the current implementation, all exchanges between nodes are synchronous. With a moderate number of cluster nodes, it does not result in high overhead. Asynchronous exchanges require more complex data processing patterns. This extension of the method will be implemented later when larger clusters become available.

The algebraic partitioning method, being simple and straightforward, is not the most optimal one. There are sophisticated domain decomposition techniques that minimize the size of the boundary areas between subdomains [9,10,22]. Figure 4 provides examples of near-optimal partitioning using advanced techniques.

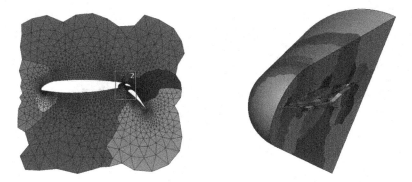

Fig. 4. Illustration of near-optimal partitioning into 8 subdomains in 2D [22] (left) and 64 subdomains in 3D [21] (right)

However, such near-optimal partitioning methods have several disadvantages. As a rule, subdomains after such a partition have a large (and different) number of neighbors with unequal sizes of the boundaries. As a result, the exchange patterns become complex, which complicates the parallelization algorithm and makes its optimal implementation more difficult. Thus, these methods are most efficient only for a large number of subdomains.

On the other hand, when using the algebraic partitioning method, each subdomain has only two neighbors, and the sizes of the boundaries between the subdomains not differ much. These properties partially compensate for the non-optimal nature of this method, especially if synchronous exchanges are used.

When using the Cuthill-McKee method, the cells are numbered layer by layer, while the character of the frontal layer propagation over the computational domain depends on variations in the grid density. In the presence of strong grid compression, the frontal layer may bend and the number of cells in the layer may increase. In the places where the domain is split, this value determines the number of boundary cells between the subdomains and, accordingly, half of the bandwidth of the matrix A, which corresponds to the amount of data exchanges between the cluster nodes.

Figure 5 shows examples of splitting domains into 8 subdomains using this method. Here you can see the bending of the boundaries between the subdomains. The number of grid cells N and half the bandwidths H for these examples are as follows: $N = 12$ million, $H = 78000$ (left), $N = 6$ million, $H = 28000$ (right). The values $H/N^{\frac{2}{3}}$ indicating the relative bandwidth are 1.48 and 0.84, respectively. This means that both grid divisions are of acceptable quality, but the second one is slightly better.

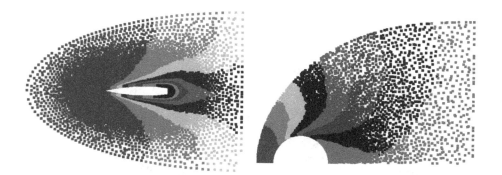

Fig. 5. Examples of grid partitioning using algebraic splitting

5.2 Structure of the Code

Technically, calculations in subdomains are organized as follows. For all data subarrays sharing a unified index space of the problem, corresponding index variables are defined, denoting the ranges of placement of internal (real) cells, dummy cells and areas of overlap. Figure 6 illustrates the structure of a subarray and the character of data exchanged between subdomains.

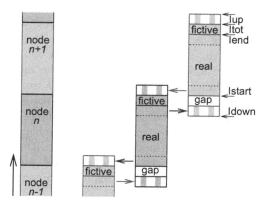

Fig. 6. Structure of subarrays in a unified index space and illustration of the exchange of overlapping areas

The main index variable ranges are Istart:Iend to handle DO-loops and Idown:Ip to allocate subarrays. The index spaces of the subarrays in the neighboring cluster nodes overlap with the following relationship between local index variables: Itot+1 for node n is equal to Istart for node $n + 1$.

Computations in the cluster nodes are performed with parallelization in the OpenMP model. After the completion of each stage of computations, exchanges are carried out with both neighboring nodes. The MPI_SENDRECV routine is used, which simultaneously receives data from a neighboring node and transfers data to this node.

Technologically, all data exchange procedures are separated from the computational code and placed in the special file XCH.F. All basic exchanges between adjacent sub-domains are implemented in the XCH subroutine, which is called once after each processing stage. The subroutine determines by itself, based on node's number and the total number of nodes, with which nodes and in what order to perform exchanges. In case the job is run on a single node, the subroutine will not do any work.

Figure 7 (left) demonstrates how the XCH subroutine works. Internally, it makes two calls of MPI_SENDRECV routine in appropriate order to exchange with both neighboring cluster nodes.

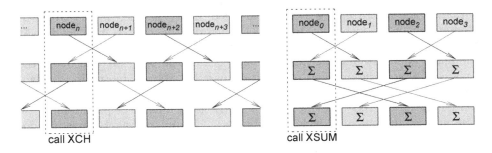

Fig. 7. Structure of basic exchanges between subdomains (left); algorithm of logarithmic complexity for reduction operations (right)

Similarly, the reduction routines are implemented to calculate the sum of the values of variables in all nodes, as well as their maximum and minimum values (XSUM, XMAX, XMIN). For these calculations, a butterfly-type algorithm is used (similar to that used in the Fast Fourier Transform) (Fig. 7, right).

Also, unified procedures are implemented for distributing data between cluster nodes after input and for collecting data from nodes before output.

Separating all calls to the MPI library routines into a special file allows the main code to be compiled with a regular compiler. A compiler that supports MPI is only required for the final build using this file. Also, the program can be compiled without the MPI library – for this case, a set of subroutines is implemented that simulate execution on one node without exchanges (XCH_NOMPI.F).

Figure 8 illustrates the evolution of computer code – sequential, OpenMP-parallel, and cluster-parallel. The principal changes of the basic DO-loop are highlighted in the picture.

```
do I=1,Nsum
  ...
  A(I)= ...
  B(K,I)= ...
  ...
enddo
```

```
!$OMP DO
  do I=1,Nsum
    ...
    A(I)= ...
    B(K,I)= ...
    ...
  enddo
!$OMP END DO
```

```
!$OMP DO
  do I=Istart,Iend
    ...
    A(I)= ...
    B(K,I)= ...
    ...
  enddo
!$OMP END DO
!$OMP SINGLE
  call XCH(...,A)
  call XCH(...,B)
!$OMP END SINGLE
```

Fig. 8. Evolution of computer code

Thus, the technology of cluster parallelization with a single mathematical address space of the problem makes it possible to implement a unified computational code for all modes – single-threaded, multi-threaded in the OpenMP shared memory model, and cluster in the hybrid OpenMP + MPI model. At the same time, parallelization of the code for the cluster mode is provided with minimal effort, without altering the main parts of the program, and under the only condition – the presence of a natural parallelization potential in the algorithm (for example, when using an explicit method of time integration). The parallelization method provides a sufficiently good quality of partitioning the computational domain, at which the number of boundary cells between subdomains does not exceed the level $O(N^{\frac{2}{3}})$.

6   Performance Results

The new parallelization algorithm was evaluated and tested on a cluster of dual-processor nodes containing Intel Xeon processors with 16 cores and a clock frequency range of 2.6 to 2.9 GHz. The cluster nodes are interconnected by Omni-Path links at a speed of 100 Gbps. The results of parallel performance when solving the aerothermodynamic problem with 33 million grid cells are shown in Fig. 9. The problem was running on up to 24 cluster nodes containing 768 processor cores (1536 parallel threads due to Hyperthreading).

Here, the computation time for one time-step is 57.8 s for serial run, 2.22 s for single-node run, and 0.135 s for 24 nodes. The loss of parallel efficiency is mainly due to the overhead of synchronous exchanges between nodes.

The above results demonstrate the good scalability of the parallelization method. The saturation effect of parallel performance begins to be observed only

#nodes	speedup	efficiency
1	1	100%
2	1.95	97.4%
4	3.79	94.7%
6	5.54	92.3%
8	7.18	89.8%
12	10.4	86.4%
16	13.1	83.1%
24	16.4	68.2%

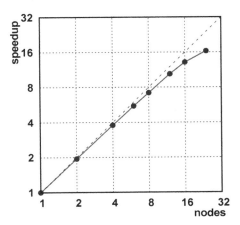

Fig. 9. Parallel speedup and efficiency of the method

at 24 nodes. Based on 26x speedup on a single cluster node versus a single thread, the total speedup across 24 nodes is as much as 425. Thus, this comparison illustrates the wide range of computing platforms on which this technique can be effectively used.

7 Conclusion

In this paper, we present a technological approach for efficient cluster parallelization of gas dynamic and aerodynamic problems using unstructured grids. This approach is based on a single mathematical address space of the problem data for all cluster nodes of the job. Due to this, parallelization of the code is carried out with minimal effort, without changing the main parts of the program. Also, grid partitioning is greatly simplified with this approach.

Additionally, we propose some techniques to simplify the work with exchanges between cluster nodes. In particular, all calls to the MPI library routines are separated from the main code.

As a result of applying the above approaches, it becomes possible to implement a single code for all computational modes – sequential, multi-threaded (shared memory), and cluster (distributed memory).

Performance measurements demonstrate good scalability of the proposed parallelization method.

Acknowledgements. This work was partially supported by the Russian State Assignment under contract No. AAAA-A20-120011690131-7. The author thanks Dr. N. Kharchenko for his help and cooperation.

References

1. Shang, J.S., Surzhikov, S.T.: Simulating nonequilibrium flow for ablative Earth reentry. J. Spacecr. Rockets **47**, 806–815 (2010). https://doi.org/10.2514/1.49923
2. Zheleznyakova, A.L., Surzhikov, S.T.: Application of the method of splitting by physical processes for the computation of a hypersonic flow over an aircraft model of complex configuration. High Temp. **51**(6), 816–829 (2013). https://doi.org/10.1134/S0018151X13050234
3. Afzal, A., Ansari, Z., Faizabadi, A.R., Ramis, M.K.: Parallelization strategies for computational fluid dynamics software: state of the art review. Arch. Comput. Methods Eng. **24**(2), 337–363 (2016). https://doi.org/10.1007/s11831-016-9165-4
4. Probst, A., Knopp, T., Grabe, C., Jägersküpper, J.: HPC requirements of high-fidelity flow simulations for aerodynamic applications. In: Schwardmann, U., et al. (eds.) Euro-Par 2019. LNCS, vol. 11997, pp. 375–387. Springer, Cham (2020). https://doi.org/10.1007/978-3-030-48340-1_29
5. Versteeg, H., Malalasekera, W.: An Introduction to Computational Fluid Dynamics: The Finite Volume Method. Prentice Hall, Harlow (2007)
6. Jost, G., Robins, B.: Experiences using hybrid MPI/OpenMP in the real world: parallelization of a 3D CFD solver for multi-core node clusters. Sci. Program. **18**, 127–138 (2010). https://doi.org/10.3233/SPR-2010-0308
7. Wang, B., Hu, Z., Zha, G.-C.: General subdomain boundary mapping procedure for structured grid implicit CFD parallel computation. J. Aerosp. Comput. Inf. Commun. **5**, 425–447 (2008). https://doi.org/10.2514/1.35498
8. Gourdain, N., et al.: High performance parallel computing of flows in complex geometries: I. Methods. Comput. Sci. Discov. **2**, 015003 (2009). https://doi.org/10.1088/1749-4699/2/1/015003
9. Karypis, G., Kumar, V.: A fast and high quality multilevel scheme for partitioning irregular graphs. SIAM J. Sci. Comput. **20**(1), 359–392 (1998). https://doi.org/10.1137/S1064827595287997
10. Karypis, G., Kumar, V.: Multilevel κ-way partitioning scheme for irregular graphs. J. Parallel Distrib. Comput. **48**, 96–129 (1998). https://doi.org/10.1006/jpdc.1997.1404
11. Cuthill, E., McKee, J.: Reducing the bandwidth of sparse symmetric matrices. In: ACM'69: Proceedings of 24th National Conference, pp. 157–172 (1969). https://doi.org/10.1145/800195.805928
12. Dagum, L., Menon, R.: OpenMP: an industry-standard API for shared-memory programming. IEEE Comput. Sci. Eng. **5**(1), 46–55 (1998). https://doi.org/10.1109/99.660313
13. Gropp, W., Lusk, E., Thakur, R.: Using MPI-2: Advanced Features of The Message-Passing Interface. The MIT Press, Cambridge (1999). https://doi.org/10.7551/mitpress/7055.001.0001
14. Shang, Z.: High performance computing for flood simulation using Telemac based on hybrid MPI/OpenMP parallel programming. Int. J. Model. Simul. Sci. Comput. **5**(04), 1472001 (2014). https://doi.org/10.1142/S1793962314720015
15. Terboven, C., Schmidl, D., Cramer, T., an Mey, D.: Task-parallel programming on NUMA architectures. In: Kaklamanis, C., Papatheodorou, T., Spirakis, P.G. (eds.) Euro-Par 2012. LNCS, vol. 7484, pp. 638–649. Springer, Heidelberg (2012). https://doi.org/10.1007/978-3-642-32820-6_63
16. Bessonov, O.: Technological aspects of the hybrid parallelization with OpenMP and MPI. In: Malyshkin, V. (ed.) PaCT 2017. LNCS, vol. 10421, pp. 101–113. Springer, Cham (2017). https://doi.org/10.1007/978-3-319-62932-2_9

17. Einfeldt, B., Munz, C.D., Roe, P.L., Sjögreen, B.: On Godunov-type methods near low densities. J. Comput. Phys. **92**, 273–295 (1991). https://doi.org/10.1016/0021-9991(91)90211-3
18. Godunov, S.: Finite difference method for numerical computation of discontinuous solutions of the equations of fluid dynamics. Mat. sbornik **47**(89) (3), 271–306 (1959). (in Russian)
19. Kharchenko, N., Kryukov, I.: Aerothermodynamics calculation of the EXPERT reentry flight vehicle. J. Phys.: Conf. Ser. **1009**, 012004 (2018). https://doi.org/10.1088/1742-6596/1009/1/012004
20. Kharchenko, N., Kotov, M.: Aerothermodynamics of the Apollo-4 spacecraft at earth atmosphere conditions with speed more than 10 km/s. J. Phys.: Conf. Ser. **1250**, 012012 (2019). https://doi.org/10.1088/1742-6596/1250/1/012012
21. Ermakov, M., Kryukov, I.: Supercomputer modeling of flow past hypersonic flight vehicles. J. Phys.: Conf. Ser. **815**, 012016 (2017). https://doi.org/10.1088/1742-6596/815/1/012016
22. Zheleznyakova, A.L.: Effective domain decomposition methods for adaptive unstructured grids applied to high performance computing for problems in computational aerodynamics. Phys.-Chem. Kinet. Gas Dyn. **18**(1) (2017). (in Russian). http://chemphys.edu.ru/issues/2017-18-1/articles/673/

Computational Aspects of Solving Grid Equations in Heterogeneous Computing Systems

Alexander Sukhinov[1] , Vladimir Litvinov[1,2(✉)] , Alexander Chistyakov[1] ,
Alla Nikitina[3] , Natalia Gracheva[2] , and Nelli Rudenko[2]

[1] Don State Technical University, Rostov-on-Don, Russia
[2] Azov-Black Sea Engineering Institute of Don State Agrarian University,
Zernograd, Russia
[3] Southern Federal University, Rostov-on-Don, Russia

Abstract. The prediction of environmental disasters, both technogenic
and natural, is currently based on advances in mathematical modeling.
The high cost and costly maintenance of computing clusters actualizes
the research in the field of heterogeneous computing. One of the directions of them is to maximize the use of all available hardware resources,
including the central processor and the video adapters (GPU). The purpose of the research is to develop an algorithm and a software module that implements it for solving a system of linear algebraic equations (SLAE) by the modified alternating-triangular iterative method
(MATM) (self-adjoint and non-self-adjoint cases) for the hydrodynamics
problem of shallow water using NVIDIA CUDA technology. The conducted experiment with the flow distribution along the Ox and Oz axes
of the computational grid at a fixed value of the grid nodes along the Oy
axis allowed reducing the implementation time of one step of the MATM
on the GPU. A regression equation was obtained at the experimental
data processing in the Statistica program, on the basis of which it was
found that the implementation time of one step of the MATM on the
GPU is affected only by the number of threads along the axis Oz. The
optimal two-dimensional configuration of threads in a computing unit
executed on a single thread multiprocessor is determined, in which the
calculation time on the GPU for one step of the MATM is minimal.

Keywords: Mathematical modeling · Parallel algorithm · Graphics
accelerator

1 Introduction

The prediction process of environmental disasters of natural and technogenic
nature requires an operational approach in order to reduce the negative consequences on the environment and the population living in the surrounding

Supported by Russian Science Foundation, project № 21-71-20050.

V. Malyshkin (Ed.): PaCT 2021, LNCS 12942, pp. 166–177, 2021.
https://doi.org/10.1007/978-3-030-86359-3_13

areas. Hydrophysical processes have a significant impact on the water shoreline, coastal protection structures and coastal constructions. Currently, the research of hydrodynamical processes in waters with complex bathymetry is one of the most important problems. This problem can be effectively solved using mathematical modeling methods.

Mathematical modeling of hydrodynamical processes is based on the Navier-Stokes motion equations, the continuity equations, as well as the heat and salt transfer equations. As a result of numerical implementation, a continuous mathematical model is transformed into a discrete one, the solution of which is reduced to the solution of a system of linear algebraic equations (SLAE).

Many Russian and foreign scientists are engaged in research and forecasting of aquatic ecosystems. Representatives of the scientific school by G.I. Marchuk study the computational aspects of atmospheric and ocean physics. Comprehensive re-searches of the environment and biota in the Azov and Black Seas are performed under the leadership of G.G. Matishov. Bonaduce A., Staneva J. proposed the mathematical models of sea level dynamicss [1]. Marchesiello P., Androsov A., etc. scientists are engaged in improving ocean models [2,3]. Developed software systems, designed for monitoring and forecasting the state of waters (SALMO, CHARISMA, MARS3D, CHTDM, CARDINAL, PHOENICS, Ecointegrator), have a number of advantages, are easy to use, and allow solving computationally labors problems for a wide range of research areas. The disadvantages include the lack of consideration of the spatially inhomogeneous transport of water environment, the lack of accuracy in modeling the vortex structures of currents, the shore and bottom topography [1–4].

The team of authors developed the AZOV3D software, which uses the spatial-three-dimensional models of the hydrodynamics of shallow waters (coastal systems). These models include the motion equations in all three coordinate directions and taking into account the wind stress, bottom friction, complex geometry of the shore and bottom of water, Coriolis force, precipitation evaporation, as well as the nonlinear character of microturbulent exchange in the vertical direction [5]. Testing of this software was performed during the reconstruction of the extreme storm surge of water on September 23–24, 2014 in the port area of Taganrog, when the level rise was more than 4 m at the average depth of the bay is about 5 m. The prediction was performed with the error of 3–5%.

The complex geometry of the computational domain requires the use of computational grids with a large number of nodes in spatial coordinates. As a result, it's necessary to solve the SLAE with dimension from 10^7, 10^9 and more [6]. The implementation of such calculations for the time interval from the occurrence of an emergency to the receipt of forecasting results, established by regulatory acts, is very difficult without the use of parallel computing and supercomputer technologies. The high cost and costly maintenance of computing clusters actualizes research in the field of heterogeneous computing, which aims to maximize the use of all available hardware resources, which include video adapters along with the central processor. Modern video adapters have a large amount of VRAM (up to 24 GB) and stream processors, the number of which can achieve the several thousand. There are software interfaces that allow you to implement the computing process on a graphics accelerator, one of which is NVIDIA CUDA. International research teams are actively conducting research in this area [7,8].

The purpose of the research is to develop algorithms for solving large-dimensional SLAE in a limited time, and their software implementation in the environment of heterogeneous computing systems.

2   Grid Equations Solving Method

Let A be is linear, positive definite operator $(A > 0)$ and in a finite-dimensional Hilbert space H it is necessary to solve the operator equation [9, 10]

$$Ax = f, A : H \to H. \tag{1}$$

For the grid Eq. (1), iterative methods are used, which in canonical form can be represented by the equation [9,10]

$$B \frac{x^{m+1} - x^m}{\tau_{m+1}} + Ax^m = f, B : H \to H, \tag{2}$$

where m is the iteration number, $\tau_{m+1} > 0$ is the iteration parameter, B is the preconditioner. Operator B is constructed proceeding from the additive representation of the operator A_0 – the symmetric part of the operator A

$$A_0 = R_1 + R_2, R_1 = R_2^*, \tag{3}$$

where $A = A_0 + A_1$, $A_0 = A_0^*$, $A_1 = -A_1^*$.

The preconditioner is formed as follows

$$B = (D + \omega R_1) D^{-1} (D + \omega R_2), D = D^* > 0, \omega > 0, \tag{4}$$

where D is the diagonal operator, R_1, R_2 are the lower- and upper-triangular operators respectively.

The algorithm for calculating the grid equations by the modified alternating-triangular method of the variational type is written in the form:

$$r^m = Ax^m - f, B(\omega_m)w^m = r^m, \tilde{\omega}_m = \sqrt{\frac{(Dw^m, w^m)}{(D^{-1}R_2 w^m, R_2 w^m)}},$$

$$s_m^2 = 1 - \frac{(A_0 w^m, w^m)^2}{(B^{-1} A_0 w^m)(Bw^m, w^m)}, k_m^2 = \frac{(B^{-1} A_1 w^m, A_1 w^m)}{(B^{-1} A_0 w^m, A_0 w^m)}, \tag{5}$$

$$\theta_m = \frac{1 - \sqrt{\frac{s_m^2 k_m^2}{(1+k_m^2)}}}{1 + k_m^2 (1 - s_m^2)}, \tau_{m+1} = \theta_m \frac{(A_0 w^m, w^m)}{(B^{-1} A_0 w^m, A_0 w^m)},$$

$$x^{m+1} = x^m - \tau_{m+1} w^m, \omega_{m+1} = \tilde{\omega}_m,$$

where r^m is the residual vector, w^m is the correction vector, the parameter s_m describes the rate of convergence of the method, k_m describes the ratio of the norm of the skew-symmetric part of the operator to the norm of the symmetric part.

The method convergence rate is:

$$\rho \leq \frac{\nu^* - 1}{\nu^* + 1}, \tag{6}$$

where $\nu^* = \nu \left(\sqrt{1 + k^2} + k\right)^2$, where ν is the condition number of the matrix C_0, $C_0 = B^{-1/2} A_0 B^{-1/2}$.

The value ω is optimal for

$$\omega = \sqrt{\frac{(Dw^m, w^m)}{(D^{-1} R_2 w^m, R_2 w^m)}} \tag{7}$$

and the condition number of the matrix is estimated C_0:

$$\nu = \max_{y \neq 0} \left(\frac{1}{2} + \frac{\sqrt{(Dy, y)(D^{-1} R_2 y, R_2 y)}}{(A_0 y, y)} \right) \leq \frac{1}{2} \left(1 + \sqrt{\frac{\Delta}{\delta}} \right) = \frac{1 + \sqrt{\xi}}{2\sqrt{\xi}}, \tag{8}$$

where $\xi = \frac{\delta}{\Delta}$, $D \leq \frac{1}{\delta} A_0$, $R_1 D^{-1} R_2 \leq \frac{\Delta}{4} A_0$.

3   Software Implementation of the Method for Solving Grid Equations

To solve the hydrodynamics problem, a computational grid is introduced as [11]

$$\bar{w}_h = t^n = n\tau, x_i = ih_x, y_i = jh_y, z_k = kh_z; n = \overline{0, n_t - 1}, i = \overline{0, n_1 - 1},$$
$$j = \overline{0, n_2 - 1}, k = \overline{0, n_3 - 1}, n_t \tau = T, n_1 h_x = l_x, n_2 h_y = l_y, n_3 h_z = l_z,$$

where τ is the time step; h_x, h_y, h_z are space steps; n_t is the time layers number; T is the upper bound on the time coordinate; n_1, n_2, n_3 are the nodes number by spatial coordinates; l_x, l_y, l_z are space boundaries of a rectangular parallelepiped in which the computational domain is inscribed.

At discretization the hydrodynamics model, we obtained a system of grid equations. Each equation of the system can be represented in a canonical form. We will use a seven-point template (Fig. 1):

$$c(m_0)u(m_0) - \sum_{i=1}^{6} c(m_0, m_i)u(m_i) = F(m_0),$$

$m_0(x_i, y_j, z_k)$ is the template center, $M'(P) = \{m_1(x_{i+1}, y_j, z_k), m_2(x_{i-1}, y_j, z_k), m_3(x_i, y_{j+1}, z_k), m_4(x_i, y_{j-1}, z_k), m_5(x_i, y_j, z_{k+1}), m_5(x_i, y_j, z_{k-1})\}$ is the

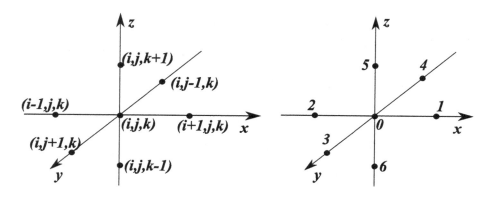

Fig. 1. Grid template for solving hydrodynamic equations.

neighborhood of the template center, $c_0 \equiv c(m_0)$ is the template center coefficient, $c_i \equiv c(m_0, m_i)$ are coefficients of the neighborhood of the template center, F is the vector of the right parts, u is the calculated vector.

The developed software module uses one-dimensional arrays. The transition from a three-dimensional representation of the grid node (i, j, k) to a one-dimensional (node number) is performed using the following formula:

$$m_0 = i + jn_1 + kn_1n_2.$$

The numbers of nodes in the neighborhood of the template center are calculated by the formulas:

$$m_1 = m_0 + 1, m_2 = m_0 - 1, m_3 = m_0 + n_1,$$
$$m_4 = m_0 - n_1, m_5 = m_0 + n_1n_2, m_6 = m_0 - n_1n_2.$$

The MATM algorithm consists of four stages:

- calculating the values of the residual vector r^m and its uniform norm;
- calculating the correction vector w^m;
- calculating the scalar products and iterative parameters based on them τ_{m+1}, ω_{m+1};
- transition to the next iterative layer.

The computational process is performed until the norm of the residual vector reaches the specified accuracy.

The most laborious part of the algorithm is the calculation of the correction vector from the equation:

$$(D + \omega R_1)y^m = r^m, (D + \omega R_2)w^m = Dy^m.$$

The algorithm fragment of solving SLAE with the lower-triangular matrix is given below (Algorithm 1).

Algorithm 1. matm(IN: $c_0, c_1, c_2, c_3, c_4, c_5, c_6, \omega$; OUT: r)

1: **for** $k \in [1; n_3 - 2]$ **do**
2:     **for** $i \in [1; n_1 - 2]$ **do**
3:         **for** $j \in [1; n_2 - 2]$ **do**
4:             $m_0 \leftarrow i + n_1 \cdot j + n_1 \cdot n_2 \cdot k$
5:             **if** $c_0[m_0] > 0$ **then**
6:                 $m_2 \leftarrow m_0 - 1;\ m_4 \leftarrow m_0 - n_1;\ m_6 \leftarrow m_0 - n_1 \cdot n_2$
7:                 $r[m_0] \leftarrow (\omega \cdot (c_2[m_0] \cdot r[m_2] + c_4[m_0] \cdot r[m_4] + c_6[m_0] \cdot r[m_6]) + r[m_0])/((0.5 \cdot \omega + 1) \cdot c_0[m_0])$
8: **for** $k \in [n_3 - 2; 1]$ **do**
9:     **for** $i \in [n_1 - 2; 1]$ **do**
10:         **for** $j \in [n_2 - 2; 1]$ **do**
11:             $m_0 \leftarrow k + n_3 \cdot j + n_2 \cdot n_3 \cdot i$
12:             **if** $c_0[m_0] > 0$ **then**
13:                 $m_1 \leftarrow m_0 + n_2 \cdot n_3;\ m_3 \leftarrow m_0 + n_3;\ m_5 \leftarrow m_0 + 1$
14:                 $r[m_0] \leftarrow (\omega \cdot (c_1[m_0] \cdot r[m_1] + c_3[m_0] \cdot r[m_3] + c_5[m_0] \cdot r[m_5]) + r[m_0] \cdot c_0[m_0])/((0.5 \cdot \omega + 1) \cdot c_0[m_0])$

The residual vector is calculated in $14N$ arithmetic operations, where N is a basic arithmetic operation such as adding, multiplying etc. The complexity of calculating the values of the correction vector is $19N$ arithmetic operations ($9N$ and $10N$ each for solving SLAE of upper-triangular and lower-non-triangular types, respectively). The transition to the next iteration will require $2N$ arithmetic operations. In total, the total number of arithmetic operations required to solve the SLAE with a seven-diagonal matrix using MATM in the case of known iterative parameters τ_{m+1}, ω_{m+1} is $35N$.

We determine the complexity of adaptive optimization of the minimum correction MATM. The calculation of $A_0 w^m$, $A_1 w^m$ and $R_2 w^m$ vectors requires $13N$, $11N$ and $7N$ operations each. The multiplication of vectors by diagonal operators D^{-1} and D will require N operations each. The conversion B to determine vectors $B^{-1} A_0 w^m$ and $B^{-1} A_1 w^m$ will require $19N$ operations each. It is also necessary to calculate 6 scalar products, each of which will require 2N operations. Thus, each adaptive optimization of the minimum correction MATM requires $83N$ arithmetic operations in the non-self-adjoint case and $49N$ in the self-adjoint case. The calculation process of iterative parameters τ_{m+1}, ω_{m+1} is laborious, but its establishment is observed quite quickly at solving grid equations in the adaptive case. As a result, these parameters do not need to be calculated at each iteration.

4   Parallel Implementation

Parallel algorithms focused on heterogeneous computing systems were developed for numerical implementation of the proposed hydrodynamics model. Each computing node of the system can contain from 1 to 2 central processing units (CPU) containing from 4 to 32 cores, and from 1 to 4 NVIDIA video accelerators with

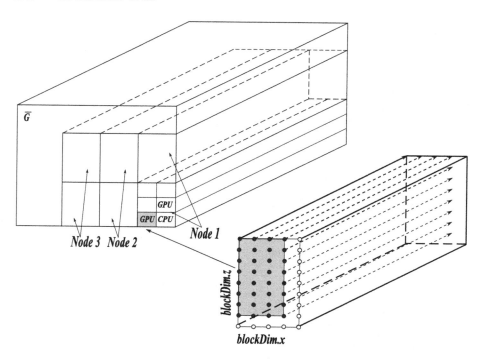

Fig. 2. Decomposition of the computational domain. $Node1$, $Node2$, $Node3$ are computational nodes; CPU, GPU are fragments of the computational domain, calculated on the CPU and GPU, respectively; $blockDim.x$, $blockDim.z$ – dimensions of the computing CUDA block.

CUDA technology (GPU), having from 192 (NVIDIA GeForce GT 710) to 5120 (NVIDIA Tesla V100) CUDA cores. Data exchange between nodes is performed using MPI (Message Passing Interface) technology. An algorithm that controls all available CPU and GPU threads performs the organization of calculations on each node. The computational domain is divided into subdomains assigned to the computational nodes. Next, each subdomain is divided into fragments assigned to each CPU core and each GPU computing unit (Fig. 2).

The solution of mathematical modeling problems using numerical methods, in particular, the finite difference method (FDM) on equal-dimensional grids, leads to the necessary to use with sparse matrices, the elements of which for internal nodes are a repeating sequence. This leads to inefficient memory consumption in the case of high-dimensional problems. Using the CSR (Compressed Sparse Row) matrix storage format avoids the necessary to store null elements. However, all non-zero elements, including many duplicate ones, are stored in the corresponding array. This disadvantage is not critical at using computing systems with shared memory. However, it can negatively affect the performance at data transferring between nodes in heterogeneous and distributed computing systems. A CSR1S modification of the CSR format was developed to improve the efficiency of data storage with a repeating sequence of elements for modeling hydrodynamic processes by the finite difference method. In this case, to change the differential operator, instead of repeatedly searching and replacing values in an array of non-zero elements, it is enough to simply change them in an array that preserves a repeating sequence.

Let's consider the conversion of a sparse matrix from CSR1S to CSR format. The input data of the algorithm is an object of the matrix class with repeated elements SMatrix1Seq, encapsulating an array of non-zero elements $Values$; the array of indexes of columns, containing non-zero elements $ColIdx$; the array of indexes of non-zero elements that are first in the rows (the last element of the array is the total number of non-zero elements) $RowIdx$; the array for storing a repeating sequence; the array for storing the indexes of columns, containing the first elements of a repeating sequence. In this case, the $Values, ColIdx and RowIdx$ arrays indicate elements that are not part of a repeating sequence. The output data – an object of the $MatrixCsr$ class – a sparse matrix in CSR format containing arrays $Values, ColIdx and RowIdx$. The data types and array assignments are similar to the corresponding arrays of the SMatrix1Seq class. We present an algorithm for converting a sparse matrix from the CSR1S to CSR format.

1. Calculation the size of the $MatrixCsr$ class arrays (output arrays).
2. Reserving of RAM for storing output arrays.
3. Saving the $resultValues$ array size value.
4. Copying the non-repeating elements from the $Values$ input array to the $resultValues$ output array.
5. Filling the $resultValues$ array with duplicate elements using CUDA.
6. Copying the column indexes of non-repeating elements.
7. Copying the column indexes of duplicate elements using CUDA.
8. Copying the indexes of rows, containing non-repeating elements.
9. Copying the indexes of rows, containing duplicate elements using CUDA.
10. Generating an output object of the $MatrixCsr$ class, containing the $resultValues$ array of non-zero elements, an array of column indexes of non-repeating elements, and an array of row indexes, containing non-repeating elements.
11. Clearing the resources, returning the result to the calling method.

Let's estimate the memory capacity in the CSR format:

$$P_{csr} = N_{nz}B_{nz} + (N_{nz} + R + 1)B_{idx},$$

in the CSR1S format:

$$P_{csr1s} = (N_{nz} - N_{seq}(R_{seq} + 1))B_{nz} + (N_{nz} - R_{seq}(N_{seq} + 1) + R + 1)B_{idx},$$

where R is the number of matrix rows; R_{seq} is the number of matrix rows, containing a repeating sequence of elements; N_{nz} is the number of non-zero matrix elements; N_{seq} is number of elements in a repeating sequence; B_{nz} is the memory capacity to store a single non-zero element; B_{idx} to store a single non-zero element to store a single index.

Let's introduce the coefficients $k_r = R_{seq}/R$ and $k_i = B_{idx}/B_{nz}$. After arithmetic transformations, we obtained the following:

$$P_{csr1s} = B_{nz}[N_{nz}(k_i + 1) - N_{seq}(k_i k_r R + k_r R + 1) - k_i(k_r R - R - 1)].$$

Efficient function libraries have been developed to solve the system of grid equations that arise during the sampling process in CSR format on GPUs using CUDA technology. The developed algorithm for solving the problem uses the modified CSR1S data storage format with further conversion to the CSR format to solve the resulting SLAE on a graphics accelerator using NVIDIA CUDA technology. In this case, there is the problem of developing a matrix conversion algorithm the from CSR1S to CSR format in the shortest possible time.

Experimental researches of the dependence of the execution time of the transformation algorithm on the number of elements of the repeated sequence N_{seq} and the ratio of the matrix rows containing the sequence to the total number of rows k_r were performed. According to the obtained results, the algorithm with using NVIDIA CUDA technology is more efficient at $N_{seq} > 7$. The point of equal efficiency decreases starting from $k_r = 0.7$. The resulting regression equation $k_r = -0.02N_{seq} + 0.08329$ with the determination coefficient 0.9276 describes the boundary of equal time consumption of the sequential algorithm and the algorithm using NVIDIA CUDA. Thus, we can calculate the minimum value k_r, by substituting a value N_{seq} into it, above which the second algorithm will be more efficient than the first.

The part of the computational load is passed to the graphics accelerator to increase the efficiency of calculations. For this, the corresponding algorithm and its software implementation on the CUDA C language were developed [12].

An algorithm for finding a solution to a system of equations with a lower-triangular matrix (straight line) on CUDA C is given (Algorithm 2).

The input parameters of the algorithm are the vectors of the coefficients of the grid equations c_0, c_2, c_4, c_6 and the constant ω. The output parameter is the vector of the water flow velocity r. Before running the algorithm, we must programmatically set the dimensions of the CUDA computing block $blockDim.x, blockDim.z$ in spatial coordinates x, z, respectively. The CUDA framework runs this algorithm for each thread; in this case, the values of the variables $threadIdx.x, threadIdx.z, blockIdx.x, blockIdx.z$ automatically initialized by the indexes of the corresponding threads and blocks. Global thread indexes are calculated in rows 1 and 2. The row index i, the layer index k, which the current thread processes, are calculated in rows 3 and 5. The variable j is initialized in row 4, representing a counter by coordinate y. The calculation pipeline is organized as a loop in line 6. The indexes of the central node of the grid template m_0 and surrounding nodes m_2, m_4, m_6 are calculated in rows 8, 10–12. The two-dimensional array $cache$ is located in the GPU shared memory and designed to store the calculation results of on the current layer by the coordinate y. This allows us to reduce the number of reads from slow global memory and accelerate the calculation process by up to 30%.

The conducted researches represent a significant dependence of the implementation time of the algorithm for calculating the preconditioner on the ratio of threads in spatial coordinates.

Algorithm 2. matmKernel(IN: $c_0, c_2, c_4, c_6, \omega$ IN/OUT: $r;$)

1: $threadX \leftarrow blockDim.x \cdot blockIdx.x + threadIdx.x$
2: $threadZ \leftarrow blockDim.z \cdot blockIdx.z + threadIdx.z$
3: $i \leftarrow threadX + 1$
4: $j \leftarrow 1$
5: $k \leftarrow threadZ + 1$
6: **for** $s \in [3; n_1 + n_2 + n_3 - 3]$ **do**
7:     **if** $(i + j + k = s) \wedge (s < i + n_2 + k)$ **then**
8:         $m_0 \leftarrow i + (blockDim.x + 1) \cdot j + n_1 \cdot n_2 \cdot k$
9:         **if** $c0[m0] > 0$ **then**
10:             $m_2 \leftarrow m_0 - 1; m_4 \leftarrow m_0 - n_1; m_6 \leftarrow m_0 - n_1 \cdot n_2$
11:             $rm4 \leftarrow 0$
12:             **if** $(s > 3 + threadX + threadZ)$ **then**
13:                 $rm4 \leftarrow cache[threadX][threadZ]$
14:             **else**
15:                 $rm4 \leftarrow r[m_4]$
16:             $rm2 \leftarrow 0$
17:             **if** $(threadX \neq 0) \wedge (s > 3 + threadX + threadZ)$ **then**
18:                 $rm2 \leftarrow cache[threadX - 1][threadZ]$
19:             **else**
20:                 $rm2 \leftarrow r[m_2]$
21:             $rm6 \leftarrow 0;$
22:             **if** $(threadZ \neq 0) \wedge (s > 3 + threadX + threadZ)$ **then**
23:                 $rm6 \leftarrow cache[threadX][threadZ - 1]$
24:             **else**
25:                 $rm6 \leftarrow r[m_6]$
26:             $rm0 \leftarrow (\omega \cdot (c2[m_0] \cdot rm2 + c4[m_0] \cdot rm4 + c6[m_0] \cdot rm6) + r[m_0]) / ((0.5 \cdot \omega + 1) \cdot c_0[m_0])$
27:             $cache[threadX][threadZ] \leftarrow rm0$
28:             $r[m_0] \leftarrow rm0$
29:     $j \leftarrow j + 1$

GeForce MX 250 video adapter was used in experimental researches; it specifications: the VRAM capacity is 4 GB, the core clock frequency is 1518–1582 MHz, the memory clock frequency is 7000 MHz, the video memory bus bit rate is 64 bits, and the number of CUDA cores is 384.

The purpose of the experiment is to determine the flow distribution along the Ox and Oz axes of the computational grid at fixed value of grid nodes along the Oy axis, equal to 10000, so that the implementation time on the GPU of one MATM step is minimal.

Two values are taken as factors: X is the number of threads on the axis Ox, Z is the number of threads on the axis Oz. The criterion function T_{GPU} is the implementation time of a single MATM step on GPU, ms.

The composition of the streams X and Z must not exceed 1024. This restriction is imposed by CUDA, since 1024 is the number of threads in a single block. Therefore, the levels of variation of the factors X and Z were chosen as shown in the Table 1.

Table 1. Experiment results.

X	Z	T_{GPU}, ms
16	64	64
32	32	65
64	16	81
128	8	109
256	4	100
512	2	103

The regression equation was obtained in the result of experimental data processing:

$$T_{GPU} = 119.797 - 9.371 \log_2 Z, \tag{9}$$

where T_{GPU} is the implementation time of a single MATM step on GPU, ms; Z is the number of threads on the axis Oz. The coefficient of determination was 0.78.

As a result of the analysis of experimental data, it was found that only the number of threads along the axis Oz affects the implementation time of one MATM step on GPU. The implementation time of one MATM step on GPU is inversely proportional to the number of nodes of the computational grid along the axis Oz. The calculation time decreases according to the logarithmic law at increasing the number of nodes along the axis Oz. Therefore, it is advisable to perform the domain decomposition in the form of parallelepipeds, in which the size on the Oz axis is maximum, and on the Ox axis is minimal.

Due to the conducted experimental researches, we established the optimal values of X and Z, which were equaled to the 16 and 64, respectively.

5   Conclusion

The algorithm and software unit that implements it were developed in the result of the conducted researches to solve the SLAE, which arises during the sampling of the hydrodynamics problem of shallow water, MATM using NVIDIA CUDA technology. The method of domain decomposition, applicable for heterogeneous computing systems, was described. The developed modification of the CSR –

CSR1S format made it possible to increase the efficiency of data storage with a repeating sequence of elements. It is determined that the algorithm using the NVIDIA CUDA technology is more effective at $N_{seq} > 7$. In this case, the point of equal efficiency decreases, starting from $k_r = 0.7$. The optimal two-dimensional configuration of threads in a computing unit, implemented on a single thread multiprocessor, was determined, in which the implementation time on GPU of a single MATM step is minimal and equaled to the 64 ms.

References

1. Bonaduce, A., Staneva, J., Grayek, S., Bidlot, J.-R., Breivik, Ø.: Sea-state contributions to sea-level variability in the European Seas. Ocean Dyn. **70**(12), 1547–1569 (2020). https://doi.org/10.1007/s10236-020-01404-1
2. Marchesiello, P., Mc.Williams, J., Shchepetkin, A.: Open boundary conditions for long-term integration of regional oceanic models. Oceanic Modell. J. **3**, 1–20 (2001)
3. Androsov, A.: Straits of the world ocean. General approach to modeling, St. Petersburg (2005)
4. Nieuwstadt, F., Westerweel, J., Boersma, B.: Turbulence. Introduction to Theory and Applications of Turbulent Flows. Springer, Cham (2016). https://doi.org/10.1007/978-3-319-31599-7
5. Sukhinov, A., Atayan, A., Belova, Y., Litvinov, V., Nikitina, A., Chistyakov, A.: Data processing of field measurements of expedition research for mathematical modeling of hydrodynamic processes in the Azov Sea. Comput. Continuum Mech. **13**(2), 161–174 (2020). https://doi.org/10.7242/1999-6691/2020.13.2.13
6. Sukhinov, A., Chistyakov, A., Shishenya, A., Timofeeva, E.: Predictive modeling of coastal hydrophysical processes in multiple-processor systems based on explicit schemes. Math. Models Comput. Simul. **10**(5), 648–658 (2018)
7. Oyarzun, G., Borrell, R., Gorobets, A., Oliva, A.: MPI-CUDA sparse matrix-vector multiplication for the conjugate gradient method with an approximate inverse preconditioner. Comput. Fluids **92**, 244–252 (2014)
8. Zheng, L., Gerya, T., Knepley, M., Yuen, D., Zhang, H., Shi, Y.: GPU implementation of multigrid solver for stokes equation with strongly variable viscosity. In: Yuen, D., Wang, L., Chi, X., Johnsson, L., Ge, W., Shi, Y. (eds.) GPU Solutions to Multi-scale Problems in Science and Engineering. Lecture Notes in Earth System Sciences, pp. 321–333. Springer, Heidelberg (2013). https://doi.org/10.1007/978-3-642-16405-7_21
9. Konovalov, A.: The steepest descent method with an adaptive alternating-triangular preconditioner. Differ. Eqn. **40**, 1018–1028 (2004)
10. Sukhinov, A., Chistyakov, A., Litvinov, V., Nikitina, A., Belova, Y., Filina, A.: Computational aspects of mathematical modeling of the shallow water hydrobiological processes. Numer. Methods Program. **21**(4), 452–469 (2020). https://doi.org/10.26089/NumMet.v21r436 https://doi.org/10.26089/NumMet.v21r436
11. Samarsky, A., Vabishchevich, P.: Numerical methods for solving convection-diffusion problems. URSS, Moscow (2009)
12. Browning, J., Sutherland, B.: C++20 Recipes. A Problem-Solution Approach. Apress, Berkeley (2020)

Optimized Hybrid Execution of Dense Matrix-Matrix Multiplication on Clusters of Heterogeneous Multicore and Many-Core Platforms

Gerassimos Barlas[✉]

Computer Science and Engineering Department, College of Engineering,
American University of Sharjah, POB 26666, Sharjah, UAE
gbarlas@aus.edu

Abstract. In this paper we analytically solve the partitioning problem for dense matrix-matrix multiplication, running on a cluster of heterogeneous multicore machines, equipped with a variety of accelerators. Closed-form solutions are provided, that can yield an optimum partitioning in linear time with respect to the number of cores in the system.

We also show that a run-time, online calculation of system parameters for the application of DLT is feasible, allowing the easy deployment of DLT frameworks without a costly a-priori benchmarking procedure.

The paper concludes with an extensive experimental study that shows that our DLT framework coupled with online parameter calculation, can outperform dynamic partitioning while leveraging existing optimized Dense Linear Algebra (DLA) libraries, such as NVidia's cuBLAS and Intel's MKL.

Keywords: Divisible load theory · Dense linear algebra · Heterogeneous parallel computing · Hybrid numerical computation

1 Introduction

Divisible Load Theory (DLT) allows the optimum load partitioning and scheduling of operations on the assumptions that the load can be arbitrary divisible, and that there is no dependence between the computations on different nodes. DLT has been previously employed for solving problems in many domains including image registration, video encoding and distribution, cloud resource allocation and scheduling [1], and others. It can be also combined with genetic algorithms [2], linear programming or other techniques for fine tuning the resources committed to solving a problem.

DLT analyses have been largely based on linear or affine cost models [3]. More recently, there have been successful attempts for the solution of problems where the computational cost is a second-order function of the problem size [4]. In this work we tackle the DLT partitioning of matrix-matrix multiplication over

© Springer Nature Switzerland AG 2021
V. Malyshkin (Ed.): PaCT 2021, LNCS 12942, pp. 178–195, 2021.
https://doi.org/10.1007/978-3-030-86359-3_14

a heterogeneous cluster, with nodes equipped with different types of accelerators. We distinguish accelerators into two categories, based on their access to main memory. The ones that do have access we call *on-chip*, and the ones that do not we call *off-chip* accelerators, the latter having their own memory space.

Matrix multiplication (MM) is an integral component of a wide variety of applications. The associated computation complexity has driven the development of multicore and many-core implementations, as delivered by software packages such as LAPACK, MAGMA, MKL and cuBLAS. CPU execution on clusters has been also targeted by ScaLAPACK.

DLT-derived schedules are typically static in the sense that they are determined a priori, although dynamic scheduling can be also accommodated in certain applications. Static schedules can result in substantial performance benefits (e.g. Song et al. report a 2.5x improvement for Cholesky factorization [5]). In this paper we employ a dynamic calculation of system parameters between runs, to enable the accurate derivation of near-optimum static schedules.

The major contributions of this paper are:

- We derive closed-form solutions for optimizing MM on a heterogeneous network with on-chip and off-chip accelerators.
- Our model accounts for every significant overhead, including both data distribution and collection times, involving network and the PCIe bus transfers.
- We examine both cases of data locality, i.e. the case where data are resident in one of the nodes and the case where the data are fetched remotely.
- We illustrate how the cost parameters needed for applying the proposed equations can be computed at run-time, without the hassle of a-priori testing.
- We prove that cubic complexity workloads can be successfully partitioned with DLT, with closed-form solutions.

In the following section we discuss related work, while Sect. 3 presents the target execution platform, notations and cost models used for our analysis. The intra-node ordering problem is examined in Sect. 4 and the closed-form solutions are presented in Sect. 5. The methodology for the online calculation of the cost parameters is described in Sect. 6. Finally, Sect. 7 presents our experimental findings, as tested on a small-scale GPU-equipped cluster.

2   Related Work

Khan et al. in [6] utilize Strassen's/Winograd's algorithm for performing dense MM on many-core architectures. A depth-first search of the recursive matrix decomposition required by Strassen's algorithm, is performed, followed by a switch to cuBLAS or MKL when the recursion depth exceeds a certain limit. While Strassen's algorithm does allow for a reduced computational complexity, the authors target only single accelerators. In this paper we utilize the "brute-force" algorithm to model our computation cost.

Kang et al. also employ Strassen's algorithm in HPMaX [7] to multiply matrices on both CPUs and GPUs. Apart from the implied limitation on matrix sizes, HPMaX cannot utilize heterogeneous networks like the proposed work.

Single multicore CPUs and GPUs are also targeted by Kelefouras et al. in [8], where the architectural characteristics of a machine, e.g. number of registers, cores, cache size and number of cache levels, are used to derive near optimal tiling arrangements for the execution of the classical dense MM algorithm.

Solomonik and Demmel [9] proposed a 2.5D block-based partitioning (an evolution of Cannon's algorithm) where additional data replication provides for optimal execution on a 2D grid of homogeneous CPUs. More recently, Lazzaro et al. [10] showed how 2.5D partitioning can be used for sparse matrix multiplication using one-sided MPI communications, on homogeneous machines equipped with GPU accelerators. In contrast, our work targets platforms mixing heterogeneous CPUs and different types of accelerators.

Hybrid GPU execution of LU factorization is examined in [11]. Tomov et al. split the matrix between the CPU cores and the GPU based on an empirically determined setting that is dependent on the input size.

KBLAS [12] provides implementations for a subset of BLAS level-2 routines, that can utilize multiple GPUs on a single node. Adbelfattah et al. are concerned with the kernel execution configuration (grid and block design) and data layout on multiple GPUs in order to obtain the highest speedup possible.

In [13] the authors use a block-based dynamic distribution of workload to the CPU and GPU computing resources of a heterogeneous cluster, while employing machine-learning for fine-tuning the GPU kernel execution configuration, based on the block dimensions.

Malik and Lastovetsky prove that three partitioning schemes are optimal under different circumstances for the case of three heterogeneous processors [14]. In [15] they also consider energy consumption due to the communication involved. The extension to a larger platform is an open problem.

A first attempt at solving the MM partitioning problem using DLT was made in [16], but that work allowed the use of only a single discrete accelerator in each system node. In this paper we significantly extend the work published in [16] to account for nodes with an arbitrary number of accelerators, and with different access to a node's main memory.

3   System Model

We assume that our computing platform is made-up of E interconnected Multicore Nodes (MNs), each hosting an arbitrary number of CPU cores and accelerators. In our treatment of the problem we assume that each CPU core and each accelerator (regardless of how many cores it may contain), can be the individual targets of computational load. In this paper we use the term core to refer to either a CPU core or an accelerator.

In order to cover the greatest possible range of setups, we assume that a system may be equipped with multiple **CPU** cores, *oN-chip* Accelerators (**NA**s), such as integrated GPUs, or the big cores in a big.LITTLE ARM machine, and *ofF-chip* Accelerators (**FA**s) such as discrete GPUs. The key attribute for distinguishing accelerator types is how they access main memory: NAs share the

Table 1. Model parameter notations and associated typical units.

a_j	oN-chip Accelerators in node j
b	communication latency (sec)
c_j	number of CPU cores in node j
E	total number of MN in the system
e_j	constant overhead for starting computation on a CPU core of node j (sec)
$e_{j,fa}$	constant overhead associated with computation on an FA of node j (sec)
$e_{j,na}$	constant overhead associated with computation on an NA of node j (sec)
f_j	number of oFf-chip Accelerators in node j
K	the number of columns of A and rows of B
l	inverse of the communication speed between LON and a MN (sec/byte)
l_{P_j}	inverse of the PCIe communication speed of MN j (sec/byte)
M	the number of rows of A and C
N	the number of columns of B and C
n_j	total computing devices/cores in node j, i.e. $n_j = c_j + a_j + f_j$
p_j	inverse of core j's computational speed (sec)
$p_{j,fa}$	inverse of FA's speed in node j (sec)
$p_{j,na}$	inverse of NA's speed in node j (sec)
$part_{j,i}$	percent of matrix C computed by core i of node j
s	bytes used in the representation of the matrices elements
$t_j^{(i)}$	total execution time of MN j, which is of type i ($i \in 1, 2, 3$)

CPU memory, while FAs are typically housed on PCIe expansion cards and have separate memory spaces. MNs are heterogeneous but we assume that within each MN the different types of cores are homogeneous.

Assuming that we have to multiply two matrices, A (MxK) and B (KxN), we assign each MN in the system a number of **rows** of the result matrix C. Computation at a node can start after the whole of B is transmitted and the corresponding rows of A are also received. Computation at a core can commence after the rows of A corresponding to its own part of the C matrix are collected. This means that computation at a MN starts in a staggered fashion. This also necessitates a core ordering, e.g. specifying which cores should start processing first in order to minimize the overall execution time.

Block-wise partitioning could certainly be a valid alternative that does not require one of the operand matrices to be broadcasted to all MNs. Block-wise partitioning in the context of DLT, is an open problem. However, if $K \cdot N > M \cdot K$, i.e. the communication cost of A is lower than B's, we can switch to column wise partitioning, where A is broadcasted and B is partitioned. The formulas that we derive would be equally applicable in that case, with slight modifications (e.g. switching where appropriate M to K and K to N).

We use affine models for the communication and computation costs. Thus, the time required for downloading B is (summary of notations in Table 1)

$$t_{comm}^{(B)} = l \ K \ N \ s + b \qquad (1)$$

where s is the size in bytes of the type used to represent matrix elements, l is the inverse of the communication speed and b is the communication latency.

We assume that $t_{comm}^{(B)}$ is a cost incurred by all MNs, except from the "load originating node" (LON), i.e. the machine originally holding A and B, and where C is ultimately collected. We also assume that it takes place in the form of a broadcast. We examine two possible configurations as far as the LON is concerned: (a) LON does not compute, and (b) LON participates in the computation.

The number of C matrix rows assigned to a core i of node j are calculated as a percent $part_{j,i}$ of M. Thus the time needed to communicate the required part of A is

$$t_{comm}^{(A)}(j, i) = l \ M \ K \ part_{j,i} \ s + b \qquad (2)$$

Enforcing the assignment of all the workload means (aka the normalization equation):

$$\sum_{j=0}^{E-1} \sum_{i=0}^{n_j-1} part_{j,i} = 1 \qquad (3)$$

where n_j is the total number of computing devices/cores in node j.

Inter-node communications are assumed to take place concurrently, thus, all MNs receive data at the same time. However each MN receives B followed by A parts in sequence.

FAs also incur an additional communication cost for having to transfer data across the PCIe bus. This is assumed to be a linear function of the transferred data:

$$t_{PCI}^{(B)} = l_{P_j} \ K \ N \ s \qquad (4)$$

$$t_{PCI}^{(A)}(j, i) = l_{P_j} \ M \ K \ part_{j,i} \ s \qquad (5)$$

The FA-to-host communications are assumed to be also sequential during matrix C parts collection (in the presence of multiple FAs). This does not apply for the dissemination of matrix B as contemporary PCIe switches have multicast capabilities, allowing the transfer of B to all the FAs in an MN in one step.

The computational cost for a CPU core i of node j is:

$$t_{comp}^{(CPU)}(j, i) = p_j \ M \ K \ N \ part_{j,i} + e_j \qquad (6)$$

where p_j is the inverse of the computational speed of core j, and e_j corresponds to its constant setup overhead. p_j is essentially the time spent per pair of input elements processed towards the calculation of C. A similar equation applies for NAs and FAs:

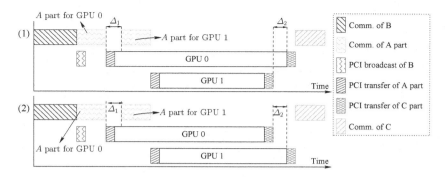

Fig. 1. Two possible timing configurations for distributing load to two FA/discrete GPUs. Timing gaps Δ_1 and Δ_2 represent arbitrary interruptions in the delivery and collection sequences that correspond to operations concerning other cores/accelerators.

$$t_{comp}^{(NA)}(j, i) = p_{j,na} \ M \ K \ N \ part_{j,i} + e_{j,na} \tag{7}$$

$$t_{comp}^{(FA)}(j, i) = p_{j,fa} \ M \ K \ N \ part_{j,i} + e_{j,fa} \tag{8}$$

It should be noted that a multicore CPU can be treated in our framework as one computational unit, the same way that a GPU is. The benefit of recognizing the individual cores is that it allows computation and communication to overlap. A CPU core can start computing as soon as its part of the matrix A arrives, instead of requiring the communication for the whole CPU as a unit to complete.

4   The Ordering Problem

The ordering problem is the determination of which core type should receive its part of the workload first. We will start by considering pairs of cores, before we attempt to generalize further. There are three pairings that need to be considered based on access to main memory:

- On-chip accelerator, CPU-core pair
- Off-chip accelerator, CPU-core pair (the latter could also be an NA)
- Pair of off-chip accelerators

For the first two cases, it can be proven that there exists no universal rule-of-thumb for ordering them.

In the case of the two FAs, the two possible orderings are shown in Fig. 1. The Δ_1 and Δ_2 time gaps represent times where communications are conducted by other cores/accelerators. In the following discussion we will assume that Δ_1 and Δ_2 are constant, effectively focusing on the relative distribution of load between the two FAs shown.

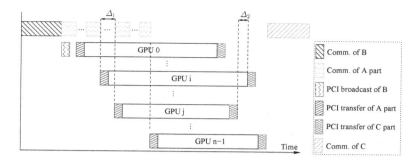

Fig. 2. A sequence that violates Lemma 1 for n off-chip accelerators.

Calculating the time difference between the two configurations produces the following

$$t_{(1)} - t_{(2)} = \frac{MKNll_Ps^2 + ((Nb + \Delta_1 N)l_P + \Delta_2 Kl)s}{((N + 2K)l_P + Kl)s + 2KNp_{fa}} > 0 \Rightarrow t_{(1)} > t_{(2)} \quad (9)$$

We can extend this result to an arbitrary set of identical off-chip accelerators:

Lemma 1. *For minimizing the processing on a set of identical off-chip accelerators, the load distribution sequence must match the result collection sequence.*

Proof. Let us assume that another optimum sequence exists that violates the lemma. This means that there must be at least one pair of accelerators i and j as shown in Fig. 2, that matches configuration (1) in Fig. 1. However, Eq. (9) dictates that we can switch the collection order for i and j, reducing the execution time for the two, without affecting the other accelerators. Hence we have a contradiction. □

Although it is impossible to pre-determine an optimum sequence for the distribution of the A matrix parts, except for the case of FAs, it can be empirically shown by calculating the execution times for alternative orderings, that for the current state-of-the-art hardware, the optimum workload assignment order is FAs, NAs, followed by CPU cores. For this reason and in order to reduce the complexity of the problem, we will assume that the distribution sequence for the cores of a node is precisely this. The collection sequence for the FAs matches their load assignment order as dictated by Lemma 1.

5 Closed-Form Solutions

5.1 Case I: LON Does Not Compute

We assume that each system node j is equipped with f_j FAs, a_j NAs, and c_j CPU cores, totaling n_j computing devices/cores. Following the notations used in the previous sections, $part_{j,0}$ to $part_{j,f_j-1}$ are the assignments to FAs, $part_{j,f_j}$ to $part_{j,f_j+a_j-1}$, are the assignments to NAs, and $part_{j,f_j+a_j}$ to $part_{j,n_j-1}$ are the CPU core assignments.

In order to derive a closed-form solution to the partitioning problem, we have to establish the timing relationships between the participating nodes. Optimality dictates that all nodes finish delivering their part of matrix C at the same time instance. Hence we must have for any two MNs j and i:

$$t_j = t_i \tag{10}$$

Figure 3 illustrates the timing of communication and computation phases for MN j. Accordingly, the total time spent by node j is:

$$
\begin{aligned}
t_j = \ & lsKN + b + \ // \text{ Communication of B} \\
& lsMK \sum_{m=0}^{n_j-1} part_{j,m} + n_j b + \ // \text{ Communication of A parts} \\
& p_j MKN part_{j,n_j-1} + e_j + \ // \text{ Computation on last CPU core} \\
& lsMN \sum_{m=0}^{n_j-1} part_{j,m} + b \ // \text{ C collection} \tag{11}
\end{aligned}
$$

We start by associating the parts assigned to the MN j's cores to the part assigned to the last CPU core $part_{j,n_j-1}$. For the last two cores in a system, (assuming they are both CPUs) it can be shown that:

$$part_{j,n_j-2} = part_{j,n_j-1}(1 + \frac{ls}{p_j N}) + \frac{b}{p_j MKN} \tag{12}$$

Using a similar logic for the third to last CPU core we can get:

$$part_{j,n_j-3} = part_{j,n_j-1}(1 + \frac{ls}{p_j N})^2 + \frac{b}{p_j MKN}(1 + (1 + \frac{ls}{p_j N})) \tag{13}$$

Which can be generalized for any CPU core $r \in [f_j + a_j, n_j)$ to:

$$part_{j,r} = part_{j,n_j-1}(1 + \frac{ls}{p_j N})^{n_j-r-1} + \frac{b}{lsMK}\left((1 + \frac{ls}{p_j N})^{n_j-r-1} - 1\right) \tag{14}$$

Equation (14) can be rewritten for $r \in [0, c_j)$ as:

$$part_{j,f_j+a_j+r} = part_{j,n_j-1}A(j,r) + B(j,r) \tag{15}$$

where A and B are shown in Table 2.

For the last-in-order NA, and in relation to the first in order CPU core we must have (expressing time span 1 in Fig. 3 as the sum of the individual operations):

$$p_{j,na}MKN part_{j,n_j-c_j-1} + e_{j,na} = lsMK part_{j,n_j-c_j} + b + p_j MKN part_{j,n_j-c_j} + e_j \Rightarrow$$

$$part_{j,n_j-c_j-1} = part_{j,n_j-c_j}\frac{ls + p_j N}{p_{j,na} N} + \frac{b + e_j - e_{j,na}}{p_{j,na}MKN} \tag{16}$$

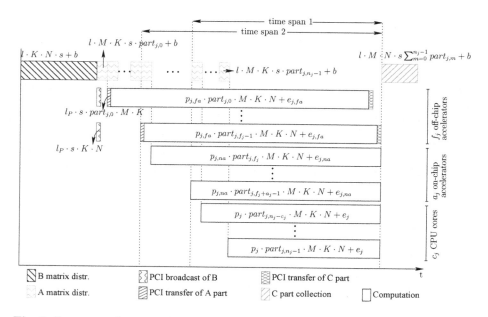

Fig. 3. Sequence of communication and computation operations for accelerators and CPU cores for a MN.

Which by utilizing Eq. (15) becomes (D and F are shown in Table 2):

$$part_{j,n_j-c_j-1} = part_{j,n_j-1}D(j) + F(j) \qquad (17)$$

For a pair of successive NAs q and $q+1$ we can show that:

$$part_{j,q} = part_{j,q+1}\left(1 + \frac{ls}{p_{j,na}N}\right) + \frac{b}{p_{j,na}MKN} \qquad (18)$$

Which can be extended to :

$$part_{j,q} = part_{j,q+r}\left(1 + \frac{ls}{p_{j,na}N}\right)^r + \frac{b}{p_{j,na}MKN}\sum_{m=0}^{r-1}\left(1 + \frac{ls}{p_{j,na}N}\right)^m \qquad (19)$$

Or in another form, with $0 \le r < a_j$ (O and P are shown in Table 2):

$$part_{j,f_j+r} = part_{j,f_j+a_j-1}O(j,r) + P(j,r) \qquad (20)$$

Given $f_j + a_j - 1 = n_j - c_j - 1$, we can combine (17) and (20) as:

$$part_{j,f_j+r} = part_{j,n_j-c_j-1}O(j,r) + P(j,r) = part_{j,n_j-1}Q(j,r) + R(j,r) \qquad (21)$$

where Q and R are shown in Table 2.

Table 2. Functions for a MN j and one of its cores r, used to abbreviate formulas. Functions $A'(r)$ to $J'(r)$ are derived from their similarly named counterparts, by setting $j \leftarrow E - 1$, and by replacing l with l' and b with b'.

CPU cores, $r \in [0, c_j)$	
$A(j,r) = (1 + \frac{ls}{p_j N})^{c_j - r - 1}$	$B(j,r) = \frac{b}{ls MK}(A(j,r) - 1)$
NA accelerators, $r \in [0, a_j)$	
$D(j) = A(j,0)\frac{ls + p_j N}{p_{j,na} N}$	$F(j) = B(j,0)\frac{ls + p_j N}{p_{j,na} N} + \frac{e_j - e_{j,na} + b}{p_{j,na} MKN}$
$O(j,r) = (1 + \frac{ls}{p_{j,na} N})^{a_j - r - 1}$	$P(j,r) = \frac{b}{ls MK}(O(j,r) - 1)$
$Q(j,r) = D(j)O(j,r)$	$R(j,r) = F(j)O(j,r) + P(j,r)$
FA accelerators, $r \in [0, f_j)$	
$G(j,r) = \left(\frac{K(s(l + l_{P_j}) + p_{j,fa} N)}{l_{P_j} s(K + N) + p_{j,fa} KN}\right)^{f_j - 1 - r}$	$H(j,r) = \frac{b}{M}\frac{G(j,r) - 1}{s(lK - l_{P_j} N)}$
$I(j,r) = Q(j,0)G(j,r)\frac{K(ls + p_{j,na} N)}{l_{P_j} s(K + N) + p_{j,fa} KN}$	
$J(j,r) = R(j,0)G(j,r)\frac{K(ls + p_{j,na} N)}{l_{P_j} s(K + N) + p_{j,fa} KN} + \frac{b + e_{j,na} - e_{j,fa}}{M(l_{P_j} s(K + N) + p_{j,fa} KN)} G(j,r) + H(j,r)$	

For the last-in-order FA, and in relation to the first in order NA (f_j) we can get by expressing time span 2 in Fig. 3 as the sum of the individual operations:

$$part_{j,f_j - 1} = part_{j,f_j} \frac{K(ls + p_{j,na} N)}{l_{P_j} s(K + N) + p_{j,fa} KN} + \frac{b + e_{j,na} - e_{j,fa}}{M(l_{P_j} s(K + N) + p_{j,fa} KN)} \tag{22}$$

We can associate two successive FAs r and $r + 1$, with $r \in [0, f_j - 1)$, with:

$$part_{j,r} = part_{j,r+1} \frac{K(s(l + l_{P_j}) + p_{j,fa} N)}{l_{P_j} s(K + N) + p_{j,fa} KN} + \frac{b}{M(l_{P_j} s(K + N) + p_{j,fa} KN)} \tag{23}$$

Thus we can express the part assigned to an arbitrary FA $0 \le r < f_j$ as:

$$part_{j,r} = part_{j,f_j - 1} \left(\frac{K(s(l + l_{P_j}) + p_{j,fa} N)}{l_{P_j} s(K + N) + p_{j,fa} KN}\right)^{f_j - 1 - r} +$$

$$\frac{b}{M(l_{P_j} s(K + N) + p_{j,fa} KN)} \cdot \sum_{m=0}^{f_j - r - 2} \left(\frac{K(s(l + l_{P_j}) + p_{j,fa} N)}{l_{P_j} s(K + N) + p_{j,fa} KN}\right)^m =$$

$$part_{j,f_j - 1} G(j,r) + H(j,r) \tag{24}$$

where G and H are shown in Table 2.

Establishing a relationship between $part_{j,r}$ in the above equation and $part_{j,n_j - 1}$, is a matter of utilizing (22) and (21), with I and J as shown in Table 2):

$$(24) \overset{via (22),(21)}{\Rightarrow} part_{j,r} = part_{j,n_j - 1} I(j,r) + J(j,r) \tag{25}$$

Table 3. Formulas for the calculation of the $U(j)$ and $V(j)$ functions for MN j. Helper functions $u(j)$ and $v(j)$ are also shown.

$$U(j) = M(p_j KN + ls(K + N)u(j))$$

$$V(j) = lsM(K + N)v(j) + (n_j + 2)b + e_j + lsKN$$

$$u(j) = \sum_{m=0}^{f_j-1} I(j,m) + \sum_{m=0}^{a_j-1} Q(j,m) + \sum_{m=0}^{c_j-1} A(j,m))$$

$$v(j) = \sum_{m=0}^{f_j-1} J(j,m) + \sum_{m=0}^{a_j-1} R(j,m) + \sum_{m=0}^{c_j-1} B(j,m)$$

Given the above, the total execution time (see (11)) for MN j is:

$$t_j = part_{j,n_j-1} M \left(p_j KN + ls(K+N) \left(\sum_{m=0}^{f_j-1} I(j,m) + \sum_{m=0}^{a_j-1} Q(j,m) + \right. \right.$$

$$\left. \left. \sum_{m=0}^{c_j-1} A(j,m) \right) \right) + lsM(K+N) \left(\sum_{m=0}^{f_j-1} J(j,m) + \sum_{m=0}^{a_j-1} R(j,m) + \sum_{m=0}^{c_j-1} B(j,m) \right) +$$

$$(n_j + 2)b + e_j + lsKN \tag{26}$$

Equation (10) allows us to establish **inter-node** part relationships:

$$t_j = t_i \stackrel{\text{via (26)}}{\Rightarrow} part_{j,n_j-1} = part_{i,n_i-1}S(i,j) + T(i,j) \tag{27}$$

where (U and V are shown in Table 3)

$$S(i,j) = \frac{U(i)}{U(j)} \text{ and } T(i,j) = \frac{V(i) - V(j)}{U(j)} \tag{28}$$

With this result, we can associate the individual parts with the part assigned to the last CPU core of MN 0:

– For the CPU cores, $0 \le r < c_j$:

$$part_{j,f_j+a_j+r} = part_{0,n_0-1}S(0,j)A(j,r) + T(0,j)A(j,r) + B(j,r) \tag{29}$$

– For the NAs, $0 \le r < a_j$:

$$part_{j,f_j+r} = part_{0,n_0-1}S(0,j)Q(j,r) + T(0,j)Q(j,r) + R(j,r) \tag{30}$$

– For the FAs, $0 \le r < f_j$:

$$part_{j,r} = part_{0,n_0-1}S(0,j)I(j,r) + T(0,j)I(j,r) + J(j,r) \tag{31}$$

We can the use the normalization Eq. (3) to produce:

$$part_{0,n_0-1} = \frac{1 - \sum_{j=0}^{E-1} T(0,j)u(j) - \sum_{j=0}^{E-1} v(j)}{\sum_{j=0}^{E-1} S(0,j)u(j)} \tag{32}$$

where $u(j)$ and $v(j)$ are shown in Table 3.

Equipped with (32), (29), (30) and (31) we can calculate the partitioning in linear time and space complexity with respect to the total number of computing devices $\Theta(\sum_{j=0}^{E-1} n_j)$. Because only the $S(0,j)$ and $T(0,j)$ terms are present in these equations, only E of each of these terms are required instead of E^2.

5.2   Case II : LON Participates in the Computation

We assume that MN $E-1$ serves as the LON, to allow us to reuse the Equations derived for Case I. The total execution time for the LON, made up of the reading of B, the "communication" of the A parts, the computation on the last CPU core and the writing of C is:

$$t_{E-1} = l'sKN + b' + l'sMK \sum_{m=0}^{n_{E-1}-1} part_{E-1,m} + n_{E-1}b' +$$

$$p_{E-1}MKN part_{E-1,n_{E-1}-1} + e_{E-1} + l'sMN \sum_{m=0}^{n_{E-1}-1} part_{E-1,m} + b' \quad (33)$$

where l' and b' correspond to the *local* data access cost. Similarly to (26), we can show that:

$$t_{E-1} = part_{E-1,n_{E-1}-1}M\left(p_{E-1}KN + l's(K+N)\left(\sum_{m=0}^{f_{E-1}-1} I'(m) + \sum_{m=0}^{a_{E-1}-1} Q'(m)\right.\right.$$

$$\left.\left.+ \sum_{m=0}^{c_{E-1}-1} A'(m)\right)\right) + l'sM(K+N)\left(\sum_{m=0}^{f_{E-1}-1} J'(m) + \sum_{m=0}^{a_{E-1}-1} R'(m) + \sum_{m=0}^{c_{E-1}-1} B'(m)\right)$$

$$+ (n_{E-1}+2)b' + e_{E-1} + l'sKN \quad (34)$$

where the A' to J' are derived from the similarly named functions shown in Table 2, by setting $j \leftarrow E-1$, and by replacing l with l' and b with b'. For example, for the CPU cores $0 \leq r < c_{E-1}$ we have:

$$part_{E-1,f_{E-1}+a_{E-1}+r} = part_{E-1,n_{E-1}-1}A'(r) + B'(r) \quad (35)$$

Following the same logic we can prove that:

$$t_{E-1}^{(1)} = t_0^{(1)} \Rightarrow part_{E-1,n_{E-1}-1} = part_{0,n_0-1}S'(0) + T'(0) \quad (36)$$

where

$$S'(0) = \frac{U(0)}{U'(E-1)} \text{ and } T'(0) = \frac{V(0) - V'(E-1)}{U'(E-1)} \quad (37)$$

Which allows the use of the normalization Eq. (3) to produce:

$$part_{0,n_0-1} = \frac{1 - \sum_{j=0}^{E-2} T(0,j)u(j)}{\sum_{j=0}^{E-2} S(0,j)u(j) + S'(0)u'(E-1)} -$$

$$\frac{\sum_{j=0}^{E-2} v(j) + T'(0)u'(E-1) + v'(E-1)}{\sum_{j=0}^{E-2} S(0,j)u(j) + S'(0)u'(E-1)} \quad (38)$$

where $v'(E-1) = \sum_{m=0}^{f_{E-1}-1} J'(m) + \sum_{m=0}^{a_{E-1}-1} R'(m) + \sum_{m=0}^{c_{E-1}-1} B'(m)$ and $u'(E-1) = \sum_{m=0}^{f_{E-1}-1} I'(m) + \sum_{m=0}^{a_{E-1}-1} Q'(m) + \sum_{m=0}^{c_{E-1}-1} A'(m))$ as per the convention mentioned above for the prime functions.

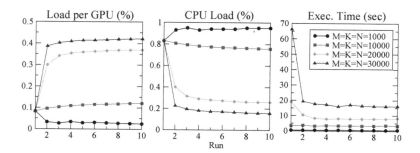

Fig. 4. Load assignment (expressed as a percent of the total) for one GPU and for the CPU, and execution time over 10 runs, for multiplying two square matrices with a size of 1000, 10000, 20000 and 30000.

6   On-Line Estimation of Cost Parameters

Typically, DLT-based partitioning requires the a-priori estimation of all the cost parameters needed. This is problematic for heterogeneous installations, and the situation is further complicated by the fact that thermal throttling is common in contemporary hardware, leading to performance variations that a fixed set of parameters fails to address properly [17].

In this paper we use a dynamic approach for updating the computation cost parameters at the end of each execution. Generic values are used to initialize these parameters when previous timing data do not exist. Then, after each execution of our program, the part of the workload assigned to a core i of a MN j, i.e. $part_{j,i} \cdot M \cdot K \cdot N$, along with the measured time it took to complete it, are fed to a routine performing linear regression in order to update the corresponding p and e parameters.

The following rules are used for the update process:

- The most recent of the time measurements are used to perform linear regression. At least two measurements are required before regression is attempted.
- The data are indexed by the M, K and N values, so that a different speed p can be calculated for each core and for each problem size.
- All the data accumulated for a particular type of core are used to derive the constant overhead e. This is attempted only if there are data for at least two sets of M, K and N values.

During the experiments described in Sect. 7, a sequence of 10 runs was adequate for getting from a generic p to a one closely matching the speed of a core.

Figure 4 illustrates how partitioning and the corresponding execution times evolved over 10 runs, when executing on the **dune-1080** node (see details in the next section). For the first run, p was set to 1 and e to 0.

7   Experimental Results

For testing we used square matrices with M, K and N ranging from 1000 to 30000. The matrices were randomly generated for data types `float`, `double` and `complex<double>` / `cuDoubleComplex`, in order to test how partitioning is influenced by larger communication costs and the relative inefficiency of GPUs in processing higher precision floating point data.

Our test platform details are shown in Table 4. Preliminary testing of the CPUs and GPUs with the MKL and cuBLAS libraries respectively, showed that they all exhibit an affine computational cost in relation to the $M \cdot K \cdot N$ product of the input matrices, validating our assumption. For example, for `float` inputs, i7-8700k had an execution time approximated by $t = 3.36 \cdot 10^{-8} \cdot M \cdot K \cdot N + 284$ msec with Pearson's $R^2 = 0.986$. A similar equation applied for Titan-X: $t = 3.9 \cdot 10^{-10} \cdot M \cdot K \cdot N + 321$ msec with $R^2 = 0.967$, and the same behavior was observed for all input types.

The communication parameters were obtained by using a "ping-pong" procedure, and they were found to be $l = 8.50 \cdot 10^{-5}$msec/byte (11.2 MB/s), $b = 0.891$msec and $l_P = 1.638 \cdot 10^{-7}$ msec/byte (5.68 GB/s).

Intel's MKL library has the capability to use multiple cores natively, without the programmer's intervention. In our tests, we discovered that performance was boosted when MKL handled the CPU workload, but only if the input was of type `float`. In the other two input type cases, launching one thread per logical core and using MKL individually from every thread (by calling `mkl_set_num_threads_local(1)`), proved a better alternative. The reason for this behavior is unknown.

In order to validate our results and quantitatively measure the performance benefits afforded by our mathematical framework, we compared it against a dynamic scheduler that breaks matrix A into blocks of fixed number of rows r. Thus, each core calculated $r \cdot N$ elements of the C matrix, before requesting more data to process, etc. As r is a critical parameter for the performance of the dynamic scheduler, we tested for values of $50, 100, 200, 500$ and 1000, and used the best overall time as the execution time.

Table 4. Hardware and software specifications of the test platform used in our experiments. Theoretical single-precision performance is also shown for each component. All machines had hexa core CPUs with hyperthreading enabled, and they were connected via 1 Gb Ethernet.

Hardware	dune-titan	dune-1080	dune-970	dune-frg
CPU	i7-5930K, 3.5GHz	i7-8700K, 3.7GHz	i7-5820K, 3.3GHz	i7-5820K, 3.3GHz
RAM	64	64	32	64
GPU	Titan-X	2xGTX-1080Ti	2xGTX-970	2xGTX-970
VRAM	12	11	4	4

Software	OS	Compiler	Libraries	
	Kubuntu 18.04	GCC 7.4	CUDA 10.2	Intel MKL 2019
			cuBLAS	MPICH 3.3

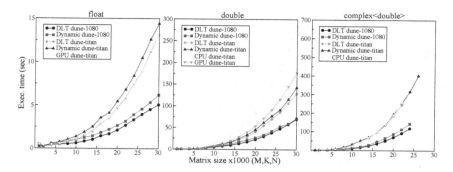

Fig. 5. Average performance for processing square input matrices. The hardware configurations of the nodes are detailed in Table 4. The best results produced for any of the tested r are used for the dynamic scheduler curves.

In terms of competing frameworks, we were unable to find a readily available one that offers hybrid computation. For example, the MAGMA library does not currently (as of v2.5.4) offer a hybrid option for their `magma_*gemm` routines, opting for either CPU or GPU based execution. Additionally ScaLAPACK offers only CPU-based execution. We still used MAGMA/cuBLAS to gauge the benefit that the addition of a CPU offers in hybrid execution.

Figure 5 shows the average performance over 10 runs for the various scenarios examined, as executed individually on our two fastest platforms. It is evident that our proposed partitioning outperforms the best a dynamic scheduler has to offer, by a good margin. Exclusive CPU and GPU results are also shown, clearly indicating that our proposed solution can leverage the best that each computing device can offer.

However, there is no improvement for relatively small inputs i.e. below roughly 5k × 5k elements. The execution time is low enough in this case, to warrant the dynamic partitioning superior. This is caused by two factors, the first being the frequency scaling characteristics of modern CPUs. Modern CPUs can maintain a maximum operating frequency for one or a handful of cores, but they have to scale back when all the cores are running at full speed. The values of r that perform best reveal this characteristic, as small matrices favor the higher values of r, leading to processing in as few cores as possible. On the other hand, bigger inputs favor modest to small values of r that lead to better load balancing.

The second factor is the delay introduced by the collection/reading of the hardware characteristics and the calculation of the partitioning. This is however a small delay (measured during the tests to 1–2 hundreds of a second), which means that further investigation on the number of cores utilized for a computation is required under the prism of frequency boosting and throttling. This is beyond the scope of this paper.

The benefits of our framework would become even more apparent if a networked platform were to be utilized. Unfortunately, due to the relatively slow

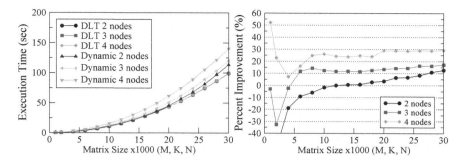

Fig. 6. (a) Performance and (b) percent improvement of the proposed partitioning method (labeled DLT), against a dynamic scheduler, for node sets comprising of 2, 3, and 4 nodes. The master/LON is not participating in the computation of the `float`-type input.

1 Gb/s Ethernet network interconnect, none of the possible multinode configurations we tested, resulted in a DLT-partitioning that utilized anything more than a single node, even when the LON was not participating in the computation.

We tested a configuration where `dune-frg` would serve as the LON node without participating in the computation, and `dune-1080`, `dune-titan` and `dune-970` would be added -in this order- in MPI node sets consisting of 2, 3 and 4 nodes respectively (`dune-frg` serving as the master). For all the tests, a single MPI process was launched per node and C++11 threads were used to utilize the available computing devices.

Our analytical framework allows us to predict the performance of all the possible configurations, disregarding nodes despite their availability, because using them would degrade performance. In our test, for all configurations only the fastest (`dune-1080`) was picked based on the predicted results, and the measured performance especially in comparison with the dynamic scheduler as shown in Fig. 6, justifies this choice. As observed in Fig. 6 adding more nodes increases the execution times as a consequence of the slow network. There is also a small degradation for DLT, caused by the initial node characteristics data collection, which is required before the partitioning is done.

8   Conclusion and Future Work

This paper presents a generic analytical framework that yields closed-form solutions to the problem of dense matrix-matrix multiplication on a heterogeneous network of machines equipped with multicore CPUs and a variety of accelerators. Our work constitutes the first successful attempt at applying DLT to the partitioning of cubic complexity workloads. We also demonstrated that the online calculation of the parameters needed for the application of DLT frameworks is feasible and accurate enough to eliminate the requirement for lengthy offline benchmarks, while also allowing for some form of adaptation to platform changes between runs.

As proven by our experimental study, our framework produces significant performance benefits for processing large matrices compared to a dynamic scheduler, optimizing the use of the available computing resources by balancing the best of what CPU and GPU domains can offer.

Arguably, a multi-installment strategy, which is analogous to the tiling used by heuristic schedulers, could boost performance even further. How the beneficial effects of this approach could be counter-balanced by the increased communication delays due to latency, is a open question. Furthermore, energy and memory constraints could be noteworthy additions to be considered.

References

1. Kang, S., Veeravalli, B., Aung, K.M.M.: Dynamic scheduling strategy with efficient node availability prediction for handling divisible loads in multi-cloud systems. J. Parallel Distrib. Comput. **113**, 1–16 (2018)
2. Wang, X., Veeravalli, B.: Performance characterization on handling large-scale partitionable workloads on heterogeneous networked compute platforms. IEEE Trans. Parallel Distrib. Syst. **28**(10), 2925–2938 (2017)
3. Barlas, G.: Multicore and GPU Programming: An Integrated Approach, 1st edn. Morgan Kaufmann, Burlington (2014)
4. Suresh, S., Run, C., Kim, H.J., Robertazzi, T.G., Kim, Y.-I.: Scheduling second-order computational load in master-slave paradigm. IEEE Trans. Aerosp. Electron. Syst. **48**(1), 780–793 (2012)
5. Song, F., Tomov, S., Dongarra, J.: Enabling and scaling matrix computations on heterogeneous multi-core and multi-GPU systems. In: ICS 2012, pp. 365–375 (2012)
6. ul Hassan Khan, A., Al-Mouhamed, M., Fatayer, A., Mohammad, N.: Optimizing the matrix multiplication using Strassen and Winograd algorithms with limited recursions on many-core. Int. J. Parallel Program. **44**(4), 801–830 (2016). https://doi.org/10.1007/s10766-015-0378-1
7. Kang, H., Kwon, H.C., Kim, D.: HPMaX: heterogeneous parallel matrix multiplication using CPUs and GPUs. Computing **102**(12), 2607–2631 (2020). https://doi.org/10.1007/s00607-020-00846-1
8. Kelefouras, V., Kritikakou, A., Mporas, I., Kolonias, V.: A high-performance matrix-matrix multiplication methodology for CPU and GPU architectures. J. Supercomput. **72**(3), 804–844 (2016). https://doi.org/10.1007/s11227-015-1613-7
9. Solomonik, E., Demmel, J.: Communication-optimal parallel 2.5D matrix multiplication and LU factorization algorithms, Technical Report UCB/EECS-2011-10, University of California at Berkeley, February 2011
10. Lazzaro, A., VandeVondele, J., Hutter, J., Schütt, O.: Increasing the efficiency of sparse matrix-matrix multiplication with a 2.5D algorithm and one-sided MPI. In: Proceedings of the Platform for Advanced Scientific Computing Conference, PASC 2017 (2017)
11. Tomov, S., Dongarra, J., Baboulin, M.: Towards dense linear algebra for hybrid GPU accelerated manycore systems. Parallel Comput. **36**, 232–240 (2010)
12. Abdelfattah, A., Keyes, D., Ltaief, H.: KBLAS: an optimized library for dense matrix-vector multiplication on GPU accelerators. ACM Trans. Math. Softw. **42**(3), 1–31 (2016)

13. Sivkov, I., Lazzaro, A., Hutter, J.: DBCSR: a library for dense matrix multiplications on distributed GPU-accelerated systems. In: 2019 International Multi-Conference on Engineering, Computer and Information Sciences (SIBIRCON), pp. 0799–0803 (2019)
14. Malik, T., Lastovetsky, A.: Optimal matrix partitioning for data parallel computing on hybrid heterogeneous platforms. In: 2020 19th International Symposium on Parallel and Distributed Computing (ISPDC), pp. 1–11 (2020)
15. Malik, T., Lastovetsky, A.: Towards optimal matrix partitioning for data parallel computing on a hybrid heterogeneous server. IEEE Access **9**, 17229–17244 (2021)
16. Barlas, G., Hiny, L.E.: Closed-form solutions for dense matrix-matrix multiplication on heterogeneous platforms using divisible load analysis. In: PDP, Cambridge, UK, pp. 376–384 (2018)
17. Ghanbari, S., Othman, M.: Time cheating in divisible load scheduling: sensitivity analysis, results and open problems. In: 6th International Conference on Smart Computing and Communications, ICSCC 2017, pp. 935–943 (2017)

Parallelization of Robust Multigrid Technique Using OpenMP Technology

Sergey Martynenko[1,2,3,4](✉) ⓘ, Weixing Zhou[5] ⓘ, İskender Gökalp[6,7] ⓘ,
Vladimir Bakhtin[8] ⓘ, and Pavel Toktaliev[1]

[1] Institute of Problems of Chemical Physics of the Russian Academy of Sciences,
Ac. Semenov avenue 1, Chernogolovka, Moscow 142432, Russian Federation
{Martynenko,Toktaliev}@icp.ac.ru

[2] Bauman Moscow State Technical University, ul. Baumanskaya 2-ya,
5/1, Moscow 105005, Russian Federation

[3] Joint Institute for High Temperatures of the Russian Academy of Sciences,
Izhorskaya st. 13 Bd.2, Moscow 125412, Russian Federation

[4] Baranov Central Institute of Aviation Motors,
2, Aviamotornaya Street, Moscow 111116, Russian Federation

[5] Harbin Institute of Technology, 92 West Dazhi Street, Nangang District,
Harbin 150001, Heilongjiang Province, China
zhouweixing@hit.edu.cn

[6] Middle East Technical University Üniversiteler Mahallesi,
Dumlupınar Bulvarı No:1, 06800 Çankaya, Ankara, Turkey
igokalp@metu.edu.tr,iskender.gokalp@cnrs-orleans.fr

[7] The Institute of Combustion, Aerothermics, Reactivity and Environment
– Centre National de la Recherche Scientifique, 1C, avenue de la Recherche
Scientifique, 45071 Cedex 2 Orléans, France

[8] Keldysh Institute of Applied Mathematics of the Russian Academy of Sciences,
Miusskaya sq., 4, Moscow 125047, Russian Federation
Bakhtin@keldysh.ru

Abstract. This article represents the parallel multigrid component analysis of Robust Multigrid Technique (RMT). The RMT has been developed for black-box solution of a large class of (non)linear boundary value problems in computational continuum mechanics. Parallel RMT can be constructed by combination of the algebraic and geometric approaches to parallelization. The geometric smoother-independent approach based on a decomposition of the given problem into 3^κ ($\kappa = 1, 2, \ldots$) subproblems without an overlap should be used to overcome the problems of large communication overhead and idling processors on coarser levels. The algebraic grid-independent approach based on a decomposition of the given problem into $C3^\kappa$ ($\kappa = 1, 2, \ldots$) subproblems with an overlap (multicoloured Vanka-type smoother) should be used for parallel smoothing on finer levels. Standard programming model for shared memory parallel programming OpenMP has been used for parallel implementation of RMT on personal computer and computer cluster.

The activity is a part of the work "Supercomputer modelling of hypervelocity impact on artificial space objects and Earth planet" supported by Russian Science Foundation (project no. 21-72-20023).

© Springer Nature Switzerland AG 2021
V. Malyshkin (Ed.): PaCT 2021, LNCS 12942, pp. 196–209, 2021.
https://doi.org/10.1007/978-3-030-86359-3_15

This paper represents parallel multigrid cycle, algebraic and geometric approaches to parallelization, estimation of the parallel RMT efficiency and parallel multigrid component analysis.

Keywords: Boundary value problems · Robust multigrid technique · OpenMP · Parallel solvers

1 Introduction

Multigrid algorithms are well known for being the fastest numerical methods for solving elliptic boundary-value problems. There are two trends with respect to the choice of multigrid components [5]:

- in optimized multigrid algorithms, one tries to tailor the components to the problem at hand in order to obtain the highest possible efficiency for the solution process;
- in robust multigrid algorithms, one tries to choose the components independently of the given problem, uniformly for as large a class of problems as possible.

At the end of the 70ies and at the beginning of the 80ies there was a real boom in research on the multigrid methods. Very interesting multigrid approach had been proposed and developed in Theoretical Division of the Los Alamos National Laboratory. In paper [1], P.O. Frederickson and O.A. McBryan studied efficiency of parallel superconvergent multigrid method (PSMG). The basic idea behind PSMG is the observation that for each fine grid there are two natural coarse grids – the even and odd points of the fine grid. Authors tries to develop optimized multigrid algorithm by combination of these coarse grid solutions for more accurate fine grid correction. The PSMG and related ideas essentially refer to massively parallel computing. To keep all processors of a massively parallel system busy especially on coarse grids, PSMG works simultaneously on many different grids, instead of working only on the standard coarse grid hierarchy [5].

Also in 1990, S.I. Martynenko (Baranov Central Institute of Aviation Motors, Moscow) suggested to use similar multiple coarse grid correction strategy for development of robust multigrid method for black-box software. To avoid terminological confusion, we define software to be black-box if it does not require any additional input from the user apart from the physical problem specification consisting of the domain geometry, boundary and initial conditions, the enumeration of equations to be solved (heat conductivity equation, Navier–Stokes equations, Maxwell equations, etc.) and mediums. The user does not need to know anything about numerical methods, or high-performance and parallel computing [2]. Developed solver is called the Robust Multigrid Technique (RMT), RMT history is given in [3]. For a theoretical description of RMT and corresponding parallel analysis, we refer to [2].

To overcome the problem of robustness, the essential multigrid principle[1] has been used in the single-grid solver. RMT has the least number of problem-dependent components, close-to-optimal algorithmic complexity and high parallel efficiency for large class of boundary value problems. Application field of the RMT is mathematical modelling of complex physical and chemical processes described by the systems of nonlinear strongly coupled partial differential equations. As a result, RMT can use not only a segregated smoothers, but also the coupled Vanka-type smoothers in the multigrid iterations. RMT can be used as efficient solver on structured grids or as a multigrid preconditioner on unstructured grids in black-box software [2]. It should be noted that RMT has close-to-optimal algorithmic complexity: the number of multigrid iterations needed for solving the boundary value problems is independent of the number of unknowns, but computational cost of each multigrid iteration is $O(N \log N)$ arithmetic operations. Loss in computational efforts compared to classic multigrid ($\sim \log N$ arithmetic operations) is a result of true robustness of RMT [2].

In [5], important aspects of parallel classic multigrid are summarized:

- on coarse grids, the ratio between communication and computation becomes worse than on fine grids, up to a (possibly) large communication overhead on very coarse grids;
- on very coarse grids we may have (many) idle processes;
- on coarse grids, the communication is no longer local.

The simplest way to construct a parallel multigrid algorithm is to parallelize all of its components. Although suitable multigrid components may be highly parallel, the overall structure of standard multigrid is intrinsically not fully parallel for two reasons. The first reason is that the grid levels are run through sequentially in standard multigrid. The second reason is that the degree of parallelism of multigrid is different on different grid levels (i.e. small on coarse grids) [5]. In addition, parallelizing the multigrid components will only allow constructing a small-scale granulated algorithm, it means small tasks can be performed in parallel.

The Chan-Tuminaro and the Gannon-van Rosendale algorithms both belong to the class of concurrent methods. The basic approach is to generate independent sub-problems for the different grid levels by projection onto orthogonal sub-spaces. The algorithms differ in the way this decomposition is performed and the way solutions are combined again.

The algorithm of Fredrickson and McBryan follows a completely different approach. Opposed to standard multigrid, the method does not employ a single grid to compute a coarse grid correction, but composes on each level several coarse grid problems. Ideally, the additional information obtained from these multiple coarse grids can be used to improve the convergence rate of the multigrid method, thus improving not only parallel efficiency, but also actual run-time [1]

---

[1] The essential multigrid principle is to approximate the smooth (low frequency) components of the error on the coarse grids. The nonsmooth (high frequency) components are reduced with a small number (independent of mesh size) of smoothing iterations on the fine grid [6].

Large-scale granularity means large tasks that can be performed independently in parallel. Goal of the paper is to analyse granularity of the basic components of parallel RMT using OpenMP technology. Now the presented approach is suitable for use only with shared memory systems.

2   Parallel RMT

Let N_{G_0} points form the computational grid G_0. We assume that it is possible to form non-overlapping subgrids $G_i \in G_0$ of N_{G_i} points, such what

$$G_0 = \bigcup_{i=1}^{I} G_i \quad \text{and} \quad G_n \cap G_m = \emptyset, \quad n \neq m.$$

All subgrids G_i, $i = 1, 2, \ldots, I$ form the first grid level and

$$\sum_{i=1}^{I} N_{G_i} = N_{G_0},$$

but the finest grid G_0 forms zero level. There are a number of the subgrid generation strategy, but we will be interested in an optimal strategy that minimizes the approximation error on the coarse grids G_i.

Let 1D uniform grid for the finite volume discretization is defined as

$$x_i^{\mathrm{v}} = \frac{i-1}{N_x^0} = (i-1)h_x, \qquad i = 1, 2, \ldots, N_x^0 + 1,$$

$$x_i^{\mathrm{f}} = \frac{x_i^{\mathrm{v}} + x_{i+1}^{\mathrm{v}}}{2} = \frac{2i-1}{2}h_x, \qquad i = 1, 2, \ldots, N_x^0,$$

where N_x^0 is a discretization parameter and $h_x = 1/N_x^0$ is a mesh size. Points x_i^{v} and x_i^{f} can be vertices or faces in the finite volume discretization of mathematical model equations. Figure 1 represents a triple coarsening used in RMT. In this case, each finite volume on the coarser grids is union of three finite volumes on the finer grid. It it easy to see that the smoothing steps on different grid of the same level are independent of each other and can be performed in parallel.

This grid hierarchy will be called a multigrid structure generated by the finest grid G_0. Note that each grid of the multigrid structure can generate own multigrid structure. The triple coarsening and the finite volume discretization lead to the problem-independent restriction and prolongation operators [2].

Multigrid cycle of RMT is smoothing on the multigrid structure as shown on Fig. 2. The multigrid schedule of RMT is sawtooth cycle, i.e. a special case of the V-cycle, in which smoothing before the coarse grid correction (presmoothing) is deleted [6].

OpenMP is an implementation of multithreading, a method of parallelizing whereby a primary thread (a series of instructions executed consecutively) forks a specified number of sub-threads and the system divides a task among them

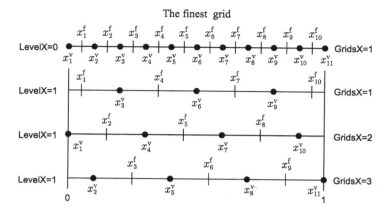

Fig. 1. Triple coarsening in RMT: the finest grid and three coarse grids of the first level.

[8]. The threads then run concurrently, with the runtime environment allocating threads to different processors. OpenMP uses a portable, scalable model that gives programmers a simple and flexible interface for developing parallel RMT for platforms ranging from the standard desktop computer to the supercomputer.

From the parallel point of view, RMT has the following attractive features:

1. All coarse grids of the same level have no common points and the smoothing iterations can be performed in parallel.
2. The number of grids (known in advance) on each level predicts the number of threads for parallel RMT.
3. Almost the same number of points on each grid of the same level leads to almost uniform load balance.
4. The most powerful coarse grid correction strategy used in RMT makes it possible to use the weak smoothers in the parallel multigrid iterations.

Each grid level of the multigrid structure consists of 3^{dl} grids (the problem dimension $d = 1, 2, 3$; $l = 0, 1, \ldots, L^+$), where L^+ is the coarsest level. Therefore we should use $p = 3^\kappa$ ($\kappa = 1, \ldots, L^+$) threads for uniform load balance. Case $\kappa = 0$ corresponds to sequential implementation.

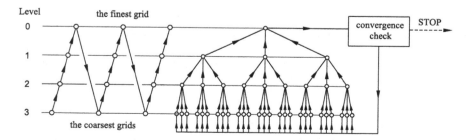

Fig. 2. Sequential multigrid cycle of RMT for solving 1D problems.

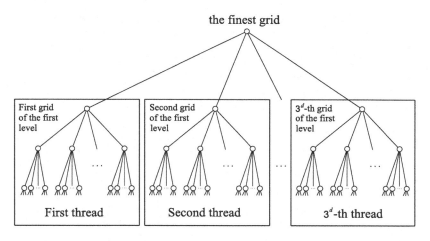

Fig. 3. Distribution of the coarse grids of the first level to $p = 3^d$ threads.

If we use $p = 3^\kappa$ ($\kappa = 1, \ldots, L^+$) threads for parallel RMT, we have to distinguish two cases:

- *algebraic parallelism* $\kappa > dl$ (finer levels). In the first case, multicoloured unknown ordering could be viewed as a specific way to parallelize a smoothing iterations independently of the computational grid [4];
- *geometric parallelism* $\kappa \leq dl$ (coarse levels). In the second case, RMT has almost full parallelism independently of the smoothing procedure. Distribution of the coarse grids of the first level to $p = 3^d$ (or $\kappa = d = 2, 3$) threads is shown on Fig. 3.

Remember the common measure of parallelism [4]:

Definition 1. *The efficiency* E *of a parallel algorithm is*

$$\mathrm{E} = \frac{1}{p} \frac{T(1)}{T(p)},$$

where $T(1)$ is an execution time for a single thread and $T(p)$ is an execution time using p threads.

Now we analyse parallel multigrid cycles. Since RMT has almost full parallelism on the coarse levels ($\kappa \geq dl$), it is possible to perform extra smoothing iterations on these levels. Figure 4 demonstrates a parallel multigrid cycle with two extra multigrid iterations on the coarse levels ($q^* = 3$). It is clear that parallel efficiency of the smoothing iterations of the finest grid will be critical for the parallel RMT efficiency.

In addition to abovementioned common measure of parallelism, we introduce measure of parallel properties of the smoothers:

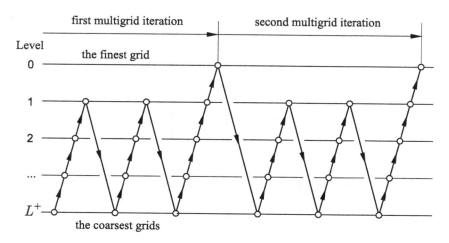

Fig. 4. Parallel multigrid cycle ($q^* = 3$).

Definition 2. *The efficiency* E_l *of a parallel smoother is*

$$E_l = \frac{1}{p}\frac{T_l(1)}{T_l(p)}, \tag{1}$$

where $T_l(1)$ *is an execution time for a single thread of the smoother,* $T_l(p)$ *is an execution time using p processors and* $l = 0, 1, \ldots, L^+$ *is serial number of the grid levels.*

Since all grids of the same level have the same number of points, it is expected the execution time for the smoothing iterations is *l*-independent: $T_l(1) = const.$ For example, the execution time of the sequential multigrid iteration of RMT becomes

$$T(1) = T_0(1) + q^* \sum_{l=1}^{L^+} T_l(1) = \left(1 + q^* L^+\right) T_0(1),$$

where q^* is the number of the multigrid iterations on coarse levels. Efficiency of the parallel multigrid cycle shown of Fig. 4 can be estimated as

$$E \approx \frac{q^* L^+ + 1}{q^* L^+ + \dfrac{1}{E_0}}.$$

This estimation predicts that $E > E_0$.

For this cycle, solution u of a boundary value problem should be represented as

$$u = c + c_d + \hat{u},$$

where c is the coarse grid correction (defined on the finest grid), c_d is the coarse grid correction (defined on the multigrid structures generated by the coarse

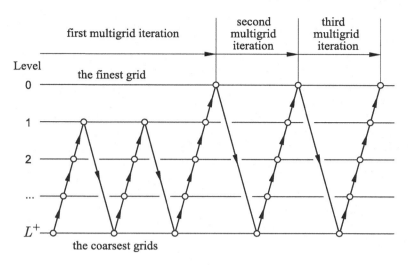

Fig. 5. Simplified parallel multigrid cycle ($q^* = 2$).

grids of the first level) and $\hat{u}$ is approximation to the solution u. This approach is difficult to use for nonlinear boundary value problems.

Simplified parallel multigrid cycle is shown on Fig. 5. This cycle makes it possible to apply standard Σ-modification of the solution used in RMT [2]. In this case, efficiency of parallel RMT depends on the number of the multigrid iterations q:

$$\mathrm{E}(+\infty) = \frac{L^+ + 1}{L^+ + \dfrac{1}{\mathrm{E}_0}} < \mathrm{E}(q) < \frac{q^* L^+ + 1}{q^* L^+ + \dfrac{1}{\mathrm{E}_0}} = \mathrm{E}(1).$$

To illustrate the simplified parallel multigrid cycle, we consider the model Dirichlet boundary value problem for the Poisson equation

$$u''_{xx} + u''_{yy} + u''_{zz} = -f(x, y, z) \tag{2}$$

in domain $\Omega = (0, 1)^3$. Substitution of the exact solution

$$u_a(x, y, z) = \exp(x + y + z), \tag{3}$$

into (2) gives the right-hand side function

$$f(x, y, z) = -3\exp(x + y + z)$$

and the boundary conditions.

Let us define error of the numerical solution as

$$\|e\|_\infty = \max_{ijk} |u_a(x_i, y_j, z_k) - \hat{u}^h_{ijk}|, \tag{4}$$

where $\hat{u}^h_{ijk}$ is approximation to the solution u.

As smoother, we use $3 \times 3 \times 3$ block Gauss–Seidel iterations (Vanka–type smoother [7]). This means that all 27 unknowns are updated collectively. As a rule, the Vanka–type smoother is used for solving systems of PDEs including saddle points problems. Of course, the discrete Poisson problem on an uniform grid does not require application of the Vanka–type smoother, since for the algebraic system that results from the seven-point discretization a point smoother is efficient, but this algorithm can be used for algorithmic complexity estimation in simulation of the coupled physicochemical phenomena.

Figure 6 represents reduction of the error (4) in the first multigrid iteration of the simplified parallel cycle (Fig. 5) starting the iterand zero. After four multigrid iterations steps on the multigrid structures generated by the coarse grids of the first level, error of approximation to the solution of (2) composed from solutions of 27 independent problems becomes small.

This error can be estimated as follows: for second-order discretization of (2), we have

$$\|u_a - u^h\| = Ch^2.$$

Second-order seven-point discretization of (2) on coarse grids of the first level with mesh size $3h$ results in

$$\|u_a - u^{3h}\| = C(3h)^2 = 9Ch^2.$$

Error estimation becomes

$$\|u^h - u^{3h}\| \leq \|u_a - u^h\| + \|u_a - u^{3h}\| = Ch^2 + 9Ch^2 = 10\,Ch^2.$$

i.e. difference between numerical solution and approximation to the solution is one significant digit.

Compared to the traditional single-grid Seidel method, RMT has a single extra problem-dependent component – the number of smoothing iterations. Numerical experiments are intended for studying efficiency of the parallel components of RMT as a function of the unknowns number (or the number N of grid points used for the boundary problem approximation).

3   Algebraic Parallelism of RMT

Efficiency of Gauss–Seidel method depends strongly on the ordering of equations and unknowns in many applications. Also, the possibility of parallel computing depends strongly on this ordering. The equations and unknowns are associated in a natural way with blocks of the unknowns. It suffices, therefore, to discuss orderings of the unknown blocks. Figure 7 represents three coloured block ordering in one dimension. Multicoloring allows the parallel execution of Gauss–Seidel relaxation. In d dimensions, such multicoloured block ordering defines the parallel block plane smoothing.

If the number of threads is less than the number of grids forming the given level, unknown blocks partitioning is a natural approach to algebraic parallelism

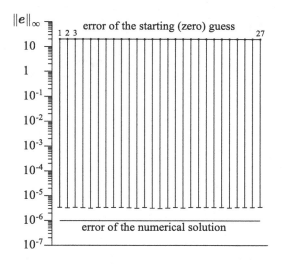

Fig. 6. Reduction of the error (4) in 27 independent subproblems (the first multigrid iteration).

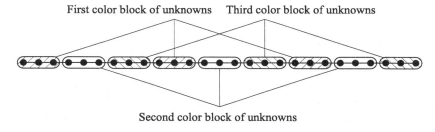

Fig. 7. 1D three coloured ordering of the unknown blocks.

of RMT. In this approach, the set of unknown blocks is split into $C3^\kappa$ ($\kappa = 1, 2, \ldots$) subsets, such that 3^κ available threads can jointly solve the underlying discrete problem. Here C is the number of colours in the used multicoloured ordering of the unknown blocks.

Personal computers (Intel(R) Core(TM) i7-4790 CPU@3.60 GHz) and computer cluster K-60 of Keldysh Institute of Applied Mathematics [9] (Russian Academy of Sciences) are used for the computational experiments for study of the parallel smoothing efficiency E_0 (1) on the finest grid.

Figure 8 represents results of the parallel computations. Reduction of the parallel RMT efficiency is observed in 27 thread implementation. Smoothing iterations on the finest grid is small-scale granulated component of RMT, but large algorithmic complexity of the Vanka-type smoother ($\sim N^3$ arithmetic operations, for each box, one has to solve a $N \times N$ system of equations to obtain corrections for the unknowns) leads to almost full parallelism.

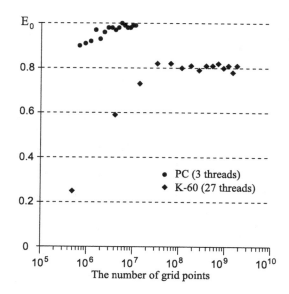

Fig. 8. Parallel smoothing efficiency E_0 (1) on the finest grid.

4   Geometric Parallelism of RMT

Remember that the problem of the very coarse grids leads to multigrid specific parallel complications which do not occur in classical single-grid algorithms. This crucial impact of the coarse grids increases, the more often the coarse grids are processed in each cycle. A parallel W-cycle, for example, has a substantially different parallel complexity from that of a parallel V-cycle [5].

An important advantage of the geometric approach to parallelization is the reduction of the discrete boundary value problem to the set of 3^κ independent problems, which can be solved in parallel without data exchange between processors for any used solver. Therefore one aim of parallel RMT is to perform as little work as possible on the finest grid and to do as much work as possible on the coarse levels. Extra multigrid iterations on the coarse levels lead to a considerably better parallel efficiency of RMT. Smoothing iterations on the coarse levels is large-scale granulated component of RMT.

Figure 9 represents results of the parallel solution of the model boundary value problem (2) on the coarse levels. We perform four multigrid iterations on the multigrid structures generated by the coarse grids of the first level. Also reduction of the parallel RMT efficiency is observed in 27 thread implementation. This test demonstrates a significant loss of efficiency for solving 27 independent subproblems. It means that the memory access pattern for computing those multigrid iterations need to be carefully designed, as their performance is very dependent on the memory bandwidth. Our study illustrates that memory bandwidth is a major bottleneck for multigrid. The role of memory and the deepening

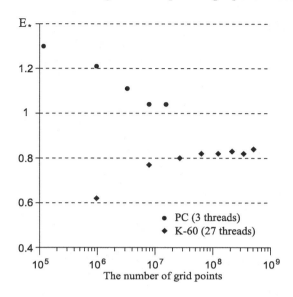

Fig. 9. Parallel smoothing efficiency on the coarse levels.

memory hierarchy on contemporary processors in the performance in numerical codes cannot be overstated.

Parallel RMT allows one to avoid a load imbalance and communication overhead on very coarse grids.

5   Parallel Multigrid Iteration

First parallel multigrid iteration of RMT shown on Fig. 5 consists of four multigrid iterations on the independent multigrid structures generated by the coarse grids of the first level (geometric parallelism of RMT) and parallel smoothing on the finest grid based on the multicoloured block Gauss–Seidel iterations (algebraic parallelism of RMT). Figure 10 represents efficiency of the first parallel multigrid iteration.

6   Remarks on Parallel Implementation

Inefficient memory access is one of the most common performance problems in parallel programs. The speed of loading data from memory traditionally lags behind the speed of their processor processing. The trend of placing more and more cores on a chip means that each core has a relatively narrower channel for accessing shared memory resources. On NUMA computers, accessing remote RAM is slower than accessing local memory. Therefore, to access the RAM of another socket it is necessary to access the QPI or HyperTransport bus, which is slower than the local RAM access bus. The program analysis performed by the

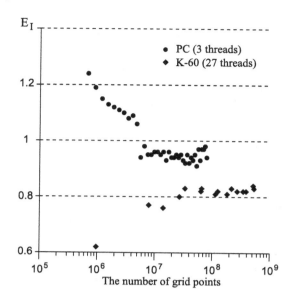

Fig. 10. Efficiency of the first parallel multigrid iteration.

Intel VTune Performance Analyzer shows that when 27 thread implementation using 2 Intel Xeon Gold 6142 v4 processors leads to 15–18% of memory accesses are accesses to remote memory. It results in reduction of the parallel program efficiency. This problem does not arise if all threads go to one processor.

Our future work is development of approaches to made memory-bandwidth efficient for parallel RMT.

7   Conclusions

The RMT has been developed for application in black-box software, because it has the least number of the problem-dependent components. This technique can solve many (non)linear problems to within truncation error at a cost of $CN \log N$ arithmetic operations, where N is the number of unknowns and C is a constant that depends on the problem. Large-scale granularity of the parallel RMT (geometric parallelism) coupled with the multicoloured Gauss-Seidel iterations (algebraic parallelism) lead to almost full parallelism of the multigrid iterations. The geometric approach, based on a decomposition of the given problem into a number of subproblems without an overlap, should be used to overcome the problems of large communication overhead and idling processors on the very coarse grids independent of the smoother. Results of numerical experiments on shared memory architectures show the high parallel efficiency of RMT components.

References

1. Frederickson, P.O., McBryan, O.A.: Parallel superconvergent multigrid. Multigrid Methods. In: McCormick, S. (ed.) Theory, Applications and Supercomputing, pp. 195–210. Marcel Dekker, New York (1988)
2. Martynenko, S.I.: The Robust Multigrid Technique: For Black-Box Software. De Gruyter, Berlin (2017)
3. Martynenko, S.I.: Sequential software for robust multigrid technique. Triumph, Moscow (2020). https://github.com/simartynenko/Robust_Multigrid_Technique_2020
4. Ortega, J.M.: Introduction to Parallel and Vector Solution of Linear Systems. Plenum, New York (1988)
5. Trottenberg, U., Oosterlee, C.W., Schüller, A.: Multigrid. Academic Press, London (2001)
6. Wesseling, P.: An Introduction to Multigrid Methods. Wiley, Chichester (1992)
7. Vanka, S.P.: Block-implicit multigrid solution of Navier-Stokes equations in primitive variables. J. Comput. Phys. 65(1), 138–158 (1986)
8. OpenMP Homepage. www.openmp.org
9. KIAM Homepage. www.kiam.ru/MVS/resourses/k60.html

Network Reliability Calculation
with Use of GPUs

Denis A. Migov[1]([✉]), Tatyana V. Snytnikova[1], Alexey S. Rodionov[1],
and Vyacheslav I. Kanevsky[2]

[1] Institute of Computational Mathematics and Mathematical Geophysics SB RAS,
prospect Akademika Lavrentjeva 6, 630090 Novosibirsk, Russia
`mdinka@rav.sscc.ru`, `snytnikovat@ssd.sscc.ru`, `rodionov@sscc.ru`
[2] LLC CityAir, Inzhenernaya st. 20, 630090 Novosibirsk, Russia

Abstract. The report discusses the problem of exact calculation and
evaluation of network reliability. We consider two common reliability
measures for networks with unreliable edges: all-terminal network reliability and diameter constrained network reliability. For each of them the
problem of its calculation is NP-hard. Parallel algorithms for calculating
and evaluating these characteristics with use of GPUs are proposed. The
results of numerical experiments are presented.

Keywords: Network reliability · Random graph · Diameter
constraint · Monte Carlo method · Associative algorithm · GPU

1 Introduction

The task of network reliability calculation is an important component of network
topology design and optimization. In practice, elements of a network can fail, as
in any other technical system. For networks for various purposes, such elements
are nodes and edges that fail due to breakage, wear, or other reasons. Typically,
such a system is represented by an undirected random graph, in which the vertices are network nodes, and the edges are network communication channels. We
study the case of unreliable edges, and the nodes are assumed to be perfectly
reliable. Thus, for each edge its presence probability is given.

The classic measure of network reliability $R(G)$ is the all-terminal network
reliability (ATR), i.e. the probability that any two nodes are connected by a
path of workable edges [1]. However, since in practice the number of transit
nodes always matters when establishing a connection [2], another measure of the
reliability $R_K(G, D)$ is also investigated: the network reliability with diameter
constraint (NRDC). NRDC is the probability that any two nodes from a given
set of terminals $K \subseteq V$ are connected by a path of workable edges and the
number of edges in each path is not greater than a given integer D [3,4]. The
problems of exact calculations of these measures are NP-hard [1,5].

To solve the problems of network reliability analysis, both sequential and
parallel methods are actively studied. There are parallel methods for exact reliability calculation [6,7] and evaluation [8–10]. Note that all of these methods were

© Springer Nature Switzerland AG 2021
V. Malyshkin (Ed.): PaCT 2021, LNCS 12942, pp. 210–219, 2021.
https://doi.org/10.1007/978-3-030-86359-3_16

designed to run on supercomputers with CPUs only. Recently, GPUs have also been used to analyze the reliability of systems (based on Markov chains) [11], but they have not yet been used to solve network reliability analysis problems. This paper introduces such methods for the first time.

We propose new methods designed to run on a GPU: a parallel matrix method for calculating NRDC and parallel methods for estimating and exact calculating ATR. Methods for calculating ATR are based on an associative algorithm for checking connectivity, which was developed for an abstract model of associative parallel processors (STAR-machine) [12]. However, for ATR calculation we need to check a large number of subgraphs. Therefore, a new modification for checking of a batch of subgraphs in parallel is designed with use of CUDA and cuRand library. Unlike the algorithm in [12], it does not use the implementation of the STAR machine on GPU.

An analysis of the numerical experiments, including experiments conducted on Siberian Supercomputer Center of SB RAS, shows the effectiveness and applicability of these methods. Due to the almost linear scalability of the reliability evaluation by Monte Carlo method, the proposed evaluation technique based on associative algorithms opens up new possibilities for solving such problems using high-performance hybrid clusters with GPUs.

The rest of the paper is organized as follows: basic notations and definitions are given in the Sect. 2, Sects. 3 and 4 describe the methods for calculation of ATR and NRDC respectively along with the results of numerical experiments. Section 5 is the brief conclusion.

2   Notations and Definitions

Let $G = (V, E)$ be an non-oriented graph; V is the set of vertices and E is the set of edges. For each edge e a real number p_e, $0 \leq p_e \leq 1$, is given. p_e value is the presence probability of edge e.

Discrete probabilistic space $W = (\Omega, P)$ is defines as follows. Here Ω is the space of elementary events (realizations), formed by various subgraphs of G, determined by the presence or absence of each edge. For the given elementary event, the present edges are named *operational* and the absent edges are referred to as *failing*. Space Ω may be represented as union of all binary vectors of length $|E|$, therefore, Ω consists of $2^{|E|}$ elements.

The probability of an elementary event $w \in \Omega$ is defined as the product of probabilities of the presence of operational edges, multiplied by the product of probabilities of the absence of failing edges:

$$P(w) = \prod_{e \in w_a} p_e \prod_{e \in w_b} (1 - p_e).$$

Here w_a is the set of operational edges, w_b is the set of failing edges of subgraph w.

All-terminal network reliability is defined by the following expression:

$$R(G) = \sum_{w \in \Omega} P(w) \cdot \mu(w), \qquad (1)$$

where μ is the connectivity indicator function:

$$\mu(w), = \left\{ \begin{array}{ll} 1, & \text{if } w \text{ is connected,} \\ 0, & \text{otherwise.} \end{array} \right.$$

Let us also assume that integer value D and subset K of nodes (terminals) are given. Network reliability with diameter constraint is used when each pair of terminals should be connected by path of length not greater than D (D-connected). Then, diameter constrained network reliability is defined as:

$$R_K(G, D) = \sum_{w \in \Omega} P(w) \cdot \chi(w), \tag{2}$$

where χ is the D-connectivity indicator function:

$$\chi(w), = \left\{ \begin{array}{ll} 1 & \text{if } w \text{ is } D\text{-connected,} \\ 0 & \text{otherwise.} \end{array} \right.$$

3   ATR Calculation Using GPU

For calculation of the all-terminal network reliability with unreliable edges, in accordance with the definition of this indicator, it is necessary to process all subgraphs (realizations) of the network graph. The number of such subgraphs grows exponentially with the number of edges of this graph. For each subgraph, the fact of its connectivity should be established. Then, the probabilities of connected realizations are summed up to obtain the network reliability value. For example, to calculate the reliability of the 4×4 grid network (16 vertices, 24 edges), it is necessary to check the connectivity of more than 2.5 million subgraphs.

Earlier [6,7], we study the possibility of supercomputers with distributed or shared memory using for exact reliability calculation. Here we carry on develop parallel methods, using GPU for faster reliability calculation. In our implementation, the subgraphs are processed in batches of 8196, which takes about 1.3 s for 4×4 grid using an Nvidia GeForce 920m graphics card. However, for the 5×5 grid network, calculation cannot be performed in a reasonable time.

Therefore, instead of the exhaustive enumeration of all subgraphs, we can generate a given number L of random subgraphs (taking into account the probabilities of the presence of edges) and, on its basis, estimate the reliability using the Monte Carlo method [10,13], which is equal to the ratio of the number of connected realizations to the number of all realizations. For ATR, the L value can be defined before the calculations in order to achieve given accuracy by the 3-sigma rule [10].

To check a random subgraph for connectivity, we represent the graph as list of edges. Since the subgraph connectivity check can be performed independently for each subgraph, each block of GPU is used to generate and to check subgraphs from a certain subset, whereas a thread corresponds to an edge. This approach allows us to minimize communications between CPU and GPU. Only the initial graph is sent from CPU to GPU.

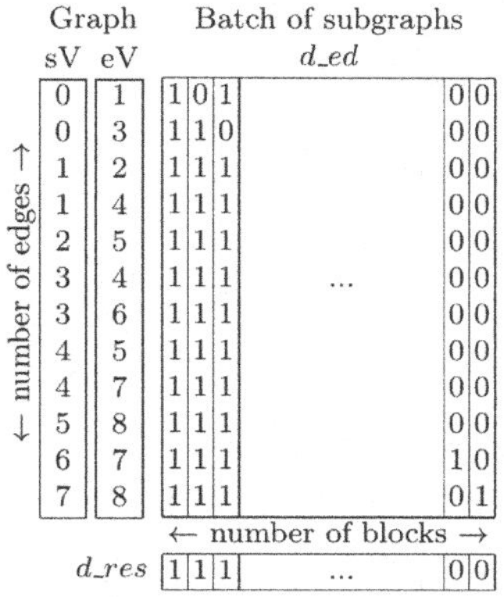

Fig. 1. Structure of data

The Fig. 1 shows the chosen structure for the data representation. Arrays sV and eV are the list of edges for 3×3 grid. Variable d_ed is an array of arrays. Its i-th element defines subgraph for checking by i-th block. The result of checking added to $d\_res[i]$.

At the first, subgraphs are generated using the cuRAND library. Then the connectivity of the each subgraph is checked. The technique used is similar to the associative parallel algorithm from [12]. Note that the each block i performs L_i generations of random subgraphs and checks their connectivity. The value of L_i depends on the initial L and the number of blocks that can be physically executed in parallel by a GPU.

The listing 1.1 shows the step of Dejkstra's algorithm for connectivity checking of a subgraph. Here the variable *iter* obtains the value only if any vertex is marked as passed on this step. Arrays $sV[M]$ and $eV[M]$ determine the original graph, which is the same for all blocks. Array $ed[M]$ specifies the processed subgraph ($ed = d\_ed[blockId.x]$). Array $d0[N]$ marks vertices passed in the previous step. Array $d1[N]$ marks vertices passing at the current step.

Listing 1.1. "Step of connectivity checking."

```
__device__ void dejkstra_step ( int it , int* iter , int *sV,
int *eV, Pointer ed, Pointer d0, Pointer d1)
 {
   int k=threadIdx.x;
   int st=sV[k];
   int en=eV[k];
   if ( ed[k]==1)
```

```
{
    if  (d0[st]<d0[en])
    {
        d1[st]=d0[en];
                *iter=it;
    }
    if  (d0[st]>d0[en])
    {
                d1[en]=d0[st];
                *iter=it;
    }
}
}
```

Figure 2 shows how the algorithm works for the connectivity checking of a subgraph of the 3 × 3 grid network. Initially, the vertex 0 is marked as passed (Fig. 2 (a)). Then the procedure dejkstra_step is called, until the variable *iter* becomes updated (Fig. 2 (b)-(f)). The subgraph is connected if all vertices are marked as passed.

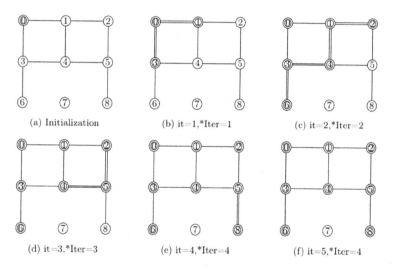

Fig. 2. An example of the connectivity checking of a subgraph

The result of the first stage is an array, where i-th element contains the number of connected subgraphs on the i-th block. Thereafter, the all elements are summed, and the obtained value is sent to CPU.

Numerical experiments were performed with use of PC with GeForce 920 m. Another series of experiments were conducted on the Kepler K40 computational cluster (Siberian Supercomputer Center of SB RAS). As the graph topology

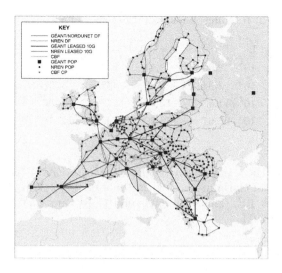

Fig. 3. Tested network

the GEANT network structure in 2009 (Fig. 3) was chosen in assumption that reliabilities of all edges are equal with each other. Let us denote this value by P. The dependence of the running time on the parameters P and L is presented in Tables 1–2 (in seconds). The difference in the computation time follows from the graphic accelerators characteristics.

Table 1. Running time of ATR calculation with use of GeForce 920M

L	P = 0.75	P = 0.9	P = 0.99	P = 0.999
10^5	22.85	23.00	23.16	23.18
10^6	27.18	29.1	30.65	30.77
10^7	69.6	88.5	103.65	104.81
10^8	493.72	680.85	833.82	843.92

Analysis of the experimental data shows that for a fixed number L of realizations, the dependence of reliability calculation time on P value is approximated by the function $t = a * P^{1.42}$. The main reason is not big difference for connectivity checking disconnected and connected subgraphs. The dependence of computational time on L for a fixed P is linear, taking into account an additive constant, which is determined by the number of edges of a graph. When calculating on the GeForce 920M card, the additive constant is approximately equal 22 s, while on Kepler K40 computational cluster it is an order of magnitude lower (2 s) for the same graph.

Table 2. Running time of ATR calculation with use of K40

L	P = 0.75	P = 0.9	P = 0.99	P = 0.999
10^5	2.09	2.11	2.16	2.16
10^6	2.94	3.39	3.70	3.73
10^7	11.38	15.69	18.93	19.09
10^8	95.67	138.69	170.83	172.70

Table 3 shows the comparison of the time of GEANT network reliability estimation by Monte Carlo method with use of GPU (GeForce 920M) and CPU in the case of $L = 10^7$. The CPU has Intel Core Duo 2.4 GHz processor; the GPU is GeForce 920M.

Table 3. Running time of ATR calculation with use of GPU and CPU

Edge reliability value	P = 0.75	P = 0.9	P = 0.99	P = 0.999
Calculation time with use of CPU	45	72	314	402
Calculation time with use of GPU	70	89	104	105

Another experiments were performed for ATR evaluation of linear wireless sensor networks. Despite their linear physical structure, the corresponding longitudinal graph may be not linear due to links between not only neighbor nodes [14]. Such graphs can be not series-parallel, so the reliability evaluation methods are in demand for the linear wireless sensor networks analysis. Numerical experiments (Table 4) were conducted on graphs obtained by the recursive procedure of joining to an already formed longitudinal graph of a complete graph on 4 vertices K_4, provided that two vertices of these graphs coincide. K_4 is also taken as an initial graph. The longitudinal graph obtained by this procedure of union of $k - 1$ graphs K_4 we denote by K_4^k (Fig. 4). Calculated reliability values with an accuracy of 5 decimal places for K_4^{100} were 0.98208 and 0.45341 for $P = 0.9$ and $P = 0.75$ respectively. For K_4^{100}, evaluated reliability was 0.989024 in case of $P = 0.9$ and 0.65684 in case of $P = 0.75$. Use crude evaluation [10] in case of $L = 10^8$, an error for the calculated values is 0.00015. The dependencies of the calculation time on P and L differ from similar dependencies for the GEANT network. This can be explained by different graph densities and different proportions of connected implementations.

The analysis of execution shows the following. On the one hand, the processing time for a connected subgraph is slightly longer than for an disconnected one. On the other hand, the proportion of connected subgraphs among the generated subgraphs significantly decreases with P decreasing. Obviously, the runtime linearly depends on L, taking into account the additive constant. As the results show, the main advantage of using GPU in comparison with CPU for network

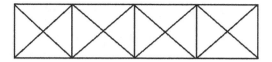

Fig. 4. Graph K_4^5

Table 4. Running time of K_4^k ATR calculation with use of GeForce 920M

L	P = 0.75 (K_4^{50})	P = 0.9 (K_4^{50})	P = 0.75 (K_4^{100})	P = 0.9 (K_4^{100})
10^5	1.33	1.41	24.27	24.91
10^6	7.43	8.19	44.69	50.52
10^7	66.48	73.86	243.62	301.87
10^8	656.8	729.94	2232.26	2810.87

reliability evaluation is the weak dependence between runtime and an edge reliability value. This advantage is especially noticeable for networks with highly reliable edges. Note that such networks are common in various real-world applications.

4 Calculating NRDC Using GPU

The next proposed method is based on the usage of GPU and CUDA for fast NRDC calculation. It is well-known, that if T is an adjacency matrix of a graph G and $W = T^D$, then $W[i,j] \neq 0$ if and only if nodes i and j are connected by some path of the length D (some edges may be used more than once). Thus, summarizing all $W_d = T^d$ for $d = 1, \ldots, D$ we obtain matrix W_D in which non-zero element $W_D[i,j]$ shows possibility of reaching node j by not more than D steps starting from the node i. If all $W_D[i,j]$ are non-zero, then the diameter of G is equal or lesser than D. If all $W_D[i,j]$, $i, j, \in K$ are non-zero, then G is reliable from point of view of NRDC (D-connected).

As it was mentioned above (2), we can obtain NRDC value of by summarizing probabilities of all D-connected graph realizations, the number of which is $2^{|E|}$. Thus, for implementing this approach we need to generate all possible realizations of G which number is very large. For example, lattice 4×4 has 16 nodes and 24 edges, thus it has $2^{24} = 16777216$ possible realizations. Unlike the approach above for ATR calculation, the realizations are prepared in CPU and loaded by parts according operative memory of GPU.

With help of GPU, we obtain for each realization all matrices $W_d = V^d$ for $d = 1, \ldots, D$, summarizing them simultaneously. Obtained matrix is sent from GPU to CPU, where it checked for D-connectivity. If a matrix, corresponding to realization under consideration, is D-connected, then its probability is calculated as production of p_e for existent edges e and $1 - p_e$ for failed ones, and added to

NRDC value, which is initially equal to 0. Thus, for NRDC calculation we use GPU just for matrices multiplication.

Experiments show that this method is effective only in case of large D values and $|K|$ values near to $|V|$. Otherwise, the sequential factoring method [1] on CPU is more effective. For example, $R_V(G, 6)$ and $R_V(G, 23)$ calculation for the mentioned lattice 4×4 take 56.0 and 56.5 s on GPU, correspondingly. Calculation time with use of CPU was 1.5 s and 24.5 min, correspondingly. It should be taken into account that the calculation on the CPU is carried out by the factorization method, which is much faster than implementation of the enumeration. Therefore, time of computation by CPU grows significantly with the diameter value increasing. The same can be said about the dependence of time on the number of terminals. At the same time, computational time with use of GPU is practically independent of the diameter value or the number of terminals. The PC with CPU intel i7 4790, 8 GB RAM, and GPU GeForce 780 Ti 3 GB was used for experiments. It seems advisable to use this approach on hybrid computers (CPUs+GPUs), which allows us significantly speed up the calculation compared to using only CPUs.

5   Conclusion

Parallel algorithms for networks reliability calculation with use of GPUs are presented. Algorithms are based on the enumeration of all graph implementations and on the analysis of a sample of graph implementations, i.e. on the Monte Carlo simulation. The complexity of the first approach is exponential. With the help of the second approach, we carry out the calculation with a given accuracy, having previously determined the required number of implementations. The main advantage of such approach is not so strong dependence between runtime and an edge reliability as for Monte Carlo network reliability evaluation with use of CPU. For diameter constrained network reliability, GPU is used for fast exponentiation of the adjacency matrix of a graph. This approach has proved quite useful for large diameter values and large proportion of terminals.

Acknowledgment. The reported study was funded by RFBR and NSFC, project number 21-57-53011, and by the state contract with ICM&MG SB RAS, project number 0251-2021-0005.

References

1. Ball, M.O.: Computational complexity of network reliability analysis: an overview. IEEE Trans. Reliab. **35**, 230–239 (1986)
2. Pandurangan, G., Raghavan, P., Upfal, E.: Building low-diameter peer-to-peer networks. IEEE J. Sel. Areas Commun. (JSAC) **21**(6), 995–1002 (2003)
3. Petingi, L., Rodriguez, J.: Reliability of networks with delay constraints. Congr. Numer. **152**, 117–123 (2001)
4. Jin, R., Liu, L., Ding, B., Wang, H.: Distance-constraint reachability computation in uncertain graphs. VLDB Endow. **4**, 551–562 (2011)

5. Canale, E., Cancela, H., Robledo, F., Romero, P., Sartor, P.: Full complexity analysis of the diameter-constrained reliability. Int. Trans. Op. Res. **22**(5), 811–821 (2015)
6. Migov, D.A., Rodionov, A.S.: Parallel implementation of the factoring method for network reliability calculation. In: Murgante, B., et al. (eds.) ICCSA 2014. LNCS, vol. 8584, pp. 654–664. Springer, Cham (2014). https://doi.org/10.1007/978-3-319-09153-2_49
7. Nesterov, S.N., Migov, D.A.: Parallel calculation of diameter constrained network reliability. In: Malyshkin, V. (ed.) PaCT 2017. LNCS, vol. 10421, pp. 473–479. Springer, Cham (2017). https://doi.org/10.1007/978-3-319-62932-2_45
8. Khadiri, M.E., Marie, R., Rubino, G.: Parallel estimation of 2-terminal network reliability by a crude Monte Carlo technique. In: Proceedings of the 6th International Symposium on Computer and Information Sciences, pp. 559–570. Elsevier (1991)
9. Martinez, S.P., Calvino, B.O., Rocco, S.C.M.: All-terminal reliability evaluation through a Monte Carlo simulation based on an MPI implementation. In: Proceedings of the European Safety and Reliability Conference: Advances in Safety, Reliability and Risk Management (PSAM 2011/ESREL 2012), pp. 1–6. Helsinki (2012)
10. Migov, D.A., Weins, D.V.: Parallel implementation and simulation of network reliability calculation by Monte Carlo method. Vestnik Tomskogo gosudarstvennogo universiteta. Upravlenie vychislitelnaja tehnika i informatika [Tomsk State Univ. J. Control Comput. Sci.] **47**, 66–74 (2018). (in Russian). https://doi.org/10.17223/19988605/47/8
11. de Freitas, I.M.A.T.: Monte Carlo Parallel Implementation for Reliability Assessment. Master Thesis. Porto University, p. 80 (2019)
12. Nepomniaschaya, A.S., Snytnikova, T.V.: Associative parallel algorithm for dynamic update of shortest paths tree after inserting an arc. Appl. Discret. Math. **46**, 58–71 (2019). https://doi.org/10.17223/20710410/46/5
13. Cancela, H., Robledo, F., Franco, R., Rubino, G., Sartor, P.: Monte Carlo estimation of diameter-constrained network reliability conditioned by pathsets and cutsets. Comput. Commun. **36**, 611–620 (2013)
14. Mohamed, N., Al-Jaroodi, J., Jawhar, I., Lazarova-Molnar, S.: Failure impact on coverage in linear wireless sensor networks. In: Proceedings of the SPECTS, 2013, pp. 188–195 (2013)

Memory-Efficient Data Structures

Implicit Data Layout Optimization for Portable Parallel Programming in C++

Vladyslav Kucher$^{(\boxtimes)}$ and Sergei Gorlatch

University of Muenster, Muenster, Germany
{kucher,gorlatch}@uni-muenster.de

Abstract. The programming process for modern parallel processors including multi-core CPUs and many-core GPUs (Graphics Processing Units) represents a significant challenge for application developers. We propose to use the widely-popular programming language C++ for parallel programming in a portable way, allowing the same program to be run on different target architectures. In this paper we extend our framework PACXX (Programming Accelerators in C++) with an additional compilation pass which simplifies data management for the programmer and makes the programming process less error-prone. These changes result in a significant reduction of execution stalls caused by memory throttling. We describe the implementation of the new data layout optimization and we report experimental results that confirm the advantages of our approach.

Keywords: GPU programming · High-performance computing · Memory access optimization · Unified parallel programming · C++

1 Introduction

Many current high-performance applications exploit modern parallel processor architectures like multi-core processors (CPUs) and so-called *accelerators*, most notably Graphics Processing Units (GPU). Computers with accelerators offer the user a higher computation power due to many cores working in parallel, but at the same time they pose new challenges with respect to their programming. One of the major difficulties in programming such systems is that two distinct programming models are required: 1) the host code for the CPU is usually written in C/C++ with a restricted, C-like API for memory management, 2) the device code for the accelerator has to be written using a device-dependent, explicitly parallel programming model, e.g., OpenCL (Open Computing Language) [9] or CUDA (Compute Unified Device Architecture) [17].

This makes the programming process complicated and error-prone and requires expert-level knowledge from the application developer.

We are developing PACXX (Programming Accelerators in C++) [7] - a novel framework that simplifies accelerator program development by allowing application developers to write programs in a widely used language, C++, with all

© Springer Nature Switzerland AG 2021
V. Malyshkin (Ed.): PaCT 2021, LNCS 12942, pp. 223–234, 2021.
https://doi.org/10.1007/978-3-030-86359-3_17

the comfortable features of the modern C++14/17 standards [3]. A mechanism of implicit memory management [12] in PACXX shortens the program code and makes the programming process less error-prone. Implicit memory management in the framework is implemented by a custom compilation pass which generates additional host code using the Clang front-end and LLVM intermediate representation (IR) [14] for data transfers between host and accelerator, transparently for the developer. However, data restructuring is still required to avoid performance problems due to memory throttling, that are caused by the limited memory transfer bandwidth of an accelerator. Especially in applications with complicated memory access patterns, this remains a significant problem which requires a manual data layout optimization.

In this paper, we extend the PACXX framework to further improve the state of the art in programming parallel systems with accelerators. Our main contribution is introducing and implementing in PACXX an implicit, i.e., automatic optimization of the data layout on the accelerator, thereby removing a major cause for execution stalls (memory throttling) and liberating the programmer from the need for manual data restructuring.

The remainder of this paper is structured as follows. In Sect. 2, we compare the existing PACXX framework to the two most popular approaches, CUDA and OpenCL, and also to related work on using C++ for parallel programming. Section 3 describes our idea of automatic data layout optimization in PACXX. In Sect. 4 we explain the design choices we make to implement this extension of the PACXX framework. Our examples and experiments throughout the paper demonstrate that our approach helps the programmer to avoid performance issues caused by memory throttling in applications with complex memory access patterns.

2   Programming for Accelerators: PACXX vs. State of the Art

Currently, OpenCL and CUDA are the most popular approaches to parallel programming for high-performance systems with accelerators. They combine two different programming models: restricted C/C++ with the corresponding OpenCL- or CUDA-specific features. This makes several programming aspects significantly more complex and tedious as compared to the modern capabilities of C++ programming for traditional computers.

In particular, memory for static and dynamic arrays must be allocated explicitly and this has to be done twice, first in the host memory and then again in the device memory. In addition, the restricted, C-like API for memory management implies significantly longer boilerplate code as compared to the modern C++. Furthermore, the developer is responsible for performing synchronization between the host and device memories and for the proper error handling, because both CUDA and OpenCL provide C-like API error codes that must be checked for each function call. This results in unnecessarily long, complicated, and error-prone program code.

Figure 1 shows a simple example code in C++ for performing the addition of two vectors; it requires only 7 Lines of Code (LOCs) in standard C++.

```
1 int main() {
2   vector<int> a(1024), b(1024), c(1024);
3   auto vadd = [&]{
4     for (size_t i = 0; i < a.size(); ++i)
5       c[i] = a[i] + b[i];
6   };
7 }
```

Fig. 1. Simple code example: Vector addition in plain C++

The simple vector addition example shown in Fig. 1 was also programmed in the tutorials on CUDA and OpenCL by the developers of these approaches, i.e., by the experts in the field. The results are quite disillusioning as the new programs are an order of magnitude longer: the CUDA version [6] requires 70+ LoCs, while OpenCL requires 110+ LoCs for this very simple example [18]. There have been efforts to extend the use of C++ for accelerator programming:

1. The C++ AMP approach [16] extends standard C++ by an explicit data-parallel construct, and so-called array_views provide functions for memory transfers. However, the developer still needs to write more than one line of code for each allocation and has to use the C++ AMP views instead of the original C++ data types if synchronization is to be handled transparently.
2. VectorPU [15] follows a similar approach by providing special classes (so-called smart containers) that handle the memory transfers, but still forces the developer to use smart-containers instead of the original C++ data types.
3. SYCL [10] is similar to VectorPU, but with altered naming convention: buffers and accessors instead of smart containers. The supported accelerator types vary based on the implementation.
4. NVIDIA Thrust [8] and AMD Bolt [2] are libraries implementing the functionality of the C++ Standard Template Library (STL) in a data-parallel way, but they are restricted to GPUs from the corresponding vendor and do not support modern C++ language features.
5. Annotation-based approaches like OpenMP [5] and OpenACC [1] expect the user to use parallelism-oriented directives in addition to C++.
6. STAPL [4] offers STL functionality which is executed in parallel by an underlying runtime system; it targets distributed-memory systems with MPI, OpenMP and PThreads, rather than GPUs.

We develop PACXX [7] - a framework for programming systems with accelerators, which offers a unified programming model based on the newest C++ standard, without any language extensions and uniformly covers both host (CPU) and accelerator (GPU) programming. It is implemented by a custom compiler

```
1  int main() {
2    auto& exec = Executor::get();
3    vector<int> a(1024), b(1024), c(1024);
4    auto vadd = [=, &c](auto &config){
5      auto i = config.get_global(0);
6      if ( i >= a.size()) return;
7      c[i] = a[i] + b[i];
8    };
9    KernelConfiguration config({1, 1, 1}, {a.size(), 1, 1});
10   exec.launch(vadd, config);
11 }
```

Fig. 2. Vector addition expressed in PACXX with our previous additions [12]

which generates accelerator code using the Clang front-end and LLVM IR [14] and modifies the host code for kernel execution transparently for the developer.

Using our previous enhancement of PACXX [12], the developer can transform the sequential, CPU-only C++ program in Fig. 1 into a portable parallel application shown in Fig. 2: it is capable of running on accelerators of different vendors with significantly fewer changes and additions (4 LOCs) when compared to CUDA (70+ LOCs) [6] or OpenCL (110+ LOCs) [18].

The new PACXX code in Fig. 2 requires only a modest additional programming effort compared to CUDA or OpenCL:

- The auto type is used throughout the program to rely on type inference of the compiler instead of specifying the complete type name by hand.
- The accelerator initialization is handled by PACXX internally. The provided Executor class allows such simplified initialization using the get() method that automatically detects available accelerator and prepares the appropriate back-end for it (line 2).
- The returned Executor object gives access to kernel management (line 10).
- The kernel is represented as the original lambda function from Fig. 1 with an additional argument generated by the PACXX runtime, which provides access to common execution thread identifying information, such as global thread id (line 5); the capture type of the kernel lambda defines the data transfer behaviour.
- To launch a kernel (line 10), the developer specifies the degree of parallelism in up to three dimensions (line 9).

Figure 3 shows PACXX code for a more complicated example of matrix multiplication which has a comparatively complex memory access pattern. The code demonstrates that, despite our previous improvements [12], difficulties remain when writing PACXX code for parallel systems with accelerators. An inherent problem in this code is that no matter how one selects the thread indices (lines 6–7), the memory addresses in line 10 will be out of order for neighboring threads.

```
 1 int main() {
 2   auto& exec = Executor::get(0);
 3
 4   double a[1024][1024], b[1024][1024], c[1024][1024];
 5   auto matmul = [=, &c](auto &config) {
 6     auto row = config.get_global(0);
 7     auto column = config.get_global(1);
 8     double val = 0.0;
 9     for (unsigned i = 0; i < width; ++i)
10       val += a[row * width + i] * b[i * width + column];
11     c[row * width + column] = val;
12   };
13
14   KernelConfiguration config({width / threads, width},
                                {threads, 1}, 0);
15   exec.launch(matmul, config);
16   exec.synchronize();
17 }
```

Fig. 3. Matrix multiplication expressed in PACXX with our previous additions [12]

Figure 4 shows the profiling results measured on an Intel Xeon E5-1620 v2 with a NVIDIA Tesla K20c on the GPU backend for the code of Fig. 3 using our profiling interface [11] that in this case relies on the CUDA Profiling Toolkit CUPTI: they demonstrate that in applications with increasingly complicated memory access patterns the execution stalls caused by memory throttling (i.e., memory access delays) also increase correspondingly (up to 72% in this example), becoming a significant factor that damages the overall application performance.

In the following, we present our automatic approach to avoid the execution stalls caused by memory throttling. We also aim at further reducing the required additional LoCs in complex applications that employ application frameworks like DUNE [13]. Altogether, our approach improves both the programmability and the performance of applications with complicated memory access patterns.

3   Data Layout Optimization: From Manual to Automatic

The currently used, low-level parallel programming approaches like OpenCL and CUDA offer only one way to transfer data between the host and the accelerator, namely a memcpy-like interface, which corresponds to what happens at the hardware level. The developer can manage (allocate, transfer, etc.) only a single continuous block of raw memory at a time. The previous additions to PACXX [12] simplify the management of complex data structures by replicating the data layout between host and accelerator.

In this section, we further improve the PACXX framework by making the data layout optimization on the accelerator fully automatic. The data on the host is usually laid out based on its meaning (for example, as shown in the matrix multiplication, line 4 in Fig. 3):

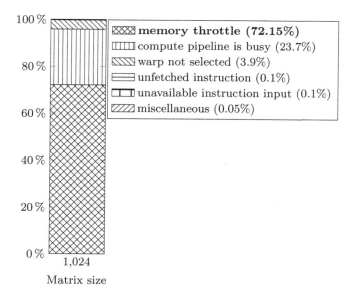

Fig. 4. Execution stalls distribution for the code in Fig. 3.

- The member fields of an object are grouped together in a structure placed in a continuous block of memory.
- Multiple structures of the same type are placed one after another in an array.
- All arrays together form the data structure of the program.

Such data layout as a structure of arrays of structures is useful for preserving the meaning of the program. However, to reduce the memory-related execution stalls during parallel data processing, we propose the following organization of the application data:

- All processing threads in a work group are executing in lock step, so loading the same member field of each object has to be completed before any of the threads resume execution. Since threads in a work group share a single local cache block, the requested data for the whole work group has to be located in a continuous block of memory. Therefore we arrange the member fields of all objects required for the work group one after another in an array.
- Multiple work groups execute independently and have private cache memory. To further increase the efficiency of each local cache block we arrange the required arrays one after another in a single structure - a data set for each work group.
- All work group data sets can be arranged one after another in an array to allow an arbitrary amount of work group processing tasks.

This proposed organization of the application data will be further discussed in Sect. 4 (Fig. 7, Fig. 8). Summarizing, our idea is to use an array of structures of arrays of application data in order to improve the efficiency of the memory controllers available on the accelerators.

Figure 5 shows an example of the additional data layout manipulations required by applications with complicated memory access patterns like the matrix multiplication in Fig. 3. By employing the new mechanism described in Sect. 4, such additional data layout manipulations are now performed automatically by the PACXX framework, thus reducing the amount of required additional LoCs and making the applications less error-prone.

```
 1 :
 2 a[1024][1024] , b[1024][1024] , c[1024][1024];(Fig. 3 line 4)
 3 a = transpose a
 4 segmented_a = a split into segments of length threads
 5 segmented_b = b split into segments of length threads
 6 data_set[2][1024][1024]
 7 foreach a_segment in a and b_segment in b:
 8     data_set.append(a_segment)
 9     data_set.append(b_segment)
10 auto matmul = [=data_set, &c](auto &config) {(Fig. 3 line 5)
11 :
12 segmented_results[1024*1024/threads/threads][threads][threads]
13 c = regroup segmented_results
```

Fig. 5. The pseudocode for the additional data layout manipulations in code of Fig. 3 for avoiding memory access stalls.

4 Implicit Data Layout: Design and Implementation

Figure 6 shows how we extend the PACXX framework by introducing an implicit data layout optimization, where parts of the figure are labelled with the numbers of steps as explained below:

1. The PACXX compiler performs the following additional steps during compilation as compared to the usual Clang front-end [7]:
   (a) Code is generated for data transfers between host and accelerator.
   (b) Data transfer code is adjusted for optimized data layout on the accelerator (grey box, Fig. 6).
   (c) LLVM Intermediate Representation (IR) for the kernels is generated.
   (d) During compilation for the host, the kernel IR is integrated into the final executable.
   (e) Calls to the PACXX runtime, additionally linked to the executable, are generated for memory and accelerator management.

2. The kernel IR code is compiled by the online compiler (PACXX runtime) for the available accelerator.
3. The PACXX runtime automatically performs all necessary steps to launch kernels.
4. Different back-ends allow to create programs for diverse hardware (e.g., AMD/NVIDIA GPUs).

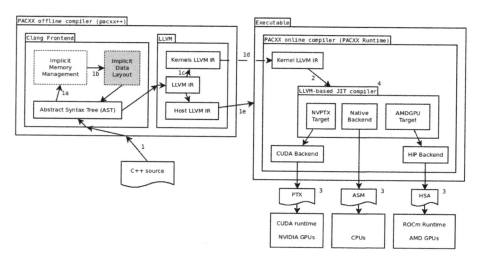

Fig. 6. The PACXX framework with our extension (grey box)

In our implementation of PACXX, performing additional data manipulations is best done as a Clang AST (Abstract Syntax Tree) pass, because it retains all the original data type information [12]. The new Clang AST Pass implementing the data layout optimization is called by the modified Clang front-end (1b above) after the implicit memory management pass and before exporting the application code to LLVM IR. This pass analyzes the provided AST and inserts new AST nodes or replaces existing AST nodes containing the memory transfer operations.

The improved PACXX framework described in this paper initially analyzes all memory load/store instructions in the kernel code. Any memory load/store instruction with addresses that are constant or depend on application data value is left as is. All remaining memory load/store instructions, with processing thread index dependent addresses, are considered for optimizing of the memory controller usage according to Sect. 3: for each referenced array element in these memory load/store instructions, if the array element's size is smaller than the distance between the addresses of two neighboring processing threads, then the memory page accessed by the memory controller contains unused data.

Figure 7 and Fig. 8 show that in order to improve the efficiency of the memory load/store instructions, the originally generated data transfer code [12] (Fig. 7a,

Fig. 8a) is split and rearranged using looped multiple per-element memory operations (Fig. 7b, Fig. 8b). A more detailed explanation of the introduced modifications is as follows:

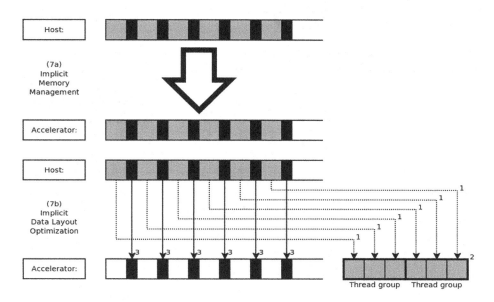

Fig. 7. Data transfer upload replacement (Numbers refer to the numbered steps described under "From host to accelerator")

- From host to accelerator (Fig. 7):
  1. All elements of the referenced array are transferred into multiple array segments of a parameterized length, which is assigned with the size of the thread group of the hardware at runtime (grey boxes in Fig. 7b).
  2. The memory address in the kernel's memory load/store instruction for the referenced array element is replaced with the new relocated address, thereby improving the efficiency of the memory page access.
  3. The transfer code for the remaining data is replaced with individual transfer operations for the data segments around the processed array elements. This preserves the expected addresses of objects which have not yet been analyzed or will not be analyzed due to having constant or application data dependent addresses (black boxes in Fig. 7b) and avoids transferring duplicate data.
- From accelerator to host for variables captured by reference by the kernel lambda (Fig. 8):
  1. All elements of the referenced array are individually transferred from multiple array segments of a parameterized length, which is assigned with the size of the thread group of the hardware at runtime, into the original memory addresses (grey boxes in Fig. 8b).

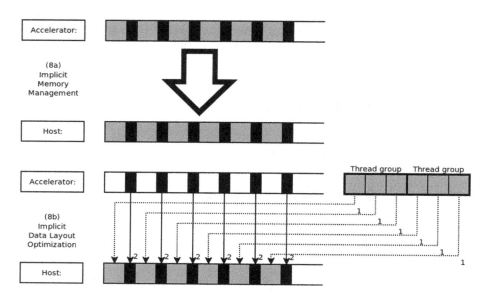

Fig. 8. Data transfer download replacement (Numbers refer to the numbered steps described under "From accelerator to host")

2. The transfer code for the remaining data is replaced with individual transfer operations for the data segments around the processed array elements. This preserves the expected addresses of objects which have not been analyzed yet or will not be analyzed due to having constant or application data dependent addresses (black boxes in Fig. 8b) and avoids transferring duplicate data.

After analyzing and adjusting the relevant memory load/store instructions, if the kernel code contains no memory load/store instructions with addresses that depend on application data value, the generated data transfer operations are analyzed. If any transferred objects of the original data structure are not referenced by the kernel code, they can be exempted from the data transfer to avoid placing unused objects in the accelerator's memory. Such exemption is done by replacing the existing memory transfer operation with individual memory transfer operations for the data segments around the exempted objects.

Finally, the generated segmented memory operations are checked for zero length segments, which are removed from the generated data transfer code.

Figure 9 shows the effect of implementing the data layout changes described above. We observe a substantial performance improvement for the code example in Fig. 3 by avoiding execution stalls caused by memory throttling. It can be seen that the throttling effect has been significantly reduced (from 72% to 0,1%), effectively removing execution stalls caused by memory throttling.

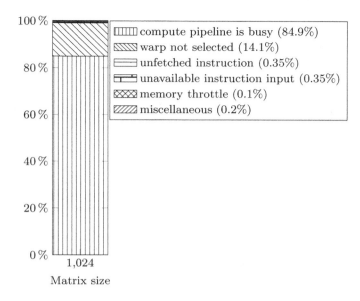

Fig. 9. Execution stalls distribution for Fig. 3, profiled after applying the new data layout optimization approach introduced in this paper

Summarizing, due to our improvements the PACXX framework can fully automatically address the following two aspects:

- The layout of the data required by the kernel is automatically adjusted on the accelerator.
- Memory no longer has to be duplicated and rearranged explicitly, but instead its layout is automatically optimized.

5   Conclusion

This paper describes the design and implementation of data layout optimization that significantly reduces the code size required to program accelerators in a portable fashion in C++, while simultaneously improving the program's performance by eliminating memory access induced stalls, which are typical for manual data layout approaches. Our extensive profiling results demonstrate that our automatic optimization of the data layout removes the major cause of execution stalling due to memory access delays.

The introduced changes to the PACXX framework relieve the developer of the necessity to optimize the accelerator's memory manually in applications with complicated memory access patterns such as matrix multiplication and real-world application frameworks like DUNE [13].

References

1. The OpenACC Application Programming Interface (2013). openacc-standard.org, version 2.0a
2. Bolt C++ Template Library, version 1.2 (2014)
3. Programming Languages - C++ (Committee Draft) (2014). isocpp.org
4. An, P., et al.: STAPL: an adaptive, generic parallel C++ library. In: Dietz, H.G. (ed.) LCPC 2001. LNCS, vol. 2624, pp. 193–208. Springer, Heidelberg (2003). https://doi.org/10.1007/3-540-35767-X_13
5. Beyer, J.C., Stotzer, E.J., Hart, A., de Supinski, B.R.: OpenMP for accelerators. In: Chapman, B.M., Gropp, W.D., Kumaran, K., Müller, M.S. (eds.) IWOMP 2011. LNCS, vol. 6665, pp. 108–121. Springer, Heidelberg (2011). https://doi.org/10.1007/978-3-642-21487-5_9
6. CUDA Vector addition example (2019). https://github.com/olcf/vector_addition_tutorials/tree/master/CUDA
7. Haidl, M., Gorlatch, S.: PACXX: towards a unified programming model for programming accelerators using C++14. In: 2014 LLVM Compiler Infrastructure in HPC, pp. 1–11, November 2014. https://doi.org/10.1109/LLVM-HPC.2014.9
8. Hoberock, J., Bell, N.: Thrust: A Parallel Template Library, version 1.6 (2014)
9. Khronos OpenCL Working Group: The OpenCL Specification, version 1.2 (2012)
10. Khronos SYCL Working Group: The SYCL Specification, version 2020 (2021)
11. Kucher, V., Fey, F., Gorlatch, S.: Unified cross-platform profiling of parallel C++ applications. In: 2018 IEEE/ACM Performance Modeling, Benchmarking and Simulation of High Performance Computer Systems (PMBS), pp. 57–62 (2018)
12. Kucher, V., Gorlatch, S.: Towards implicit memory management for portable parallel programming in C++. In: Proceedings of the 2020 ASSE, pp. 52–56. ACM, New York (2020). https://doi.org/10.1145/3399871.3399881
13. Kucher, V., Hunloh, J., Gorlatch, S.: Toward performance-portable finite element methods on high-performance systems. In: 2019 SigTelCom, pp. 69–73, March 2019. https://doi.org/10.1109/SIGTELCOM.2019.8696146
14. Lattner, C.: LLVM and Clang: next generation compiler technology. In: The BSD Conference, pp. 1–2 (2008)
15. Li, L., Kessler, C.: VectorPU: a generic and efficient data-container and component model for transparent data transfer on GPU-based heterogeneous systems. PARMA-DITAM 2017, pp. 7–12. ACM, New York (2017). https://doi.org/10.1145/3029580.3029582
16. Microsoft: C++ AMP: Language and Programming Model, version 1.0 (2012)
17. Nvidia: CUDA C Programming Guide, version 6.5 (2014)
18. OpenCL Vector addition example (2019). https://github.com/olcf/vector_addition_tutorials/tree/master/OpenCL

On Defragmentation Algorithms for GPU-Native Octree-Based AMR Grids

Pavel Pavlukhin[1,2]([✉]) and Igor Menshov[1]

[1] Keldysh Institute of Applied Mathematics, Moscow 125047, Russia
{pavelpavlukhin,menshov}@kiam.ru
[2] Research and Development Institute "Kvant", Moscow 125438, Russia

Abstract. The GPU-native CFD framework with dynamical adaptive mesh refinement (AMR) requires periodical execution of memory compaction operations to relieve memory expenses. The present paper addresses several different parallel GPU memory defragmentation algorithms for octree-based AMR grids. These algorithms are tested on benchmark CFD problems typical for AMR transformation. The results show that the memory defragmentation algorithm based on the prefix scan procedure is not only 1–2 order faster compared to algorithm based on space filling curve (z-curve) but also surprisingly and dramatically impacts on CFD solver performance by reducing total GPU runtime for some problems up to 37%.

Keywords: AMR · CUDA · CFD · Octree

1 Introduction

The CFD framework we develop for the last few years was initially based on three-dimensional Cartesian type grids. It efficiently performed on supercomputers equipped with hundreds of GPUs due to a simple static domain decomposition [1]. However structured Cartesian grids have very limited flexibility for managing different local mesh resolutions inside the computational domain and often lead to excessive computational cost increase when applied to solving spatial CFD problems with features of much different length scales. To overcome this difficulty, one can implement one of adaptive mesh refinement (AMR) approaches. In our opinion, the octree-based AMR is the most promising approach for massively-parallel scalable CFD solver implementations since one efficiently supports simple dynamic load balancing by exploiting (Morton or Hilbert) space filling curve linearly ordering all grid cells. For pure unstructured AMR dynamic load balancing is the main factor limiting scalability on supercomputers.

Exploiting Cartesian grids with recursive refinements forming quad-/octree graphs requires adaptation of CFD numerical methods and AMR metadata management for corresponding grid structures arising as the result of the cell coarsening and refinement procedures. In conventional approaches for GPU-accelerated

© Springer Nature Switzerland AG 2021
V. Malyshkin (Ed.): PaCT 2021, LNCS 12942, pp. 235–244, 2021.
https://doi.org/10.1007/978-3-030-86359-3_18

CFD codes, the AMR function is partially or fully performed on CPU. This requires regular GPU↔CPU data transfers which lead to GPU stalls. As opposed to that, in the CFD framework we develop one of main features is the AMR management module natively executing on GPU without back and forth grid data transferring to/from CPU RAM. For this module, all AMR procedures including cells refinement and coarsening, neighbors (cells with common face) searching, cell geometry extracting have been developed to efficiently perform on GPU. The comparison with the P4est (MPI octree-based AMR framework) [2] has shown clear advantages of our implementation for GPU-native AMR [3].

Combined with the Godunov-type CFD solver, our AMR module brings only 2% runtime overhead on Nvidia Tesla V100 in test problems with grid coarsening and refinement performed every timestep iteration [4]. During such grid transformations newly coarsen and refined cells are consequently stored in the free memory region whereas the memory for corresponding parent cells is not used anymore. That leads to fragmentation and exhaustion of available GPU memory. To fight this issue in our AMR module, a memory defragmentation procedure is implemented on GPU. In fact, this procedure occurs to be the most time-consuming part of the AMR module. The main purpose of this paper is to optimize the GPU memory defragmentation kernel and to investigate the influence of this procedure on the overall CFD solver performance.

In conventional approach, as one mentioned before, CFD solver is performed on GPU while AMR procedures are fully or partially executed on CPU since they are hardly implemented in multi-threaded programming model. To the author knowledge there are only couple of papers describing fully-native GPU AMR management [5,6] which are limited by employment only of two-dimensional triangular grids. Unfortunately, the details of the memory management during GPU grid refinement and coarsening aren't provided in [5]. Authors of [6] solved the memory fragmentation and exhaustion problem by tracking and reusing free spaces (left after cell refinement/coarsening) directly in grid modification kernels while we always add newly modified cells in the end of the corresponding arrays and then separate defragmentation kernel is launched. In case when total grid resolution is considerably decreased original in-place compaction algorithm is used in [6] which actually represents some sort of memory defragmentation. It would be interesting to compare our approach with [6] but that paper doesn't provide implementation details about direct tracking and reusing free memory spaces.

The paper is organized as follows. In Sect. 2, the baseline numerical method on octree-based grids in our framework is described. In Sect. 3, different optimizations for GPU memory defragmentation are considered. In Sect. 4, numerical results for different memory defragmentation implementations and CFD problems are presented.

2   Discrete Model on a Cartesian Grid with AMR

The finite volume method is used for discretizing the system of governing gas dynamics equations on AMR Cartesian grid. For increasing the order of approximation, the piece-wise linear sub-cell reconstruction of the solution is implemented with the MUSCL interpolated scheme [7]. The Godunov method is employed to approximate the function of numerical flux with implementing the exact solution to the face-based Riemann problem [8].

Time integration of the system of semi-discrete equations is fulfilled with the explicit second-order predictor-corrector scheme that ensures the max norm diminishing property, if the time step meets the CFL condition. This scheme is known in literature as Godunov-Kolgan-Rodionov scheme. Details of the scheme can be seen in [9], for example.

A limited gradient of the primitive state vector defined in each computational cell is used for calculating face-related interpolated state vectors. This gradient represents a linear cell reconstruction of the solution vector that maintains the strong monotonicity condition: $(u_j - u_i)(u_j^\sigma - u_i^\sigma) \geq 0$, where the subscripts i and j denote neighboring cells, the superscript σ indicates the face interpolated value. Commonly, the limited gradient for Cartesian grids is introduced by applying a 1D slope limiter procedure to each coordinate direction. We extend this approach to non-conformal locally adaptive octree-based grids. It should be noted that the proposed extension is reduced to the conventional directional limited gradient for Cartesian grids without adaptation.

The limiting algorithm is based on the least square method (LSM) applied to a stencil consisting of the current cell and all neighboring cells having a common face (or its part) with the current face. Let $\boldsymbol{x}_i^j = \boldsymbol{x}_j - \boldsymbol{x}_i$ is the radius-vector of the j-th neighbor. The subscript i is used here for the current cell. $\Delta_j u_i = u_j - u_i$ is the difference between the values of a function u in the neighbor and current cells. First, we calculate the cell LSM gradient (unlimited gradient) that is reduced to central difference approximations for regular grids with no adaptation. Then, we consider the derivative along the direction to j-th neighbor, $\partial_j u_i = \frac{(\nabla u_i, \boldsymbol{x}_i^j)}{|\boldsymbol{x}_i^j|}$, and introduce the following one-sided derivatives

$$\partial_j^+ u_i = \frac{\Delta_j u_i}{|\boldsymbol{x}_i^j|}, \partial_j^- u_i = 2\partial_j u_i \frac{\Delta_j u_i}{|\boldsymbol{x}_i^j|} \tag{1}$$

Finally, we define the limited derivative along the direction to j-th neighbor as follows:

$$\bar{\partial}_j u_i = \phi(\theta_i^j)\partial_j^+ u_i, \theta_i^j = \frac{\partial_j^- u_i}{\partial_j^+ u_i} \tag{2}$$

where $\phi(x)$ is the limiter function. There are several options for choosing this function, which in 1D calculations guarantee the monotonicity condition, e.g., MINMOD, MC, SUPERBEE, NOTVD (for definition see [4]). The limited gradient is calculated with the LSM in the similar way as unlimited one.

The linear subcell reconstruction of the solution with the limited gradient defined by derivatives (2) is used to calculate face-related values required for numerical flux approximations and also for projecting the solution from the coarse to fine grid in the refinement grid procedure.

The AMR procedure is done with accordance to the refinement criterion proposed in [10]. This criterion is local, just avoiding any global operations, bounded by 0 and 1 that permits preset tolerances, and also non-dimensional, so that specific physical parameters can be used without problems of dimensioning. Its extension to multidimensional case and grids of arbitrary topology can be suggested in the following way:

$$\chi(u)_i = \sqrt{\sum_{i,j}(A_{ij})^2 / \sum_{i,j}(B_{ij} + \varepsilon C_{ij})^2}. \tag{3}$$

where the summation is taken over the number of space dimension, and

$$A_{ij} = \sum_{\sigma}\left(\frac{\partial u}{\partial x_i}\right)_{\sigma} n_{\sigma,j} S_{\sigma}, B_{ij} = \sum_{\sigma}\left|\left(\frac{\partial u}{\partial x_i}\right)_{\sigma}\right|\left|n_{\sigma,j} S_{\sigma}\right|, \tag{4}$$

$$C_{ij} = \frac{\sum_{\sigma}|u|\left|n_{\sigma,j}\right|S_{\sigma} \cdot \sum_{\sigma}|u|\left|n_{\sigma,j}\right|S_{\sigma}}{Vol}.$$

where Vol is the cell volume. The derivatives at cell interfaces in (4) are calculated by averaging the gradients with cell volumes. The threshold values for cell refinement and coarsening are generally model dependent. However, numerical experiments show that these values can be chosen as $\chi_{rf} \sim [0.2, 0.25]$, $\chi_{cr} \sim [0.05, 0.15]$ [11]. The constant ε is given a value of 0.01; it switches off unnecessary refinement in regions of small oscillations appearing due to loss of monotonicity.

The value assigned to the cell at the coarsening procedure is computed on the base of conservative vectors, $\boldsymbol{q}_{cr} = \sum_i \boldsymbol{q}_{cr(i)}/8$ (or/4 for 2D grids), the index $cr(i)$ indicates children cell to be coarsen into the cr cell. In the refinement procedure, the values assigned to the arising children cells are computed on the base of the conservative and the limited gradient of the refined cell: $\boldsymbol{q}_{cr(i)} = \boldsymbol{q}_{cr} + \frac{\partial \boldsymbol{q}}{\partial \boldsymbol{z}}\overline{\nabla}\boldsymbol{z}_{cr}(\boldsymbol{x}_{cr(i)} - \boldsymbol{x}_{cr})$. This ensures the conservative property in both the refinement and coarsening procedure.

3   Parallel Defragmentation Algorithms

The computational domain is initially discretized by a Cartesian grid which we refer to as the base grid (Fig. 1). The cells (shown as "□") of the grid are the roots for local octrees when recursively refined. The dangling or leaf nodes (shown as "○") in octrees are physical cells actually used in the CFD solver. Other octree cells are called anchored or virtual cells; they are shown in Fig. 1 as "●". Three arrays of base, anchored, and dangling cells with corresponding links between their elements represent full tree structure of the AMR grid.

Once a cell of the AMR grid is refined, its position in the dangling cell array is marked as "empty" ("×" on Fig. 1) and new cells resulted from the refinement are atomically added to the end of the array. Corresponding updates are

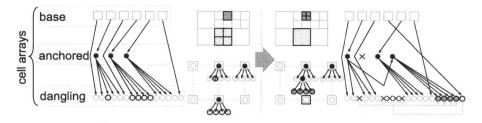

Fig. 1. Grid, corresponding graph and arrays before (left) and after (right) coarsening/refining.

performed in the anchored node array with relinking between the octree cells to represent changes in the updated grid structure. The cell coarsening procedure is performed in the same way (see Fig. 1). One may see that multiple procedures of refining and coarsening lead to "holes" (denoted as "×") in the arrays of dangling and anchored cells and fast exhaustion of free space in the end of these arrays. Therefore, one needs to periodically perform memory defragmentation (compaction) procedure. In this paper, we propose solutions to the defragmentation problem for dangling cells; the proposed solutions are easy implemented also for anchored cells in nearly the same way.

The first defragmentation algorithm which was initially implemented in our framework we refer to as *octree-centered*. In this algorithm, each n-th CUDA warp is assigned one base cell with $S_n \geq 1$ – number of dangling cells in its local octree and prefix scan operation is performed:

$$B_n = \sum_{i=1}^{n} S_i. \tag{5}$$

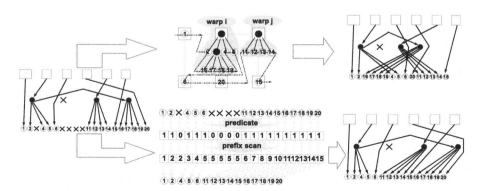

Fig. 2. Initial grid arrays (left); octree-based (center, above) and cell-based (center, bottom) defragmentation algorithms; defragmented arrays (right).

After that all dangling cells of the local octree are consequently copied to a new array in the place starting from the index B_n in the order defined by "Z" space filling curve (SFC) (z-curve linearly orders all leaf nodes in the quad-/octrees), Fig. 2. Since the size of the base grid in tests conducted is not so large (about $10^3 - 10^4$ cells) and octree traversing over z-curves (to copy cells in new arrays) in different threads of warp (if each thread is assigned to separate octree) may lead to warp divergence problems, it is decided to perform all operations in only one of 32 warp threads. Obviously this reduces GPU resources utilization but permits to spread octrees traversing over all GPU Streaming Multiprocessors (SMs) rather than to concentrate them into one or few SMs letting other ones to stall.

Prefix scan for (5) was naively implemented via polling and consequential incrementing volatile variables inside each warp with appropriate _threadfence() synchronization calls. This led to partial serialization of warp execution and, consequently, to overall performance decrease. Test results on different GPUs show that the defragmentation procedure is the most time-consuming one compared to another AMR kernels [3]; hence, it becomes the primary target for optimization.

The first attempt to increase the defragmentation performance is to exploit the CUDA-optimized prefix scan implementation [12] so that the operation (5) is performed in separate kernels preceding octree traversing. Having optimized prefix scan in our framework, we can easily implement the new defragmentation algorithm which we called *cell-centered*. It is based on the stream compaction operation which exploits prefix scan. Stream compaction accepts the array of dangling cells as the input vector and its predicate is just the cell flag (actual physical cell ("○") or "empty" ("×")). Prefix scan is performed over the array generated by predicate, which consists of 0 or 1. An i-th dangling non-"empty" cell is then copied to a new defragmented array at the index equals to "i"-th result of the previous prefix scan, Fig. 2. In other words, the cell-centered algorithm compacts the array of dangling nodes by just eliminating "holes" without changing the ordering of nodes.

The octree-centered algorithm is initially developed since it orders dangling nodes in accordance with the z-curve thus preserving in some degree the memory locality i.e. when the reference to geometrically nearby physical cells in the CFD solver is performed at nearby addresses. Our assumption is that the octree-centered algorithm despite significant runtime due to octrees traversing nevertheless possibly may allow to gain a higher solver performance thanks to better memory locality in the defragmented array compared to the cell-centered algorithm, which, in general, doesn't aware of grid cells geometrical neighborhood. However, the tests carried out (see below) show that the cell-centered algorithm is not only significantly faster than the octree-centered one, but also unexpectedly leads to considerable increase in the solver performance, so that further optimizations in the octree-centered algorithm lose edge and importance.

4   Numerical Results

The test problems are calculated on NVidia Tesla K20c (CUDA 7.5). To validate our AMR framework, 2D Riemann problems from [13] are considered, namely, the configuration 6 and 16. The computational domain for these problems is a $[0;1] \times [0;1]$ quad and the base grid resolution is $50 \times 50 \times 2$ (since 2D solution is actually calculated on a 3D grid); 1- and 2-level maximum adaptations are used. Refinement and coarsening are performed each timestep whereas the defragmentation procedure is executed every 20-th timestep.

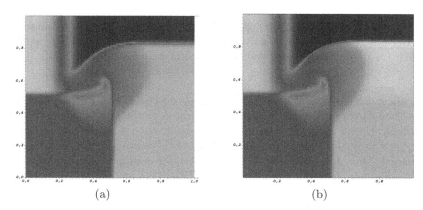

(a)   (b)

Fig. 3. Density fields for configuration 6 problem, 1- (left) and 2-level (right) adaptation of 50×50 base grid.

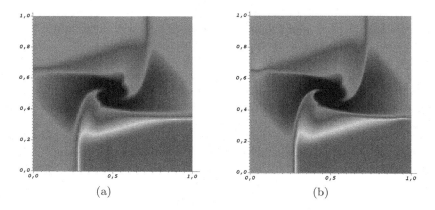

(a)   (b)

Fig. 4. Density fields for configuration 16 problem, 1- (left) and 2-level (right) adaptation of 50×50 base grid.

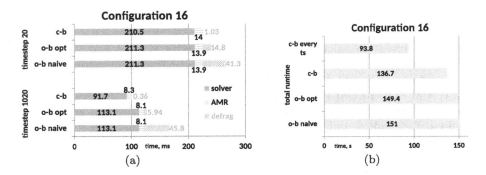

Fig. 5. (a) times per timestep 20 (233998 cells), 1020 (108747 cells) for cell-based (c-b) and octree-based algorithm with naive prefix scan (o-b naive), optimized prefix scan (o-b opt); (b) total runtime for different defragmentation algorithms.

2D slices for numerical solutions are presented in Fig. 3, 4. The AMR code correctly captures eddy formations in the center of the computational domain for configuration 6 and tracks specific low-density zone near the $(0.5; 0.6)$ point; the higher adaptation level permits to obtain more detailed and precise solution, as expected. One should note that the grid size for configuration 6 is only about 2×10^4 cells, which is not enough for full GPU SMs utilization, so the performance results only for configuration 16 are presented in Fig. 5. As one can see, the optimized scan for the octree-centered algorithm reduces the defragmentation time in 2.8–7.7 times compared to naive scan version but almost doesn't impact on the total runtime. In contrast, the cell-centered algorithm is 14.3–16.5 times faster than the octree-centered one and leads to 8.5% reduction of the total runtime. Since the cell-centered defragmentation runtime is negligible compared to solver iterations, one additional solver execution is performed with the defragmentation procedure executed each timestep ("c-b every ts" in Fig. 5). This leads to a further performance improvement; one able to gain 37% total runtime reduction compared to the octree-centered algorithm.

The second test problem is the Sedov blast wave problem [14]. It is solved on $20 \times 20 \times 20$ base grid; other parameters are identical to ones used in the previous paper [4]. Numerical solution is also described in [4]. Here we only discuss the performance results obtained with different defragmentation algorithms (Fig. 6). Refinement and coarsening are performed each timestep, whereas defragmentation is executed every 40-th timestep. Here we can see that optimized scan for the octree-centered algorithm reduces the defragmentation time by 7–42% but the cell-centered algorithm is 19–215 times faster and, more importantly, leads to total runtime reduction by 23.5%. Performing the cell-centered defragmentation every timestep (instead of every 40-th timestep) does not introduce additional performance overhead but even brings further reducing of the total runtime by 26.7%.

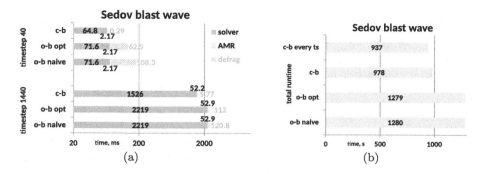

Fig. 6. (a) times per timestep 40 (70496 cells), 1440 (2733576 cells) for cell-based (c-b) and octree-based algorithm with naive prefix scan (o-b naive), optimized prefix scan (o-b opt); (b) total runtime for different defragmentation algorithms.

5   Conclusions

Different defragmentation algorithms for dynamically updated GPU-native octree-based grids used in the AMR CFD framework have been considered. Numerical tests performed with these algorithms have shown that the simple cell-centered algorithm is 14.3–215 times faster than the octree-centered one. A more important and unexpected result obtained is that the cell-centered algorithm which, in general, doesn't aware of grid cells geometrical neighborhood leads to better memory locality and decrease in the total runtime for the test CFD problems considered by 26.7–37% compared to the octree-centered algorithm, which actually preserves memory locality in some degree thanks to cells ordering in accordance with Z-SFC.

Acknowledgments. This work was supported by Moscow Center of Fundamental and Applied Mathematics, Agreement with the Ministry of Science and Higher Education of the Russian Federation, No. 075-15-2019-1623.

References

1. Pavlukhin, P., Menshov, I.: On implementation high-scalable CFD solvers for hybrid clusters with massively-parallel architectures. In: Malyshkin, V. (ed.) PaCT 2015. LNCS, vol. 9251, pp. 436–444. Springer, Cham (2015). https://doi.org/10.1007/978-3-319-21909-7_42
2. Burstedde, C., et al.: Extreme-scale AMR. In: Proceedings of the 2010 ACM/IEEE International Conference for High Performance Computing, Networking, Storage and Analysis, pp. 1–12. IEEE Computer Society (2010)
3. Pavlukhin, P., Menshov, I.: GPU-aware AMR on octree-based grids. In: Malyshkin, V. (ed.) PaCT 2019. LNCS, vol. 11657, pp. 214–220. Springer, Cham (2019). https://doi.org/10.1007/978-3-030-25636-4_17
4. Menshov, I., Pavlukhin, P.: GPU-native gas dynamic solver on octree-based AMR grids. J. Phys.: Conf. Ser. **1640**, 012017 (2020). https://publishingsupport.iopscience.iop.org/questions/how-to-cite-an-iop-conference-series-paper/

5. Alhadeff, A., Leon, S.E., Celes, W., Paulino, G.H.: Massively parallel adaptive mesh refinement and coarsening for dynamic fracture simulations. Eng. Comput. **32**(3), 533–552 (2016). https://doi.org/10.1007/s00366-015-0431-0
6. Giuliani, A., Krivodonova, L.: Adaptive mesh refinement on graphics processing units for applications in gas dynamics. J. Comput. Phys. **381**, 67–90 (2019)
7. VanLeer, B.: Towards the ultimate conservative difference scheme V: a second-order sequel to Godunov's method. J. Comp. Phys. **32**, 101–136 (1979)
8. Godunov, S.K., Zabrodin, A.V., Ivanov, M.V., Kraiko, A.N., Prokopov, G.P.: Numerical solution of multidimensional problems of gas dynamics. Nauka, Moscow (1976)
9. Menshov, I.S., Pavlukhin, P.V.: Efficient parallel shock-capturing method for aerodynamics simulations on body-unfitted cartesian grids. Comput. Math. Math. Phys. **56**(9), 1651–1664 (2016). https://doi.org/10.1134/S096554251609013X
10. Lohner, R.: An adaptive finite element scheme for transient problems in CFD. Comput. Methods Appl. Mech. Eng. **61**, 323–338 (1987)
11. Dumbser, M., Zanotti, O., Hidalgo, A., Balsara, D.S.: ADER-WENO finite volume schemes with space-time adaptive mesh refinement. J. Comput. Phys. **248**, 257–286 (2013)
12. Nguyen, H.: GPU Gems 3, 1st edn. Addison-Wesley Professional, Boston (2007)
13. Lax, P.D., Liu, X.-D.: Solution of two-dimensional Riemann problems of gas dynamics by positive schemes. SIAM J. Sci. Comput. **19**(2), 319–340 (1998)
14. Sedov, L.: Similarity and Dimensional Methods in Mechanics. Academic Press, NewYork (1959)

Zipped Data Structure for Adaptive Mesh Refinement

Anton Ivanov[1] and Anastasia Perepelkina[1,2]

[1] Keldysh Institute of Applied Mathematics, Moscow, Russia
mogmi@narod.ru
[2] Kintech Lab Ltd., Moscow, Russia

Abstract. Adaptive mesh refinement (AMR) is a dynamic approach to non-uniform grids which is commonly used to cut the simulation costs of multiscale problems in mathematical modeling of physical phenomena.

In this work, we propose a new dynamic data structure for AMR implementations which is based on a Z-order curve and tiles with variable size. It is a generalization of classical octree and various tile-based octrees, which can be seen as special cases of it. The tree height is dynamically decreased wherever possible by adjusting the number of children of nodes, increasing the size of tiles. Thus, the events of access to neighboring tiles become less frequent, and the complexity of access becomes less. Trivial data serialization presents another advantage of the data structure. In a specific case where the refinement level is constant over some region, the sub-tree height is equal to one, thus the neighbor access is just as simple as in a uniform multidimensional mesh. The structure inherits the locality properties of the Z-order space-filling curve.

In the text, the detailed description of the structure, algorithms for traversal, random access, neighbor search, and mesh adaptation are described.

Keywords: AMR · Grid refinement · Data structure · Z-curve · aiwlib

1 Introduction

Adaptive mesh refinement (AMR) [2] is a dynamic approach to non-uniform grids which is commonly used to cut the simulation costs of multiscale problems in mathematical modeling of physical phenomena. The implementation of simulation codes which use AMR is, however, a complicated problem. The efficiency of the simulation algorithms is very important in practical large-scale applications. Many numerical simulation schemes are memory-bound problems [24], thus the efficiency of the algorithm depends on the data structure, which should support efficient traversal, neighbor access, serialization, and load balancing [6].

In problems with complex geometry, the cell data is stored with all necessary data for the mesh topology [9,12]. This way, the cell coordinates, the cell size and shape parameters, the address of the neighbor cell data, are all stored along

© Springer Nature Switzerland AG 2021
V. Malyshkin (Ed.): PaCT 2021, LNCS 12942, pp. 245–259, 2021.
https://doi.org/10.1007/978-3-030-86359-3_19

with the cell in an AoS (array of structure) fashion. Such method provides means for free generation of meshes of any type, and any type of subdivision, however, the data overhead for the storage of the mesh topology data is high.

By imposing rules on the structure of the mesh and its subdivision rules, some of the information in the cell data may be omitted. In the block structured AMR [1,3,13,20,25] the data is divided into several blocks, which may be represented as separate meshes. The absolute and relative position, shape and size parameters are stored not for each cell, but for each patch, this way the data overhead for topology is less. However, at the regions where coarse and fine patches overlap, often only one patch is relevant, that is, the patch with a finer grid. Thus, data overhead for spurious storage of cell data is possible.

The hierarchical tree data structure is especially efficient for storage of adaptively refined meshes. The cell data are associated with a tree node, and the cells, produced in the refinement, are associated with the children of the node. If, in a refinement, a node is subdivided into two parts in each direction the tree is a binary tree, quadtree or octree for 1D, 2D and 3D mesh subdivision correspondingly. Thus, for generality, we talk about 2^D-binary trees [19] (D is the number of space dimensions).

In the traditional pointer-based implementation [5], each node contains pointers to its parent and children. It is relatively simple and the mesh refinement operation is cheap [14,22]. Its shortcoming is low data access locality. Mesh traversal and search for the nearest neighbors that is required for application of a numerical scheme stencil are difficult, since they require a large number of conditional tests. This leads to lower efficiency of the simulation codes that are built with such structures. The mean number of operations required for the neighbor cell access increases with the tree height.

Various tile-based octrees [21], in which the tree nodes contain a block of cells instead of just one tree node, show higher data locality since many neighbor access operations address cells in the same tile. However, storage of larger tiles in the tree nodes produce large overhead in the memory footprint.

Many optimizations for the tree-based structures exist in the computer graphics [19]. However, the specifics of image processing and physical simulation are different. The tree volume data storage popular in computer graphics [16] is optimal for sparse volume representation, while the volume data for mathematical modelling are often dense and the priority is in the reducing the cost of neighbor access.

Thus, the tree structures for mathematical modelling may also contain pointers to neighbors. This way, the stencil data access is faster at the cost of larger storage overhead. The pointers to the node children may be reused to point to the node neighbors if the children do not exist [8,18].

Another approach is the use of 'heavy' and 'light' tiles [11]. In it, the tiles which exist in the tree structure but their cells are not used in the simulation (e.g. they are overshadowed by a finer grid) store the data of another tile instead of the superfluous information. The data of the 'light' tiles is stored in the unused space of the 'heavy' tiles. That approach shows significant improvement in reducing storage space in multi-dimensional simulation, shows excellent access locality, and allows implementation of the advanced parallel algorithms [10]. However, the efficient implementation of such structure remains a complex problem.

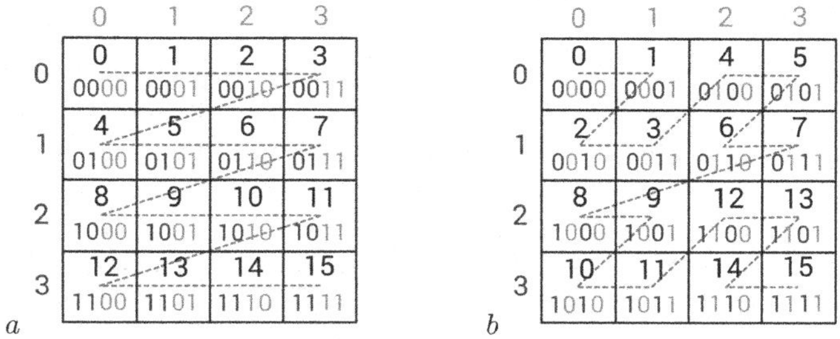

Fig. 1. The traditional indexation of a 2D array (*a*) and indexation with the Morton Z-curve (*b*). In each cell the binary representation of its number is shown. The color distinguishes bits representing different coordinate axis.

Linear representation of octree [19] stores only the leaf nodes, each one with a locational code. This kind of implementation reduces the required storage space, which is the priority for GPU simulations [4,17]. The order in which the leaves are stored is defined with a space-filling curve. The traditional storage order, where the data is looped over in the axes direction one axis after another is not efficient for many-dimensional stencil codes. Hierarchically defined space-filling curves are more suitable for AMR data representation [4,19,20,23].

In this work, a new AMR data structure based on the Morton Z-curve [15] is proposed. It is a forest of 2RD-binary trees, where R is a variable parameter which depends on the current mesh refinement structure. A traditional 2^D-binary tree hierarchy and its tile-based variants may be described as its special cases. The use of the Z-order curve allows to decrease the tree height by dynamically increasing the size (number of children) of a node from 2^D to 2RD.

The properties of the Z-order curve help to increase the search complexity to $O(1)$ for any node. Data organization into tiles with size 2RD provides high data access locality by simplifying mesh traversal and neighbor access in a scheme stencil. The memory storage overhead is observed to be less than in the traditional 2^D-tree data structures.

2   Data Structure

A traditional indexation of a D-dimensional array allows to compute the index f of a cell in an array with size $N_i \times N_j \times N_k \times \ldots$ from its coordinates $i, j, k\ldots$ as

$$f = i + N_i j + N_i N_j k + \ldots$$

For a cube-shaped array with an edge length equal to 2^R a change in any coordinate leads to the change in the corresponding group of R bits of the number

f (Fig. 1 a). The Morton Z-curve [15] is an alternative indexation. In it, the bits representing each coordinate axis are interleaved (Fig. 1 b).

Efficient algorithms are known for computing f from $i, j, k...$ and back [15]. The computation of the nearest neighbors is not difficult as well.

Z-curve indexation leads to a higher data access locality in comparison with the traditional indexation, and this is especially important in dealing with big memory-bound problems [24]. For the implementation of AMR-compatible data structures, it is important for Z-curve to provide the same data traversal order as a 2^D tree. This property allows compressing segments of the tree that implements the AMR data. This leads to advantages in all aspects of the code implementation.

In Fig. 2 a the traditional AMR data structure based on the octree is shown [19]. The application of the Z-curve indexation to the tree leaves allows enlarging the data block, which, in turn, improves data locality, simplifies traversal, random access, and the search of neighbor cells inside one block (Fig. 2 b).

The next step is the application of a similar compression to the tree nodes which are not leaves. The change in the data block size and the number of sub-nodes in a node results in different options for data organization (Fig. 3 a, b, c), while the cell traversal remains unchanged.

We propose the following data structure. The mesh is built as a forest of trees. The tree nodes are placed in the cells of a rectangular grid with a traditional indexation. Some cells may be vacant. With this approach the simulation areas with complex non-rectangular shapes can be described.

Let S be the level of refinement. The size of the cells at level S is 2^S times smaller than the size of the cells at level $S - 1$. At each level a separate Z-curve may be introduced. Let the tile of rank R be a cube-shaped mesh segment with the edge length in 2^R cell so that the cells of this segment are positioned in a sequence on the Z-curve. This means that the binary representations of the Z-curve indexes of cells in a tile have equal bits in all positions higher than RD.

A 64-bit structure ptile_t is introduced. It contains one signal bit, six bits for the tile rank R, and 57 bits for the integer index f. For example,

```
struct ptile_t{
    uint64_t blob;
    void set(bool sign, uint32_t rank, uint64_t off)
    {    blob = uint64_t(sign)|((rank&0x3F)<<1)|(off<<7); }
    bool get_sign() const { return blob&1; }
    uint32_t get_rank() const { return (blob>>1)&0x3F; }
    uint64_t get_off() const { return blob>>7; }
};
```

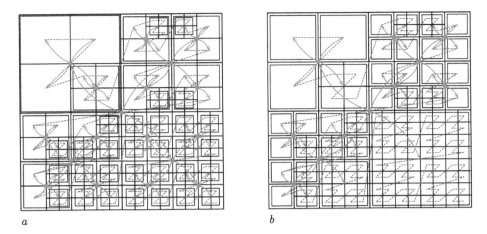

Fig. 2. AMR data organization. (a)—traditional D-binary tree, (b)—Z-curve. Blue color represent data blocks. Green lines show tree branches. The cell traversal rule remains the same (red dotted line). (Color figure online)

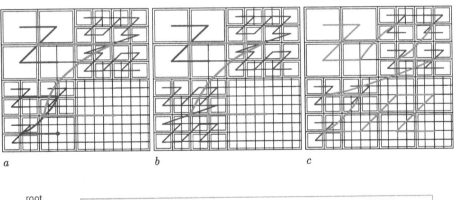

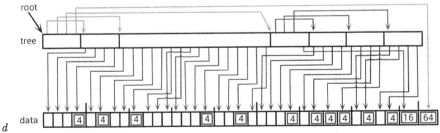

Fig. 3. The three options of implementations of a tree-based AMR data structure with the Z-curve (a, b, c) and linking structure for option a (d)

The tree data structure is

```
struct tree_t {
    ptile_t root;                    // tree root
    std::vector<ptile_t> tree;       // tree description
    std::vector<T> data;             // data
    int Smin, Smax;                  // minimal and maximal
};                                   // levels of refinement
```

If the signal bit is set, the `ptile_t` structure instance addresses a tile (tree node) that is in the `tree` array. Otherwise, the tile in the `data` array is addressed (Fig. 3 d). The tile size is determined from the rank as 2RD. The tiles are placed one by one in the array. The beginning of the tile is pointed at by the index f. The elements of a tile are indexed with a Z-curve. Each tile is a tree node with 2RD children. The elements of a data tile are the tree leaves.

The data structure that is described here is a generalization of 2^D-binary trees, where the nodes have a dynamically variable degree 2RD. The increase in R leads to a decrease in the tree height, which leads to better data access locality. With lower trees, the number of pointer access operations is lower for neighbor search in the most non-local access events. The tree control structure is more compact in comparison to the traditional tree implementation. This increases caching ability which may lead to better performance.

The new data structure is named ZAMR, where Z emphasizes the compactness and the use of a Z-order curve. ZAMR is better than the data structure proposed in [11] both in terms of data locality and implementation simplicity. The 'light' tiles in [11] require one additional pointer access operation to address their data. The simple implementation of ZAMR allows for the use of efficient algorithms for better performance in mathematical simulation.

For unrestricted use, the AMR implementation has to provide the following capabilities:

1. random access to the cells by their coordinates (to implement initial conditions and sources);
2. sequential traversal of the cells in a mesh segment, and the traversal of the cells on a specific level of refinement (to implement cell updates according to a numerical scheme).
3. access to the cell neighbors according to the scheme stencil;
4. mesh adaptation;
5. data serialization (read/write data on disc).

The trivial data serialization/deserialization is one of the advantages of ZAMR. The structure does not contain pointers and can be saved and loaded in three parts (`root`, `tree`, `data`) by a byte-to-byte copy operation.

2.1   Random Access and Sequential Traversal

Let f_S be the position f on a Z-order curve on the level S. For a transition to the Z-curve on the level S' only a bit shift in $(S' - S)$ bits is required. A positive

$(S' - S)D$ is a left shift, a negative value is a right shift, and some bits of f are lost in this case. For addressing a specific cell its index f_S on a uniform Z-curve on the level S is sufficient. The search algorithm is as follows:

1. f_S value is transformed into $f_{S\,max}$;
2. the `ptile_t cursor = root` and `int Sres = 0`; (final refinement level) variables are created;
3. while `cursor` is indexing the tree node (the signal bit is set):
   (a) the current refinement level is incremented
   ```
   Sres += cursor.get_rank();
   ```
   (b) the tile cell position under the pointer is computed
   ```
   uint64_t c = f>>D*(Smax-Sres);
   ```
   (c) the remaining bits of the shift f are computed:
   ```
   f -= c<<D*(Smax-Sres);
   ```
   (d) `cursor` is set to the next cell, and this corresponds to a descent in a tree by one node
   ```
   cursor = tree[cursor.get_off()+c];
   ```
4. the resulting refinement level is computed, the lower bits of f are truncated, and the cell index is computed

```
Sres += cursor.get_rank();
f >>= (Smax-Sres)*D;
T &cell = data[curosr.get_off()+f];
```

This algorithm allows an arbitrary definition of the original level of refinement S, while the f_S index should be consistent with it. If S is less than `Stot` then the cell at the beginning of the Z-curve is found. If S is larger than `Stot`, the lower bits of f are discarded.

For the implementation of a numerical scheme, all cells have to be traversed in sequential order. The sequence is according to the data placement in memory, thus it amounts to an increment in a cell address. However, the cell position in space should be accessible. The algorithm for extracting x, y, z from uniform Z-curve indexes are known, thus, the Z-curve index should be computed from its address. To find it, the cell position in a tile (defined in the previous section) has to be known. If the incremented index value is inside the tile the position of the cell in the Z-curve and the cell index are incremented. Otherwise, the cell with an incremented position is found with the random access algorithm. However, such misses are sufficiently rare.

The traversal of only the cells on a specified level of refinement S is a separate problem. For this, a field is added into the `ptree_t` data structure:

```
std::vector<std::vector<std::pair<ptile_t, uint64_t> > > index;
```

For each level of refinement S an array of pairs, that is, a pointer to a tile, position of a tile on the Z-curve of a tree, is created. This is enough for an efficient traversal. The index is updated in each mesh adaptation.

2.2 Algorithms for Neighbor Search

Let us consider the problem of the neighbor search for a cell $c_{i,j,k...}^S \equiv c$, where $i, j, k, ...$ are the coordinates, S is the degree of refinement. Let f_S be the position of c on the Z-curve at level S. The neighbor cell $c_{i+di,j+dj,k+dk...} \equiv c'$ may be addressed with a tuple of D numbers $di, dj, dk, ...$, which defines its shift relative to the c position in the units of length that are 2^{-S} smaller than the cell size of the initial, unrefined mesh. The algorithm for the neighbor search is as follows.

1. The position f'_S of a cell c' on the Z-curve is found. If it is in another tree, f'_S may be negative. To overcome this problem, D higher bits from the position SD and upward are added to the f beforehand. This provides positive f'_S values with any reasonable shift into the current and neighboring trees.
2. If the higher f_S and f'_S bits match, that is, `(f>>R*D)==(f'>>R*D)`, then c' is in the same tile as c and the problem is solved.
3. If f'_S is inside the tree, that is, `(f>>S*D)==(f'>>S*D)`, then the random access of the cell in the current tree is used. The cell is looked up at the position f'_S without its higher D bits.
4. Otherwise, the neighboring tree is found with the higher D bits of f'_S, and, inside it, the cell is looked up with random access at the position f'_S without its higher D bits.

The constructed algorithm works correctly if the neighbor cells are of the same size as c or larger than it. Otherwise, the search would hit a smaller cell (with a larger S). Depending on the chosen numerical scheme, several methods may be used to resolve the issue.

1. The use of a buffer zone at the refinement boundary, which guarantees the presence of the cells of the same size in the scheme stencil. The data transfer in the buffer zone between the levels of refinement is a separate task.
2. Generation of a cell with the same size on the fly, e.g. computation of the mean value across the closest cells. This solution corresponds to a virtual buffer zone.
3. For numerical schemes with fluxes, it is often enough to access the directly adjacent cells which share a boundary with the current one. Here, several smaller cells may share one boundary with the current cell.

The implementation of the first option is not considered in the current work. For the implementation of the second option, the following modification of the random access algorithm is proposed.

1. The f'_S position is converted into $f_{S\,max}$;
2. the variables for the resulting level of refinement `int Sres = 0;`, the remainder level of refinement `dS = 0;`, and `ptile_t cursor = root` are created;
3. while the `cursor` variable points to a tree node (the signal bit is set) and `Sres<S`:
   (a) the new level of refinement is computed

```
dS = cursor.get_rank() - std::min(S-Sres, cursor.get_rank());
Sres += std::min(S-Sres, cursor.get_rank());
```

(b) the cell position to which the `cursor` is pointing is found
```
uint64_t c = f'>>D*(Smax-Sres);
```
(c) the remaining bits of the f' shift are found:
```
f' -= c<<D*(Smax-Sres);
```
(d) the `cursor` is switched to the next cell, and this corresponds to a descent in a tree by one node
```
cursor = tree[cursor.get_off()+c];
```

$...b_y^{j-1}b_x^{i-1}$	$...b_y^{j-1}b_x^i 10$	$...b_y^{j-1}b_x^i 11$	$...b_y^{j-1}b_x^{i+1}$
$...b_y^j b_x^{i-1} 01$			$...b_y^j b_x^{i+1} 00$
	$...b_y^j b_x^i$		
$...b_y^j b_x^{i-1} 11$			$...b_y^j b_x^{i+1} 10$
$...b_y^{j+1}b_x^{i-1}$	$...b_y^{j+1}b_x^i 00$	$...b_y^{j+1}b_x^i 01$	$...b_y^{j+1}b_x^{i+1}$

Fig. 4. Sample Z-curve shifts for looping over the boundary. Blue denotes bits corresponding the y axis, green denotes bits corresponding to the x axis. (Color figure online)

After this, the recursive algorithm is started. In a loop for $i \in [f'_S : f'_S + 2^{dS\,D})$:

1. if the signal bit of the cursor is set, the function for the sum for the cell in `data[cursor.get_off()+i]` is called;
2. otherwise, the same algorithm for the parameter `cursor2 = data[cursor.get_off()+i]`, `dS = cursor2.get_rank()` is started.

The traversal of the adjacent cells is implemented similarly (Fig. 4). The boundary is determined by the number of the coordinate axis perpendicular to it. The bitmask is created. For the neighbors on the right side, the bits corresponding to that axis should be equal to zero, and, for the neighbors on the left side, these bits should be equal to one. When the level S is reached, the recursive algorithm is started. The function for the boundary processing is called for the cells in which the higher bits for the axis corresponding to the boundary match the corresponding bits of the bitmask.

2.3   Mesh Adaptation Algorithms

The algorithms for mesh adaptation are often the most complicated issue in the implementation of AMR data structure.

The mesh refinement is infrequent since it is required in the synchronization instants [10]. Three functions are required. The function `f_check` takes a cell as an argument and returns -1 if the cell should be coarsened, 0 if the cell remains the same, and 1 if the cell should be refined. The `f_join`/`f_split` functions perform the coarsening and refinement of cells.

In the mesh adaptation, the data consistency in the `f_check`, `f_join`, `f_split` function calls should be provided.

The cell move and cell delete operations may cause undetermined behavior if the attempt to access the cell as a neighbor is made after them. Moreover, for some numerical schemes it may be required to impose rules on mesh nesting [2]. For example, two cells may be prohibited from sharing an edge if their levels of refinement differ in more than one.

For the parallel implementation of the mesh adaptation it is distributed by trees. Each tree is processed with a separate thread.

In one tree, the adaptation algorithm consists of the following steps.

1. Execution of the f_check function on each cell. The output of f_check is saved into the bit array. Two bits per cell encode three possible states: the cell persists, the cell is refined or coarsened. Since the coarsening is a non-local operation and the f_check function does not analyze the state of the neighbor cells, only the cubes of 2^D cells in which all cells are ready for coarsening are to be coarsened.
2. The analysis of the bit array and its correction according to the imposed rules (such as prohibition of large jumps in the refinement level), if necessary.
3. Creation of a new tree data structure. The unchanged cells are copied bit-to-bit. The f_join/f_split functions are called for the cells that have to be changed. The new data structure is created separately from the old one, and the old structure remains operable at this stage.
4. The new data structure replaces the old one.

The described algorithm traverses the forest of trees as a wavefront along the longest dimension of the simulation area. At each cross-section of the simulation area, initially, the first step is executed. After that, the second and the third steps are executed with a delay in one step in the forest of trees. Finally, the fourth step is executed with a delay in one more step.

Let us discuss the third step as the most complicated one. Here, the data (leaves) are merged into tiles and the structure is organized in a way that is optimal both for random access and cell traversal. The solution to this problem is not unique. Let us consider an algorithm that results in a minimal tree height (Fig. 3 c). A requirement for maximal data tile size presents a polar opposite (Fig. 3 a), however, it would result in a higher tree, thus the neighbor search outside large tiles becomes more expensive, and the relative size of those tiles may be small at the same time. In any case, the tree structure affects neither the data placement in the memory storage nor their locality.

At the beginning of the third step, the new data array and the new data tile array are created based on the bit array which was built in the first step and modified at the second step. The data may be efficiently processed in blocks of 32 cells. If the cells should not be changed according to the bitmask, the basic data copy is executed. Otherwise, the f_join/f_split functions are called. To describe a tile, the structure of the following type is used.

```
struct rebuild_tile_t{
    ptile_t p;    // the start and the rank of the tile in the data array
    uint64_t z;   // position of the tile in the Z-curve in a tree
    int S;        // level of the refinement of the Z-curve
};
```

The array of such structures describe the created tiles. In the refinement and coarsening of the cells from the old tile, the tile is separated into several tiles of the smaller size. A segment of a tree may be considered. For its definition, it is sufficient to define a range on a Z-curve and the level of refinement of the Z-curve S. To address a whole tree, $S = 0$ and the Z-curve range $[0, 1)$ can be taken as arguments. With a ΔS increase in S the beginning and the end of the range are multiplied by $2^{D\Delta S}$. After that, the following recursive algorithm is started.

1. The range of tiles the data of which are inside the range is found for the tree segment. In this range, the minimal and the maximal levels of refinement $S_{\min}$ and $S_{\max}$ are computed.
2. If $S_{\min} = S_{\max}$, all data tiles are merged into one and become a resulting data tile. The index of the data tile is complemented.
3. Otherwise, a tree node is constructed as a tile with rank $S - S_{\min}$. S is increased up to $S_{\min}$, and the same algorithm is called for each branch.

3   Implementation

The described data structure is implemented in the `aiwlib` [7] library as a class

```
template <typename T, int D>  class ZipAdaptiveMesh;
```

where T is a type of the cell, D is the number of space dimensions.
    To access a cell the following structure is defined

```
struct ZipAdaptiveMesh<T, D>::Cell;
```

For the data access, the following operations are overloaded

```
T& operator * ();
T* operator -> ();
```

For the access to neighbors, the following operations are overloaded

```
const Cell& operator [](Ind<D> delta);
template <typename F> void cell(Ind<D> delta, F &&f);
template <typename F> void face(int axe, F &&f);
```

where f is the user function. In its call, it takes one neighbor cell with type const Cell& as its argument. axe is an axis perpendicular to the chosen boundary. The $-1, -2, -3...$ values define neighbors to the left of the x, y, z axis correspondingly, and $1, 2, 3...$ define the neighbors to the right.
    For cell traversal in the ZipAdaptiveMesh class the following methods are implemented

```
template <typename F>
size_t foreach(F &&f, int S=-1, bool parallel=false);
template <typename F1, typename F2>
size_t foreach2Xdt(F1 &&f1, F2 &&f2, bool parallel=false);
```

the methods return the number of processed cells. Here `f`, `f1`, `f2` are the functions applied to the cells with type `Cell` of the level of refinement S, and to all cells if $S = -1$. The `parallel` flag switches the parallel processing on. The `foreach2Xdt` method implements a two-step numerical scheme with a stepwise traversal rule and with the time step refined in correspondence to the mesh step [10].

The following method is implemented for mesh adaptation

```
template <typename F_CHECK, typename F_SPLIT, typename F_JOIN>
void rebuild(F_CHECK &&f_check, F_SPLIT &&f_split, F_JOIN &&f_join);
```

All required Z-order curve operations are implemented in the header file `aiwlib/zcube`.

The class has methods for mesh initialization, random access, various traversal rules, mesh adaptation, save/load of the mesh to the disc in a binary format. The `uplt` viewer from the `aiwlib` library is used for result diagnostics.

4   Benchmarks

For the study of the efficiency of the proposed method we considered the sample function in 2D and 3D (Fig. 5).

$$f(\mathbf{r}) = \exp\left[-(|\mathbf{r}| - 0.5)^2/0.2^2\right].$$

This type of scalar field is relevant for multiscale implementations, for example, in shock propagation in computational fluid dynamics. The cell refinement level has been set to $S = 5 + (R_{\max} - 5)f$.

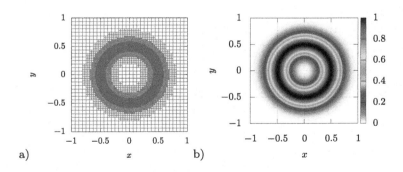

Fig. 5. The sample function f (a) and mesh refinement with $R_{\max} = 8$ (b).

The dependency of the main efficiency characteristics of the proposed data structure is presented (Fig. 6). It is observed that in data access according to the cross stencil, which is relevant is many Finite-Difference, Finite-Element or Finite Volume numerical schemes, it provides significantly fewer tile misses

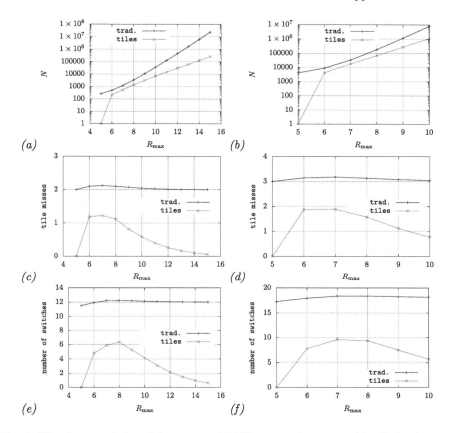

Fig. 6. The characteristics of the proposed AMR versus the maximal level of refinement R_{max} in 2D (a, c, e) and 3D (b, d, f). The same results for the traditional octree are presented for reference (**trad.**). (a, b) data tiles; (c, d) tile misses, when the data access with a cross stencil requires the data from another tile; (e, f) switches, i.e. the average number of pointers that have to be dereferenced and conditional choices for nearest neighbor access in a cross stencil.

(when the access to another tile is required) in comparison with a traditional octree implementation which is taken here for reference. The access complexity here is measured as the average number of pointers that have to be dereferenced and instances conditional processing when tile miss is encountered.

5   Conclusion

In this work, we proposed a new data structure ZAMR for the implementation of AMR in physical simulations. It is a generalization of a traditional octree structure to an octree with tiles of variable rank. The variable rank allows adapting the tree structure to current mesh refinement.

The tree height is dynamically decreased wherever possible. Thus, the events of access to neighboring tiles become less frequent, and the complexity of access becomes less. Trivial data serialization presents another advantage of the data structure. In a specific case where the refinement level is constant over some region, the sub-tree height is equal to one, thus the neighbor access is just as simple as in a uniform multidimensional mesh. The structure inherits the locality properties of the Z-order space-filling curve.

On the flip side, mesh adaptation is relatively computationally complex. However, the implementation is lightweight, and the performance of mesh adaptation may be further optimized. Nevertheless, mesh adaptation is supposed to be infrequent and performed only in the synchronization instants [10], thus this shortcoming is minor. At the same time, higher data access locality and higher efficiency in neighbor search should increase the performance of stencil simulation problems.

References

1. Adams, M., et al.: Chombo software package for AMR applications design document. Lawrence Berkeley National Laboratory Technical Report LBNL-6616E (2015)
2. Berger, M.J., Colella, P.: Local adaptive mesh refinement for shock hydrodynamics. J. Comput. Phys. **82**(1), 64–84 (1989)
3. Bryan, G.L., et al.: Enzo: an adaptive mesh refinement code for astrophysics. Astrophys. J. Suppl. Ser. **211**(2), 19 (2014)
4. Burstedde, C., Wilcox, L.C., Ghattas, O.: p4est: scalable algorithms for parallel adaptive mesh refinement on forests of octrees. SIAM J. Sci. Comput. **33**(3), 1103–1133 (2011)
5. Cormen, T.H., Leiserson, C.E., Rivest, R.L., Stein, C.: Introduction to algorithms second edition. The Knuth-Morris-Pratt Algorithm (2001)
6. Dubey, A., et al.: A survey of high level frameworks in block-structured adaptive mesh refinement packages. J. Parallel Distrib. Comput. **74**(12), 3217–3227 (2014)
7. Ivanov, A.V., Khilkov, S.A.: Aiwlib library as the instrument for creating numerical modeling applications. Sci. Vis. **10**(1), 110–127 (2018). https://doi.org/10.26583/sv.10.1.09
8. Khokhlov, A.M.: Fully threaded tree algorithms for adaptive refinement fluid dynamics simulations. J. Comput. Phys. **143**(2), 519–543 (1998)
9. Kirk, B.S., Peterson, J.W., Stogner, R.H., Carey, G.F.: libmesh: a c++ library for parallel adaptive mesh refinement/coarsening simulations. Eng. Comput. **22**(3–4), 237–254 (2006). https://doi.org/10.1007/s00366-006-0049-3
10. Ivanov, A., Levchenko, V., Korneev, B., Perepelkina, A.: Management of computations with LRnLA algorithms in adaptive mesh refinement codes. In: Voevodin, V., Sobolev, S. (eds.) RuSCDays 2020. CCIS, vol. 1331, pp. 25–36. Springer, Cham (2020). https://doi.org/10.1007/978-3-030-64616-5_3
11. Ivanov, A., Perepelkina, A., Levchenko, V., Pershin, I.: Memory-optimized tile based data structure for adaptive mesh refinement. In: Voevodin, V., Sobolev, S. (eds.) RuSCDays 2019. CCIS, vol. 1129, pp. 64–74. Springer, Cham (2019). https://doi.org/10.1007/978-3-030-36592-9_6

12. Lutsky, A.E., Severin, A.V.: Numerical study of flow x–43 hypersonic aircraft using adaptive grids. Keldysh Institute Preprints (102) (2016)

13. MacNeice, P., Olson, K.M., Mobarry, C., De Fainchtein, R., Packer, C.: PARAMESH: a parallel adaptive mesh refinement community toolkit. Comput. Phys. Commun. **126**(3), 330–354 (2000)

14. Menshov, I., Sheverdin, V.: A parallel locally-adaptive 3D model on cartesian nested-type grids. In: Malyshkin, V. (ed.) PaCT 2017. LNCS, vol. 10421, pp. 136–142. Springer, Cham (2017). https://doi.org/10.1007/978-3-319-62932-2_12

15. Morton, G.M.: A computer oriented geodetic data base and a new technique in file sequencing (1966)

16. Museth, K.: VDB: high-resolution sparse volumes with dynamic topology. ACM Trans. Graph. (TOG) **32**(3), 1–22 (2013)

17. Pavlukhin, P., Menshov, I.: Parallel implicit matrix-free CFD solver using AMR grids. In: Journal of Physics: Conference Series, vol. 1141, p. 012035. IOP Publishing (2018)

18. Popinet, S.: Gerris: a tree-based adaptive solver for the incompressible Euler equations in complex geometries. J. Comput. Phys. **190**(2), 572–600 (2003)

19. Samet, H.: The quadtree and related hierarchical data structures. ACM Comput. Surv. (CSUR) **16**(2), 187–260 (1984)

20. Schive, H.Y., ZuHone, J.A., Goldbaum, N.J., Turk, M.J., Gaspari, M., Cheng, C.Y.: GAMER-2: a GPU-accelerated adaptive mesh refinement code - accuracy, performance, and scalability. Mon. Notices Royal Astron. Soc. **481**(4), 4815–4840 (2018). https://doi.org/10.1093/mnras/sty2586

21. Stout, Q.F., De Zeeuw, D.L., Gombosi, T.I., Groth, C.P.T., Marshall, H.G., Powell, K.G.: Adaptive blocks: a high performance data structure. In: Proceedings of the 1997 ACM/IEEE Conference on Supercomputing. SC 1997, pp. 1–10. ACM, New York (1997). https://doi.org/10.1145/509593.509650

22. Wahib, M., Maruyama, N., Aoki, T.: Daino: a high-level framework for parallel and efficient AMR on GPUs. In: SC 2016: Proceedings of the International Conference for High Performance Computing, Networking, Storage and Analysis, pp. 621–632 (2016). https://doi.org/10.1109/SC.2016.52

23. Weinzierl, T.: The Peano software–parallel, automaton-based, dynamically adaptive grid traversals. ACM Trans. Math. Softw. (TOMS) **45**(2), 1–41 (2019)

24. Williams, S., Waterman, A., Patterson, D.: Roofline: an insightful visual performance model for multicore architectures. Commun. ACM **52**(4), 65–76 (2009)

25. Zhang, W., et al.: AMReX: a framework for block-structured adaptive mesh refinement. J. Open Source Softw. **4**(37), 1370–1370 (2019)

Automatic Parallel Tiled Code Generation Based on Dependence Approximation

Wlodzimierz Bielecki$^{(\boxtimes)}$ and Maciej Poliwoda

Faculty of Computer Science and Information Systems, West Pomeranian University of Technology in Szczecin, Zolnierska 49, 71210 Szczecin, Poland
{wbielecki,mpoliwoda}@wi.zut.edu.pl
http://www.wi.zut.edu.pl

Abstract. The paper results concern automatic parallel program generation based on a program dependence approximation technique. Dependence approximation allows us to directly form linear time partition constraints necessary for extracting loop nest statement instance schedules to be used for parallel tiled code generation. The proposed approach was implemented in the publicly available DAPT optimizing compiler, which takes on its input the C program and automatically generates parallel tiled code in the OpenMP C/C++ standard. Empirically, we discovered that for some dynamic programming codes, DAPT generates tiled code whose tiles are of a larger dimension than that of tiles yielded with popular closely related optimizing compilers based on Farkas' lemma. This allows us to considerably increase code locality for such applications.

Keywords: Automatic code optimization · Tiling · Program dependence · Scheduling · OpenMP

1 Introduction

This paper deals with automatic parallel tiled code generation based on deriving loop nest statement instance schedules. Program loop tiling is a well-known compiler transformation for both sequential and parallel program optimization. It allows us to generate parallel high-performance code running on modern architectures by increasing code granularity and data locality.

State-of-the-art techniques of loop tiling are based on the affine transformation framework (ATF) [2,5,6,9], which is implemented in many optimizing compilers.

ATF envisages that the following tasks should be realized. For a given loop nest, dependences are extracted. They are used to form time partition constraints [6], which can be resolved to find as many as possible linearly independent solutions. Those solutions are used to form statement instance schedules applied for parallel tiled code generation.

© Springer Nature Switzerland AG 2021
V. Malyshkin (Ed.): PaCT 2021, LNCS 12942, pp. 260–275, 2021.
https://doi.org/10.1007/978-3-030-86359-3_20

In general, time partition constraints are represented with a system of non-linear integer equations and inequalities. Resolving an integer non-linear system is NP-hard. Feautrier suggested the application of Farkas' lemma to linearize time partition constraints [3,4]. Many state-of-the-art optimizing compilers successfully apply Farkas' lemma to linearize non-linear time partition constraints. However, such a linearization leads to growing the size of a resulting system of affine constraints. That increases the computational complexity of target code generation.

In this paper, we propose an alternative way to form linear time partition constraints. We suggest to approximate irregular (non-uniform) dependences with regular (uniform) ones. Then uniform dependences are used for forming linear time partition constraints.

Under an approximation of a distance vector, we mean deriving a set of uniform vectors that can be used instead of that vector to form time partition constraints.

We compare schedules obtained by means of the proposed approximation with those achieved with the classic linearization of time partition constraints [3,4] for Polybench benchmarks [8].

The main contributions of the paper are the following.

- Proposition of an approximation of irregular (non-uniform) dependences with regular (uniform) ones and proving its validity.
- Suggestion of a technique to extract linearly independent solutions to time partition constraints.
- Implementation of the propositions in a publicly available optimizing compiler, DAPT (Dependence Approximation for Parallelism and Tiling) and its usage for carrying out experiments.

2 Background

The loop nest can expose dependences among statement instances. A dependence is a situations when two statement instances access the same memory location and at least one of these accesses is write. Each dependence is represented with its source and destination provided that the source is executed before the destination.

To extract dependences available in the loop nest, we use the polyhedral model returned with PET [15] and apply the ISL library to implement calculations on polyhedral sets and relations in a way presented in paper [14] as well as the iscc calculator that is an interactive interface to the barvinok counting library available at http://barvinok.gforge.inria.fr/barvinok.pdf and PET.

For the dependence, a distance vector is the difference between the iteration vector of its destination and that of its source. If dependences are uniform then for each distance vector, all its elements are constant.

In this paper, we deal with both perfectly and arbitrarily nested loops. For imperfectly nested loops, for statements surrounded with the different number of loops, the length of iteration vectors is different that prevents calculating the distance vector when the source and destination of a dependence are represented with iteration vectors of different length. To resolve that problem, we apply the dependence relation normalization algorithm presented in paper [1]. This allows us to extract distance vectors such that the length of each of them is the same. Such a conversion is carried out automatically by means of our DAPT optimizing compiler whose details are presented in Sect. 7.

We call so calculated distance vectors as normalized ones. Dependences available in the loop nest can be uniform or non-uniform. Dependences are uniform when all the distances between dependent loop nest statement instances in the iteration space are constants; otherwise they are non-uniform.

Automatic tiling and parallelization of loop nests with non-uniform dependences is much difficult than those exposing uniform ones. The reason is that in general, for irregular dependences, constraints formed to extract affine transformations (to be next applied to tile and parallelize loops) are non-constant, this considerably increases the computational complexity of extracting affine transformations.

To optimize a loop nest (parallelize and tile), we should form *time-partition constraints* [6] that imply that if one statement instance is dependent upon the other, then in target code, the dependence destination must be executed no earlier than the corresponding dependence source. If they are assigned in the same iteration, then it is understood that the destination will be executed after than the source within the iteration. Details can be found in papers [2, 6]. A solution to time partition constraints is used to form schedules for each statement instance of a loop nest.

The schedule defines a strict partial order, i.e., an irreflexive and transitive relation, on the statement instances that specifies the order in which they are or should be executed. Obtained schedules are applied to generate parallel code, details are presented in papers [2, 6].

When dependences are non-uniform, time-partition constraints are non-linear too. Usually, they are linearized by means of applying Farcas' lemma that leads to increasing the constraints size and the computation complexity of code generation.

This is why, in this paper, to directly form affine constraints, we suggest a way to approximate original non-uniform dependences to uniform ones and prove the correctness of such an approximation.

We should resolve time partition constraints so to extract as many as possible linearly independent solutions. The more independent solutions are extracted, the larger target code parallelism degree and tile dimension are.

In this paper, we examine an alternative way to extract independent solutions to time partition constraints that is simpler than that used in the PLUTO compiler [2].

Using obtained linear independent solutions, parallel tiled code can be easily generated [2].

3   Idea of Dependence Approximation and Code Generation

The first task of the proposed approach is to approximate each normalized distance vector whose one or more elements are non-constant with vectors whose all elements are constants. This allows us to form affine time partition constraints, i.e., constraints without any non-linear term and to resolve affine constraints applying the ISL library [14].

3.1   Dependence Approximation

We distinguish the following cases of normalized distance vector elements: i) element is a constant, for example, 10; ii) element is an expression with a known lower bound and an unlimited upper bound, for example, $n \geq 2$; iii) element is an expression with a known upper bound and an unlimited lower bound, for example, $n \leq 2$; iv) element is an expression with unlimited upper and lower bounds, for example, n.

We denote the lower and upper bounds of iterator i as $lb(i)$ and $ub(i)$, respectively.

Let m normalized distance vectors be given $D_k, k = 1, 2, ..., m$, and the elements of vector D_k be $d_k^j, j = 1, 2, ..., n$, where n is the vector length. For each normalized distance vector, D_k, we form vector, V_k, of length n according to the following rule.

Rule 1

$v_k^j = d_k^j$ if d_k^j is constant,

$v_k^j = lb(d_k^j)$ if d_k^j is an expression with a known lower bound and unlimited upper bound, i.e., $lb(d_k) \leq d_k^j \leq \infty$,

$v_k^j = ub(d_k^j)$ if d_k^j is an expression with a known upper bound and unlimited lower bound, i.e., $-\infty \leq d_k^j \leq ub(d_k^j)$,

$v_k^j = 0$ if d_k^j is an expression with unlimited lower and upper bounds, i.e., $-\infty \leq d_k^j \leq \infty$.

Let us consider vector $D_k = (2, n \geq 4, m \leq 10, l)^T$. Then vector V_k is the following $V_k = (2, 4, 10, 0)^T$.

In addition to vectors $V_k, k = 1, 2, ..., m$, we form a set of vectors X whose number is the sum of the number of the non-constant elements of vectors $D_k, k = 1, 2, ..., m$ with an unlimited lower or upper bound.

Elements $x_p^i, i = 1, 2, ..., n$, of vector X_p are formed as follows.

Rule 2. If there exists an element d_k^j with an unlimited upper bound, i.e., $lb(d_k) \leq d_k^j \leq \infty$, we form the elements of vector X as follows.

$$x_p^i = \begin{cases} 0 \text{ if } i \neq j \\ 1 \text{ if } i = j. \end{cases}$$

Rule 3. If there exists an element d_k^j with an unlimited lower bound, i.e., $-\infty \leq d_k^j \leq ub(d_k^j)$, we form the elements of vector X_p as below.

$$x_p^i = \begin{cases} 0 & \text{if } i \neq j \\ -1 & \text{if } i = j. \end{cases}$$

Let us again consider vector $D_k = (2, n \geq 4, m \leq 10, l)^T$. According to Rule 2, we form vectors $X_1 = (0, 1, 0, 0)^T$, $X_2 = (0, 0, 0, 1)^T$ and according to Rule 3, we form vectors $X_3 = (0, 0, -1, 0)^T$, $X_4 = (0, 0, 0, -1)^T$.

So, for vector $D_k = (2, n \geq 4, m \leq 10, l)^T$, all vectors formed according to Rules 1, 2, and 3 are the following.
$V_k = (2, 4, 10, 0)^T$, $X_1 = (0, 1, 0, 0)^T$, $X_2 = (0, 0, 0, 1)^T$, $X_3 = (0, 0, -1, 0)^T$, and $X_4 = (0, 0, 0, -1)^T$.

It is worth noting that all elements of those vectors are constants.

Applying Rules 1–3 to a normalized distance vector whose all elements are constants results in forming the same vector while their applying to a normalized distance vector whose one or more elements are non-constant results in an approximation of this vector with a set of uniform vectors.

For a given normalized distance vector with non-constant elements, D, extracted for a loop nest of depth dp, a set including vector V and vector(s) X approximates vector D so that it can be represented as a linear combination of vector V and vectors X, i.e., each element $d_i, i = 1, 2, ..., dp$, of vector D can be represented as follows.

$$d_i = v_i + \begin{cases} a_i * (x_i^1(d_i) = 1) & \text{if } d_i \geq v_i \\ b_i * (x_i^2(d_i) = -1) & \text{if } d_i < v_i \\ c_i * (x_i^3(d_i) = 1) + e_i * (x_i^4(d_i) = -1) \text{ if} \\ d_i \text{ is unbounded} : -\infty \leq d_i \leq \infty, \end{cases}$$

where $a_i, b_i, c_i, e_i \geq 0$ are constants, $x_i^1(d_i), x_i^2(d_i), x_i^3(d_i), x_i^4(d_i)$ are the $i - th$ elements of the corresponding vectors X. For example, vector $D = (2, n \geq 4, m \leq 10, l)^T$ can be represented as follows

$$D = (2, n \geq 4, m \leq 10, l)^T = (2, 4, 10, 0)^T + a * (0, 1, 0, 0)^T + b * (0, 0, -1, 0)^T + c * (0, 0, 0, 1)^T + e * (0, 0, 0, -1)^T, \exists a, b, c, d \geq 0.$$

We form time partition constraints according to paper [6], which state that if iteration I of statement $S1$ depends on iteration J of statement $S2$, then I must be assigned to a time partition that executes no earlier than the partition containing J, i.e., schedule(I) $\leq$ schedule(J), where schedule(I) and schedule(J) denote the discrete execution time of iterations I and J, respectively.

For the normalized loop nest, the general form of time partition constraints is the following.

$$\wedge_{i=1}^n H \bullet D_i \geq 0,$$

i.e., each scalar product of vectors H and $D_i, i = 1, 2, ..., n$ should be greater or equal to 0, where H is the unknown vector, D_i is the normalized distance vector.

For example, the time partition constraint for the vector $D_k = (2, n \geq 4, m \leq 10, l)^T$ is the following.

$2 * h_1 + n * h_2 + m * h_3 + l * h_4 \geq 0.$

That constraint is non-linear. To form affine time partition constraints, instead of the vector $D_k = (2, n \geq 4, m \leq 10, l)^T$, we use a set of vector V and vectors X and obtain:

$$2 * h_1 + 4 * h_2 + 10 * h_3 + l * h_4 \geq 0$$
$$h_2 \geq 0$$
$$h_3 \leq 0$$
$$h_4 \geq 0$$
$$h_4 \leq 0$$

or

$$2 * h_1 + 4 * h_2 + 10 * h_3 + l * h_4 \geq 0$$
$$h_2 \geq 0$$
$$h_3 \leq 0$$
$$h_4 = 0.$$

It is worth noting that the usage of vectors X for forming time partition constraints results in simple constraints of the form $h_i \geq 0, h_i \leq 0$, or $h_i = 0$.

To prove the correctness of such a substitution, we fulfil the following transformations. Because $h_4 = 0$, we have.

$$2 * h_1 + n * h_2 + m * h_3 + l * h_4 =$$
$$2 * h_1 + n * h_2 + m * h_3 \geq 0.$$

Since $h_2 \geq 0$ and $n \geq 4$, we obtain.

$$2 * h_1 + n * h_2 + m * h_3 \geq$$
$$2 * h_1 + 4 * h_2 + m * h_3 \geq 0.$$

Finally, because $h_3 \leq 0$ and $m <= 10$, we get

$$2 * h_1 + 4 * h_2 + m * h_3 \geq$$
$$2 * h_1 + 4 * h_2 + 10 * h_3 \geq 0.$$

So, we have

$$2 * h_1 + n * h_2 + m * h_3 + l * h_4 \geq$$
$$2 * h_1 + 4 * h_2 + 10 * h_3 \geq 0$$

and consequently we may conclude that if the constraint

$$2 * h_1 + 4 * h_2 + 10 * h_3 \geq 0$$

is satisfied, then the original non-linear constraint

$$2 * h_1 + n * h_2 + m * h_3 + l * h_4 \geq 0$$

is also satisfied.

So, to form time partition constraints, instead of vector D_k we may use a set of vector V and vectors X.

3.2   Time Partitions Constrains and Finding Linearly Independent Solutions to Them

In general, given a set of n vectors V and a set of m vectors X, we form linear time partition constraints as follows.

$$\wedge_{i=1}^{n} H \bullet V_i \geq 0 \ \wedge \ \wedge_{i=1}^{m} H \bullet X_i \geq 0, \tag{1}$$

where $H \bullet V_i$ and $H \bullet X_i$ are the scalar products of vectors H, V and H, X, respectively, H is an unknown vector.

For example, given vectors $V_1 = (1,0)^T$, $V_2 = (1,1)^T$, $X_1 = (1,0)^T$, $X_2 = (0,-1)^T$, we build the following constraints.

$$1 * h_1 + 0 * h_2 \geq 0 \wedge 1 * h_1 + 1 * h_2 \ \geq \ 0 \ \wedge$$
$$1 * h_1 + 0 * h_2 \geq 0 \wedge 0 * h_1 + (-1) * h_2 \geq 0.$$

There may exist many solutions to the time partition constraints above. We should avoid the zero solution to prevent mapping all statement instances to the same execution time that averts any parallelism. With this aim, we apply the following additional constraint:

$$h_1 \neq 0 \ \vee \ h_2 \neq 0 \ \vee \ ... \vee \ h_{dp} \neq 0, \tag{2}$$

where $h_i, i = 1, 2, ..d$ are the unknown elements of vector H, dp is the loop nest depth. I.e., we require that at least one element of H should be non-zero.

In addition to the requirement above, we strive to find such vector H that allows us to obtain the value of $|H \bullet D_i|$ for each of n normalized distance vectors $D_i, i = 1, 2, ..., n$, extracted for a given loop nest that is as close as possible to its minimal value. The goal is to enhance target code locality, details can be found in paper [2]. Striving to that goal, we use the following two heuristic: i) the minimization of the value

$$\sum_{i=1}^{dp} |h_i|$$

and ii) the maximization of the number of positive elements of vector H.

We experimentally discovered that satisfying conditions (1) and (2) as well as applying heuristics i) and ii) allow us to 1) to reduce the number of synchronization events in target code (the number of iterations of the outermost loop), 2) extract linearly independent solutions to time partition constraints for all loop nests examined by us without applying the mathematical method implemented in the PLUTO compiler [2].

To implement those heuristics, we introduce two additional variables b_0 and c_0 as well as the corresponding constraints that allow us to calculate the values of b_0 and $|c_0|$ as the minimum of

$$\sum_{i=1}^{dp} |h_i|$$

and the maximum of the number of positive elements of vector H, respectively. Then, we form a set whose tuple elements are $b_0, c_0, h_1, h_2, ..., h_{dp}$ and the constraints are formed as the conjunction of the time partition constraints defined above and the constraints imposed on variables b_0 and c_0.

Finally, we find the lexicographically minimal vector $(b_0, c_0, h_1, h_2, ..., h_{dp})^T$ represented with the tuple of the set constructed as described above.

Let us consider the following example: $V = (1, 2)^T$, $X = (0, -1)^T$. The resulting constraints for those vectors provided that $h_1 \neq 0$ are the following.

$\{[b_0, c_0, h_1, h_2] : h_1 \neq 0 \land \exists \, b_1, c_1, b_2, c_2 \; s.t \; ((b_1 = h_1 \land h_1 \geq 0) \lor (b_1 = -h_1 \land h_1 < 0)) \land ((c_1 = -1 \land h_1 \geq 0) \lor (c_1 = 0 \land h_1 < 0)) \land ((b_2 = h_2 \land h_2 \geq 0) \lor (b_2 = -h_2 \land h_2 < 0)) \land ((c_2 = -1 \land h_2 \geq 0) \lor (c_2 = 0 \land h_2 < 0)) \land b_0 > 0 \land b_0 = b1 + b2 \land c_0 = c_1 + c_2 \land h_2 <= 0 \land 1 * h_1 + 2 * h_2 \geq 0\}.$

Applying the iscc *lexmin* operator to the set above, we obtain the following solution $\{[1, -2, 1, 0]\}$, i.e., $b_0 = 1, c_0 = -2, h_1 = 1, h_2 = 0$.

$b_0 = 1$ means that the sum of $|h_1|$ and $|h_2|$ is 1, $c_0 = -2$ or $|c_0| = 2$ implies that the number of non-negative solutions is 2, $h_1 = 1, h_2 = 0$ are elements of vector H_1, i.e., $H_1 = (1, 0)^T$.

Replacing in the set above the term $h_1 \neq 0$ for the term $h_2 \neq 0$, we get the following solution $\{[3, -1, 2, -1]\}$. So, vector H_2 is the following.

$H_2 = (2, -1)^T$.

Let us remind that we try to find as many solutions as the depth of the original loop nest, dp. We seek for dp solutions applying in the constraints consequentially $h_i \neq 0, i = 1, 2, ..., dp$.

In general, the procedure above does not guarantee that all solutions would be linearly independent. So, finally we check whether each solution is not a linear combination of the remanding ones. If so, we discard each linearly dependent solution.

For this purpose, we check whether there exist constants $C_1, C_2, ..., C_{dp}$ such that the equation below is satisfied.

$$C_i * H_i = \sum_{j \neq i} C_j * H_j, C_i \neq 0, i = 1, 2, ..., dp.$$

If so, vector H_i is discarded.

3.3   Parallel Tiled Code Generation

To form parallel tiled code, we apply the well-known algorithm presented in paper [2] based on affine schedules and wave-fronting technique. To generate target code with m parallel loops enumerating tile identifiers, we use $m+1$ vectors $H_i, i = 1, 2, ..., m+1$. Let the serial tiled code has iteration vector $IT = (ID, I)^T$, where $ID = (id_1, id_2, ..., id_{dp})^T$, $I = (i_1, i_2, ..., i_{dp})^T$. Then, the first iteration index of a target loop nest is formed as $id_1 = H_1 + H_2 + ... + H_{m+1}$, loops defined with indexes $id_2, id_3, ..., id_{m+1}$ are marked as parallel, the reminding loops are marked as serial.

4   Formal Algorithm

Below we present a formal algorithm to approximate original normalized distance vectors and extract affine transformations.

Algorithm 1. Extracting affine transformations via dependence approximation.

Input: n normalized distance vectors $D_k, k = 1, 2, ..., n$, with the lower and upper bounds of each non-constant element d_k^j, $lb(d_k^j)$ and $ub(d_k^j)$, respectively, dp is the loop nest depth.
Output: Affine transformations presented with a set of vectors H.
Method:

1. For each normalized distance vector D_k with elements $d_k^j, j = 1, 2, ..., dp$ form vector, V_k, of length dp as follows

   $v_k^j = d_k^j$ if d_k^j is constant,

   $v_k^j = lb(d_k^j)$ if d_k^j is an expression with a known lower bound and unlimited upper bound, i.e., $lb(d_k) \leq d_k^j \leq \infty$,

   $v_k^j = ub(d_k^j)$ if d_k^j is an expression with a known upper bound and unlimited lower bound, i.e., $-\infty \leq d_k^j \leq ub(d_k^j)$,

   $v_k^j = 0$ if d_k^j is an expression with unlimited lower and upper bounds, i.e., $-\infty \leq d_k^j \leq \infty$.

2. In addition to vectors $V_k, k = 1, 2, ..., m$, form a set of vectors X as follows. If there exists an element d_k^j with an unlimited upper bound, i.e., $lb(d_k) \leq d_k^j \leq \infty$, form the elements of vector X as follows

$$x_i = \begin{cases} 0 \text{ if } i \neq j \\ 1 \text{ if } i = j. \end{cases}$$

If there exists an element d_k^j with an unlimited lower bound, i.e., $-\infty \le d_k^j \le ub(d_k^j)$, form the elements of vector X as below

$$x_p^i = \begin{cases} 0 & \text{if } i \ne j \\ -1 & \text{if } i = j. \end{cases}$$

3. Form time partition constraints as follows

$$\wedge_{i=1}^n H \bullet V_i \ge 0 \ \wedge \ \wedge_{i=1}^m H \bullet X_i \ge 0,$$

where m and n are the numbers of vectors V and X, respectively, $H \bullet V_i$ and $H \bullet X_i$ mean the scalar product of vectors H, V and H, X, respectively.

4. Try to find dp solutions to the time partition constraints formed in the previous step satisfying the following requirements i) the minimization of the value

$$\sum_{i=1}^{dp} |h_i|,$$

ii) the maximization of the number of positive elements of vector H; iii) $h_i \ne 0$ for each sequential solution $i = 1, 2, ..., dp$.

5. Check whether the solutions obtained in the previous step are linearly independent. If not, discard each dependent solution. For this purpose, check whether for each $H_i, i = 1, 2, ..., dp$, there exists a solution to the equation below

$$C_i * H_i = \sum_{j \ne i} C_j * H_j, C_i \ne 0, i = 1, 2, ..., dp.$$

If so, discard vector H_i.

To prove the correctness of the presented algorithm, let us consider the time partition constraint built on the basis of an original normalized distance vector D_k:

$\sum_{i=1}^{dp} d_i * h_i \ge 0,$

where $h_i, i = 1, 2, ..., dp$ are the elements of vector H.

Let i) the corresponding vector V be the following

$V = (v_1, v_2, .., v_{dp})^T;$

ii) set x_1 include the positions of elements of vector X such that there exist two elements in the same position, say i, and one element is equal to 1 while the second is equal to -1; such vector positions imply that element h_i is 0;

iii) sets x_2, x_3 include the element positions of vector X whose values are 1 and -1, respectively, that is equivalent to $h_2 \ge 0$ and $h_3 \le 0$, respectively.

Because each $h_i, i \in x_1$ is zero, it is evident that

$$\sum_{i}^{dp} d_i * h_i = \sum_{i,i \notin x_1}^{dp} d_i * h_i.$$

Since each element $h_i, i \in x_2$ is greater or equal to 0 ($h_i \geq 0$) and $d_i \geq v_i$, we obtain

$$\sum_{i,i \notin x_1}^{dp} d_i * h_i \geq \sum_{i,i \notin x_1, i \in x_2}^{dp} v_i * h_i + \sum_{i,i \notin x_1, i \in x_3}^{dp} d_i * h_i.$$

Because each element $h_i, i \in x_3$ is less than or equal to 0 ($h_i \leq 0$) and $d_i \leq v_i$, we get

$$\sum_{i,i \notin x_1, i \in x_2}^{dp} v_i * i_i + \sum_{i,i \notin x_1, i \in x_3}^{dp} d_i * h_i \geq$$
$$\sum_{i,i \notin x_1, i \in x_2}^{dp} v_i * h_i + \sum_{i,i \notin x_1, i \in x_3}^{dp} v_i * h_i = \sum_{i,i \notin x_1}^{dp} v_i * h_i \geq 0.$$

So,

$$\sum_{i}^{dp} d_i * h_i = \sum_{i,i \notin x_1}^{dp} d_i * h_i \geq \sum_{i,i \notin x_1}^{dp} v_i * h_i \geq 0.$$

This means that solutions h_i satisfying the time partition constraints based on vectors V and X also satisfy the time partition constraints formed on the basis of vector D_k. This is true for each of n vectors $D_k, k = 1, 2, ..., n$ representing all loop nest dependences. Hence, to form time partition constraints, instead of vectors D_k, we may use a set of vectors V and X. That proves the validity of the substitution of vectors D_k for vectors V and X with the aim of forming affine time partition constraints.

5   Related Work

In state-of-the-art techniques, the linearization of time partition constraints is based on Farkas' lemma [3,4]. Let us consider the following dependence relation.

$$R := N \rightarrow \{ (i,j) \rightarrow (j,i) \mid 0 < i \leq N \wedge j - i > 1 \wedge 2 \leq j \leq N \}.$$

For that relation, the time partition constraints are the following

$$(h_1 j + h_2 i) - (h_1 i + h_2 j) \geq 0, 2 \leq j \leq N, 1 \leq i \leq N, j - i \geq 1.$$

Applying Farkas' lemma [11], we obtain.

$(h_2 - h_1)i + (h_1 - h_2)j \equiv \lambda_0 + \lambda_1(N - i) + \lambda_2(N - j) + \lambda_3(j - i - 1) + \lambda_4(i - 1) + \lambda_5(j - 2),$

where $\lambda_0, \lambda_1, \lambda_2, \lambda_3, \lambda_4, \lambda_5 \geq 0$ are the Farkas multipliers.

In the equation above, the LHS and RHS coefficients for i, j, N and the constants are equated to obtain:

$$h_2 - h_1 = -\lambda_1 - \lambda_3 + \lambda_4$$
$$h_1 - h_2 = -\lambda_2 + \lambda_5 + \lambda_3$$
$$\lambda_1 + \lambda_2 = 0$$
$$\lambda_0 - \lambda_3 - \lambda_4 - 2\lambda_5 = 0$$
$$\lambda_0 \geq 0$$
$$\lambda_1 \geq 0$$
$$\lambda_2 \geq 0$$
$$\lambda_3 \geq 0$$
$$\lambda_4 \geq 0$$
$$\lambda_5 \geq 0.$$

Eliminating the Farkas multipliers through Gaussian and Fourier-Motzkin eliminations [11] yields:

$$h_1 - h_2 \geq 0.$$

So, to obtain the constraint above, we should i) apply the Farkas' lemma to non-linear constraints, ii) form additional linear constraints to eliminate Farkas multipliers, iii) eliminate Farkas multipliers from the additional constraints.

Applying the approach presented in this paper, we first calculate the normalized distance vector using the iscc *deltas* operator to relation R and obtain

$$D = N \rightarrow \{ (i, -i) \mid 0 < i < N \}.$$

Next according to Algorithm 1, we approximate the vector above to the following vector

$$V = (1, -1)^T.$$

Using that vector, we directly obtain the following time partition constraint

$$h_1 - h_2 \geq 0$$

without applying Farkas' lemma.

I.e., obtaining linear time partition constraints by means of the proposed approach is characterized by a lower computational complexity in comparison with that based on Farkas' lemma.

Converting non-uniform dependences to a set of uniform ones is considered in many papers, for example, [10,12,13]. All proposed approaches use the same basic idea of routing data in a uniform manner through the computation domains of the system of recurrence equations. They, however, differ in the class of recurrence equations considered and the requirements that are imposed in the technique. Although many uniformization techniques are proposed, their implementation into publicly available design tools has been limited.

Paper [7] unifies previously published uniformization techniques and also highlights some of the problems that need to be overcome. The authors demonstrate how auxiliary transformations can be used to enhance a uniformization framework by avoiding pipelining and routing under certain conditions. The authors have also implemented their algorithms and integrated them into the MMAlpha system, which is freely available under the GNU public license. But that paper does not consider any parallel tiled code generation based on dependence uniformization.

So, to our best knowledge, using dependence uniformization to form time partition constraints with the aim to generate parallel tiled code never was presented in publications.

6   Experiment

The proposed approach was implemented in the publicly available optimizing compiler, DAPT (Dependence Approximation for Parallelism and Tiling) by means of the ISL library [14]. The DAPT sources are available at https://sourceforge.net/projects/dapt/files/. For a given loop nest, DAPT extracts dependence relations and next normalizes them to calculate normalized distance vectors (each vector has the same length), approximates irregular dependences to regular ones, forms linear time partition constraints, resolves those constraints to obtain the maximum number of linearly independent solutions to be used for schedule extraction and generates parallel tiled code in the OpenMP C/C++ standard.

6.1   Comparison of Schedules Generated for Polybench Benchmarks

Using DAPT as well as the ISL scheduler [14], we experimented with PolyBench (http://www.cs.ucla.edu/pouchet/software/polybench) – a benchmark suite of 30 numerical computations with static control flow, extracted from operations in various application domains (linear algebra computations, image processing, physics simulation, dynamic programming, statistics, etc.).

The goal of experiments was forming schedules with the proposed approach as well as with the ISL scheduler based on deriving affine schedules in the classic way – linearization with Farkas' lemma.

Schedules returned with the ISL and DAPT schedulers were obtained for the same set of normalized dependence relations for each Polybench benchmark. Those schedules are presented at https://sourceforge.net/projects/dapt/files/.

Analysing schedules returned with ISL and DAPT, we conclude that they are the same for each benchmark. This means that for examined normalized loops, target codes generated with those compilers are the same provided that the same code generation algorithm is used.

6.2   Examining Tile Dimensions for Dynamic Programming Codes

We experimented with various dynamic programming applications to discover tile dimensions generated with DAPT and PLUTO using Farkas' lemma to linearize time partition constraints. We observed that for many dynamic programming applications that expose irregular dependences, DAPT generates tiles whose dimension is larger than that of tiles yielded with PLUTO. In this paper, we present only one example shown below.

```
for (int c0 = 2; c0 < n; c0 += 1)
 for (int c1 = 1; c1 <= n - c0; c1 += 1)
  for (int c2 = c0 + c1; c2 <= min(n, 2 * c0 + c1 - 2);
       c2 += 1) {
   if (2 * c0 + c1 >= c2 + 3)
     c[c1][c2]=min(c[c1][c2]),w[c1][c2]+
     c[c1][-c0 + c2 + 1]+c[-c0 + c2 + 1][c2]);
     c[c1][c2] = min(c[c1][c2]),w[c1][c2]+
     c[c1][c0 + c1 - 1]+c[c0 + c1 - 1][c2]);
```

That code implements an optimal binary search tree dynamic programming algorithm. For that code, DAPT generates the following parallel tiled code representing 3D tiles of size $16 \times 16 \times 16$.

```
for (int c1=2; c1<=floord(n - 2, 8) + 2; c1 +=1) {
  #pragma omp parallel for
  for (int c2 = max(-((n + 15) / 16),
          -c1 - (n + 14) / 16 + 2);
          c2 <= -c1 + c1 / 2; c2 += 1) {
    for (int c3 = max(0, -c1 - c2 + (c1 + 1) / 3);
             c3 <= -c1 - c2 + c1 / 2; c3 += 1) {
      for (int c5 = max(max(2, 16 * c1 +
         16 * c2 + 16 * c3 - 15), -8 * c2 -
         8 * c3 - 14);   c5 <= min(min(min(n - 1,
         -16 * c2 - 1), 16 * c1 + 16 * c2 + 16 * c3),
         n - 16 * c3);   c5 += 1) {
      for (int c6 = max(max(-16 * c3 - 15, -n + c5),
             16 * c2 + c5);
             c6 <= min(min(-1, -16 * c3),
             16 * c2 + 2 * c5 + 13); c6 += 1) {
        for (int c7 = max(-16 * c2 - 15, c5 - c6);
             c7 <= min(min(n, -16 * c2),
             2 * c5 - c6 - 2); c7 += 1) {
          if (2 * c5 >= c6 + c7 + 3) {
          c[-c6][c7] = ((c[-c6][c7] <
          ((w[-c6][c7] + c[-c6][-c5 + c7 + 1]) +
          c[-c5 + c7 + 1][c7])) ? c[-c6][c7] :
          ((w[-c6][c7] + c[-c6][-c5 + c7 + 1]) +
          c[-c5 + c7 + 1][c7]));}
        c[-c6][c7] = ((c[-c6][c7] < ((w[-c6][c7] +
        c[-c6][c5 - c6 - 1]) +
```

```
c[c5 - c6 - 1][c7])) ? c[-c6][c7] :
((w[-c6][c7] +  c[-c6][c5 - c6 - 1]) +
c[c5 - c6 - 1][c7]));}}}}}}
```

The PLUTO compiler (version 0.11.4) generates code representing only 2D tiles. That code is much more complex (includes many lines), so we inserted it at the website https://sourceforge.net/projects/dapt/files/ in the PLUTO catalog, that code represents tiles of size 16×16. This means that for the time partition constraints formed for that loop nest on the basis of dependence approximation, DAPT finds and applies for code generation three independent solutions, while for the time partition constraints formed due to applying Farkas' lemma, PLUTO extracts only two independent solutions.

We carried out a performance analysis of those codes on a multicore computer: Intel(R) Xeon(R) CPU X5570 @ 2.93 GHz, 16 cores, 24 GB, 8192 KB Intel® Smart Cache and discovered that the 3D tiled code is 3 to 7 times faster than the 2D PLUTO tiled code dependent on the number of threads varied by us from 2 to 32. Such a performance improving is due to better code locality of the 3D tiled code in comparison with that of the 2D tiled one.

7   Conclusions

We presented a simple way to approximate non-regular dependences available in the loop nest to regular ones and proved its correctness. This allows us to directly form linear time partition constraints necessary for extracting loop nest statement instance schedules. The approach is implemented in the publicly available DAPT compiler. We discovered that for some dynamic programming codes, DAPT generates code representing 3D tiles while closely related compilers using Farkas' lemma for constraint linearization, yield code exposing only 2D tiles.

References

1. Bielecki, W., Palkowski, M., Skotnicki, P.: Generation of parallel synchronization-free tiled code. Computing **100**(3), 277–302 (2017). https://doi.org/10.1007/s00607-017-0576-3
2. Bondhugula, U., Hartono, A., Ramanujam, J., Sadayappan, P.: A practical automatic polyhedral parallelizer and locality optimizer. In: Proceedings of the 29th ACM SIGPLAN Conference on Programming Language Design and Implementation, pp. 101–113 (2008)
3. Feautrier, P.: Some efficient solutions to the affine scheduling problem. I. One-dimensional time. Int. J. Parallel Prog. **21**(5), 313–347 (1992). https://doi.org/10.1007/BF01407835
4. Feautrier, P.: Some efficient solutions to the affine scheduling problem. Part II. Multidimensional time. Int. J. Parallel Prog. **21**(6), 389–420 (1992). https://doi.org/10.1007/BF01379404
5. Irigoin, F., Triolet, R.: Supernode partitioning. In: Proceedings of the 15th ACM SIGPLAN-SIGACT Symposium on Principles of Programming Languages, pp. 319–329 (1988)

6. Lim, A.W., Cheong, G.I., Lam, M.S.: An affine partitioning algorithm to maximize parallelism and minimize communication. In: Proceedings of the 13th International Conference on Supercomputing, pp. 228–237 (1999)
7. Manjunathaiah, M., Megson, G.M., Rajopadhye, S., Risset, T.: Uniformization of affine dependence programs for parallel embedded system design. In: 2001 International Conference on Parallel Processing, pp. 205–213. IEEE (2001)
8. Pouchet, L.N., et al.: Polybench: The polyhedral benchmark suite (2012). http://www.cs.ucla.edu/pouchet/software/polybench
9. Ramanujam, J., Sadayappan, P.: Tiling multidimensional iteration spaces for multicomputers. J. Parallel Distrib. Comput. **16**(2), 108–120 (1992)
10. Rapanotti, L., Megson, G.M.: Uniformisation techniques for reducible integral recurrence equations. In: Algorithms and Parallel VLSI Architectures III, pp. 283–294. Elsevier (1995)
11. Schrijver, A.: Theory of Linear and Integer Programming. Wiley, Hoboken (1998)
12. Shang, W., Hodzic, E., Chen, Z.: On uniformization of affine dependence algorithms. IEEE Trans. Comput. **45**(7), 827–840 (1996)
13. Tzen, T.H., Ni, L.M.: Dependence uniformization: a loop parallelization technique. IEEE Trans. Parallel Distrib. Syst. **4**(5), 547–558 (1993)
14. Verdoolaege, S.: isl: an integer set library for the polyhedral model. In: Fukuda, K., Hoeven, J., Joswig, M., Takayama, N. (eds.) ICMS 2010. LNCS, vol. 6327, pp. 299–302. Springer, Heidelberg (2010). https://doi.org/10.1007/978-3-642-15582-6_49
15. Verdoolaege, S., Grosser, T.: Polyhedral extraction tool. In: Second International Workshop on Polyhedral Compilation Techniques (IMPACT 2012), Paris, France, pp. 1–16 (2012)

Experimental Studies

Scalability Issues in FFT Computation

Alan Ayala[1]([✉]) [ID], Stanimire Tomov[1], Miroslav Stoyanov[2],
and Jack Dongarra[1,2,3]

[1] Innovative Computing Laboratory - UT, Knoxville, TN 37996, USA
aayala@icl.utk.edu
[2] Oak Ridge National Laboratory, Oak Ridge, TN 37830, USA
[3] University of Manchester, Manchester M13 9PL, UK

Abstract. The fast Fourier transform (FFT), is one the most important
tools in mathematics, and it is widely required by several applications
of science and engineering. State-of-the-art parallel implementations of
the FFT algorithm, based on Cooley-Tukey developments, are known
to be communication-bound, which causes critical issues when scaling
the computational and architectural capabilities. In this paper, we study
the main performance bottleneck of FFT computations on hybrid CPU
and GPU systems at large-scale. We provide numerical simulations and
potential acceleration techniques that can be easily integrated into FFT
distributed libraries. We present different experiments on performance
scalability and runtime analysis on the world's most powerful supercomputers today: Summit, using up to 6,144 NVIDIA V100 GPUs, and
Fugaku, using more than one million Fujitsu A64FX cores.

Keywords: Scalability · Parallel FFT · Hybrid systems

1 Introduction

The fast Fourier transform (FFT) is a key mathematical tool and widely used
in a variety of fields in science and engineering. In essence, the FFT of x, an m-
dimensional vector of size $N := N_1 \times N_2 \times \cdots \times N_m$ is defined by $y := FFT(x)$,
which is obtained as follows,

$$\tilde{y} := \sum_{n_1=0}^{N_1-1} \sum_{n_2=0}^{N_2-1} \cdots \sum_{n_m=0}^{N_m-1} \tilde{x} \cdot e^{-2\pi i \left(\frac{k_1 n_1}{N_1} + \frac{k_2 n_2}{N_2} \cdots + \frac{k_m n_m}{N_m} \right)}, \qquad (1)$$

where $\tilde{y} = y(k_1, k_2, \ldots, k_m)$, and $\tilde{x} := x(n_1, n_2, \ldots, n_m)$.

From Eq. 1, we see that the FFT could be directly computed by a tensor
product. However, this would cost $\mathcal{O}(N \sum_{i=1}^{m} N_i)$, while the advantage of the
FFT is that the cost can be reduced to $\mathcal{O}(N \log_2 N)$ operations.

This research was supported by the Exascale Computing Project (ECP), Project Number: 17-SC-20-SC, a collaborative effort of two DOE organizations (the Office of Science
and the National Nuclear Security Administration) responsible for the planning and
preparation of a capable exascale ecosystem.

© Springer Nature Switzerland AG 2021
V. Malyshkin (Ed.): PaCT 2021, LNCS 12942, pp. 279–287, 2021.
https://doi.org/10.1007/978-3-030-86359-3_21

The parallel FFT is implemented by a sequence of 1-D or 2-D FFTs, see e.g., [13], which are computed using efficient intra-node optimized libraries, such as FFTW [11] and CUFFT [1]. Figure 1 shows the steps to perform a 3-D FFT, typical in molecular dynamics, c.f., [14,17]. For some applications the input data has a shape ready to perform one-dimensional (*pencils*) or two-dimensional (*slabs*) FFTs and no initial nor final reshaping is needed. In [5], authors showed that saving one reshape step can accelerate the runtime around 25%, since, asymptotically, the multi-dimensional FFT runtime is dominated by the number of data-reshapes.

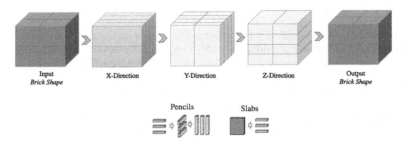

Fig. 1. Sequence for the computation of 3-D FFTs. If slab decomposition is possible, then an extra reshape step is saved.

In the current state-of-the-art, many authors have reported the impact of multi-process communication on distributed FFT performance [4,12,16,18], using both binary and collective MPI communication schemes that are available in current libraries. In this paper, we study these performance impacts from a numerical perspective, with a focus on architecture and algorithmic tuning for better performance. We analyze the effects of the communication bottleneck on scalability and provide techniques to maintain linear scaling. In Sect. 2, we make evident how FFT computation halts scaling even using latest efforts on MPI communication and their ability to perform CUDA-aware communication and specialized MPI for accelerators such as the NCLL library from NVIDIA [2]. This is critical for upcoming exascale system with millions of cores [6]. When addressing how network topology issues break scalability, we also provide techniques to prevent them.

Finally, the FFT is a key component for applications ranging from electronics to molecular dynamics. It is used at small and large scale; as within software targeting exascale (e.g., LAMMPS [14] and HACC [10]) and those from the machine learning community [15]. Such applications are being prepared for very large computing systems, with hybrid components and complex topologies. Therefore, it is critical to ensure parallel FFT scalability at large-scale.

2   Parallel FFT Performance Bottleneck

A major issue with distributed hybrid FFTs is that, due to the sheer compute capabilities of today's supercomputers, the algorithm quickly becomes communication bound. Such type algorithms, where already studied and authors in [8] warned of their effect on upcoming large-scale clusters. In [7], authors performed an extensive theoretical analysis on hybrid systems targeting exascale and realized that the FFT computation itself would take only a small fraction of the total run time, while the communication between processors would be the bottleneck where most of the run-time is spent. Nowadays, computing systems have greatly increased their computation power but their communication features have not been increased in the same proportion. For example, Summit supercomputer uses powerful nodes with two IBM POWER9 processors and six Nvidia V100 GPUs capable of reaching 42,000 GFlop/s in double precision, but the interconnect between the nodes is supported by a bandwidth of just 25 GB/s. Another supercomputer, the Sunway TaihuLight, has SW26010 processors with 260 cores, and 1 execution thread per core, with a unidirectional bandwidth of 8 GB/s between nodes and $1\,\mu$ of latency [9]. It therefore becomes critical to develop techniques and methodologies that help us of dealing with limited communication capabilities, together with an ecosystem of integrated tuning techniques for better communication frameworks. Such approaches are crucial in general and are paramount for the FFT, where communication can take more than 95% of total run-time on the latest GPU-accelerated supercomputers [4,5].

2.1   Scalability Issues

The recent developments of parallel FFT libraries capable of handling CPU and GPU components at the same time, has allowed considerable speedups in computation. However, this is highly limited by the communication bottleneck which has a considerable impact even for small-scale problems (due to latency issues) [5]. The bottleneck behaves different for every architecture and no general conclusion can be given on optimization criteria. For instance, experiments from [4,12,18] show that MPI All-to-All communication, was, in general, the best behaving methodology for data exchange in Summit-like architectures; however, as it can be seen in Fig. 2, in some systems, such as Fugaku, at large-scale, All-to-All (A2A) communication drastically fails to scale. An alternative for this case is to switch to binary MPI communication (P2P) which helps to keep a linear scaling. Note, however, that for a given problem size, if it is too small compared with the number of resources, then the scalability will also start to break, due to increased latency, see for example the P2P curve for the 256^3 problem. The experiment was performed using *heFFTe* library [3].

2.2   Peak Performance Model

When making a software contribution on parallel implementation, it is important to see how well the performance approaches to the machine theoretical peak.

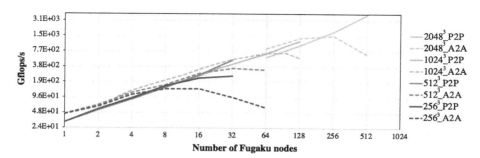

Fig. 2. Strong scaling of a 3-D FFTs. Using 48 A64FX cores per node. Comparing the scalability of A2A and P2P approaches using double-complex precision data.

Since the bandwidth injection between a single node is, in general, very high compared to inter-node injection. We developed a mathematical model for the theoretical performance peak on a supercomputer, c.f., [5, Sec. 3], given as:

$$\Phi := \frac{5P \log(N) B}{\alpha r} (GFlops/s), \qquad (2)$$

where, the parameters are explained in Table 1.

Table 1. Parameters for communication model

Symbol	Description
N	Size of FFT
P	Number of nodes
r	Number of reshapes (tensor transpose, c.f., Fig. 1)
α	Size of datatype (Bytes)
B	Theoretical inter-node bandwidth (GB/s)

In Fig. 3, we show the roofline model for *heFFTe* v.2.0 on Summit and Fugaku, which have, respectively, 25 and 40.8 GB/s of inter-node theoretical bandwidth injection.

2.3 Choosing the Fastest FFT Parallel Algorithm

In Fig. 1 we see that there exists different ways to implement the parallel FFT, and it also depends on the user's data arrangement at input and output. For the sake of simplicity, let us consider a 3-D FFT, where the possible reshape combinations are (*B: Bricks, P: Pencils, S: Slabs*):

- **Pencils:** $B2P \rightarrow P2P \rightarrow P2P \rightarrow P2B$; this approach is the one available in libraries such as AccFFT [12] and FFTMPI [17].

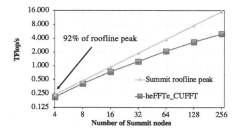
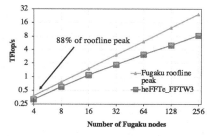

Fig. 3. Roof-line performance model—heFFTe performance on a 3-D FFT of size 1024^3 using 6 MPI/node, 1 GPU-Volta 100 per MPI for Summit, and 48 A64FX per node on Fugaku.

- **Slabs:** $B2P \rightarrow P2S \rightarrow S2B$; this approach uses a combination of pencils and slabs, and it is included in *heFFTe* library [3].

The choice of a given reshape sequence will depend on the type of architecture. Note that, for example, the number of messages for a P2P reshape is of the order of $P^{2/3}$, where P is the number of processors involved in the communication, c.f., Fig. 1. Hence, assuming 3-D double-complex data—and using Eq. 2 and the asymptotic number of messages sent by each of the reshape types, with $B = 25$ GB/s and $L = 1\,\mu$s—Fig. 4 is a phase diagram for Summit, which allows to choose the theoretical fastest decomposition to use. This offline pre-processing tuning strategy can help users to identify which 3-D decomposition to use for the FFT parallel algorithm. The proposed methodology can easily be extended to other supercomputers and higher dimension transforms.

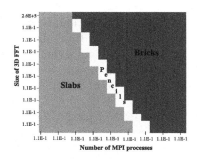

Fig. 4. Selection of the best reshape approach based on the 3-D FFT size and the number of resources.

3   Experimental Results

In this section, we present numerical experiments to support our analysis from previous sections. Since this paper targets large-scale computation, our results were obtained using the two world's most powerful supercomputers today, with the following architectures:

- *Summit* at ORNL - USA, having 4,608 nodes, each consisting of 2 IBM POWER9 CPUs and 6 NVIDIA V100 GPUs. These 6 GPU accelerators provide a theoretical double-precision capability of approximately 40 TFLOP/s. Within the same node, processors have two NVIDIA NVLink interconnects, each having a peak bandwidth of 25 GB/s (in each direction), hence V100 and P900 can communicate at a peak of 50 GB/s (100 GB/s bi-directional).
- *Fugaku* at RIKEN - Japan, currently at testing stage, and has 158,976 nodes, each consisting of Fujitsu A64FX CPU. We use the maximum amount of number of resources currently allowed with 48 cores per node.

Experiments on this paper where performed using a state-of-the-art library for parallel FFTs: *heFFTe* version 2.0 [3], which reportedly provides considerable speedups with respect to its peers [4]. If not stated otherwise, our results display average values of 10 experiments (5 forward and 5 backward 3D-FFT computations) using double-complex precision random data and 4 data reshapes per direction (Input → X → Y → Z → Output).

3.1   Strong and Weak Scalability

Several authors have shown that parallel FFT runtime on large problems are highly due to MPI communication, which asymptotically takes more than 95% of runtime on hybrid systems, c.f., [4,5,12]. Hence, it is critical to select the fastest MPI (binary or collective) communication for the data exchanges required by parallel FFT distributions. In Fig. 5, we present weak and strong scalability on up to one million A64FX cores on Fugaku, this experiment clearly shows the effect on scalability of the selection of the Point-to-Point (MP2P) and All-to-All (A2A) communication frameworks, and its relationship with the number of resources. When dealing with hybrid systems, such as Summit supercomputer, the percentage of time spend on communication exploits, making the performance scaling highly dependent on the underlying MPI library, we explore the MPI selection in next subsection.

The strong scalability plot from Fig. 5, sheds light on how P2P communication is faster for large number of resources, and we verified this for medium sized allocation and employed the P2P approach for our largest experiments on the weak scalability plot. Figure 6 shows a weak scaling using AlltoAll communication on Summit, and using SpectrumMPI 10.3 with data striping enabled, we can get good linear scaling and this is faster than the P2P approach, which is the opposite situation compared to Fugaku.

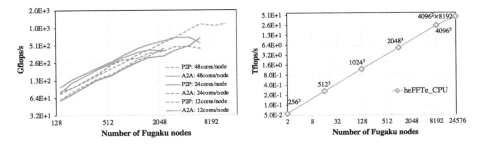

Fig. 5. *Left:* Comparison of strong scaling for a 3-D FFT of size 1024^3, using different node count. *Right* Weak scalability for different 3-D FFT sizes. For both experiments we use *heFFTe* with FFTW backend and 48 MPI processes (1 MPI processes per A64FX core) per node.

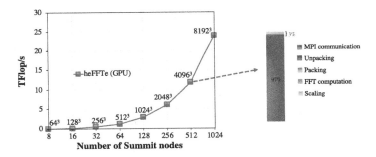

Fig. 6. Weak scalability for different 3-D FFT sizes on a hybrid architecture (Summit). Using NVIDIA CUFFT backend and 6 MPI processes (1 MPI processes per Volta 100 GPU) per node.

3.2   MPI Selection for Further Acceleration

In Sect. 3, we showed how the right reshaping algorithm can provide speedups of over 25% compared to default implementations. Next, assuming that the algorithm is fixed and properly chosen, we observed that to achieve linear scalability, it is very important to figure out how to optimally use the computational resources and architecture tools from manufactures to manually tune the port inter-connections to achieve maximum bandwidth injection. Therefore, let us analyze the parallel computing technologies in both, Summit and Fugaku, supercomputers:

– Fugaku uses a TofuD network topology, with three different types of options: *torus, mesh, noncont*. For our experiments we used MPIFCC provided with the Fujitsu compiler, and we enabled auto-parallelization using the *Kparallel* flag. We observe that the *torus* and *noncont* networks provided the best injection bandwidth. In theory, using the 6 available TofuD ports we can get a total of 40.8 GB/s theoretical bandwidth injection.

– Summit inter-node connections are not as fast as the NVLINKS available intra-nodes, and they are arranged on a non-blocking fat tree topology with dual-rail EDR InfiniBand network that provides a theoretical bandwidth of 25 GB/s. For our experiments we use IBM SpectrumMPI, which is optimized for this architecture.

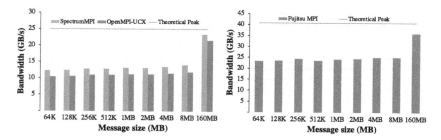

Fig. 7. Comparison of bandwidth injection obtained for different MPI implementations on Fugaku and Summit.

Information about the interconnections have to be obtained in advance and can be integrated to state-of-the-art libraries for auto-tuning, this feature is not, currently, supported by libraries covered in Sect. 1. Next, for a given FFT computation we can find the message size that will be transferred between nodes and Fig. 7 shows which MPI implementation offers the best bandwidth injection. For instance, for a $256 \times 256 \times 256$ double-complex (16 Bytes) precision FFT on 128 nodes, each processor communicates around 2 MB of data.

4   Conclusion

In this paper, we studied performance and scalability limitations of large-scale FFT computation on state-of-the-art CPU and GPU distributed systems. We provided methodologies to further accelerate parallel FFT by targeting software improvements on critical algorithm bottlenecks and making them aware of the underlying architecture. Our numerical studies and bounds on performance scalability can be generalized to all type of architectures (e.g., those from grid computing) and can be employed to make performance predictions. We finally presented experiments on today's top supercomputers, showing how carefully chosen system-aware parameters and algorithms can lead to very good linear strong and weak scalability.

References

1. cuFFT library (2018). http://docs.nvidia.com/cuda/cufft
2. NCLL library (2019). https://github.com/NVIDIA/nccl

3. heFFTe library (2020). https://bitbucket.org/icl/heffte
4. Ayala, A., et al.: Impacts of Multi-GPU MPI collective communications on large FFT computation. In: 2019 IEEE/ACM Workshop on Exascale MPI (ExaMPI) (2019)
5. Ayala, A., Tomov, S., Haidar, A., Dongarra, J.: *heFFTe*: highly efficient FFT for exascale. In: Krzhizhanovskaya, V.V., et al. (eds.) ICCS 2020. LNCS, vol. 12137, pp. 262–275. Springer, Cham (2020). https://doi.org/10.1007/978-3-030-50371-0_19
6. Balaji, P., et al.: MPI on a million processors. In: Ropo, M., Westerholm, J., Dongarra, J. (eds.) EuroPVM/MPI 2009. LNCS, vol. 5759, pp. 20–30. Springer, Heidelberg (2009). https://doi.org/10.1007/978-3-642-03770-2_9
7. Czechowski, K., McClanahan, C., Battaglino, C., Iyer, K., Yeung, P.K., Vuduc, R.: On the communication complexity of 3D FFTs and its implications for exascale (2012). https://doi.org/10.1145/2304576.2304604
8. Demmel, J.: Communication-avoiding algorithms for linear algebra and beyond. In: 2013 IEEE 27th International Symposium on Parallel and Distributed Processing (2013)
9. Dongarra, J.: Report on the sunway TaihuLight system. Technical report (2016)
10. Emberson, J., Frontiere, N., Habib, S., Heitmann, K., Pope, A., Rangel, E.: Arrival of first summit nodes: HACC testing on phase I system. Technical report, MS ECP-ADSE01-40/ExaSky, Exascale Computing Project (ECP) (2018)
11. Frigo, M., Johnson, S.G.: The design and implementation of FFTW3. Proc. IEEE **93**(2), 216–231 (2005). Special issue on "Program Generation, Optimization, and Platform Adaptation'
12. Gholami, A., Hill, J., Malhotra, D., Biros, G.: AccFFT: a library for distributed-memory FFT on CPU and GPU architectures. CoRR abs/1506.07933 (2015)
13. Grama, A., Gupta, A., Karypis, G., Kumar, V.: Accuracy and Stability of Numerical Algorithms, 2nd edn. Addison Wesley, Boston (2003)
14. Large-scale atomic/molecular massively parallel simulator (2018). https://lammps.sandia.gov/
15. Lin, S., Liu, N., Nazemi, M., Li, H., Ding, C., Wang, Y., Pedram, M.: FFT-based deep learning deployment in embedded systems. In: 2018 Design, Automation Test in Europe Conference Exhibition (DATE), pp. 1045–1050 (2018)
16. Parallel 2d and 3d complex FFTs (2018). http://www.cs.sandia.gov/~sjplimp/download.html
17. Plimpton, S., Kohlmeyer, A., Coffman, P., Blood, P.: fftMPI, a library for performing 2d and 3d FFTs in parallel. Technical report, Sandia National Lab. (SNL-NM), Albuquerque, NM, USA (2018)
18. Takahashi, D.: Implementation of parallel 3-D real FFT with 2-D decomposition on Intel Xeon Phi Clusters. In: 13th International Conference on Parallel Processing and Applied Mathematics (2019)

High Performance Implementation of Boris Particle Pusher on DPC++. A First Look at oneAPI

Valentin Volokitin[1], Alexey Bashinov[2], Evgeny Efimenko[2], Arkady Gonoskov[3], and Iosif Meyerov[1]([✉])

[1] Mathematical Center, Lobachevsky University, Nizhni Novgorod 603950, Russia
meerov@vmk.unn.ru
[2] Institute of Applied Physics, Russian Academy of Sciences, Nizhny Novgorod 603950, Russia
[3] Department of Physics, University of Gothenburg, 41296 Gothenburg, Sweden

Abstract. New hardware architectures open up immense opportunities for super-computer simulations. However, programming techniques for different architectures vary significantly, which leads to the necessity of developing and supporting multiple code versions, each being optimized for specific hardware features. The oneAPI framework, recently introduced by Intel, contains a set of programming tools for the development of portable codes that can be compiled and fine-tuned for CPUs, GPUs, FPGAs, and accelerators. In this paper, we report on the experience of porting the implementation of Boris particle pusher to oneAPI. Boris particle pusher is one of the most demanding computational stages of the Particle-in-Cell method, which, in particular, is used for supercomputer simulations of laser-plasma interactions. We show how to adapt the C++ implementation of the particle push algorithm from the Hi-Chi project to the DPC++ programming language and report the performance of the code on high-end Intel CPUs (Xeon Platinum 8260L) and Intel GPUs (P630 and Iris Xe Max). It turned out that our C++ code can be easily ported to DPC++. We found that on CPUs the resulting DPC++ code is only ~10% on average inferior to the optimized C++ code. The code is compiled and run on new Intel GPUs without any specific optimizations and shows the expected performance.

Keywords: Parallel computing · HPC · Heterogeneous computing · oneAPI

1 Introduction

The development of computational architectures in the last decades has led to the emergence of new possibilities for supercomputer simulations. However, the appearance of devices with fundamentally different architectures required the development of appropriate approaches to programming and code optimization. It turned out that the development of a universal framework that allows implementing a single code that can be

This study is supported by Intel (oneAPI Center of Excellence program) and by the Ministry of Science and Higher Education of the Russian Federation, project no. 0729-2020-0055.

© Springer Nature Switzerland AG 2021
V. Malyshkin (Ed.): PaCT 2021, LNCS 12942, pp. 288–300, 2021.
https://doi.org/10.1007/978-3-030-86359-3_22

compiled and, no less important, work efficiently on different hardware is not straight-forward. Such frameworks and libraries, in particular, include OpenCL [1], OpenACC [2], Alpaka [3], Kokkos [4], and many others. In 2020, Intel introduced oneAPI – a new unified open model for heterogeneous programming, which includes a wide set of tools and a new DPC++ language [5] for heterogeneous programming based on the SYCL language. The DPC++ language allows using various computing devices in calculations, in particular, CPUs, GPUs, and FPGAs.

In this paper, we report on the experience of porting the algorithm of Boris pusher to DPC++. The Boris pusher is a frequently used algorithm for advancing the classical state of a charged particle under the action of a given electromagnetic field. It is one of the main computational cores of the High-Intensity Collisions and Interactions (Hi-Chi) framework [6, 7], which is an open-source collection of Python-controlled tools for performing simulations and data analysis in the research area of strong-field particle and plasma physics. In particular, we address the following questions. Firstly, we demonstrate how such code can be ported to DPC++. Secondly, we analyze the performance of the DPC++ code on high-end Intel CPUs versus the baseline C++ implementation and show how the key code optimization techniques affect performance in different simulation scenarios. Finally, we assess the performance of the DPC++ code on new Intel GPUs versus CPUs without any additional optimizations for GPUs.

2 Method

In this work we employ the commonly used Particle-in-Cell (PIC) method; a detailed description is given in [8–12]. The method is used to model the interaction of an elec-tromagnetic field with plasma using kinetic description. It operates on two distinct sets of data: grid field data and particle data. The values of electric and magnetic fields are defined on a spatial grid. The plasma is represented as an ensemble of particles, each with a charge, mass, position and momentum. Each particle used in simulation is in fact a macroparticle that represents a cloud of real particles, whose distribution is described by a fixed localized shape function, also referred to as the form factor of a macroparticle. A notable feature of the method is that particles do not interact with each other directly; instead each particle interacts with a set of nearby grid values of the electromagnetic field, depending on the form factor.

This paper concerns one of the main parts of the PIC method: the integration of particle motion in electromagnetic fields. This stage, usually called the Particle push, is of particular interest for performance optimization [13–19], because this stage becomes the most time consuming for realistic problems due to a large number of macroparticles (as compared to the number of grid nodes). The numerical code Hi-Chi implements the commonly used and *de-facto* standard Boris method [12]. Further we discuss porting and optimization of this method using the DPC++ language.

3 Data Structures and Algorithm

The developments reported in this paper are a part of the Hi-Chi project [6]. The project Hi-Chi is an open-source collection of Python-controlled tools for performing simula-tions and data analysis in the research area of strong-field particle and plasma physics.

The tools are being developed in C++ and provide high performance using either local or supercomputer resources. The project is intended to offer an environment for testing, benchmarking and aggregative use of individual components, ranging from basic routines to supercomputer codes.

A `Particle` class is the key data structure used in our simulations. For each particle, we store position and momentum vectors of 3 floating point numbers each, as well as scalar floating point values of the particle weight and the Lorenz factor γ. Additionally, we store an integer value of the particle type to determine its mass and charge. These parameters, corresponding to particles of different types, are stored in a separate table in a single copy.

The code is implemented so that we can easily switch between using single and double precision data types. To do this, we abstracted the floating point data type as `FP`, which can be `float` or `double` depending on the settings. Similarly, the `FP3` data type describes a vector of 3 `float` or `double` components. In the case of single precision, storage of each particle requires 34 bytes of memory (36 bytes after memory alignment), in the case of double precision, each particle takes 66 bytes of memory (72 bytes after memory alignment). The investigation of the possibility of performing calculations in single and double precision is beyond the scope of this study. Here we are only comparing the performance of calculations in single and double precision. We should also note that in the considered benchmarks, we did not observe any inaccuracies caused by the use of single precision.

The way of organizing an ensemble of particles deserves special attention. For example, in programs for supercomputer modeling of laser plasma by the particle-in-cell method, two main approaches of representing an ensemble of particles are commonly used. The first method assumes that each cell stores its own array of particles. This representation has many advantages, but it requires handling the movement of particles between cells, which causes an additional overhead when parallelizing computations. The second way is to store the entire ensemble of particles in a single array. In this case, we do not need to handle the movement of particles between cells, but we have to periodically sort the array of particles in order to improve cache locality. In the Hi-Chi code, we employ the second method.

The next question that arises when choosing data structures to represent an ensemble of particles is which of the common patterns is better to use: an *array of structures* (AoS) or a *structure of arrays* (SoA). This issue has been studied for a long time as applied to various problems. It is known that both approaches of data representation have their pro et contra. For example, the AoS pattern allows us to preserve memory locality. However, this scheme is not very efficient in the case of code vectorization, since it entails non unit-stride access to the data of different particles. On the contrary, the SoA pattern is less efficient in utilizing cache memory, but it allows efficiently loading data for vector computations and does not use time-consuming scatter/gather operations. In the general case, none of the schemes is unconditionally better. Everything is determined by the properties of the algorithm, problem, and target architecture. Therefore, Hi-Chi allows using any of these patterns. Next, we will compare how the choice of data structures affects the performance.

Note that in order to use different ways of storing data, we implement the `Parti-cleProxy` class, which completely repeats the functionality of the `Particle` class, but stores *references* to objects. This approach allows us to effectively employ the C++ templates and use the single code regardless of the storage structure.

4 Exploiting Parallelism Using the OneAPI Technology

4.1 Reference Implementation of the Boris Pusher

As a reference implementation, we consider a parallel version implemented using the OpenMP technology. Parallelism in this version is exploited at the level of particle processing, and the loop over particles is *parallelized* and *vectorized* as follows:

```
// Numerical integration loop over numSteps time steps
for (int step = 0; step < numSteps; step++)  {
    // Run the Pusher for every particle in an ensemble
    #pragma omp parallel for simd
    for (int ind = 0; ind < numParticles; ind++)  {
        // Run the Boris pusher for particle #ind
        ...
    }
}
```

4.2 Porting the Pusher to DPC++

Smart memory management is a key factor to achieving good performance and scalability of codes. In the case of using accelerators, this issue becomes even more important. DPC++ provides two ways to manage memory and access/share/move data between devices. The first method involves the use of special concepts – buffers, which allow us to define regions of memory that can be used on the device (buffers), and accessors, which allow us to plan access to data and their movement between devices. The second method (Unified Shared Memory, USM) is more low-level and allows us to work in a style similar to working with C++ pointers. This model is quite convenient for codes that have been already based on C++ pointers. In this case, porting to DPC++ requires just minimal modifications to memory allocation instructions.

We employ the USM model. It is the simplest, but quite functional option for shared memory allocation providing data access on a device and a host. We also rely on oneAPI runtime for memory management. This approach allowed us to quickly port the code to DPC++, with only minimal changes and reasonable performance. Compared to the reference implementation, our DPC++ code is quite similar:

```
// Numerical integration loop over numSteps time steps
for (int step = 0; step < numSteps; step++) {
    // Create a "kernel" function
    auto kernel = [&](sycl::handler& h) {
        // Work with particles in parallel
        h.parallel_for(sycl::range<1>(numParticles),
                       [=](sycl::id<1> ind) {
            // Run the Boris pusher for particle #ind
            ...
        }
    }
    // Submit the kernel
    device.submit(kernel).wait_and_throw();
}
```

The code, as before, processes the movement of particles in parallel. Unlike typical C++ code, for processing particles, we create a kernel using special C++ lambda expression (supported since the C++ 11 standard). This kernel employs a special DPC++ mechanism parallel_for, which calls the Boris pusher in parallel for particles from the ensemble. Code vectorization is also automatically provided by the compiler. Since the Boris pusher is implemented as a lambda expression that captures objects *by copy*, these objects must have a default copy constructor that will create full copies of objects with the same addresses in memory. Therefore, we could not use the standard vector class to implement an array of particles. Instead, we use a C-style pointer to a buffer, which is copied without actually copying the contents of the buffer when capturing objects to the kernel. Such copying is usually a mistake for C++ classes, but in this case it is exactly the required behavior.

4.3 Improving Scaling Efficiency

DPC++ runtime on a CPU employs the widely used Threading Building Blocks (TBB) library for parallel computations. Compared to OpenMP, TBB always uses dynamic scheduling, which can substantially improve performance in complex unbalanced problems. However, in balanced applications, the overhead of dynamic scheduling may not be justified. However, a small overhead is a reasonable price to pay for the versatility of the code that can be compiled and run on different architectures.

Appropriate use of platforms with Non-Uniform Memory Access (NUMA) architecture deserves a separate discussion. Thus, on modern supercomputers, a configuration with several (often two) CPUs is typical. In such cases the access of the cores to the local memory of their processor is much faster than access to the memory of another processor installed on the same node. This is especially important for memory-bound applications, in particular for the considered pusher.

In codes parallelized with OpenMP, we can often achieve that the data is localized in the cache memory of the CPU that will process it. In the case of using TBB (recall that DPC++ uses this scenario), we can also work with memory in a *NUMA-friendly* manner. In this regard we use the DPCPP_CPU_PLACES environment variable with the value numa_domains. In this case, the iteration space is divided into NUMA domains, and TBB performs dynamic scheduling of parallel execution of tasks only within the corresponding NUMA arena. This ensures that the same particles are processed on the same CPU at every time step. It will be shown below that this significantly improves performance and scaling efficiency of the code. In what follows, we will refer to such launches as 'DPC++ (NUMA)'.

5 Numerical Results

5.1 Computational Infrastructure

The computational experiments were performed at a node of the supercomputer Endeavour with 2x Intel Xeon Platinum 8260L (Caskade Lake, 24 cores each), 48 cores overall, 192 GB RAM, RedHat 4.8.5, Intel C++ Compiler and Intel DPC++ Compiler from the Intel OneAPI Toolkit Base and HPC (Gold Release 2020) suite. All tests on Intel P630 and Iris Xe Max GPUs were executed on Intel DevCloud. Some preliminary tests were executed on the Lobachevsky supercomputer at Lobachevsky University. The hardware parameters are presented in Table 1.

Table 1. Hardware parameters

Parameter	2x Intel Xeon Platinum 8260L	P630	Iris Xe Max
Number of CPU cores/GPU execution units	48	24	96
Clock frequency	2.4 GHz (3.9 GHz Boost)	0.35 GHz (1.15 GHz Boost)	0.3 GHz (1.65 GHz Boost)
RAM	DDR4 192 GB	DDR4 32 GB (CPU RAM)	LPDDR4X 4 GB
Memory bandwidth	250 GB/sec	40 GB/sec	68 GB/sec
Peak performance (single precision)	3.6 TFlops	0.441 TFlops	2.5 TFlops

5.2 Benchmarks

We considered *two simulation scenarios* as benchmarks for analyzing performance. In the *first scenario*, all field values are precalculated and stored in the corresponding array. This scenario allows excluding all operations from measurements except for particle motion. The *second scenario* assumes that the fields are specified analytically. In this case, we do not have to store a large data array. On the contrary, field values are computed using analytical formulas when they are directly needed in calculations. Both scenarios are in demand in practice and, hypothetically, can lead to different conclusions regarding code optimization, since in the first case, we store much more data, and in the second, we perform much more calculations.

In order to test our implementations we consider the motion of electrons in the tightly focused fields in the form of a standing magnetic dipole (m-dipole) wave [20]. This study is necessary to determine the optimal parameters of a seed target for the vacuum breakdown in multi-petawatt m-dipole wave [21]. Tight focusing allows decreasing of the threshold power of this phenomenon [22] that is favourable for upcoming experiments at 10-PW laser facilities [23]. For this reason, we consider ultimate focusing [24] in a form of the dipole wave.

The pulsed multi-PW incoming m-dipole wave can ionize matter at its leading edge and pull unbound electrons to the wave focus. When the wave passes through the focus the diverging wave appears and electrons start to oscillate in the standing wave. In order to trigger the vacuum breakdown a number of particles should remain in the focus when the instantaneous wave power becomes greater than 10 PW [21]. However, due to strong field inhomogeneity, particles can rapidly escape the focal region while instantaneous power is not high enough. With the help of simulations of the particle motion in the standing m-dipole wave the rate of particle escape from the focal region can be obtained. Based on these results the optimal parameters of the seed target can be chosen.

Particle escape is fastest in the range of powers from approximately 4 GW to 1 PW when fields are relativistic, but radiative trapping effects [25] are absent. For the test we consider the wave power P = 0.1 PW. In the simulation the electric and magnetic field components are set analytically as follows:

$$E_x = -\frac{2A_0 y}{R(x,y,z)}\cos(\omega_0 t)f_1(R(x,y,z))$$
$$E_y = \frac{2A_0 x}{R(x,y,z)}\cos(\omega_0 t)f_1(R(x,y,z))$$
$$E_z = 0$$
$$B_x = -\frac{2A_0 xz}{R^2(x,y,z)}\sin(\omega_0 t)f_2(R(x,y,z))$$
$$B_y = -\frac{2A_0 xy}{R^2(x,y,z)}\sin(\omega_0 t)f_2(R(x,y,z))$$
$$B_z = -\frac{2A_0 z^2}{R^2(x,y,z)}\sin(\omega_0 t)\left(\frac{z^2}{R^2(x,y,z)}f_2(R(x,y,z)) + f_3(R(x,y,z))\right),$$

where t is time, x, y, z are Cartesian coordinates, $R(x,y,z) = \sqrt{x^2 + y^2 + z^2}$, $A_0 = k\sqrt{3P/c}$, c is the light velocity, $\omega_0 = 2.1 \times 10^{15}$ s^{-1} is the wave frequency corresponding to the wavelength $\lambda = 0.9\,\mu$m, $k = \omega_0/c$,

$$f_1(R) = \frac{\sin(kR)}{(kR)^2} - \frac{\cos(kR)}{kR}$$
$$f_2(R) = \left(\frac{3}{(kR)^3} - \frac{1}{kR}\right)\sin(kR) - \frac{3\cos(kR)}{(kR)^2}$$
$$f_2(R) = \left(\frac{1}{kR} - \frac{1}{(kR)^3}\right)\sin(kR) + \frac{\cos(kR)}{(kR)^2}.$$

Initially ($t = 0$), electrons are at rest and distributed uniformly within the sphere with radius $r = 0.6\lambda$. The experimental setup is as follows. In each experiment, 10^7 particles were simulated. The equations of motion were integrated over 10^3 time steps, which we further refer to as 'iteration'. During the experiment, 10 successive iterations were measured. To compare the performance results, we used the **NSPS** metric (*nanoseconds per particle per step*) calculated as the average time of one iteration in nanoseconds, divided by the number of particles (10^7) and by the number of steps in one iteration (10^3).

5.3   Results and Discussion

Experiments on CPUs

First of all, it is necessary to take into account the following fact. When profiling computational codes, we often observe that the first iteration of a method can take work slower than the rest. This is usually explained by the fact that at the first iteration, the data has to be loaded from RAM, while at the next iterations, part of the data is loaded from a cache. In NUMA systems, this effect is sometimes even more pronounced if the code does not implement a NUMA-friendly memory usage policy. In the case of DPC++ codes, this effect is manifested in an even more explicit form, since when the kernel is first launched, it is compiled from an intermediate representation for a specific hardware, which can take some time. In our benchmark, the first iteration takes 50% longer time than the subsequent ones, which is the cumulative effect of the reasons described above. Considering that we perform a lot of iterations, this effect does not have a significant impact on the results.

We collected the results on CPUs employing available 48 cores (2 CPUs with 24 cores each). The comparison involves implementations parallelized on OpenMP, or DPC++, or DPC++ with the NUMA-friendly memory usage policy described before. For each of these implementations, we tried using SoA and AoS memory layout patterns. As stated earlier, two simulation scenarios were considered. We refer them to as 'Precalculated Fields' and 'Analytical Fields'. For OpenMP versions, it was found that employing 96 threads is empirically the best, that is, the use of hyperthreading technology improves performance. For DPC++ implementations, the number of threads is selected by the TBB runtime. All experiments were executed both in single and in double precision (Table 2).

The results lead to the following conclusions:

1. Using the NUMA-friendly memory usage policy leads to a significant performance gain due to the elimination of the overhead of remote access to the memory of another CPU installed on the same node. Note that in the OpenMP code, similar tricks did not lead us to performance improvement, since in this case remote access occurs only at the first time steps of the method, then the data is localized within the corresponding NUMA domains. The conclusions are confirmed by profiling using Intel VTune. Note also that although such a significant effect of NUMA on performance is specific to the considered memory bound benchmark, it can be important for optimizing other DPC++ applications as well.

2. The performance of the optimized DPC++ implementation is only slightly inferior to the OpenMP implementation. The difference is usually only ~10% on average due to some overhead and a different approach to parallelization. We think this is an excellent result for DPC++ taking into account the portability of the code.

Table 2. Performance results (NSPS, nanoseconds per particle per step) on CPU for 6 implementations and 2 simulation scenarios.

Pattern	Parallelization	Precalculated fields		Analytical fields	
		Float	Double	Float	Double
AoS	OpenMP	0.53	0.98	0.58	0.84
	DPC++	0.78	1.54	1.02	1.48
	DPC++ NUMA	0.54	0.99	0.54	0.89
SoA	OpenMP	0.50	1.06	0.43	0.76
	DPC++	0.85	1.49	0.77	1.31
	DPC++ NUMA	0.58	1.20	0.60	0.90

3. The choice of the AoS or SoA patterns has almost no effect on the performance in the current benchmark. This is due to the fact that the main factor limiting performance is not loading data into vector registers, but working with RAM.

4. When going from single to double precision, the running time changes as expected, because it requires twice the memory bandwidth and doubles the amount of computation. In the problem with precomputed fields, the difference is almost twofold; in the case of analytical fields, it is slightly less due to the specifics of the calculations. Note also that in the case of DPC++, code vectorization occurs with full use of AVX-512 instructions, as it was earlier in OpenMP.

5. Since the problem is memory bound, working with memory dramatically affects performance. The two considered simulation scenarios are fundamentally different in working with memory, since in the 'Precalculated Fields' problem, we additionally store an array of field values comparable in size to the ensemble of particles. On the contrary, in the 'Analytical Fields' problem, we do a lot more resource-intensive calculations of mathematical functions. The main motivation for considering these two scenarios was to find out how these differences affect the overall simulation time. The results showed that calculations using analytical formulas and loading pre-calculated data from memory turned out to be, on the whole, comparable in terms of time consumption. At the same time, in the case of calculations in double precision, the scenario with the analytical computations of field values runs a little faster. It is noteworthy that this result does not depend on the choice of parallel programming technologies (OpenMP or DPC++).

To evaluate the efficiency of parallelization, we calculate the speedup when using 1–48 cores relative to runs on a single core. Considering that hyperthreading is enabled, we start 2 threads on each core, binding threads to cores. As an example, single precision calculations in the problem with precalculated fields are considered. The results (Fig. 1) show that in the implementation on OpenMP, a close to linear speedup is observed until the code fully utilizes memory bandwidth of the first socket. When we start using of the second socket, the run time begins to scale linearly again. For DPC++ NUMA implementations, super-linear acceleration is observed at the beginning. This is because the DPC++ single core version is quite slow. Further experiments demonstrate reasonable scaling, approaching to 63% of strong scaling efficiency when using 48 cores. As shown earlier, the overall run times for OpenMP and DPC++ NUMA versions are close to each other.

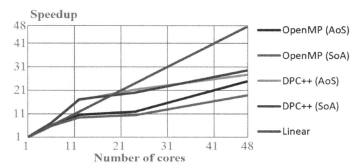

Fig. 1. Speedup of parallel computations of the OpenMP and DPC++ NUMA implementations employing the AoS and SoA data layouts in the 'Precalculated Fields' problem. Computations are performed in single precision on 1–48 cores. Single core run time is used as a reference.

Experiments on GPUs

DPC++ is the universal development tool for portable programs. However, achieving *performance portability* is even much more complex problem due to fundamental differences in computing architectures. Apparently, when porting DPC++ codes to specific architectures, some fine-tuning or even new implementations of the computational kernels can be required. One of the goals of the present work was to study how the DPC++ code, built on the basis of the C++ code optimized for Intel CPUs, will work on the new Intel GPUs without any specific optimizations. The results obtained should not be taken as a fair comparison of CPUs vs. GPUs, they only demonstrate how much performance we can get without additional work. We carried out such experiments on Intel devCloud, using currently available devices, the parameters of which are shown earlier in Table 1. Since for the Iris Xe Max, double precision operations occur only in an emulation mode, we present the results in single precision only. The results are shown in Table 3.

If for the CPUs different particles memory layouts were comparable in performance due to various factors described earlier, then on Intel GPUs the run time may differ by more than half (Table 3). This is due to a different organization of the memory subsystem in the GPUs. We should also note the lack of additional optimizations for the GPUs. Probably, the performance of the AoS version of the code can be improved, however, in any case, the importance of choosing a layout on GPUs must be taken into account when such porting. A direct comparison of the run time on the CPUs and GPUs is also of great interest. As stated earlier, this comparison is not fully objective due to the lack of GPU optimizations. However, it provides an answer to the question of whether we can expect the GPU run time to be appropriate after such porting. In the problems we are considering, we can give a positive answer to this question. Indeed, as compared to the considered Xeon CPUs, the performance of P630 and Iris Xe Max is lower by a factor of about 8 and 1.5, respectively. At the same time, the code on P630 works slower only by a factor of 3.5–4.5, and the code on Iris Xe Max is slower by a factor of 1.7–2.6, compared to 2 high-end CPUs. This comparison does not give a complete picture, since GPUs have a different memory organization, the problem is not compute- but memory-bound, and utilization of GPUs is often much harder compared to CPUs. Nevertheless, we can conclude that even without additional optimizations, we got reasonable performance on GPUs, which, probably, can be further improved.

Table 3. Performance results (NSPS, nanoseconds per particle per step) on GPUs for DPC++ implementations in 2 simulation scenarios. Computations are performed in single precision.

Pattern	Precalculated fields			Analytical fields		
	CPU	P630	Iris Xe Max	CPU	P630	Iris Xe Max
AoS	0.54	4.76	2.10	0.54	4.45	2.10
SoA	0.58	2.43	1.42	0.60	1.93	1.00

6   Conclusion

The paper presents a new DPC++ implementation of the Boris Pusher algorithm for the movement of particles in a given electromagnetic field. The implementation is obtained by porting the CPU-optimized C++ implementation in the Hi-Chi code by replacing the way of organizing parallel computations. It turned out that this porting can be done quickly enough. After running the program on the high-performance server with 2 high-end CPUs, we found that the performance of the resulting DPC++ implementation significantly depends on the run settings customization in terms of optimal use of the NUMA architecture, while the SoA and AoS patterns of the data layout have almost no effect on performance. As a result, it was found that, regardless of the simulation scenario, the DPC++ implementation is only slightly inferior to the C++ code, while it became possible to run it on Intel GPUs.

Our experiments on Intel GPUs showed that even though we did not optimize the code for the GPU, the performance results compared to the optimized code on the CPUs exceed our expectations. So, it turned out that 2 Xeon CPUs are ahead of desktop GPUs only in accordance with the difference in peak performance capabilities. We expect that the performance of the GPU implementation can be improved. This is one of the directions for further research. The code is publicly available [6].

References

1. OpenCL: open standard for parallel programming of heterogeneous systems. https://www.khronos.org/opencl/

2. OpenACC. https://www.openacc.org/
3. Matthes, A., Widera, R., Zenker, E., Worpitz, B., Huebl, A., Bussmann, M.: Tuning and optimization for a variety of many-core architectures without changing a single line of implementation code using the Alpaka library. In: Kunkel, J.M., Yokota, R., Taufer, M., Shalf, J. (eds.) ISC High Performance 2017. LNCS, vol. 10524, pp. 496–514. Springer, Cham (2017). https://doi.org/10.1007/978-3-319-67630-2_36
4. Edwards, H.C., Trott, C.R., Sunderland, D.: Kokkos: enabling manycore performance portability through polymorphic memory access patterns. J. Parallel Distrib. Comput. **74**(12), 3202–3216 (2014). https://doi.org/10.1016/j.jpdc.2014.07.003
5. Reinders, J., et al.: Data Parallel C++: Mastering DPC++ for Programming of Heterogeneous Systems Using C++ and SYCL. Apress, Berkeley (2021). https://doi.org/10.1007/978-1-4842-5574-2
6. Hi-Chi framework. https://github.com/hi-chi/pyHiChi
7. Panova, E., et al.: Optimized computation of tight focusing of short pulses using mapping to periodic space. Appl. Sci. **11**(3), 956 (2021). https://doi.org/10.3390/app11030956
8. Birdsall, C.K., Langdon, A.B.: Plasma Physics via Computer Simulation. CRC Press, Hoboken (2004)
9. Taflove, A., Hagness, S.C., et al.: Computational Electrodynamics: the Finite-Difference Time-Domain Method, 2nd edn. Artech House, Norwood (1995)
10. Tajima, T.: Computational Plasma Physics: With Applications to Fusion and Astrophysics. CRC Press, Hoboken (2018)
11. Ripperda, B., et al.: A comprehensive comparison of relativistic particle integrators. Astrophys. J. Suppl. Ser. **235**(1), 21 (2018). https://doi.org/10.3847/1538-4365/aab114
12. Boris, J.P.: Relativistic plasma simulation-optimization of a hybrid code. In: Proceedings of Fourth Conference on Numerical Simulations of Plasmas, pp. 3–67 (1970)
13. Decyk, V.K., Singh, T.V.: Particle-in-cell algorithms for emerging computer architectures. Comput. Phys. Commun. **185**(3), 708–719 (2014). https://doi.org/10.1016/j.cpc.2013.10.013
14. Fonseca, R.A., et al.: Exploiting multi-scale parallelism for large scale numerical modelling of laser wakefield accelerators. Plasma Phys. Control. Fusion **55**(12), 124011 (2013). https://doi.org/10.1088/0741-3335/55/12/124011
15. Germaschewski, K., et al.: The plasma simulation code: a modern particle-in-cell code with patch-based load-balancing. J. Comput. Phys. **318**, 305–326 (2016). https://doi.org/10.1016/j.jcp.2016.05.013
16. Surmin, I., Bastrakov, S., Matveev, Z., Efimenko, E., Gonoskov, A., Meyerov, I.: Co-design of a particle-in-cell plasma simulation code for Intel Xeon Phi: a first look at knights landing. In: Carretero, J., et al. (eds.) ICA3PP 2016. LNCS, vol. 10049, pp. 319–329. Springer, Cham (2016). https://doi.org/10.1007/978-3-319-49956-7_25
17. Surmin, I., et al.: Particle-in-cell laser-plasma simulation on Xeon Phi coprocessors. Comput. Phys. Commun. **202**, 204–210 (2016). https://doi.org/10.1016/j.cpc.2016.02.004
18. Vay, J.L., et al.: Modeling of a chain of three plasma accelerator stages with the WarpX electromagnetic pic code on GPUs. Phys. Plasmas **28**(2), 023105 (2021). https://doi.org/10.1063/5.0028512
19. Vshivkov, V., Kraeva, M., Malyshkin, V.: Parallel implementation of the particle-in-cell method. Program. Comput. Softw. **23**(2), 87–97 (1998)
20. Gonoskov, I., et al.: Dipole pulse theory: maximizing the field amplitude from 4 π focused laser pulses. Phys. Rev. A **86**(5), 053836 (2012). https://doi.org/10.1103/PhysRevA.86.053836
21. Bashinov, A., et al.: Dense e− e+ plasma formation in magnetic dipole wave: vacuum breakdown by 10-pw class lasers. arXiv preprint arXiv:2103.16488 (2021)
22. Bulanov, S., Mur, V., Narozhny, N., Nees, J., Popov, V.: Multiple colliding electromagnetic pulses: a way to lower the threshold of e+ e− pair production from vacuum. Phys. Rev. Lett. **104**(22), 220404 (2010). https://doi.org/10.1103/PhysRevLett.104.220404

23. Danson, C.N., Haefner, C., Bromage, J., et al.: Petawatt and exawatt class lasers worldwide. High Power Laser Sci. Eng. **7** (2019). https://doi.org/10.1017/hpl.2019.36

24. Bassett, I.M.: Limit to concentration by focusing. Optica Acta Int. J. Opt. **33**(3), 279–286 (1986). https://doi.org/10.1080/713821943

25. Gonoskov, A., et al.: Anomalous radiative trapping in laser fields of extreme intensity. Phys. Rev. Lett. **113**(1), 014801 (2014). https://doi.org/10.1103/PhysRevLett.113.014801

Evaluating the Performance of Kunpeng 920 Processors on Modern HPC Applications

Ilya Afanasyev[✉][iD] and Dmitry Lichmanov[iD]

Moscow Center of Fundamental and Applied Mathematics, Moscow 119991, Russia
afanasiev_ilya@icloud.com

Abstract. Nowadays, ARM processors are widely used in various HPC applications. With ARM popularity rapidly increasing, there is still a significant lack of detailed performance evaluation of such systems on various workloads. Unlike other existing approaches to the performance evaluation, this paper covers the methodology of creating a full and comprehensive benchmarking set, which allows us to present a detailed performance comparison of Kunpeng 920–6426 and Intel Xeon 6140 processors. The developed benchmarks are based on relatively simple fragments of code, frequently used in many scientific and real-world applications. For each benchmark we provide a detailed scalability and performance analysis, based on the top-down and roofline performance models, which allow to identify bottlenecks and implementation efficiency for each benchmark. The evaluation results demonstrate that Kunpeng 920 outperform Intel Xeon 6140 processors on various cache-bound and memory-bound applications, such as stencil kernels, operations with dense matrices and vectors. At the same time, Kunpeng 920 demonstrate lower performance on compute-bound problems which can be vectorised or problems, involving indirect memory accesses, such as graph algorithms.

Keywords: Performance evaluation · Arm · Kunpeng · Benchmarking · Stencil · Graph algorithms

1 Introduction

The Kunpeng 920 (formerly known as Hi1620) is HiSilicon's fourth generation server processor announced in 2018, launched in 2019. It makes this processor the newest in Hisilicon product line, as Kunpeng 930 is not launched yet at the moment of writing this paper (the first half of 2021). Due to the novelty of this product line (first Kunpeng processor was launched in 2016) and Kunpeng 920 CPU model in particular, its potential is still poorly researched. In this paper, we use the following approach aimed to evaluate the performance of Kunpeng 920 processors. First, we run several benchmarking tools on Kunpeng 920 and Intel Xeon 6140 processors aimed to estimate various hardware characteristics. With the help of LMbench benchmark [1], Empirical Roofline Toolkit (ERT)

© Springer Nature Switzerland AG 2021
V. Malyshkin (Ed.): PaCT 2021, LNCS 12942, pp. 301–321, 2021.
https://doi.org/10.1007/978-3-030-86359-3_23

[2] and Intel Vtune Profiler we discovered that Kunpeng 920 CPU has higher memory subsystem bandwidth and smaller DRAM and cache latencies than Intel Xeon has, which potentially makes Kunpeng 920 more relevant platform to run various memory-bound applications. After that we have designed a new micro-benchmark package, which is based on relatively simple fragments of code, frequently used in many scientific and real-world applications. As an example, stencil kernels are widely used in finite difference methods, like in [3] and [4]. For each benchmark we provide a detailed scalability and performance analysis. In case of Intel CPU plaforms we used top-down model, built with Intel Vtune, while for ARM CPUs it was necessary to collect data from CPU core and uncore events to obtain figures for top-down model. All these tools allowed us to estimate the efficiency and existing bottlenecks for each benchmark. The wide range of benchmarks in the developed package helped us to get a full view on Kunpeng 920 performance behaviour compared to well-known systems like Intel Xeon.

2 Target Architectures

The Chinese company Huawei started developing ARMv8-based Kunpeng processors in 2016, which have been publicly available on market in 2019. Kunpeng processors have several modifications developed for both high performance computing and desktop computers, for example the number of cores ranges from 24 to 64 with clock speeds ranging from 2.4 to 3 GHz.

In this work, we evaluated the performance of TaiShan 200 two-socket system equipped with two Kunpeng 920–6426 processors. The performance has been compared to dual-socket system based on Intel Xeon Gold 6140 processors of the Skylake microarchitecture. The comparative hardware characteristics of these processors are provided in Table 1.

In order to obtain additional characteristics of the evaluated systems (such as cache bandwidth and latency values represented in Table 1), we used several benchmarking tools. Latency of different levels of memory subsystem has been measured with LMbench benchmark, using lat_mem_rd option. Stride parameter was equal to 1024 in order to avoid accessing data in the same cache line. Cache bandwidth values for Intel-based processors have been measured using Empirical Roofline Tool.

3 Related Work

A large number of benchmarks have been developed to compare various performance aspects of modern processors and GPUs. For example, STREAM [5] benchmark estimates the achievable DRAM bandwidth using operations over dense vectors (add, multiply on scalar etc.), HPL [6] benchmark estimates the achievable peak performance when solving a system of linear equations, HPCG [7] and Graph500 [7] benchmarks – the performance of memory subsystem during processing of indirect memory accesses in sparse matrix vector

Table 1. Main hardware characteristics of Kunpeng 920 and Intel Xeon Gold 6140 processors

Hardware characteristic	Kunpeng 920	Intel Xeon Gold 6140
Frequency	2.6 GHz	2.3 GHz (average), 3.7 GHz (with boost)
Sockets	2	2
SIMD instructions, width	ARM Neon, 128-bit	AVX-512, 512-bit
NUMA nodes	2	2
Cores per socket	64	18
Threads per core	1	1(2 with hyperthreading)
Theoretical peak performance (SP)	1.331 TFlop/s	2.649 TFlop/s
Theoretical peak performance (DP)	665 TFlop/s	1.324 TFlop/s
Theoretical scalar performance	332 GFlops	81 GFlops
Max OpenMP threads	128	36
L1d cache size	64 KB	32 KB
L1i cache size	64 KB	32 KB
L2 cache size	512 KB	1 MB
L3 cache size	65 MB	25 MB
L1 bandwidth/latency per socket	5248 GB/s, 1.539 ns	2802 GB/s, 1.083 ns
L2 bandwidth/latency per socket	4163 GB/s, 3.077 ns	1810 GB/s, 3.789 ns
L3 bandwidth/latency per socket	1397 GB/s, 14.418 ns	573 GB/s, 21.42 ns
DRAM bandwidth	187 GB/s	117 GB/s
Memory type	DDR4-2933	DDR4-2666
Number of channels	8	6

multiplication (SPMV) and breadth-first search (BFS) graph algorithm. However, an essential problem is that many existing benchmarks are not sufficiently adapted and optimized for newly developed architectures, such as Kunpeng 920. In addition, many existing benchmarks (for example HPL) are extremely complex and use third-party libraries (in the case of HPL – BLAS, Atlas, MKL). At the same time, the more complex the benchmark, the more difficult it is to understand hardware bottlenecks and peculiarities of the target CPU. For these reason in this paper we used a combination of existing benchmarks (such as HPL) and simple code fragments of real-world scientific algorithms and applications.

At the moment of this writing only a single work [8] is devoted to the comparative performance evaluation of Kunpeng processors. Authors compare Kunpeng 916 processors with Intel Xeon E5-2680v3 and Intel Xeon E5-2680v4 processors (of Haswell microarchitecture) using a combination of benchmarks (HPL, STREAM, etc.) and scientific applications (SpMV, SNAP, etc.). In addition, authors use various hardware performance counters to explain the behaviour of different applications on both systems. Similar reports are frequently prepared for other recently released architectures. Reviewing these reports is interesting in order to see which applications and microbenchmarks are used for the performance evaluation. For example, in [9] the performance of NEC SX-Aurora TSUBASA vector architecture is evaluated on a wide range of scientific applications: Landmine, Earthquake, Turbulent Flow, Antenna, etc. In [10] the performance of Intel Broadwell EP and Cascade Lake SP processors is compared using DGEMM and sparse matrix-vector multiplication benchmarks. In [11] the performance of Marvell's ThunderX2 processors (ARM-based, similar to Kunpeng) is evaluated on HPCG and COSA benchmarks, as well as for GROMACS, OpenSBLI and Nektar++ scientific applications.

4 Developed Benchmarking System

During this research we have developed a benchmarking system, which consists of a benchmarking framework and a set of computational kernels. The computational kernels are based on different relatively simple fragments of real-world programs and computational algorithms, such as matrix-matrix multiplication, stencil kernels, and many others. Both framework and its computational kernels are implemented in C++, while OpenMP programming model is used in order to exploit parallelism inside computational kernels.

The benchmarking framework contains of a unified set of rules for compiling and executing computational kernels, and, in addition, allows to easily integrate various profiling and performance analysis tools. The developed system has the following advantages:

- the developed framework allows to easily vary input data size, datatypes on other input parameters for all computational kernels, such as matrix sizes, input graph scale and type, stencil radius, etc.;
- each benchmark is automatically executed multiple times with accurate averaging of the obtained results; even though in most situations all benchmarks runs have approximately the same execution time, this allows to eliminate random operating system overheads;
- between different runs automatic cache annihilation is implemented;
- possibility to choose from various code compilers. We tested GNU g++/gcc, clang, icpc, Kunpeng g++ and nvcc compiler from NVIDIA HPC Toolkit. Though, the best performance on Intel was obtained with icpc, and with Kunpeng g++ on Kunpeng;
- for each computational kernel the estimated sustained bandwidth (in GB/s) and sustained performance (in GFlop/s) is collected, which allows to roughly estimate the performance of memory-bound and compute-bound benchmarks respectively;
- linux perf [12] performance analysis tool is integrated into the benchmarking system, which allows to obtain hardware performance counters and thus calculate different dynamic characteristics of the evaluated benchmarks, such as

L1/L2/L3 hit rate, the number of instructions per cycle, the number of flops executed, etc.;
- using the collected hardware performance counters the roofline [13] and top-down performance analysis models are automatically generated.

Top-down performance analysis demonstrates which activities take up most of the execution time of the evaluated benchmark. For example a program can be classified as L1-, L2-, L3- or DRAM-bound, which indicates that most of the time the program spends loading information from a specific cache or DRAM, and thus can potentially benefit from faster and lower latency bandwidth caches of target architecture.

The roofline model (shown in Fig. 1) allows to estimate execution efficiency of each benchmark. Using the roofline model each benchmark can be classified either as compute-bound or memory-bound based on its arithmetic intensity ratio. Then the efficiency of compute-bound benchmarks is calculated as the percentage of theoretical peak performance (scalar or vector depending on properties of the program), while of memory-bound – as the percentage of the one of memory roofs, which are based on cache and DRAM bandwidths. Further in the paper we will say "benchmark demonstrates higher performance on platform A compared to platform B" meaning one of these values depending on the type of the evaluated benchmark.

5   Implemented Benchmarks

The following subsections are devoted to a detailed performance comparison of various developed microbenchmarks, executed on the previously described Arm Kunpeng and Intel Xeon platforms. At the first place we implemented multi-threaded version for each benchmark, which is executed on either 64 (on Kunpeng 920) or 18 (on Intel Xeon 6140) OpenMP threads (so they occupy all available cores). This allows to provide a more fair comparison, since per-core performance of Kunpeng 920 will be in many situations significantly lower compared to Intel Xeon 6140 (due to Kunpeng having 3.5 times higher number of cores). However, since for many people per-core performance is also very important, in the final subsection we will evaluate scaling of different benchmarks, which belong to either compute-bound, L1, L3 cache bound or DRAM bound classes. In addition, we vectorized each benchmark using either AVX-512 on Intel Xeon or ARM NEON on Kunpeng 920, either by providing necessary compiler flags (which in all cases was enough for Intel), or using corresponding C++ ARM NEON API (Table 2).

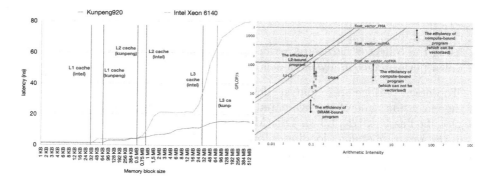

Fig. 1. Latency values (in ns) obtained using LMbench benchmark on the evaluated platform (left), the roofline model generated for multiple benchmarks on the Kunpeng 920 processors and the approach to calculating benchmark efficiency (right)

5.1 Triada Variations

Triada benchmark is designed to evaluate the performance of different memory access patterns to DRAM memory when operating with dense vectors. In contrast to the triada implementation in STREAM benchmark, we developed 10 different memory access patterns, frequently used in real-world programs. These memory access patterns (in C++ notation) are listed in Fig. 2 along the X axis. Benchmarks (1) – (4) are designed to evaluate the efficiency of triada operations for a different number of input vectors (from 2 to 4). Benchmarks (5) – (7) are designed to evaluate the efficiency of triada operation, when some arrays are indirectly accessed using indexes which have linear (equal to i) values, since on some architectures (for example NEC SX-Aurora TSUBASA) such indirect memory accesses leads to a significant performance degradation due to using gather/scatter instructions instead of load/store. In benchmarks (8) – (10) indexes are set randomly to generate as many cache line misses as possible.

The obtained results of benchmarking are demonstrated in Fig. 2, 3, 4, which allow to make the following observations. First, benchmarks (1)–(7) demonstrate 1.5–1.79 times higher performance (in terms of sustained memory bandwidth) on Kunpeng 920 compared to Intel Xeon, which is caused by 1.47 times higher memory bandwidth of Kunpeng 920 processors. In addition, Kunpeng 920 processors also demonstrate slightly higher efficiency (61–74% against 52–63%) calculated based on the roofline model. Second, according to the top-down analysis benchmarks (1)–(7) are 87% and 90% memory-bound on both platform; however, benchmarks executed on Intel Xeon are in average 11% L1-bound, 11% store-bound and 54% DRAM-bound, while on Kunpeng – 93% DRAM-bound, 2% L1 bound and 0.02% store bound. This indicates that benchmarks (1)–(7) running on Intel spend significantly more time on working with L1 cache and doing DRAM stores, while on Kunpeng benchmarks interact with DRAM almost all the time, resulting in Kunpeng demonstrating higher efficiency. Third, benchmarks (5)–(7) have only 10–15% lower efficiency compared to benchmarks

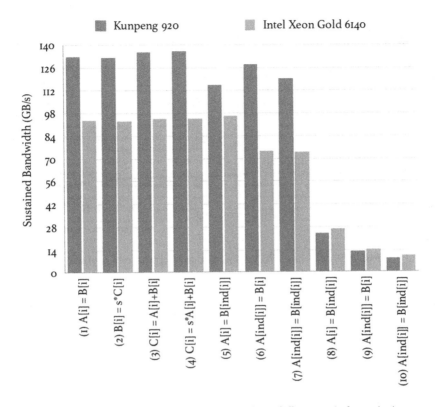

Fig. 2. The sustained bandwidth achieved on different triada variations

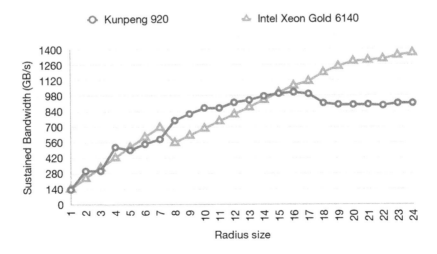

Fig. 3. The sustained bandwidth achieved on one-dimensional stencil kernels

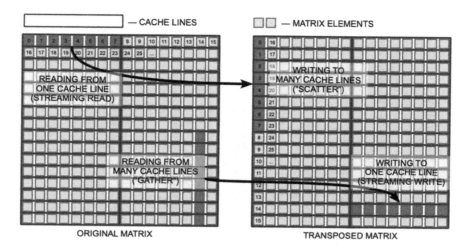

Fig. 4. Cache lines utilization in matrix transpose problem

(1)–(4) on both platforms, having approximately the same top-down analysis metrics. Finally, benchmarks (8)–(10), demonstrate the comparable sustained memory bandwidth on both platforms (Intel Xeon being 1–3% faster) and twice higher efficiency on Intel Xeon. This means that such indirect memory accesses reduce the efficiency of triada on Kunpeng 920 much more drastically (4.9 times against 3.1).

Table 2. Runtime contribution of the typical computational kernel into the whole program package

Test name	Front-end Bound	Bad Spec.	Retiring	Back-end Bound	Memory Bound	L1 Bound	L2 Bound	L3 Bound or DRAM	Core Bound	LL hit rate
triada	0.06	0.01	3.65	96.28	93.10	1.19	0.16	89.13	6.89	3.33
stencil 1D	0.40	0.71	36.80	62.08	30.48	24.79	2.34	2.83	69.51	2.85
stencil 2D	0.04	0.01	18.78	81.16	49.90	13.19	36.30	0.15	50.09	42.74
stencil 3D	0.24	0.23	39.90	59.60	44.90	18.43	2.37	23.78	55.09	27.18
n_body	0.16	0.01	28.22	71.59	26.43	21.68	4.22	0.53	73.56	95.38
random access (0)	0.04	0.01	2.32	97.62	93.20	0.25	0.03	92.41	6.79	69.55
random access (1)	0.11	0.01	1.36	98.53	93.86	0.20	0.19	93.46	6.13	5.10
rand generator	0.03	0.01	44.95	55.01	7.36	7.22	0.00	0.13	92.63	86.72
matrix transp (0)	0.19	0.01	1.58	98.21	90.06	1.32	0.57	86.92	9.93	71.70
matrix transp (2)	0.41	0.05	5.65	93.86	34.35	1.85	0.05	31.11	65.64	8.64
page rank (0)	8.83	0.98	23.58	66.59	82.37	8.93	0.68	72.75	17.62	95.98
page rank (1)	7.02	0.30	19.66	73.00	82.82	38.34	0.48	43.98	17.17	92.72

5.2   Matrix Transpose

This group of benchmarks is designed to evaluate the efficiency of various types of matrix traversals: by row or by column. Since all benchmarks are implemented in C++, all matrices are stored as one-dimensional arrays in row-major order. Accessing matrix elements by column is often refereed as "strided", since memory is accessed with stride equal to the size of matrix row. In this paper we implemented 3 variations of matrix transpose operation. First two variations are "naive", and are based on two nested loops. The first variation (1) accesses input matrix by columns and output matrix by rows, while (2) – in the mirrored opposite way (Fig. 4). The third variation is blocked transpose, which is based on four nested loops and block size 32. Blocked transpose variation performance was compared to the performance of similar kernel, written with the use of Eigen template library. Developed kernel launched on ARM Kunpeng is 1.3 faster than ARM version of library kernel, so all following performance details were obtained with our benchmarking system.

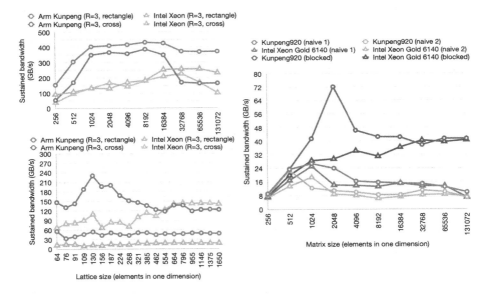

Fig. 5. The sustained memory bandwidth two-dimensional stencil (top left), three-dimensional stencil (bottom left) and matrix transpose (right) benchmarks

The obtained results of benchmarking are demonstrated in Fig. 5(right), which allow to make the following observations. First, from the two naive implementations the one writing to the same cache line and reading from different cache lines is preferred for both evaluated architectures for large matrices, with the performance difference between these 2 variations being identical for both evaluated architectures (1.5 – 2 times). Second, for both architectures a blocked implementation is approximately equivalent to a naive implementation on the small matrices (less then 1024^2), since 64 KB or 32 KB L1 cache can completely contain 8 columns of the processed matrices (the size of cache line is equal to 64 bytes, element size is 8 bytes, and information about next 7 lines is prefetched into the cache). When the size of matrix is larger than 1024 and 8 columns can not be fitted into L1 cache, the performance of blocked implementations for both architectures is significantly higher (2–3 times). Third, Kunpeng 920 processors demonstrate up to 1.18 times higher performance on naive 1 version for medium-sized matrices, and up to 2 times higher performance on blocked version with a large spike at 2048 sized matrix. At the same time with the increase of matrix size to 32768 Kunpeng 920 processors start demonstrating lower performance compared to Intel Xeons. The spike of blocked implementation at 2048 sized matrix on Kunpeng 920 can be explained by high L2 utilization: on 2048 sized matrix it is 82% core bound, 8% L2 bound and 8% L3 or DRAM bound, while on large matrices it is 5% core bound and 93% L3 or DRAM bound. Finally, the theoretical maximum performance of matrix transposition should be equal to the memory copy bandwidth, because the matrix data must be read once and written once to a different location. Both architectures demonstrate significantly lower sustained bandwidth values even on blocked implementations, however, on large matrices Kunpeng 920 having lower efficiency (20% against 30%). This indicates on Intel Xeons being better suited for problems, which have memory access pattern with very large stride (and thus having poor cache usage).

5.3 One Dimensional Stencil

In order to evaluate how cache memory hierarchy influences the performance of applications, we decided to implement several stencil kernels (one-dimensional, two-dimensional, and three-dimensional). One-dimensional (1D) stencil kernel is based on processing two linear arrays (input and output) in the following way: each element with index i of the output array is calculated as a function (the sum in our implementation) of $2 * R + 1$ elements of the input array with indices from $i - R$ to $i + R$. R is the main benchmark input parameter called "the neighborhood radius", and is varied between 1 and 12, while the size of the input and output arrays is set to 800M. Despite the fact that neighborhood radius values larger than 3 are rarely used in real-world applications, varying this parameter in a wider range allows to better evaluate caching effects of target architectures. One-dimensional stencil benchmark is characterized by high data reuse: almost all values of the input array loaded at i-th iteration (except the first element with index $i - R$) will be used at the next iteration, which potentially results into relatively efficient cache usage. Stencil-based benchmark YASL(Yet Another Stencil Kernel) shows perfect performance results on platforms with Intel processors, when is compiled with well compatible Intel compiler. Despite the fact that performance of stencil kernels in developed benchmarking system is slightly less, it is fairer to compare more general stencil kernels, so further results are based on stencil kernel from developed system.

The sustained memory bandwidth values for 1D stencil benchmark are listed in Fig. 3, from which the following observations can be made. First, on both architectures 1D stencil demonstrates high utilization of compute and cache resources. According to the top-down analysis, on Kunpeng 920 and Intel Xeon the retiring ratio (percent of time the processor spends on executing useful instructions) is very high: 18% and 30% on radius equal to 3, while increasing up to 49% and 43% on radius equal to 10. At the same time, the percentage of stalls related to requesting data from memory hierarchy is relatively small: 21% on Kunpeng 320 and 10% – 64% on Intel Xeon. Among these stalls the largest part is related to L1 requests, which increases with the radius size (8% against 18% for R=3 and R=10 on Kunpeng 920), which correlates to better data reuse and thus L1 utilization on large radius values. Second, 1D stencil on Kunpeng 920 demonstrates 1.6 times higher performance for small radius values (< 7) and up to 1.95 times higher performance for large radius values (> 8). This is mainly caused by 1.8 times higher bandwidth of L1 cache on Kunpeng 920 architecture. Third, the highest sustained bandwidth achieved by Kunpeng 920 is 588 GB/s and 238 GB/s by Intel Xeon, which is only a fraction of available L1 cache bandwidth, but 2–3 times higher compared to DRAM bandwidth. Thus accurate performance tuning of 1D stencil (vectorization and manual loop unrolling) may potentially result into higher performance on both architectures.

5.4   Two- and Three- Dimensional Stencil

In order to evaluate the performance of stencil-based real-world applications, we have implemented multiple variations of two-dimensional (2D) and three-dimensional (3D) stencil kernels. These benchmarks are a generalisation of previously introduced one-dimensional stencil, and have two types of neighborhood form: "rectangle-shaped" and "cross-shaped" with radius equal to either 1 or 3 (as used in most real-world scientific applications). This way 2D stencils have 4 elements for cross-shape neighbourhood and 8 elements for rectangle-shape neighbourhood when radius is equal to 1. When radius is equal to 3 it's 12 elements for cross-shaped and 48 for rectangle shaped – a significantly larger values compared to 1D stencil. For 3D stencil, processing each elements requires loading information about up to 342 points, which still should theoretically fit into L1 cache of both target architectures.

The key difference between 2D/3D and 1D stencils is the fact that the lattice is stored as a two-dimensional or three-dimensional array. This results into elements requested by y-offset or z-offset inside neighbourhood being loaded from different columns of matrix (in 2D case) – a situation similar to matrix transpose. However, unlike matrix transpose these elements are used right on the next iteration of innermost loop (instead of next $matrix\_size$ iteration), which results into significantly better locality and cache utilization. This is the main reason why for 2D/3D stencils we also vary the size of lattice together with radius.

The sustained memory bandwidth values for different neighborhood shapes, lattice sizes and radius of 3 (as the largest frequently used in stencil applications) are provided in Fig. 5, from which the following observations can be made. First, as discussed in the previous paragraph the sustained bandwidth of 2D and 3D stencils on both platforms does not change with the increase of lattice size, unlike matrix transpose (which significantly decreases for large matrices). The only exception is the performance decrease on 32K lattice for Kunpeng 920, which is explained by much lower L2 cache usage: according to the top-down analysis, for 16K sized lattice the benchmark is 22% L2 bound and 21% DRAM and L3 bound (L3 cache hit rate is low at 20%). For 32K sized lattice the benchmark is primary 70% DRAM and L3 bound. Second, 2D and 3D stencils with cross-shaped and rectangle-shaped neighbourhood of size 3 are 1.7–2 times faster on Kunpeng 920, with this difference being similar to the case of 1D stencil and proportional to L1 bandwidth difference. Finally, both architectures process cross-shaped borders more efficiently: the retiring ratio for rectangle-shaped neighbourhood is 18–25% range, while for cross-shaped – in 44–65%. This indicates that loading information via diagonal neighbourhood offsets is harmful on both architectures.

5.5   LCopt

This benchmark is based on the simulations software for liquid crystals [4]. The computational problem in this software belongs to the class of stochastic optimization of a functional defined on finite space cubic lattice. Namely, the solver is based on Markov chain Monte Carlo with Metropolis algorithm, paralleled by sparse checkerboard decomposition. Checkerboard decomposition uses a specific memory access patter (shown in Fig. 6 bottom left), which is interesting and important to evaluate on the investigated architectures. On different algorithm iterations pivot elements are selected based on random tick values, which can be either 0 or 1 among each dimension. The implemented benchmark (as well as the original program package) uses cubic lattice. Each pivot element is updated based on its of 26 neighbouring elements (in 3D case).

The sustained bandwidth values obtained on lcopt benchmark are provided in Fig. 6 (top left). Similar to other stencil benchmarks, lcopt demonstrates up to 1.7 times higher performance on Kunpeng 920. However lcopt benchmark is characterised by lower data reuse since gaps between pivot elements allow to reuse only 9 adjacent elements located between different pivot values in the 3D case. This observation is confirmed by the top-down analysis, which classifies lcopt benchmark as primary DRAM bound (31% on Intel Xeon) and (63% on Kunpeng 920), while L2 cache is the most heavily used on both platforms. This also results into lcopt benchmark demonstrating significantly lower achieved sustained memory bandwidth values compared to other stencil kernels.

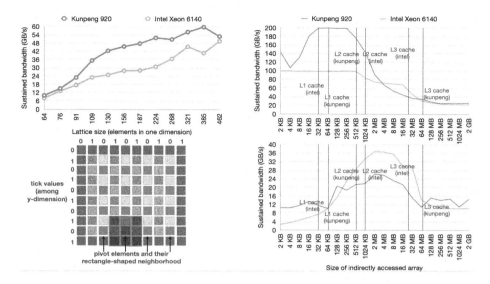

Fig. 6. The sustained bandwidth of lcopt benchmark (top left), the lattice configurations in checkerboard decomposition (bottom left), the sustained memory bandwidth values of random memory access loads (top right) and stores (bottom right) benchmarks

5.6   HPL (solving a SLE)

So far we have studied the performance only of memory-bound benchmarks. In order to compare the performance of compute-bound applications, we decided to use HPL benchmark, which is based on solving a system of linear equations. On Kunpeng 920 we compiled HPL using gcc v8.3 and openBLAS v0.3.10. On Intel Xeon Gold 6140 we also used gcc 8.3 compiler and Intel MKL. The maximum performance values obtained among different sizes of SLE are provided in Table 3: 605 GFlop/s for Kunpeng 920 (90% efficiency) and 843 GFlop/s for Intel Xeon 6140 (63% efficiency). Even though Kunpeng 920 higher efficiency (calculated based on theoretical peak performance) Intel Xeon 6140 still demonstrates 1.4 times higher performance, which is mainly caused by significantly wider SIMD instructions available (512 against 128 bit). This results into many other vectorizable compute-bound problems also being significantly slower on Kunpeng 920. This class of problems is relatively broad, since it also includes dense matrix-matrix and other linear algebra algorithms, which are frequently used in real-world applications, such as deep learning.

5.7   Random Number Generation

This benchmark is based on filling of a fixed-sized array with random numbers in parallel. Parallel random number generation is implemented via calls of rand_r function. This benchmark also belongs to the compute bound type, since for each element of the output array rand_r function is called, which executes around two dozen arithmetic operations. This is confirmed by top-down analysis, showing random generation benchmark being 44% retiring and only 7% memory-bound on both architectures. However, the important difference is that the main parallel loop (in which rand_r function is called) can not be vectorized by gcc compiler on both evaluated platforms, due to the inner complexity of rand_r function.

Runtime differences for an output array which contains 800M double-precision elements are shown in Fig. 3: Kunpeng 920 demonstrates 3.4 times higher performance compared to Intel Xeon 6140, which is explained by 3–4 times higher theoretical scalar performance of Kunpeng 920 processors. This example demonstrates that non-vectorizable compute-bound applications in most situations will be significantly faster on Kunpeng 920 platform. In addition, generating random-numbers on its own is a very important problem, frequently required in different numerical algorithms and applications: Monte-Carlo, random walk graph algorithm, etc.

Table 3. The performance of compute-bound benchmarks: random numbers generation and HPL

Benchmark	Kunpeng 920	Intel Xeon Gold 6140
rand_generator(double), L = 800M	0.072492 ms (3.45x)	0.250812 ms (1x)
HPL (max perf, double)	605 GFlop/s (1x)	843 GFlop/s (1.39x)
N-body, (max perf, float)	130 GFlop/s (2x)	64 GFlop/s (1x)

5.8   N-Body

Finally, we implemented N-body benchmark, which is well-known compute-bound problem, since its complexity is equal to $O(N^2)$, while most of the memory accesses are directed to L1 or L2 caches. During experiments we did not manage to force gcc 8.3 compiler to use ARM NEON instructions, thus we decided to completely disable vectorization on both systems using -fno-tree-vectorize flag. As shown in Table 3 Kunpeng 920 demonstrates 2 times higher performance compared to Intel Xeon 6140. This difference is lower compared to theoretical scalar performance ratio due to Intel Xeon still using SSE instructions for floating-point calculations (which was not the case with random generation, since all the calculations there were on integer).

5.9   Random Memory Access

An important characteristic of any modern processor is its ability to efficiently process random memory accesses, which are essential for many algorithms: operations with sparse matrices (SPMV), graph algorithms, etc. Random memory accesses typically imply the situation when some data arrays are accessed by indexes with random or pseudo-random nature. In programs with random memory access patterns modern CPUs still load data using cache lines; however, most of the cache line data (with the exception of the requested element) is not used on the further stages of algorithm, which results into memory bus being used very inefficiently. In addition, indirect memory accesses can be requested either for reading or writing, which may have very different performance (in terms of the sustained bandwidth). As shown by matrix transpose benchmark, writing to different cache lines can be up to two times less efficient compared to reading from different cache lines.

In order to evaluate the effects of random memory accesses we have prepared two benchmark variations. Both variations operate with arrays of two different lengths: "large" (6 GB) and "small" (from 2 KB to 2 GB). The first variation implies doing the following random loads in a loop iterating over large array: $large[i] = small[indexes[i]]$, while the second – the following random stores: $small[indexes[i]] = large[i]$. This way index and sequentially accessed arrays have "large" size, while indirectly accessed arrays – "small" size. The sustained memory bandwidth values for both implemented benchmarks are show in Fig. 6, where L1, L2 and L3 cache sizes of the evaluated processors are additionally listed. First, when data fits into L1, both benchmarks are 1.3–2.6 times faster on Kunpeng 920. Second, "random loads" benchmark demonstrates twice higher sustained bandwidth on Kunpeng 920 processors when indirectly accessed data fully fits into L2 cache, which is caused by Kunpeng 920 L2 cache having twice higher latency.

Third, Intel Xeon Gold 6140 demonstrate up to 1.5 times higher sustained bandwidth on 4–32 MB arrays, which fit into L3 cache of both processors, but do not fit into LLC partition owned by a single CPU core. This indicates that cores of Kunpeng 920 processors have problems when collectively working with data, shared via L3 cache; top-down model shows that for indirectly accessed array of 1 MB size "random loads" benchmark is 50% L3 and DRAM Bound, while on 16 MB array – 93%, which means that L3 cache can not provide data to cores fast enough when array size is growing, but still fits into L3 cache.

Finally, when indirect memory accesses are directed to DRAM, both evaluated processors demonstrate roughly equal sustained bandwidth (25 GB/s against 22 GB/s), with Kunpeng 920 having lower efficiency (13% against 17%) calculated based on theoretical DRAM bandwidth. This situation looks very similar to strided memory access wit hlarge stride, evaluated in transpose benchmark.

Thus, in general both platforms have comparable sustained bandwidth values when arrays are located in DRAM (can not be cached), while very different bandwidth values when data fits into some level of cache hierarchy (generally Kunpeng 920 being faster due to caches with higher bandwidth).

5.10   Bellman-Ford and Page Rank Graph Algorithms

In order to evaluate the performance of Kunpeng 920 architecture on graph algorithms, we have integrated VGL (Vector Graph Library) graph-processing framework into the developed benchmarking system. VGL framework provides high-performance implementations of several graph algorithms for modern multicore and vector processors [14] and NVIDIA GPUs [15]. VGL includes several optimisations (load balancing, graph-preprocessing, vectorization of graph algorithms) with the aim to maximise the sustained memory bandwidth on target architectures.

The performance evaluation has been conducted using Page Rank and Bellman-Ford graph algorithms, launched on synthetic RMAT [16], which simultaneously resemble properties of various real-world graphs (such as web-graphs and social networks) and can be easily scaled to verify the effects of caching. The smallest graph we used for testing (of scale 12) has 4 thousand input vertices, which easily can be placed into LLC of both platforms, while the larges (of scale 23) has 8 M vertices, thus significantly exceeding LLC size.

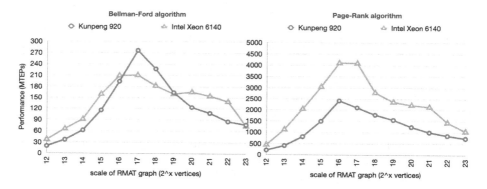

Fig. 7. The performance (in MTEPS) of Bellman-Ford (left) and page rank (right algorithm)

The obtained performance in terms of Traversed Edges Per Second (TEPS) is shown in Fig. 7: Intel Xeon 6140 processors demonstrate up to 1.6 times higher performance on both algorithms, which is caused by previously discussed lower efficiency of processing indirect memory accesses to arrays, which size exceeds L2 and L3 caches.

5.11 OpenMP-Based Benchmarks

All parallel versions of designed benchmarks are implemented via OpenMP. Naturally, the impact of various OpenMP constructs can vary in different systems, so we used EPCC OpenMP micro-benchmark suite [17] to estimate the impact of some frequently used OpenMP clauses on each target platform.

Table 4. Evaluation time of benchmarks with some OpenMP constructs on target architectures

OpenMP clause	Kunpeng 920	Intel Xeon Gold 6140
Parallel for	75.4 μs	5.21 μs
Barrier	16.8 μs	3.10 μs
Critical	18.4 μs	0.401 μs
Atomic	0.24 μs	0.145 μs
Trim step graph benchmark	1.67 ms	0.54 ms

As shown in Table 4 Kunpeng 920 requires significantly more time for each of OpenMP construct, which is possibly caused by a higher number of cores installed in Kunpeng 920. In order to further evaluate these effects in real-world programs, we have implemented a graph benchmark, which is based on calculating number of in-degree and out-degree edges for each graph vertex based on edges list representation (trim step). This operation is frequently used in various graph algorithms, such as connected components on strongly connected components in order to detect trivial components of small size. In order to perform computations in parallel, using atomic operations is required for incrementing in_degree[src_id] and out_degree[dst_id] values. According to the provided in Table 4 results this benchmark is 3.1 times slower on Kunpeng 920. Thus, the performance difference between Kunpeng 920 and Intel Xeon further increases from 1.6 times (on no-atomic algorithms) up to 3.1 times (on algorithm with atomic operations).

5.12 Scaling of Multi-threaded Benchmarks

Due to the fact that all previously discussed benchmarks are multi-threaded and have been executed on either 64 or 18 cores (all available in single socket of each evaluated processor), it is also important to compare single-core (or per-core) performance. To achieve this goal we selected three benchmarks: N-body problem, which is mainly compute-bound, 1D stencil with large radius, which is L1-bound, page rank on medium-scale graph, is primary which is L3 bound and triada, which is primary DRAM bound, and evaluated their scaling on 1–64 threads on Kunpeng 920 and 1–18 threads on Intel 6140 (Fig. 8). Selecting benchmarks, which stress different hardware resources, either private (compute, L1 cache), or shared (L3, DRAM) is important, since their scaling varies a lot.

Figure 8 demonstrates that N-body and 1D stencil benchmarks have almost linear scaling on Kunpeng (61 and 59 times acceleration on 64 threads compared to their sequential versions). This can be easily explained by the fact, that these benchmarks stress private resources of each core. At the same time, page rank and triada benchmarks, which stress shared L3 cache and DRAM, do not scale linearly (12 and 27 times acceleration when using 64 cores). This allows to conclude that per-core performance of Kunpeng is 12 times lower for DRAM-bound

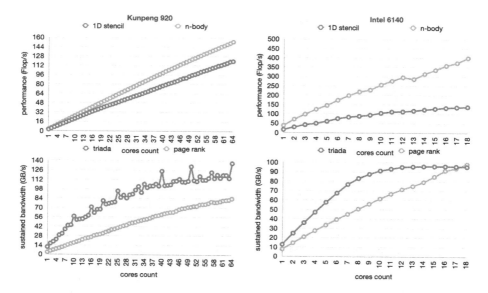

Fig. 8. Multi-threaded scaling of triada, 1D stencil, N-body and page rank benchmarks.

applications, 27 times lower for L3 bound applications and 64 times lower for L1 and compute-bound applications compared to multi-threaded versions running on 64 cores. At the same time for Intel Xeon per-core performance for these benchmarks is 7–10 times lower compared to multi-threaded versions running on 18 cores, which results into Kunpeng 920 having 1–2.5 times lower per-core performance compared to Intel Xeon.

6   Analysis of Hardware Features

In this section we are going to provide an analysis of Kunpeng 920 hardware features, which result into this architecture achieving higher or lower performance on different reviewed benchmarks.

Higher theoretical memory bandwidth (187 GB/s against 127 GB/s) allows Kunpeng 920 processors to achieve up to 1.8 times higher performance on various DRAM-bound applications, which has been demonstrated on benchmarks based on operations with dense liner vectors (triada) and lcopt benchmark, which uses checkerboard memory access pattern.

Four times higher theoretical scalar performance caused by a significantly larger number of cores allows Kunpeng 920 to outperform Intel Xeon 6140 on compute-bound non-vectorizable applications, as shown on random number generation benchmark or n-body problem (Kunpeng 920 being 3.4 and 2 times faster).

Due to Kunpeng 920 having caches with up to 2 times higher bandwidth and in most situations with lower latency, this processor achieves up to 1.8 times higher performance on 1D, 2D or 3D stencils with different radius and shape of neighbourhood.

At the same time 4 times longer vector length results into Kunpeng 920 demonstrating lower performance on vectorizable compute-bound applications, as has been shown on HPL benchmark.

We also observed a relatively poor performance of Kunpeng 920 on applications, which are typically L3 cache-bound. When some frequently accessed data arrays can be fully stored in L3 cache (for example distances or page ranks in the case of graph algorithms), according to hardware events and top-down analysis Kunpeng demonstrates low LLC hit rate and thus high DRAM utilization. This results Kunpeng demonstrating up to 1.6 times lower performance on various developed benchmarks, such as random access for several segment sizes and graph algorithms working with medium-sized graphs. Unfortunately due to top-down analysis on Kunpeng 920 classifying L3 and DRAM-bound applications into the same category, currently we struggle to provide a more detailed explanation on this problem.

In addition, we observed a significant performance degradation of Kunpeng 920 when working with indirect memory accesses to a large array exceeds L3 cache. In this case despite having lower DRAM latency and higher DRAM bandwidth Kunpeng 920 achieve comparable or lower performance with Intel Xeon 6140 on various benchmarks: random memory access, transpose of large matrices and graph algorithms.

We also observed a significant performance degradation of Kunpeng 920 processors when working with large datasets (16 GB and higher) on 2D, 3D stencil and matrix transpose problems, when memory access stride is larger than 128 KB. According to the top-down analysis, it is caused by much lower L1, L2 and L3 cache utilization of Kunpeng compared to Intel Xeons.

Finally, we have observed Kunpeng 920 spending more time on various OpenMP parallel constructs, such as atomics, criticals, barriers, and others, which is most probably caused by higher number of cores, and thus larger synchronization/atomic overheads, and confirmed its significant impact on real-world applications using a fragment from strongly connected component graph algorithm (trim step).

7   Conclusions

In this paper we have proposed a benchmarking system, which is aimed to compare the performance of Kunpeng 920 and Intel Xeon 6140 processors. According to the conducted research Kunpeng 920 processors demonstrate higher performance on various memory-bound applications, including 1D, 2D and 3D stencil kernels, lcopt benchmark, operations with dense vectors, matrix transpose. In addition, Kunpeng 920 also allow to achieve higher performance on scalar compute-bound problems, such as random number generation or N-body problem.

At the same time, Kunpeng 920 demonstrate lower performance on vectorizable compute-bound applications, such as HPL, or algorithms which involve indirect memory accesses to large arrays or strided memory accesses with a large stride, as has been shown for multiple graph algorithms and random access benchmark.

Acknowledgments. The reported study presented in Sects. 5.7 and 5.8 concerning evaluating the performance of VGL framework was is supported by Russian Ministry of Science and Higher Education, agreement No. 075-15-2019-1621. The work presented in all sections except 5.7 and 5.8 was supported by Huawei Technologies Co., Ltd. (Project No. OAA20100800391587A).

References

1. McVoy, L.W., Staelin, C., et al.: Lmbench: portable tools for performance analysis. In: USENIX Annual Technical Conference, pp. 279–294, San Diego, CA, USA (1996)
2. Lo, Y.J., et al.: Roofline model toolkit: a practical tool for architectural and program analysis. In: Jarvis, S.A., Wright, S.A., Hammond, S.D. (eds.) PMBS 2014. LNCS, vol. 8966, pp. 129–148. Springer, Cham (2015). https://doi.org/10.1007/978-3-319-17248-4_7
3. Roten, D., Olsen, K., Day, S., Cui, Y., Fäh, D.: Expected seismic shaking in los angeles reduced by san andreas fault zone plasticity. Geophys. Res. Lett. **41**(8), 2769–2777 (2014)
4. Rudyak, V.Y., Emelyanenko, A.V., Loiko, V.A.: Structure transitions in oblate nematic droplets. Phys. Rev. E **88**(5), 05250 (2013)
5. McCalpin, J.D.: Stream benchmark, vol. 22 (1995). http://www.cs.virginia.edu/stream/ref.html# what
6. Luszczek, P.R., et al.: The hpc challenge (hpcc) benchmark suite. In: Proceedings of the 2006 ACM/IEEE conference on Supercomputing, vol. 213, pp. 1188455–1188677. Citeseer (2006)
7. Marjanović, V., Gracia, J., Glass, C.W.: Performance modeling of the HPCG benchmark. In: Jarvis, S.A., Wright, S.A., Hammond, S.D. (eds.) PMBS 2014. LNCS, vol. 8966, pp. 172–192. Springer, Cham (2015). https://doi.org/10.1007/978-3-319-17248-4_9
8. Wang, Y.-C., et al.: An empirical study of hpc workloads on huawei kunpeng 916 processor, pp. 360–367 (2019)
9. Komatsu, K., et al.: Performance evaluation of a vector supercomputer sx-aurora tsubasa. In: SC18: International Conference for High Performance Computing, Networking, Storage and Analysis, pp. 685–696. IEEE (2018)
10. Alappat, C.L., Hofmann, J., Hager, G., Fehske, H., Bishop, A.R., Wellein, G.: Understanding HPC benchmark performance on intel Broadwell and cascade lake processors. In: Sadayappan, P., Chamberlain, B.L., Juckeland, G., Ltaief, H. (eds.) ISC High Performance 2020. LNCS, vol. 12151, pp. 412–433. Springer, Cham (2020). https://doi.org/10.1007/978-3-030-50743-5_21
11. Jackson, A., Turner, A., Weiland, M., Johnson, N., Perks, O., Parsons, M.: Evaluating the arm ecosystem for high performance computing. In: Proceedings of the Platform for Advanced Scientific Computing Conference, pp. 1–11 (2019)
12. De Melo, A.C.: The new linux'perf'tools. Slides Linux Kongr. **18**, 1–42 (2010)
13. Williams, S., Waterman, A., Patterson, D.: Roofline: an insightful visual performance model for multicore architectures. Commun. ACM **52**(4), 65–76 (2009)

14. Afanasyev, I.V., Voevodin, V.V., Komatsu, K., Kobayashi, H.: Vgl: a high-performance graph processing framework for the nec sx-aurora tsubasa vector architecture. J. Supercomput. 1–22 (2021)
15. Afanasyev, I.V.: Developing an architecture-independent graph framework for modern vector processors and nvidia gpus. Supercomput. Front. Innov. **7**(4), 49–61 (2021)
16. Chakrabarti, D., Zhan, Y., Faloutsos, C.: R-mat: a recursive model for graph mining. In: Proceedings of the 2004 SIAM International Conference on Data Mining, pp. 442–446. SIAM (2004)
17. Bull, J.M., Reid, F., McDonnell, N.: A microbenchmark suite for OpenMP tasks. In: Chapman, B.M., Massaioli, F., Müller, M.S., Rorro, M. (eds.) IWOMP 2012. LNCS, vol. 7312, pp. 271–274. Springer, Heidelberg (2012). https://doi.org/10.1007/978-3-642-30961-8_24

Job Management

Optimization of Resources Allocation in High Performance Distributed Computing with Utilization Uncertainty

Victor Toporkov[✉], Dmitry Yemelyanov, and Maksim Grigorenko

National Research University "MPEI", ul. Krasnokazarmennaya, 14, Moscow 111250, Russia
{ToporkovVV,YemelyanovDM,GrigorenkoMO}@mpei.ru

Abstract. In this work, we study resources co-allocation approaches for a dependable execution of parallel jobs in high performance computing systems with heterogeneous hosts. Complex computing systems often operate under conditions of the resources availability uncertainty caused by job-flow execution features, local operations, and other static and dynamic utilization events. At the same time, there is a high demand for reliable computational services ensuring an adequate quality of service level. Thus, it is necessary to maintain a trade-off between the available scheduling services (for example, guaranteed resources reservations) and the overall resources usage efficiency. The proposed solution can optimize resources allocation and reservation procedure for parallel jobs' execution considering static and dynamic features of the resources' utilization by using the resources availability as a target criterion.

Keywords: Computing · Grid · Resource · Scheduling · Uncertainty · Dynamic · Availability · Probability · Job · Allocation · Optimization

1 Introduction and Related Works

Today, Grid and cloud computing systems are used universally. Due to their commercial reach and low entry threshold, they attract users with different technical skills, who solve a wide range of computational tasks (time- and volume-wise) and require different quality of service.

It usually takes certain economic costs to build and manage the necessary computing infrastructure, including the purchase and installation of equipment, the provision of power supply, and user support. Thus, when a budget for job performance is limited, it becomes important to allocate suitable resources efficiently in accordance with both technical specification and a constraint on the total cost [1–3].

The system's resources may include computational nodes, storage devices, data communication links, software, etc. Each resource has a set of characteristics, their values determine its suitability for performing a specific job. Generally, computational nodes have the widest set of characteristics. For example, a virtual machine is the main computing resource in the commonly used CloudSim simulator [2, 3], its characteristics

© Springer Nature Switzerland AG 2021
V. Malyshkin (Ed.): PaCT 2021, LNCS 12942, pp. 325–337, 2021.
https://doi.org/10.1007/978-3-030-86359-3_24

include overall performance, number of CPU cores, size of RAM and disk memory, bandwidth limit of the data link.

It is worth mentioning the dynamic utilization issue of available resources and computational nodes at time. High performance and distributed computing systems (HPDCS) are the dynamic systems, in which the following processes take place: execution of parallel jobs from multiple users, utilization with local jobs, maintenance works, a physical shutdown of nodes (both scheduled and unscheduled). To procure the reliability and dependability of such systems, an advance allocation mechanism is used [4–7]. This mechanism allows one to pre-allocate resources for a specific job and, thereby, prevents possible contention between jobs. Thus, a utilization schedule for each resource can be obtained: a list of utilization intervals (allocated time, scheduled maintenance, and outages) and downtime periods. Downtime periods can be used to perform other jobs and to allocate the resources for the execution of user jobs.

The problem of scheduling and co-allocating resources for executing parallel jobs in a distributed computing system with non-dedicated resources is stated as follows.

1. The set R of the computing system resources, as a rule, is heterogeneous and includes resources r_i of several types with different sets of characteristics C_i. The values of these characteristics for the resources of the same type may also differ. Among the most important characteristics of a resource, one can single out its performance, which affects the execution time of a job, as well as the cost required to allocate the resource. Besides, at any specific time, some subsets of the resources may be unavailable for a user job. Therefore, available resources, as a rule, are represented in classical models as a set of slots - intervals of availability of each resource [5–7].
2. On the other hand, a job typically requires the parallel allocation of multiple resources with types and characteristics defined by the user who is running the job. The resource request for the job execution includes the number of concurrently required resources n, the minimum suitable values of the characteristics C_i, the volume of task V (the number of calculations/instructions) or the ordered resource allocation time T, as well as the total execution budget C [1, 2, 7].

However, as a rule, the structure and specifics of submitted jobs in HPDCS imply some uncertainty, primarily in the execution time and load of the allocated resources. So, users can only roughly estimate the execution time of their jobs, while special expert systems for predicting the execution time of user programs or the level of resource load (based on the use of machine learning, statistics, and big data) present the results in the form of probabilities of outcomes [4, 8, 9].

Existing systems of distributed computing provide the job flow execution according to the First Fit [2] principle or are based on deterministic models of resource scheduling [5–7, 10–12]. In the first case, pre-allocation mechanisms are not realized, user jobs in the queue wait for higher priority jobs to be executed and may *hang* with no guaranteed start time. In the second case, the efficiency and accuracy of scheduling are reduced by inaccurate estimates of the job execution time by users or experts. Late job completion time requires rescheduling of all the subsequent jobs or shutting down the job with possible loss of results. Early release of resources also requires rescheduling to minimize the resource downtime. For example, according to an existing approach [9], a scheduler may double the user's runtime estimates to improve the efficiency of the job flow. Thus, reliable and controlled scheduling of user jobs, which considers the uncertainty of resource load [4, 8], can be used both to increase the overall efficiency of the resources usage and to guarantee the priority-based execution of jobs with pre-allocated resources.

In this paper, we propose proactive algorithms for resources allocation and reservation in heterogeneous market-based computing environments considering static and dynamic resources availability uncertainties. The uncertainties are formalized with the availability probability functions as a natural way of statistical and machine learning predictions presentation. The novelty of the proposed solution is in general knapsack-based resources selection procedure performing resources availability maximization according to the parallel job requirements.

The paper is organized as follows. Section 2 presents a formal model of the resources' utilization and a general procedure for the dynamic resources' allocation optimization. Additional details are provided for the subset selection and time scan algorithms. Section 3 provides details about the simulation experiment setup, simulation results and analysis. Section 4 summarizes the paper and describes further research topics.

2   Resource Selection Algorithm

2.1   Resources Utilization Model

We consider a set R of heterogeneous computing nodes with different performance p_i and price c_i characteristics.

The probabilities (predictions) of the resource's availability and utilization for the whole scheduling interval L are provided as input data. Dynamic job execution uncertainties are modeled as a sequence of *allocation, occupation* (actual execution) and *release* events with the *occupation* probability $P_o \leq 1$. Global (static) resources utilization uncertainties, such as maintenance works or network failures, are modeled as a continuous *occupation* events with $P_o \ll 1$ during the whole considered scheduling interval.

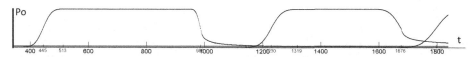

Fig. 1. Example of a resource utilization probability schedule.

Figure 1 shows an example of a single resource occupation probability P_o schedule. With two jobs already assigned to the resource, there are two resources allocation events (with expected times of allocation at 445 and 1230 time units), two resources occupation events (starting at 513 and 1319 time units) and two resources release events (expected release times are 986 and 1676 time units respectively). Gray translucent bar at the bottom of the diagram represents a sum of global utilization events with a total resource occupation probability $P_o = 0.05$.

A detailed analysis of the main utilization characteristics of real HPDCS systems was made to design and simulate an adequate resources utilization model. As the basis for modeling the availability and utilization probability of computational nodes, the log files of the ForHLR II supercomputer from the Karlsruhe Institute of Technology in Germany were taken for the analysis [13, 14]. The available files contain information on the execution of jobs from June 2016 to January 2018.

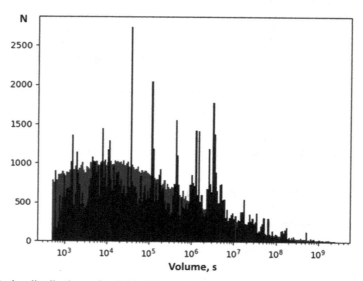

Fig. 2. Job size distributions of real (black) and simulated (blue) job-flows. (Color figure online)

After carrying out many experiments, the normal distribution on a logarithmic scale (lognormal) was chosen as the most suitable for modeling the jobs' length and size characteristics. The main parameters of the distribution (mathematical expectation and variance) were selected experimentally to achieve an acceptable accuracy. As a result, the generated distribution by form largely replicates the original one (Fig. 2). More formal comparison gives 0.14 value by the Kolmogorov - Smirnov test.

Thus, the resources *allocation* events are modeled by random variables with a normal distribution. Resources *release* events are modeled with lognormal distribution and expose heavy tails [14]. Expected allocation and release times are derived from the job's replication and execution time estimations.

2.2 Resources Allocation Under Uncertainties

To execute a parallel job a set of simultaneously idle nodes (a window) should be allocated ensuring user requirements from the resource request. The resource request usually specifies number n of nodes required simultaneously, their minimum applicable performance p, job's total computational volume V and a maximum available resources allocation budget C.

These parameters constitute a formal generalization for resource requests common among distributed computing systems and simulators [2, 7, 12, 13].

Common allocation and release times for all the window resources ensure the possibility of inter-node communications during the whole job execution. In this way, the occupation and availability probabilities should be estimated for each resource during the scheduling interval L. For the job scheduling, values $P_a^{r_i}(t; t + T)$ may be derived, representing a probability that resource r_i will be available for the whole job execution interval T starting at time t.

When a set of n resources is required for a job execution for a period T, the total window availability P_a^w during the expected job execution interval can be estimated as a product of availability probabilities of each independent window nodes:

$$P_a^w(t) = \prod_i^n P_a^{r_i}(t; t + T).$$

If any of the window nodes will be occupied during the expected job execution interval T, the whole parallel job will be postponed or even aborted. Therefore, a common resources allocation problem is a maximization of a total resources' availability probability.

Based on the model above we consider the following job resources allocation problem in heterogeneous computing environment with non-dedicated resources and utilization uncertainties: during a scheduling interval L allocate a set of n nodes with performance $p_i \geq p$ for a time T, with common allocation and release times and a restriction C on the total allocation cost. As a target optimization criterion, we assume maximization of a whole window availability probability P_a^w.

The solution for this problem may be divided into two sub-problems.

1. Static sub-problem. Given the time t_k and values $P_a^{r_i}(t_k; t_k + T)$ of the resources' availability for the following period T, allocate a subset of n resources according to the job requirements with the maximum probability $P_a^w(t_k)$.
2. Dynamic generalization. Perform time scan and execute the first sub-problem for each time moment $t_k \in [0; L]$. The solution is then obtained as a maximum from all the intermediate solutions: $P_a^w = \max_{t_k} P_a^w(t_k)$.

Thus, further in this paper we study different approaches for these two sub-problems implementation.

As an example, Fig. 3 shows maximum values of function $Z = P_a^w(t)$ obtained for a parallel job on the interval $[0; 1200]$ with the maximum availability probability reaching 0.93 at $t^{max} = 834$.

2.3 Near-Optimal Resources Allocation

Let us discuss in more details the procedure which allocates an optimal (according to the probability criterion P_a^w) subset of n resources at some static time moment t_k.

We consider the following total resources availability criterion $P_a^w = \prod_i^n P_a^{r_i}$, where $P_a^{r_i} = P_i$ is an availability probability of a single resource r_i on the interval $[t_k; t_k + T]$.

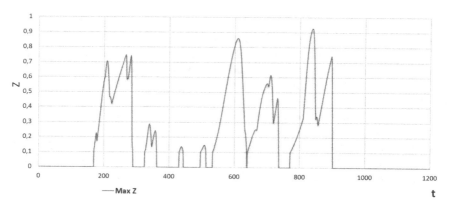

Fig. 3. An example of $\max P_a^w(t)$ function for a parallel job resources allocation.

In this way we can state the following problem of an n - size window subset allocation out of m available nodes in the system:

$$P_a^w = \prod_{j=1}^{m} x_j P_j, \tag{1}$$

with the following restrictions:

$$\sum_{j=1}^{m} x_j c_j \leq C,$$

$$\sum_{j=1}^{m} x_j = n,$$

$$x_j \in \{0, 1\}, j = 1..m,$$

where c_j is total cost required to allocate resource r_j for a time T, x_j - is a decision variable determining whether to allocate resource r_j ($x_j = 1$) or not ($x_j = 0$) for the current window.

In [15] based on a classical 0–1 *Knapsack* problem solution we proposed the following dynamic programming recurrent scheme to solve problem (1):

$$f_j(c, v) = \max\{f_{j-1}(c, v), f_{j-1}(c - c_j, v - 1) * P_j\}, \tag{2}$$

$$j = 1, .., m, c = 1, .., C, v = 1, .., n,$$

where $f_j(c, v)$ defines the maximum availability probability value for a v-size window allocated from the first j considered resources for a budget c. After the forward induction procedure (2) is finished the maximum availability value $P_{a\ max}^w = f_m(C, n)$. x_j values are then obtained by a backward induction procedure. Further in this paper we will refer to this algorithm simply as *Knapsack*.

An estimated computational complexity of the presented recurrent scheme is $O(m * n * C)$, which is n times harder compared to the original *Knapsack* problem $(O(m * C))$.

2.4 Greedy Resources Allocation Algorithm

Another approach for the static subset allocation sub-problem is to use more computationally efficient greedy algorithms. We outline four main greedy algorithms to solve the problem (1).

1. *MaxP* selects first n nodes providing maximum availability probability P_j values. This algorithm does not consider total usage cost limit and may provide infeasible solutions. Nevertheless, *MaxP* can be used to determine the best possible availability options and estimate a budget required to obtain them.
2. An opposite approach *MinC* selects first n nodes providing minimum usage cost c_j or an empty list in case it exceeds a total cost limit C. In this way, *MinC* does not perform any availability optimization, but always provides feasible solutions when it is possible. Besides, *MinC* outlines a lower bound on a budget required to obtain a feasible solution.
3. Third option is to use a weight function to regularize nodes in an appropriate manner. *MaxP/C* uses $w_j = P_j/c_j$ as a weight function and selects first n nodes providing maximum w_j values. Such an approach does not guarantee feasible solutions but performs some availability optimization by implementing a compromise solution between *MaxP* and *MaxC*.
4. Finally, we consider a joint approach *GreedyJnt* for a more efficient greedy-based resources allocation. The algorithm consists of three stages.

   a. Obtain *MaxP* solution and return it if the constraint on a total usage cost is met.
   b. Else, obtain *MaxP/C* solution and return it if the constraint on a total usage cost is met.
   c. Else, obtain *MinC* solution and return it if the constraint on a total usage cost is met.

This combined algorithm is designed to perform the best possible greedy optimization considering restrictions on total resources allocation size and cost.

Estimated computational complexity for the greedy resources' allocation step is $O(m * \log m)$.

2.5 Time Scan Optimization

Dynamic generalization of the static resources' allocation problem requires a full-time scan performed over all the considered scheduling interval L. In general, this leads to a significant increase in the computational cost of the dynamic scheduling algorithm (especially, when a full knapsack-based optimization should be performed for all time moments $t_k \in [0; L]$).

To optimize the performance of the proposed resources allocation procedure during the time scan we consider a computational method which performs search for the maximum from a set of starting time points. Assuming, that the functions $P_a^{r_i}(t)$ for each resource are continuous in time (see Fig. 1), then their product $P_a^w(t)$ will be continuous as well. This means that certain computational algorithms are applicable for $P_a^w(t)$

function study and the extrema search. Figure 2 shows an example of $P_a^w(t)$ function calculated by the resources allocation algorithm after scanning all time points if [0; 1200] interval.

A general procedure for $\max P_a^w(t)$ search optimization during the scheduling interval L can be presented as follows.

1. A set of starting time points is allocated on the interval L. Their particular locations can be given as 1) uniform, 2) randomized, 3) a combination of options 1 and 2.
2. At each starting time point t_i^s the value of $P_a^w(t_i^s)$ is calculated by the static resources' allocation algorithm (*Knapsack* or *GreedyJnt*) based on actual resources state at t_i^s.
3. The gradient value is determined for each starting point by calculating and comparing neighbor values $P_a^w(t_i^s + 1)$ and $P_a^w(t_i^s - 1)$ with $P_a^w(t_i^s)$.
4. From each starting point t_i^s an incremental movement is performed in the direction of increasing the gradient by the sequential calculation of $P_a^w(t_i^s \pm \delta * k) = P_a^w(t_i^s, k)$, where k is a step number. The movement is stopped if the maximum is reached (when $P_a^w(t_i^s, k) < P_a^w(t_i^s, k - 1)$) and, thus, can be found on the interval $[t_i^s \pm \delta * (k - 1); t_i^s \pm \delta * k]$. Besides, the search movement stops if any other starting points t_{i+1}^s or t_{i-1}^s are reached. In this case, the search will be continued independently, starting from the corresponding points.

It should be noted that the above optimization procedure does not guarantee an exact solution: scenarios of finding local maxima or missing abrupt function changes are possible. Improving the accuracy is possible by increasing the set of starting points and by decreasing the search step length δ. On the other hand, the performance of this procedure is significantly increased compared to the full-time scan: the calculation of function $P_a^w(t)$ is performed on a limited set of time points, guaranteed to be smaller than the whole interval L.

3 Simulation Study

3.1 Simulation Environment

We performed a series of simulations to study optimization properties of the proposed dynamic resources allocation approaches. An experiment was prepared as follows using a custom distributed environment simulator [11, 12, 15]. For our purpose, it implements a heterogeneous resource domain model: nodes have different usage costs and performance levels. A space-shared resources allocation policy simulates a local queuing system (like in CloudSim [2]) and, thus, each node can process only one task at any given simulation time. Additionally, each node supports a list of active global and local job utilization events.

Global static uncertainty events represent resources failure or shutdown susceptibility and keep a constant occupation probability during the whole scheduling interval L. Static utilization is generated for each resource based on a random variable P_o of occupancy probability with a normal. System-wide global-load parameter defines a standard deviation for P_o and is used to set an average global utilization for the whole computing environment. Thus, for example, when global load = 0.05, about 68% of the resources on average have global occupancy probability $P_o^{r_j} < 0.05$. More detailed study of a static resources' allocation problem under global utilization uncertainties was provided in [15].

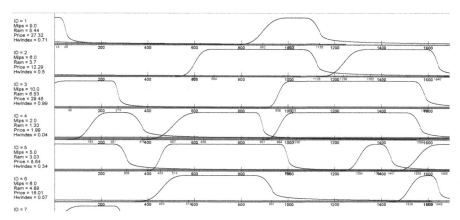

Fig. 4. An example of static and dynamic load generated for system resources.

Dynamic job-based utilization uncertainty is generated based on a preliminary job-flow scheduling simulation. For each resource a list of single-node jobs is generated with random jobs' submit times, lengths, start time and finish time uncertainty estimations. The jobs are ordered by their submit time and are scheduled in advance starting either at the submit time, or after the previous job is finished. During this scheduling, a chain of the resource *allocation*, *occupation* and *release* events is generated for each job. Corresponding expected times and standard deviations are defined by the job length and uncertainty parameters. More details regarding the simulated job-flow properties provided in Sect. 2.1. A total length of jobs generated for each resource is determined by a system wide job-load parameter. For example, when job-load = 0.1, a total length of locally generated jobs constitutes nearly 10% of the considered scheduling interval L.

Figure 1 shows a single resource utilization schedule with global and dynamic utilization events generated based on the procedures described above. Figure 4 shows an example of global and dynamic utilization uncertainties generated for a subset of the system resources in simulator [15].

3.2 Dynamic Resources Allocation

To solve the dynamic resources allocation problem for a parallel job, it is necessary to consider the available resources' schedule and utilization events which change over time Thus, the scheduling problem requires allocation of a set of suitable resources not at some static moment t_k, but during a given time interval.

Since the computational complexity and working time of the algorithms under consideration increase in proportion to the size of the considered scheduling interval, the following parameters of the scheduling problem were chosen to minimize the simulation time. It is required to maximize the probability of simultaneous availability of 6 concurrently available nodes to perform a job with a volume of 200 computational units on a time interval [0; 800] in a computing environment that includes 64 heterogeneous computational nodes. Initial load of computational nodes with global events global load = 0.05. The dynamic load of the computing system changed during the simulation within the limits of job-load $\in$ [0; 1].

The obtained results indicating the availability of the resources selected by the *Knapsack* (Sect. 2.3) and *GreedyJnt* (Sect. 2.4) algorithms depending on the dynamic load *job-load* values, are presented in Fig. 5. To obtain these results, more than 10,000 independent scenarios of scheduling and resources allocation were performed by each of the considered algorithms.

It should be noted that with *job-load* = 0 the advantage of the *Knapsack* algorithm is about 9%, and the probability of simultaneous availability of the selected resources is 0.96 for *Knapsack* and 0.87 for *GreedyJnt*.

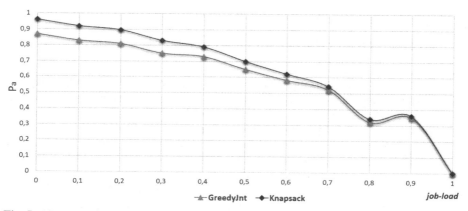

Fig. 5. Simulation results: P_a^w resources availability obtained by *Knapsack* and *GreedyJnt* algorithms depending on the resources utilization level.

With an increase in the dynamic load of the system (*job-load* > 0), the highest achievable probability P_a^w of simultaneous resource availability, as expected, sensibly monotonically decreases. The local maximum at *job-load* = 0.9 is explained by the fact that under conditions of extra high dynamic load, the number of experiments in which it was possible to find six concurrently available resources, turned out to be statistically insignificant (about 10 results). At the same time, when the value of job-load = 1 (full initial utilization of the system) was reached, a suitable set of resources ($P_a^w = 0$) was not found in any of the experiments at any time instant $t_k \in [0; 800]$.

Also, *Knapsack* provided a higher availability probability P_a^w of the required set of resources at all the considered values of the dynamic load (*job-load* < 1) in comparison to *GreedyJnt*. However, the relative advantage decreases from about 9% to almost 0% as the *job-load* increases. This decrease in relative efficiency is explained by a decrease in the dimensionality and variability of problem (1) with an increase of the resources load.

For example, when *job-load* = 0 all 64 resources are available at every instant with a probability of at least 0.95 (due to global-load = 0.05). Then as the *job-load* increases, many resources fall out of consideration due to a high probability of being utilized by other jobs (see Fig. 4). Thus, for large *job-load* values, the static algorithms often solved the degenerated problem of selecting a set of 6 concurrently available resources from 6 resources in the system that remained unloaded.

3.3 Time Scan Optimization

The time and accuracy characteristics of the proposed time scan optimization procedure (Sect. 2.5) were studied based on a resource's allocation problem in computing environment with dynamically changing utilization level. Figure 4 presents an example of a utilization schedule generated for a few computational nodes in the simulation environment [15].

To obtain reliable results, we performed more 1000 independent simulation scenarios of resources allocation for a single parallel job. The computing environment consisted of 64 heterogeneous computing nodes of varying cost and performance characteristics with dynamically changing occupation function $P_o(t)$: *job-load* = 0.5, *global-load* = 0.05. The job scheduling problem required allocation of six nodes for 200 units of time on the interval $L \in [0; 800]$. The target optimization criterion P_a^w is a simultaneous availability of the selected resources. As an additional criterion, a total algorithm working times was measured.

The time scan optimization procedure described in Sect. 2.5 was implemented with a different number of the starting points: {1, 5, 10, 20, 50, 100}.

Table 1 shows the relative results in terms of working time (performance acceleration) and accuracy in comparison with the full scan approach.

As expected, with an increase in the number of starting points the accuracy of the approximate procedure tends to 1 (i.e., to the optimal solution obtained with a full scan search). Already with 50 starting points (on an interval of 801 points) the accuracy of the optimized solution reaches 99%, while the calculation time is accelerated by almost 7 times.

On the other hand, full scan procedure with *GreedyJnt* algorithm achieves 95% accuracy with a 143× speedup! Thus, it is advisable to apply *Knapsack* with this time scan optimization technique if it is necessary to achieve a high accuracy in the presence of the light computation time restrictions. In this case, it is possible to speed up the work time by about an order of magnitude. With tighter time constraints, additional speedup can be achieved by using a greedy counterpart. In addition, the time scan optimization is applicable to *GreedyJnt* algorithm as well: for example, running *GreedyJnt* algorithm from 50 starting points allows you to speed up the computation time by 1000 times, while the solution accuracy will decrease only to 94%.

Table 1. Algorithms' efficiency comparison in terms of accuracy and performance optimization

Algorithm	Accuracy	Acceleration
Full scan (*Knapsack*)	1	1
1 starting point (*Knapsack*)	0,8	65
5 starting points (*Knapsack*)	0,93	17
10 starting points (*Knapsack*)	0,96	10,5
20 starting points (*Knapsack*)	0,97	8,3
50 starting points (*Knapsack*)	0,99	6,8
100 starting points (*Knapsack*)	0,99	3,7
Full scan (*GreedyJnt*)	0,953	143

4 Conclusion and Future Work

In this work, we presented procedure for a reliable resources' allocation in high performance computing systems with heterogeneous hosts considering utilization uncertainty. The uncertainties are formalized with probability functions as a natural way of statistical and machine learning predictions representation. The proposed solution uses an availability criterion to optimize resources allocation under static and dynamic utilization features. *Knapsack*-based and greedy algorithms were implemented and compared in a dynamic procedure performing optimized time scan over a specified scheduling interval. Both approaches were able to successfully optimize availability of the selected resources.

We considered several types of static and dynamic job-based resources utilization events with different load levels.

The simulation study addressed two main criteria: optimization efficiency and algorithms working time. *Knapsack*-based solution advantage over the greedy approach by the resources availability criterion at average reaches 5% but requires nearly 100 times more time for the calculations. Considering a relatively high computation complexity of the *Knapsack*-based solution, several optimization options were proposed to provide 99% accuracy 10 times faster or almost 94% accuracy 1000 times faster.

In our further work, we will refine the resource utilization model to simulate different types of global and local utilization events closer to real systems.

References

1. Lee, Y.C., Wang, C., Zomaya, A.Y., Zhou, B.B.: Profit-driven scheduling for cloud services with data access awareness. J. Parallel Distrib. Comput. **72**(4), 591–602 (2012)
2. Calheiros, R.N., Ranjan, R., Beloglazov, A., De Rose, C.A.F., Buyya, R.: CloudSim: a toolkit for modeling and simulation of cloud computing environments and evaluation of resource provisioning algorithms. J. Softw.: Pract. Exp. **41**(1), 23–50 (2011)
3. Samimi, P., Teimouri, Y., Mukhtar, M.: A combinatorial double auction resource allocation model in cloud computing. J. Inf. Sci. **357**(C), 201–216 (2016)

4. Ramírez-Velarde, R., Tchernykh, A., Barba-Jimenez, C., Hirales-Carbajal, A., Nolazco-Flores, J.: Adaptive resource allocation with job runtime uncertainty. J. Grid Comput. **15**(4), 415–434 (2017)
5. Nazarenko, A., Sukhoroslov, O.: An experimental study of workflow scheduling algorithms for heterogeneous systems. In: Malyshkin, V. (ed.) PaCT 2017. LNCS, vol. 10421, pp. 327–341. Springer, Cham (2017). https://doi.org/10.1007/978-3-319-62932-2_32
6. Srinivasan, S., Kettimuthu, R., Subramani, V., Sadayappan, P.: Characterization of backfilling strategies for parallel job scheduling. In: Proceedings of the International Conference on Parallel Processing, ICPP 2002 Workshops, pp. 514–519 (2002)
7. Jackson, D., Snell, Q., Clement, M.: Core algorithms of the Maui scheduler. In: Feitelson, D.G., Rudolph, L. (eds.) JSSPP 2001. LNCS, vol. 2221, pp. 87–102. Springer, Heidelberg (2001). https://doi.org/10.1007/3-540-45540-X_6
8. Tchernykh, A., Schwiegelsohn, U., El-ghazali, T., Babenko, M.: Towards understanding uncertainty in cloud computing with risks of confidentiality, integrity, and availability. J. Comput. Sci. **36** (2016)
9. Tsafrir, D., Etsion, Y., Feitelson, D.G.: Backfilling using system-generated predictions rather than user runtime estimates. IEEE Trans. Parallel Distrib. Syst. **18**(6), 789–803 (2007)
10. Rodriguez, M.A., Buyya, R.: Scheduling dynamic workloads in multi-tenant scientific workflow as a service platform. Futur. Gener. Comput. Syst. **79**(P2), 739–750 (2018)
11. Toporkov, V., Yemelyanov, D.: Optimization of resources selection for jobs scheduling in heterogeneous distributed computing environments. In: Shi, Y., et al. (eds.) ICCS 2018. LNCS, vol. 10861, pp. 574–583. Springer, Cham (2018). https://doi.org/10.1007/978-3-319-93701-4_45
12. Toporkov, V., Yemelyanov, D., Toporkova, A.: Coordinated global and private job-flow scheduling in Grid virtual organizations. Simul. Model. Pract. Theory **107**, 102228 (2021)
13. https://www.cse.huji.ac.il/labs/parallel/workload/ (2021)
14. Feitelson, D.G.: Workload Modeling for Computer Systems Performance Evaluation. Cambridge University Press, Cambridge (2015)
15. Toporkov, V., Yemelyanov, D.: Availability-based resources allocation algorithms in distributed computing. In: Voevodin, V., Sobolev, S. (eds.) RuSCDays 2020. CCIS, vol. 1331, pp. 551–562. Springer, Cham (2020). https://doi.org/10.1007/978-3-030-64616-5_47

Influence of Execution Time Forecast Accuracy on the Efficiency of Scheduling Jobs in a Distributed Network of Supercomputers

Boris Shabanov, Anton Baranov, Pavel Telegin, and Artem Tikhomirov$^{(\boxtimes)}$

Joint Supercomputer Center, Russian Academy of Sciences, 119334 Moscow, Russia
{shabanov,ptelegin}@jscc.ru

Abstract. Supercomputer users when submitting jobs often overestimate walltime. These inaccuracies lead to the jobs completion before schedule and hence the decreased efficiency of job scheduling. Machine learning, using various characteristics of user jobs, can provide job walltime forecasts before the job starts. The use of forecasts by the supercomputer job management system makes it possible to increase the efficiency of scheduling and executing jobs. In this paper, we study the efficiency of using the forecasted execution time of jobs in a geographically distributed network of supercomputer centers with de-centralized management. The execution time of a job on the computing resources of different supercomputer centers may vary. The threshold value of forecast accuracy is evaluated when scheduling jobs in a supercomputer network becomes efficient. Estimations of scheduling efficiency are made, taking into account the forecasts of job walltime.

Keywords: Supercomputer centers · Job scheduling · Machine learning · Performance forecast · Distributed network of supercomputers

1 Introduction

A stable trend in the scientific supercomputer centers (SCC) development is the integration of supercomputing resources into a single geographically distributed network (GDN) [1]. The GDN is primarily aimed at increasing the efficiency of using the SCC computing resources. The typical computing unit (CU) connected into a GDN is a separate high-performance cluster (supercomputer). Typically, the GDN includes CUs with different performance, integrated with communication channels of different bandwidth. A supercomputer job is an elementary workload object both on a separate CU and in the GDN as a whole. The user creates a job as an object containing a computational program, initial data, and the following resource requirements: number of processors (cores), amount of RAM and disk space, time required for job execution (walltime request), etc.

Each CU runs a local job management system [2], like SLURM, PBS, LSF. The basic functions of the local job management system are maintaining job queue, scheduling, submitting, and controlling the job execution process on a single CU. Local management

© Springer Nature Switzerland AG 2021
V. Malyshkin (Ed.): PaCT 2021, LNCS 12942, pp. 338–347, 2021.
https://doi.org/10.1007/978-3-030-86359-3_25

systems on different CUs form the lower (local) level of computing jobs management. Management at the top (global) level, which is the level of distributed resources, is performed by the global job management system. To increase the reliability and scalability of GDN, we examine a decentralized global job management system, where management is done by a team of peer dispatchers operating locally on each CU in GDN. The interaction of dispatchers is asynchronous in a single global job queue organized using a distributed DBMS [3]. The consistency of dispatchers decisions is achieved using reverse auction planning, where dispatchers on different CUs compete with each other for the right to process a job from the global queue. While competing for a job, dispatchers propose rates that reflect the readiness of the CU to process the job with a given quality of service, e.g., the walltime. The job is assigned to the CU, the dispatcher of which has offered the best rate. Note that the auction type may be different depending on the goal of planning [4].

The walltime request is one of the most important resource requirements for the job. It determines the time for which the Local Resource Management System (LRMS) will allocate the resources. Like other resource requirements, the walltime request is estimated by the job owner. If the actual walltime does not exceed the requested one, then allocated computing resources are freed immediately after job completion. If the walltime exceeds the requested time, then after the requested time has elapsed, the job is aborted by the LRMS. Note that some SCCs set a default value for the walltime request, this often motivates users not to change this value [5].

To avoid early termination of the job, users often deliberately exceed the time estimates when requesting the walltime. On the one hand, this ensures that the job owner will receive the results of the execution, even if the execution of the job is slowed down by any technical obstacles, for example, waiting for the completion of a write to the shared file system. On the other hand, excessive additional time can lead to delaying jobs in the queue, since due to some features of scheduling algorithms, jobs that require less time usually spend less time in the queue [6]. For instance, setting the walltime request to the maximum disables the backfilling of short jobs [7].

According to the statistics from the MVS-10P OP supercomputer [8], in the Joint Supercomputer Center (JSCC RAS), in 2020 the walltime request exceeded the actual walltime for 95%, of jobs. The median value of excess was 2.3 times, and the average excess was in 22.6 times (see Table 1). The actual walltime for 60% of jobs exceeds the requested value 2 times and more.

Table 1. Job flow statistics for MVS-10P OP in 2020.

Statistic name	Value
Number of jobs	43946
Proportion of jobs for which the walltime request exceeds the actual walltime	0.95 (41743 jobs)
Median excess (average excess)	2.3 (22.6)
Proportion of jobs completed abnormally (out of requested time)	0.003 (131 jobs)

The error in the accuracy of the walltime request does not depend on any specific CU, it is typical for most shared facility systems, as described in papers [9–12].

It is obvious, that the walltime request inaccuracy negatively affects the local resources scheduling efficiency: due to the early completion of jobs, the processing order planned by LRMS is changed. As for the global resources scheduling, the scheduler not always is able to change the order of jobs processing, or reschedule it. This happens because scheduling at the global level is associated with the transfer of initial job data between CUs. Note that, firstly, in some cases, the transfer time of the initial data can be comparable or even exceed the time of its execution, and secondly, the initial data must be copied before the job execution. So, the copying process must be proactive. In other words, we must predict the dynamically changing workload of computing resources. All this shows the error in the walltime request is critical for the distributed resources.

Analysis of recent papers has shown that it is possible to use machine learning algorithms to build forecasting systems, which, based on the job metadata, can adjust the job walltime request before submission. The article [9] shows the efficiency of using the methods for adjusting the walltime request based on modeling the real flow of computational jobs arriving at the COARE (Computing and Archiving Research Environment) SCC. The article [10] presents the results of a study implemented in the widely used Portable Batch System Professional (PBS Pro) [11] mechanism for predicting the job execution time; scenarios are given when the adjustment of the walltime request can be used. The article [5] examines the accuracy of the forecast based on the amount of initial data of the job and the amount of required computing resources. In [5], a wider list of job parameters is analyzed, which makes it possible to refine the forecast of execution time. The paper [12] is devoted to the study of the characteristics of the jobs flow coming to the SCC, as well as to modeling and comparison of methods for adjusting the walltime requests. The works [13, 14] investigate the efficiency of forecasts using the Alea 4 simulator.

The study of methods and means for the development of an information and computing environment for scientific experiments based on the federative principle of management was held within the Russian-Chinese MC2E (Meta Cloud Computing Environment) project [15]. Two new approaches to the program execution time prediction problem were proposed [16]. The first one uses CU grouping based on the Pearson correlation coefficient. The second one is based on vector representations of CU and MPI programs, so-called embeddings. The paper [16] describes applying of embeddings technique to the execution time prediction of an MPI program on a set of CUs. The papers [17, 18] are devoted to machine learning methods and algorithms for supercomputer walltime prediction. Forecasting is using submitted jobs classification based on supercomputer job management statistics. The supercomputer RIKEN Integrated Cluster of Clusters (RICC) statistics [17] and the JSCC RAS supercomputers statistics [18] were initial data. There were made probability estimates of correct predictions for well-known machine learning algorithms.

The review showed that there are good results on the efficiency of using predicted walltime for the local scheduling, while there is a lack of relevant results for the globally distributed resources. In this paper, we do not consider specific machine learning methods for walltime prediction. We assume that machine learning methods can predict walltime on different CUs with a given accuracy. The walltime may vary on different CUs since the CUs can be inhomogeneous. We study how using the actual walltime, instead of the walltime request will increase the efficiency of job scheduling on distributed resources.

In this paper, we estimate the threshold value of the prediction accuracy when the optimal assignment of jobs on the GDN is achieved. We also evaluate the performance of the auction-based scheduling algorithm with forecasting.

2 Experiments

2.1 GDN Testbed

GDN is a sophisticated system with numerous connections between its elements. As you can see in [19] the real system is influenced by many random features. Considering these features with the use of known mathematical methods is limited by a number of classical distribution functions. Consequently, the analytical model of a complex system becomes oversimplified, this inevitably leads to a decrease in the reliability of the results. We used simulation modeling for research. Currently, simulation is associated [20] with the complexity of reproducing the results by other researchers and using their models. The main obstacles are model unavailability, lack of input or output data (when average metrics are published, but not all of the outputs are available), the impossibility of access to the experimental environment. To carry out a series of experiments, we created an own GDN testbed consisting of several CUs and a single global job queue.

Each CU runs a dispatcher, LRMS, and a job execution module. In addition, there were implemented a job launcher, which placed the test set of jobs into the global queue, and a forecasting module, which replaced the walltime request with the actual one for the given proportion of the jobs from the test set. To prevent the early abortion of jobs due to a prediction error, we use the soft walltimes approach discussed in [10]. A distributed document-oriented Elasticsearch DBMS [21] was used to organize a global queue.

Each dispatcher uses an auction-based jobs scheduling algorithm. During the experiment, the first price sealed bid auction was used. At this auction, the participating dispatchers are not aware of the bids made by other participants and cannot adjust their previously proposed bid. Participants' bids for a separate job are accepted within a parametrically specified time interval – the duration of the auction. Changing the duration of the auction it is possible to adjust the number of participants who succeeded to take part in the auction.

The operation of the LRMS was simulated by the RabbitMQ message queue broker [22]. We chose this message broker due to the ease of its installation and configuration, as well as the ability of one broker to simulate the operation of several LRMS instances at once. The FCFS (First Come, First Serve) algorithm was used as a scheduling algorithm implemented in the LRMS model.

The job launchers, prediction and processing modules (PM) were also implemented in the Python programming language using the multiprocessing library (Fig. 1). RabbitMQ and Elasticsearch services functioned in docker containers and are united in a swarm. This made it possible to start the testbed quickly.

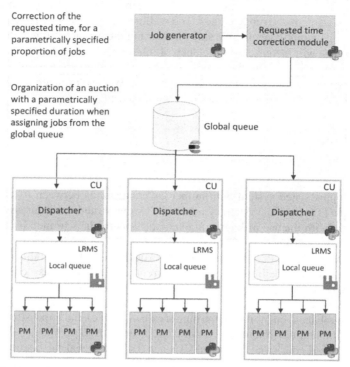

Fig. 1. A model of the GDN for researching the efficiency of walltime prediction.

2.2   The Approach for Preparing Test Set

The test set of jobs was a stream coming in over 6 days. The test set was based on the job logs of the MVS-10P supercomputer in 2020. The set was formed the following way. Using the job logs we identified the month with the maximum intensity of the incoming jobs. From the selected month array there were excluded jobs received by the CU on the days of the CU maintenance, as well as jobs with walltime equal to zero (incorrect jobs). The following information was used for all remaining job submissions: time of submission into local CU queue, walltime request, actual walltime.

As the test set was obtained from the job logs on a single CU, to simulate jobs flow on several CUs the intensity of incoming jobs was increased by merging several weeks of the month into one. In other words, one day of the week of the test set contains jobs of several days of the same name in different weeks, while the information about the time of the job submission did not change. Let's consider an example. We take information on the execution of jobs in November 2020 (11-2020) from the job logs for the CU. We consider the first week of the month from 01-11-2020 to 07-11-2020. On the first-day jobs from dates 01, 08, 15 will enter, on the second-day jobs from dates 02, 09, 16 will enter, while the time of submission, that is, hours, minutes, and seconds, does not change in any way.

To adapt the test set, it was required to determine the walltime for each job for each CU. As mentioned above the processing time of a job on different CUs may vary, therefore, the actual walltime and walltime request for different CUs were randomly determined within a given range. These values were added to the passport of each job, in each series of the experiment. Thus, for each job in the experiment, we define the optimal execution time as the minimum of all walltimes for all CUs.

In the experiment the model time was used, which is 60 times faster than the real time. This is correct since the LRMS time quantum of a job on the CU is 1 min.

2.3 Assumptions and Methodology of the Simulation

Since the study examines the efficiency of job scheduling for distributed resources, the authors make the following assumptions during the experiment for the local resource scheduling. First, the FCFS service algorithm was used as a scheduling algorithm for the LRMS, while in real systems more complex scheduling algorithms are used, for example, the Backfill algorithm. The assumption is based on the fact that the scheduling system of the distributed resource level should not interfere with the scheduling of jobs at the local level, as was already said above. The second assumption made by the authors relates to the job resource requirements. The resource requirements of each job contain only information about the walltime request. It is assumed that all jobs require the same number of computational resources for their execution. It is possible to neglect the requested number of computing resources because, during the experiment, CUs of equal performance are simulated, that is, the maximum possible number of jobs processed by them is the same.

In the series of experiments, we varied the duration of the auction and the time prediction accuracy. The forecast accuracy was used in the dispatcher's rates. The dispatcher's rate for the job will be the walltime request. That is, CU dispatchers compete with each other and the winner is the one which is ready to spend the least amount of time on the job. Since the execution time on different CUs is different, the dispatcher's rates are different. But each time each dispatcher seeks a job from the global queue with minimal walltime, as it is more reasonable to process jobs that require more processing time on other CUs in GDN.

The following metrics are evaluated: the average value of the jobs proportion assigned using the auction, the average value of the optimally scheduled jobs proportion, the average value of the jobs assignment efficiency. We will say that a job is optimally scheduled if its execution time in the experiment is equal to the optimal execution time. For every job, there exists a CU where actual walltime is optimal. There can be different fastest CUs for different jobs. This results from the fact that different CUs have different architecture and other characteristics influencing the performance of a certain job. This is not a degenerate case when there is a single CU that can execute all jobs in optimal time. We will say that a job was assigned with an auction if more than one CU dispatcher made a bid.

2.4   Simulation Results

Note that we did not observe repeatability of the results with a prediction accuracy of less than 0.2. In addition, the advantage of forecasting with such accuracy was extremely small, and the results are not of scientific interest. Experimental results are presented in Fig. 2.

Let us start with two extreme cases. In the first extreme case the proportion of jobs assigned as the auction result is 1, that is, all jobs of the test set were assigned with the auction. Note that with a prediction accuracy of 1, the entire test set is assigned optimally; at the same time note that, when forecast accuracy is 0.5, the proportion of jobs assigned optimally is more than 0.6 for the entire test set. In the second extreme case the share of jobs assigned as the auction result is 0. This case corresponds to the FCFS algorithm. It can be seen that the efficiency of scheduling the jobs, even with a forecast accuracy of 1, does not exceed 0.55. The reason for the low efficiency of job scheduling is the lack of consistency between dispatchers when making a job assignment decision since dispatchers do not take into account the interests of each other.

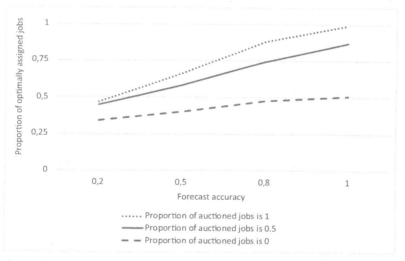

Fig. 2. Dependence of the proportion of optimally distributed jobs in the GDN on the forecast accuracy

Further, we consider an intermediate case which is reflecting reality. Let us consider the case when the share of jobs assigned with the auction is 0.5. One can see in Fig. 2 that with a prediction accuracy of more than 0.8, the proportion of optimally scheduled jobs is 0.75. This allows us to conclude that the threshold value for the prediction accuracy is 0.8. It can be seen from the graph (Fig. 2) that a breaking point is observed for this value, that is, after this value, the optimality of the distribution increases slower.

The experiments showed that the discovered dependence retains for a larger number of dispatchers. In particular, the experiment was carried out with 10 dispatchers.

Now we can look at how the distribution efficiency is achieved. Let us introduce 3 main classes of jobs: the first class (I_b) includes optimally scheduled jobs. The second class ($I_{b,p}$) consists of optimally assigned jobs when the forecasting was used in the scheduling. The third class ($I_{b,p,a}$) includes jobs of the second class ($I_{b,p}$) for which the auction was held. As one can see, all three classes are dependent, and the jobs set of a larger class includes jobs of smaller classes. Other classes of jobs can be distinguished, but they are not considered, since they are rather random.

One can see on the graph above that almost all jobs with predicted walltime were scheduled with the auction in an optimally. Results presented above (see Fig. 2) show that the percentage of jobs of class $I_{b,p}$, is almost equal both to the accuracy of prediction, and the percentage of jobs of class $I_{b,p,a}$. At the same time, the percentage of jobs of class I_b is higher than all the others, since it includes randomly assigned jobs. Graph limits are the following: with the proportion of jobs assigned with the auction number 1, for each value of the prediction accuracy all three columns are approximately equal. When the proportion of jobs assigned with the auction equal to 0, there are no jobs of the third class (see Fig. 3).

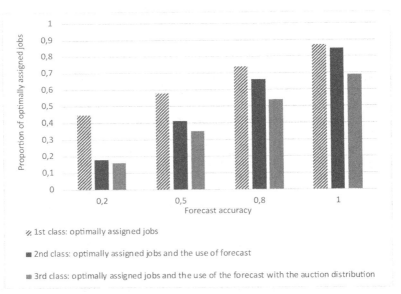

Fig. 3. Dependence of the proportion of optimally scheduled jobs on the job classes for different forecast accuracy.

3   Conclusion

A testbed has been developed to study the efficiency of job assignment to computational units with varying forecast accuracy and varying duration of the auction. It was experimentally determined that the threshold value for the prediction accuracy is 0.8. With this value, the proportion of optimally scheduled jobs is 75%. The efficiency of scheduling grows slower with larger values of the prediction accuracy.

Experiments showed that the use of the predicted job execution time makes it possible to increase the efficiency of job scheduling up to a factor of 1.6. It is impossible to use only the walltime request, it is necessary to take into account other important characteristics of GDN: communicational and computational heterogeneity.

Acknowledgments. The study was carried out within state assignment project 0580-2021-0016 and was partially supported by RFBR project No. 18-29-03236. Supercomputer MVS-10P in JSCC RAS was used in research.

References

1. Shabanov, B., Ovsiannikov, A., Baranov, A., Leshchev, S., Dolgov, B., Derbyshev, D.: The distributed network of the supercomputer centers for collaborative research. Program. Sist. Teor. Prilozh. **8**:4(35), 245–262 (2017). https://doi.org/10.25209/2079-3316-2017-8-4-245-262
2. Reuther, et al.: Scalable system scheduling for HPC and big data. J. Parallel Distrib. Comput. **111**, 76–92 (2018). https://doi.org/10.1016/j.jpdc.2017.06.009
3. Baranov, A.V., Tikhomirov, A.I.: Methods and tools for organizing the global job queue in the geographically distributed computing system. Vestn. Yuzh. Ural. Univ. Ser. Vychisl. Mat. Programm. **6**(4), 28–42 (2017). https://doi.org/10.14529/cmse170403
4. Baranov, A., Telegin, P., Tikhomirov, A.: Comparison of auction methods for job scheduling with absolute priorities. In: Malyshkin, V. (ed.) PaCT 2017. LNCS, vol. 10421, pp. 387–395. Springer, Cham (2017). https://doi.org/10.1007/978-3-319-62932-2_37
5. Gaussier, E., Glesser, D., Reis, V., Trystram, D.: Improving backfilling by using machine learning to predict running times. In: Proceedings of the International Conference for High Performance Computing, Networking, Storage and Analysis (SC 2015). Article 64, pp. 1–10 (2015). https://doi.org/10.1145/2807591.2807646
6. LUNARC Documentation pages. https://lunarc-documentation.readthedocs.io/en/latest/batch_system/. Accessed 10 Feb 2021
7. Tsafrir, D., Etsion, Y., Feitelson, D.G.: Backfilling using system-generated predictions rather than user runtime estimates. IEEE Trans. Parallel Distrib. Syst. **18**(6), 789–803 (2007). https://doi.org/10.1109/tpds.2007.70606
8. Savin, G.I., Shabanov, B.M., Telegin, P.N., Baranov, A.V.: Joint supercomputer center of the Russian academy of sciences: present and future. Lobachevskii J. Math. **40**(11), 1853–1862 (2019). https://doi.org/10.1134/S1995080219110271
9. Guo, J., Nomura, A., Barton, R., Zhang, H., Matsuoka, S.: Machine learning predictions for underestimation of job runtime on HPC system. In: Yokota, R., Wu, W. (eds.) SCFA 2018. LNCS, vol. 10776, pp. 179–198. Springer, Cham (2018). https://doi.org/10.1007/978-3-319-69953-0_11
10. Klusáček, D., Chlumský, V.: Evaluating the impact of soft walltimes on job scheduling performance. In: Klusáček, D., Cirne, W., Desai, N. (eds.) Job Scheduling Strategies for Parallel Processing. JSSPP 2018. Lecture Notes in Computer Science, vol. 11332, pp. 15–38. Springer, Cham (2019). https://doi.org/10.1007/978-3-030-10632-4_2
11. Nitzberg, B., Schopf, J.M., Jones, J.P.: PBS Pro: grid computing and scheduling attributes. Grid Resour. Manag. **64**, 183–190 (2004). https://doi.org/10.1007/978-1-4615-0509-9_13
12. Rubio, J.C., Villapando, A., Matira, C., Aborot, J.: Correcting job walltime in a resource-constrained environment. In: Panda, D.K. (ed.) SCFA 2020. LNCS, vol. 12082, pp. 118–137. Springer, Cham (2020). https://doi.org/10.1007/978-3-030-48842-0_8

13. Klusáček, D., Tóth, V., Podolníková, G.: Complex job scheduling simulations with Alea 4. In: Ninth EAI International Conference on Simulation Tools and Techniques (SimuTools 2016), pp. 124–129. ACM (2016)

14. Klusáček, D., Soysal, M.: Walltime prediction and its impact on job scheduling performance and predictability. In: Klusáček, D., Cirne, W., Desai, N. (eds.) JSSPP 2020. LNCS, vol. 12326, pp. 127–144. Springer, Cham (2020). https://doi.org/10.1007/978-3-030-63171-0_7

15. Smeliansky R., Mei, H.: MC2E – meta-cloud computing environment. In: 2020 International Scientific and Technical Conference Modern Computer Network Technologies (MoNeTeC), pp. 1–2 (2020). https://doi.org/10.1109/MoNeTeC49726.2020.9258124

16. Chupakhin, A., Bahmurov, A., Antonenko, V., Ishelev, G.: Application of recommender systems approaches to the MPI program execution time prediction. In: 2020 International Scientific and Technical Conference Modern Computer Network Technologies (MoNeTeC), pp. 1–7 (2020). https://doi.org/10.1109/MoNeTeC49726.2020.9258345

17. Baranov, A., Nikolaev, D.: Machine learning to predict the supercomputer jobs execution time. Softw. Syst. (2), 218–228 (2020). https://doi.org/10.15827/0236-235X.130.218-228

18. Savin, G.I., Shabanov, B.M., Nikolaev, D.S., et al.: Jobs runtime forecast for JSCC RAS supercomputers using machine learning methods. Lobachevskii J. Math. **41**, 2593–2602 (2020). https://doi.org/10.1134/S1995080220120343

19. Devyatkov, V.: Methodology and Technology of Simulation Studies of Complex Systems: Current State and Prospects of Development. INFRA-M Publishing House, Moscow (2013). ISBN 978-5-9558-0338-8

20. Dutot, P.-F., Mercier, M., Poquet, M., Richard, O.: Batsim: a realistic language-independent resources and jobs management systems simulator. In: Desai, N., Cirne, W. (eds.) JSSPP 2015-2016. LNCS, vol. 10353, pp. 178–197. Springer, Cham (2017). https://doi.org/10.1007/978-3-319-61756-5_10

21. Vohra, D.: Using elasticsearch. In: Pro Couchbase Development. Apress, Berkeley, CA (2015).https://doi.org/10.1007/978-1-4842-1434-3_7

22. Christudas, B.: Install, configure, and run RabbitMQ cluster. In: Practical Microservices Architectural Patterns. Apress, Berkeley, CA (2019). https://doi.org/10.1007/978-1-4842-4501-9_21

Performance Estimation of a BOINC-Based Desktop Grid for Large-Scale Molecular Docking

Natalia Nikitina[1]([envelope]) [iD], Maxim Manzyuk[2] [iD], Črtomir Podlipnik[3] [iD],
and Marko Jukić[4,5] [iD]

[1] Institute of Applied Mathematical Research,
Karelian Research Center of the Russian Academy of Sciences,
185910 Petrozavodsk, Russia
nikitina@krc.karelia.ru
[2] Internet Portal BOINC.Ru, Moscow, Russia
[3] Faculty of Chemistry and Chemical Technology, University of Ljubljana,
1000 Ljubljana, Slovenia
[4] Faculty of Chemistry and Chemical Engineering, University of Maribor,
2000 Maribor, Slovenia
[5] Faculty of Mathematics, Natural Sciences and Information Technologies,
University of Primorska, 6000 Koper, Slovenia

Abstract. This paper addresses the performance evaluation of a heterogeneous distributed computing environment (Desktop Grid) for large-scale medicinal chemistry experiments in silico. Dynamic change of the set of computational nodes, their heterogeneity and unreliability impose difficulties on task scheduling and algorithm scaling. We analyze the performance, provide efficiency metrics, statistics and analysis of the volunteer computing project SiDock@home.

Keywords: Distributed computing · Volunteer computing · Desktop Grid · Task scheduling · BOINC · Virtual screening · Molecular docking

1 Introduction

Desktop Grids are an essential tool for performing computationally-demanding scientific research. They combine non-dedicated geographically distributed computers (typically, desktop ones) connected to the central server by the Internet or a local access network and performing computations for the Desktop Grid in their idle time. The resources may be provided either by the volunteer community or by individuals and organizations related to the work. Many of the world's leading research institutions run large-scale computational projects based on Desktop Grids (e.g., Washington University: Rosetta@home [19], Folding@home [22]; CERN: LHC@home [14]; University of Oxford: Climateprediction.net [5]).

© Springer Nature Switzerland AG 2021
V. Malyshkin (Ed.): PaCT 2021, LNCS 12942, pp. 348–356, 2021.
https://doi.org/10.1007/978-3-030-86359-3_26

To organise and manage Desktop Grids, a number of software platforms are used. The most popular one is BOINC (Berkeley Open Infrastructure for Desktop Computing) [2]. Among the 157 active largest projects on volunteer computing, 89 are based on BOINC [10]; that is, BOINC can be considered a *de-facto* standard for the operation of volunteer computing projects. The BOINC platform is an actively developing Open Source software and provides rich functionality.

BOINC has a server-client architecture. The server generates a large number of tasks that are mutually independent parts of a computationally-intensive problem. When a client computer is idle, it requests work from the server, receives tasks, and independently processes them. Upon finishing, it reports results back to the server. The results are then stored in the database for further usage.

Unlike computational clusters and supercomputers, Desktop Grids are devoid of characteristics such as the high-speed interconnection between computational nodes, homogeneity, reliability, and availability of nodes during defined periods of time. These disadvantages restrict the class of computational problems that one can solve on Desktop Grids to the type of bag-of-tasks problems.

Nonetheless, Desktop Grids serve as an affordable, quickly deployable tool. An essential feature of the Desktop Grid technology is the ability to attract a large number of inexpensive computing resources for a temporary or long-term mission, providing a quick response to emerging scientific problems.

A recent example of employing Desktop Grids at the early stages of solving urgent scientific problems is the fight against novel coronavirus at the beginning of 2020. The structure of the SARS-CoV-2 spike protein (key role in pathogenesis) was accurately predicted by the Rosetta@home project [15] several weeks before its description with cryo-electron microscopy [24]. It allowed to speed up the research by many academic groups: design of novel vaccines and antivirals.

The Folding@home project is also developing an antiviral agent against SARS-CoV-2. In early 2020, more than 700,000 new participants joined the project, and its performance exceeded one exaflops, making Folding@home the first world's exascale system, more powerful than the top 100 supercomputers combined [26].

Conventional supercomputers have also been employed to fight against SARS-CoV-2 from the very early stages COVID-19 pandemic onset (see, e.g., [3] for a detailed overview). Nation-wide and cross-nation research initiatives provide scientists with supercomputer resources.

However, not every computational problem is designed to take all advantages of a supercomputing environment. The problems of bag-of-tasks type do not utilize the high-speed interconnection of supercomputer nodes. The resources are shared among many computational applications; setting a supercomputer implies high costs. For these reasons, research groups (especially of a small/medium size) typically do not have immediate, on-demand access to supercomputing resources. Desktop Grids serve as a tool that can complement and, when needed, substitute conventional high-performance computing systems.

In this paper, we discuss the performance characteristics of a BOINC-based Desktop Grid on an example of a real project performing large-scale molecular docking. In Sect. 2, we overview the related works on performance measurement in BOINC-based Desktop Grids. In Sect. 3, we describe the BOINC-based project SiDock@home. In Sect. 4, we analyze and discuss performance metrics. Finally, Sect. 5 concludes the paper.

2   Related Works

When planning the workflow, time and budget of a research, one needs to evaluate the performance of available computational resources. The owners of BOINC client computers independently determine their contribution to the computing process: the amount of resources provided and the time of their availability. Thus, they create uncertainty in the project's operation, particularly in the estimation of the project's performance. A number of works are devoted to performance measurement in BOINC-based Desktop Grids.

An up-to-date description of the BOINC platform and its credit system serving for measuring the computational performance is provided in [2]. The credit system allows to unify and rank the contributions within a BOINC project, across projects or the overall BOINC performance and potential.

The throughput may be expressed in the number of active tasks as in the LHC@Home project [4]. Such a simple characteristic is illustrative, in particular, when the workflow of BOINC-based computer simulations is being compared to the actual physical experiments, on the one side, and to the similar workflow on supercomputers, on the other side.

Another option is to measure the performance in flops (the number of floating point operations per second). This approach inherits the ambiguity of benchmark choice and is complicated by heterogeneity of the Desktop Grid. However, it is universal and common in scientific community.

Flops metric is widely used to unify participant computers by a reference CPU/GPU [26] or to match Desktop Grid performance with conventional HPC systems [25]. In many workflows, BOINC credits are being awarded according to a flops estimate of the work done by a task.

The author of [21] addresses more complex performance estimation in terms of throughput, scalability, latency and reliability, and exemplifies on two large Desktop Grid projects. Among the considered metrics, reliability is concluded to be crucial as it influences the performance consistency of a system. To increase reliability, the author suggests to designate a number of reliable nodes as a backup subsystem.

Emulation and simulation systems, such as ComBoS [1], SimGrid [18] and EmBOINC [12], hold a specific place among frameworks for performance measuring in BOINC-based Desktop Grids. They allow one to evaluate and compare scheduling algorithms under various scenarios, using generated and/or real data without interference into the operation of real computational projects; predict possible bottlenecks and limits of a BOINC project.

3   The Project SiDock@home

Since the beginning of the COVID-19 pandemics in late 2019-early 2020, many citizen science projects attracted the public's attention at analyzing viral life-cycle and working on potential therapeutic targets. Due to the world-wide character of the COVID-19 pandemic, such projects gathered large amounts of traction and computational resources. It helped many research teams boost their research, developing drugs and vaccines, sharing the results with the scientific community and broadening fundamental scientific disciplines.

In March 2020, a citizen science project *"Citizen science and the fight against the coronavirus"* (COVID.SI) [13] was initiated in the field of drug design and medicinal chemistry. The project is aimed at drug discovery, first of all, against coronavirus infection, using high-throughput virtual screening (HTVS) [17] on a small molecules library developed by the team. In the next months, SiDock@home [20], a BOINC-based extension of COVID.SI, was created and grew into a sizable, independent and competent research project for general drug design.

The project's server part was deployed in an Ubuntu 18.04 LTS-based machine under system configuration of 2 Xeon 6140 cores, 8 Gb RAM, 32 Gb SSD and 512 Gb HDD. The molecular docking application CmDock [7] was adapted for execution in the BOINC environment using a wrapper program for Windows, Linux and MacOS 64-bit operating systems.

The project was announced to the public on October 23, 2020 and attracted the interest of the BOINC volunteer community. Consequently, in the testing phase of two months, 240 participants joined the project and provided the computational resources of more than 500 computers. Such a capacity allowed us to perform virtual screening (VS) on a part of the library, to elaborate on the optimal parameters of task distribution and results processing, and to evaluate the developed CmDock software in a heterogeneous distributed environment.

HTVS on the complete library was completed for five targets in eight months. In the next section, we discuss the performance metrics obtained based on gathered statistics of project operation.

4   Performance Analysis of a Desktop Grid

The most popular performance metrics for Desktop Grids are [16] throughput (the number of tasks completed per a time unit), makespan (the time interval to complete a set of tasks), turnaround time of a task (the time from its creation to obtaining its result), reliability (the probability of the server to receive a correct result, a valid result, or any result once a task instance has been sent out) and availability (the ratio of results returned to the server to the total number of tasks sent out to the node). Other metrics have been addressed as well (total load of the computational servers, overhead due to replication or failures etc.)

To calculate the desired metric, a number of task and computer characteristics can be used. In this work, we will consider the following ones calculated by the BOINC middleware and stored in a database:

1. Computer:
   - create_time: timestamp of the moment the host registered to the project;
   - rpc_time: timestamp of the last RPC of the host to the server;
   - total_credit: the total BOINC credit earned by the host;
   - on_frac: the fraction of total time the host runs the BOINC client;
   - active_frac: of the time the host runs the BOINC client, the fraction it is enabled to use CPU;
   - p_ncpus: the number of available CPU cores;
   - p_fpops: estimate peak performance of a single CPU core, flop/s.
2. Task:
   - sent_time: timestamp of the moment the task was sent by the server;
   - received_time: timestamp of the moment the result was received by the server. We use these two values to discretise the workflow by hours;
   - elapsed_time: actual time of the task execution on the host computer;
   - flops_estimate: estimated peak flop/s of the host computer.

Note that the value p_ncpus can be edited by the host's owner to alter the number of received tasks. To estimate the performance bounds, we adjusted the apparent outliers. Specifically, we did not consider hosts with p_ncpus ≥ 1000 and manually corrected several values of typical configuration computers.

As the Whetstone benchmark [9] is generally considered representative for CPU-intensive applications performing numeric operations (which is the case for many scientific applications), it is used in BOINC [2] to estimate the peak performance (theoretical achievable performance) of a CPU in flops. The calculated value p_fpops=flops_estimate is used to obtain the amount of BOINC credits for a task which, consequently, expresses the theoretical number of flops that could have been done during the task execution on this computer.

Computer data allow to estimate the scale of the available computational resources and design the workflow. We consider a BOINC-based Desktop Grid of a computational project operating without any knowledge of other work performed by the nodes and other projects present at them.

As a theoretical upper bound of a BOINC-based Desktop Grid performance, we use its aggregated theoretical peak performance accounting availability periods of the computers and the user-imposed restrictions on resources usage by BOINC. Let us call it the theoretical available performance. This value can be considered when planning workflow at a short distance.

Let H be the set of project's hosts. For each $h \in H$, let us denote h.field the corresponding value of the field at database entry h. $I(x)$ is the indicator of an event x. Using the database, one can calculate the theoretical available performance at a specified interval of time $[t, t+1]$ as follows:

$$P_{avail}(t) = \sum_{h \in H} I(h.\text{create\_time} \leq t) \cdot I(h.\text{rpc\_time} \geq t+1)$$
$$\cdot I(h.\text{total\_credit} > 0) \cdot h.\text{p\_ncpus} \cdot h.\text{p\_fpops} \cdot h.\text{on\_frac} \cdot h.\text{active\_frac}. \quad (1)$$

Task data allows estimating the actual performance achieved by a BOINC project. Note that we consider one-core tasks, which is the case of a number of other BOINC projects including SiDock@home.

Some of the tasks end up with errors of different types or may not be needed because, at the moment of their return to the server, the quorum has been already met. Otherwise, the tasks are awarded with BOINC credit [8] which, in the most common case, is calculated as follows.

Each host $h \in H$ is assigned a value a_h, a peak performance of its CPU flops, estimated with an internal BOINC benchmark. When a task τ has been executed on host h, BOINC registers the elapsed time $T_{h\tau}$. The amount of credit the host would get for the task is $C_{h\tau} = T_{h\tau} \cdot a_h \cdot CS$. If the result passes a validity check on the server and the quorum has been met, the host is awarded $C_{h\tau}$ (or an appropriately adjusted value if quorum exceeds 1). Note that in Eq. (1), the term $I(h.\text{total\_credit} > 0)$ serves to filter out computers that haven't done any work for the project.

Here, $CS = \frac{200}{86\,400} \times 10^9$ (a Cobblestone) is a constant unifying the effective work of heterogeneous computers with a reference one that would do one gigaflops based on the Whetstone benchmark and receive 200 credits a day.

In Fig. 1, we provide the performance dynamics of the project's Desktop Grid in comparison with the performance bound. We observe that the actual performance varies in between 10–50% of the theoretical available performance of active computers. This is partly explained by the background character of the Desktop Grid computations that, by definition, use only idle time of a computer. The value is also influenced by the presence of other BOINC projects at a host computer. The positivity of the performance value corresponds to the normal operation of the Desktop Grid, while periods of zero value indicate pauses in the server work or a temporary absence of tasks.

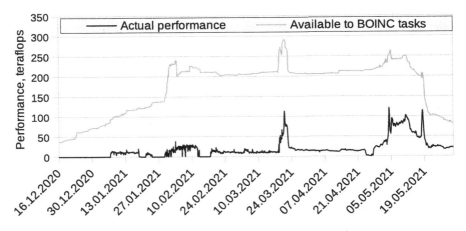

Fig. 1. Performance dynamics of the project's Desktop Grid.

4.1   Alternative Performance Metrics of a Desktop Grid

The project's performance can also be estimated in terms of active threads: the amount of work in CPU days per day of a Desktop Grid operation. The actual performance strongly depends on the specifics of the solved computationally-intensive problem. For this reason, it is complicated to compare Desktop Grids and traditional high-performance computing systems. However, the number of active threads allows one to find an approximation of a Desktop Grid's scale.

During the first 8 months of SiDock@home operation, the average number of active threads during non-zero workload periods had the order of 4 000, comparable with supercomputers in the last 10% of Top-50 supercomputers of Russia and CIS [23]. The practice has shown that for short periods of high load, such as BOINC competitions, $T(t)$ value exceeds 18 000, which is comparable with a significantly more powerful supercomputer available on demand.

To compare the scale of the Desktop Grid with traditional high-performance systems, one may also adapt an existing benchmark such as LINPACK [11]. Let us illustrate a variant of such adaptation on an example of a sample desktop computer participating in SiDock@home, available 24/7, AMD Ryzen 9 3900X 12-Core Processor, earning 22 500 BOINC credits a day and producing 500 gigaflops according to the LINPACK benchmark. Calculated using this reference computer, the average performance of SiDock@home is, again, comparable with the 10% tail of Top-50 supercomputers of Russia and CIS.

5   Conclusion

Performance measurement is of relevance when selecting a high-performance computing system, designing the experiments and planning research. In general, it is complicated to compare Desktop Grids with supercomputers and clusters directly. The actual performance of any system strongly depends on the specifics of the solved computationally-intensive problem; moreover, Desktop Grids' peculiar features impose additional difficulties.

In this work, we propose several performance metrics for Desktop Grids. We formalize them for a BOINC-based volunteer computing project, illustrate them on the project SiDock@home and discuss their practical application scope.

The results show that the actual efficient load of a Desktop Grid of a volunteer computing project is several-fold lower than its potential performance. It is partly explained by the background character of the Desktop Grid computations, but leaves open the problem of optimizing the workload.

The performance bounds based on BOINC benchmarks represent the relative scale of the BOINC project among the others regarding community involvement and theoretical peak performance. Such metrics may be helpful when planning the further development of a project. For instance, one may decide between joining an existing umbrella project or running a separate one, extending the server hardware base, targeting new client platforms or processor types.

Analysis of the Desktop Grid-based project SiDock@home has shown the applicability of the proposed metrics and their combinations for quantitative and qualitative conclusions about the project's performance.

In the future, we aim to work on Desktop Grids' performance in three main directions. Firstly, we will develop and implement a prognosis system for a Desktop Grid for estimation of the time to complete the project or a computational experiment. A possible application of such a prognosis is the long-term planning of research. Secondly, we will develop new methods for optimising a Desktop Grid utility, targeting the fastest discovery of practically valuable results. Thirdly, we will implement the developed methods in the long-running volunteer computing projects SiDock@home and RakeSearch to increase their utility.

Acknowledgements. The initial library (one billion of compounds) was prepared with the generous help of Microsoft that donated computational resources in the Azure cloud platform [6]. COVID.SI team is grateful and looking forward to future collaborations.

We wholeheartedly thank all BOINC participants for their contributions.

Funding Information. This work was partly supported by the Scholarship of the President of the Russian Federation for young scientists and graduate students (project SP-609.2021.5); the Slovenian Ministry of Science and Education infrastructure project grant HPC-RIVR; the Slovenian Research Agency (ARRS) programme P2-0046 and J1-2471, the Physical Chemistry programme grant P1-0201; Slovenian Ministry of Education, Science and Sports programme grant OP20.04342.

References

1. Alonso-Monsalve, S., García-Carballeira, F., Calderón, A.: ComBos: a complete simulator of volunteer computing and desktop grids. Simul. Modell. Pract. Theory **77**, 197–211 (2017). https://doi.org/10.1016/j.simpat.2017.06.002. https://www.sciencedirect.com/science/article/pii/S1569190X17301028
2. Anderson, D.P.: BOINC: a platform for volunteer computing. J. Grid Comput. **18**(1), 99–122 (2020)
3. By Editorial Team: The history of supercomputing vs. COVID-19. https://www.hpcwire.com/2021/03/09/the-history-of-supercomputing-vs-covid-19. Accessed 18 Mar 2021
4. Cai, Y., et al.: LHC@Home: a BOINC-based volunteer computing infrastructure for physics studies at CERN. Open Eng. **7**, 379–393 (2017)
5. Climateprediction.net—the world's largest climate modelling experiment for the 21st century. https://www.climateprediction.net. Accessed 10 June 2021
6. Cloud Computing Services—Microsoft Azure. https://azure.microsoft.com/en-us/. Accessed 10 June 2021
7. CmDock. https://gitlab.com/Jukic/cmdock. Accessed 10 June 2021
8. CreditNew - BOINC. https://boinc.berkeley.edu/trac/wiki/CreditNew. Accessed 10 June 2021
9. Curnow, H.J., Wichmann, B.A.: A synthetic benchmark. Comput. J. **19**(1), 43–49 (1976)

10. Distributed Computing – Computing Platforms. http://distributedcomputing.info/platforms.html. Accessed 10 June 2021

11. Dongarra, J.J., Luszczek, P., Petitet, A.: The LINPACK benchmark: past, present and future. Concurr. Comput.: Pract. Exp. **15**(9), 803–820 (2003). https://doi.org/10.1002/cpe.728. https://onlinelibrary.wiley.com/doi/abs/10.1002/cpe.728

12. Estrada, T., Taufer, M., Anderson, D.P.: Performance prediction and analysis of BOINC projects: an empirical study with EmBOINC. J. Grid Comput. **7**(4), 537 (2009)

13. Home – COVID.SI. https://covid.si/en. Accessed 10 June 2021

14. Home—LHC@home. https://lhcathome.web.cern.ch. Accessed 10 June 2021

15. Institute for Protein design, University of Washington: Rosetta's role in fighting coronavirus. https://www.ipd.uw.edu/2020/02/rosettas-role-in-fighting-coronavirus. Accessed 10 June 2021

16. Ivashko, E., Chernov, I., Nikitina, N.: A survey of desktop grid scheduling. IEEE Trans. Parallel Distrib. Syst. **29**(12), 2882–2895 (2018)

17. Jukič, M., Janežič, D., Bren, U.: Ensemble docking coupled to linear interaction energy calculations for identification of coronavirus main protease (3CLpro) non-covalent small-molecule inhibitors. Molecules **25**(24), 5808 (2020)

18. Legrand, A.: Scheduling for large scale distributed computing systems: approaches and performance evaluation issues. Habilitation à diriger des recherches, Université Grenoble Alpes, November 2015. https://tel.archives-ouvertes.fr/tel-01247932

19. Rosetta@home. https://boinc.bakerlab.org. Accessed 10 June 2021

20. SiDock@home. https://sidock.si/sidock. Accessed 10 June 2021

21. Suhail, M.: Performance analysis of distributed systems using BOINC. Int. J. Comput. Appl. **975**, 8887 (2016)

22. Together We Are Powerful - Folding@home. https://foldingathome.org. Accessed 10 June 2021

23. Top50—Supercomputers [in Russian]. http://top50.supercomputers.ru/list. Accessed 10 June 2021

24. Wrapp, D., et al.: Cryo-EM structure of the 2019-nCoV spike in the prefusion conformation. Science **367**(6483), 1260–1263 (2020)

25. Zaikin, O., Manzyuk, M., Kochemazov, S., Bychkov, I.V., Semenov, A.A.: A volunteer-computing-based grid architecture incorporating idle resources of computational clusters. In: Dimov, I., Faragó, I., Vulkov, L.G. (eds.) Numerical Analysis and Its Applications. Lecture Notes in Computer Science, vol. 10187, pp. 769–776. Springer, Heidelberg (2016). https://doi.org/10.1007/978-3-319-57099-0_89

26. Zimmerman, M.I., et al.: SARS-CoV-2 simulations go exascale to capture spike opening and reveal cryptic pockets across the proteome. bioRxiv (2020). https://doi.org/10.1101/2020.06.27.175430. https://www.biorxiv.org/content/early/2020/10/07/2020.06.27.175430

Essential Algorithms

Consensus-Free Ledgers When Operations of Distinct Processes are Commutative

Davide Frey[1], Lucie Guillou[1], Michel Raynal[1,2(✉)], and François Taïani[1]

[1] Univ Rennes, IRISA, CNRS, Inria, 35042 Rennes, France
raynal@irisa.fr
[2] Department of Computing, The Hong Kong Polytechnic University,
Kowloon, Hong Kong

Abstract. Considering asynchronous message-passing systems in which any number of processes may crash, this article addresses the construction of ledger objects where (i) the append operations issued from distinct processes commute, while (ii) the append operations issued from the same process do not. In a very interesting way, it appears that the implementation of such ledgers does not need consensus, which makes them both attractive and efficient. Their underlying formalization rests on Mazurkiewicz's traces.

Keywords: Asynchronous system · Blockchain · Commutative operations · Consensus-freedom · FIFO-based synchronization · Immutable memory · Mazurkiewicz's traces · Message-passing · Process crash failures · Reliable broadcast

1 Introduction

1.1 Context of the Study

Once upon a Time the Blockchain... Since its introduction, more than ten years ago in the context of cryptocurrencies [14,17], blockchains have receiving more and more attention. A blockchain is nothing more than a technology to implement ledger objects [5,8,16], i.e., an object providing its users with a list of items (also called blocks, elements, cells, etc. according to the application context) that can be accessed by two operations only, namely an operation append() which allows to add a new element at the head of the list and an operation query() which allows to obtain the current value of the full list. The important point of a ledger lies in the fact that the elements previously added cannot be modified (immutability property).

Basically, and according to the upper layer application, the implementation a ledger object involves two main domains of informatics: synchronization and fault-tolerance (which includes cryptography). The main issue consists then in allowing the processes to agree on the very same order in which items are added to the ledger, i.e., in one way or another, the processes have to solve a consensus problem.

© Springer Nature Switzerland AG 2021
V. Malyshkin (Ed.): PaCT 2021, LNCS 12942, pp. 359–370, 2021.
https://doi.org/10.1007/978-3-030-86359-3_27

Do all the Ledgers Need Consensus? It has recently been shown that not all the ledgers need consensus. This actually depends on the application. In an amazing way, it has been show that the synchronization part of cryptocurrencies does not need consensus [2–4,6,7]. So, the important point of such *weak* ledgers is immutability. This is the topic addressed in this article in the context of asynchronous message-passing systems where any number of process may commit unexpected crash failures.

1.2   Computing Model

Process Model. The system comprises a set of n sequential asynchronous processes, denoted $p_1, ..., p_n$. Sequential means that a process invokes one operation at a time, and asynchronous means that each process proceeds at its own speed, which can vary arbitrarily and always remains unknown to the other processes. Any number of processes may crash. A crash is a premature definitive halt. Hence, a process behaves correctly (i.e., executes its algorithm) until it possibly crashes.

From a terminology point of view, when considering an execution, a process that does not crash is *correct*. Otherwise it is *faulty*.

Communication. The processes communicate through an underlying message-passing point-to-point network in which there exists a bidirectional channel between any pair of processes. For simplicity, in writing the algorithms, we assume that a process can send messages to itself. Each channel is reliable and asynchronous. Reliable means that a channel neither lose, duplicates, nor corrupts messages. Asynchronous means that the transit delay of each message is finite but arbitrary.

1.3   When the Operations append() of Different Processes Commute

Considering the previous computing model, the problem addressed in this article is the following.

– Impose the same view of each local order. Given any process p_i, ensure that all the processes see all the invocations of append() by p_i in the order in which p_i issued them),
– Allow global disorder. This means that the processes may see the appends issued by distinct processes in different order. Let op() denote an append invocation. This means that, if p_i issues op()1 and p_j issues op2(), a process p_k can see first op1() and then op2() while another process sees first op2() and then op1().

One can see that these constraints are the ones required by money transfer: to prevent double spending from occurring, two transfers issued by a process must be seen in their sending order, while transfers from distinct processes may be seen in different order by different processes. Other applications such as work stealing [12] and the distributed simulation of Petri nets belong to the same family of problems. In the context of failure-free systems such an approach based on commutative operations been investigated for about 10 years [10,18].

From an implementation point of view, it is easy to see that, this requires to implement FIFO channels between each pair of processes, which is a simple peer-to-peer problem not requiring consensus. Before presenting a distributed algorithm satisfying the two previous properties, the next section presents a formal definition of the problem based on Mazurkiewicz's traces, which turns to be the theoretical basis on which relies the specification of this family of problems.

2 Underlying Formalization

2.1 A Quick Look at Mazurkiewicz's Traces

A trace monoid (or free partially commutative monoid, also known as Mazurkiewicz's Traces [13,15]) is a generalization of the notion of words (finite sequence over an alphabet Σ, which allow us to capture the independence and the conflicts on operations (represented as letters of Σ).

More formally, a trace monoid over an alphabet Σ is defined by a symmetric independence relation $I \subseteq \Sigma \times \Sigma$ between the letters (operations) of Σ. $(a, b) \in I$ means that the operations a and b commute, i.e. the effect of ab and ba are equivalent.

Two (finite) words $u, v \in \Sigma^*$ are said to be *equivalent under I*, noted $u \overset{I}{\sim} v$, if and only if one can transform u into v (and reciprocally) by exchanging adjacent operations that are independent within u.

Relation $\overset{I}{\sim}$ is an equivalence relation over Σ^*, and a (finite) trace is simply an equivalence class of $\overset{I}{\sim}$, which is a congruence with respect to the concatenation operator (note $\oplus$, but generally omitted), i.e. if $x \overset{I}{\sim} y$ and $u \overset{I}{\sim} v$ then $xu \overset{I}{\sim} yv$. As a result, the concatenation over words translates to the set of traces. More precisely, $[u]_I [v]_I = [uv]_I$, where $u, v \in \Sigma^*$ are words over Σ, $[u]_I$ is the trace represented by u (equivalence class of u under the relation $\overset{I}{\sim}$). The resulting structure $\left(\Sigma^* / \overset{I}{\sim} \right)$ is called *free partially commutative monoid*, denoted $\mathbb{M}(\Sigma, I)$. A subset of $\mathbb{M}(\Sigma, I)$ is called a *trace language*.

2.2 Problem Formalization

We consider a ledger with the two types of operations defined below.

- Type A denotes append operations that allow processes to add elements to the ledger. Each append operation returns the symbol $\perp$ (which informs the invoking process it can continue its execution). Let A_i be the bounded set of the append operations invoked only by p_i. Each set A_i is thus attached to a process p_i in the sense that only this process can invoke the operations it contains. The operations in any given A_i do not commute with each other, with respect to the content of the ledger at a given instant, while any operation in A_i and any operation in A_j, with $j \neq i$, commute. In the following, op_i denotes an operation of A_i.

- Type Q denotes query operations that do not modify the ledger and return a value that depends on the current content of the ledger as seen by the invoking process. A query operation can be invoked by any process. In the following, query denotes an operation of Q independently of the process that invokes it.

Process-Commutative Ledger (PC-Ledger) Specification. Mazurkiewicz's traces allow us to capture the correct behaviors of a ledger. More precisely, a PC-ledger specification is a triple $((A_i)_i, L, Q)$ such that:

- Each set A_i is the set of append operations that p_i can invoke. We define $\Sigma = \bigcup_i A_i$ because the content of the ledger only depends on append operations. Then we leverage the fact that two operations $\mathsf{op}_i \in A_i$ and $\mathsf{op}_j \in A_j$ commute if and only if $i \neq j$ to define an independence relation I over Σ, namely

$$ I = (\Sigma \times \Sigma) \setminus \bigcup_{1 \leq i \leq n}(A_i \times A_i). $$

- L is a trace language defined on the monoid $\mathbb{M}(\Sigma, I)$ that satisfies a *forward acceptability* property defined as follows. let t be a trace in $\mathbb{M}(\Sigma, I)$. In the following $\mathsf{mset}(t)$ denotes the multiset of the operations appearing in the trace t. *Forward acceptability* states that for any two traces $u, v \in L$, and any operation $\mathsf{op}_i \in A_i$, we have

$$ u \oplus \mathsf{op}_i \in L \wedge \left(\exists\, \mathsf{op}_k \in \bigcup_{j \neq i} A_j : \mathsf{mset}(v) = \mathsf{mset}(u) \cup \{\mathsf{op}_k\}\right) \Rightarrow v \oplus \mathsf{op}_i \in L. $$

Forward acceptability means that an append operation op_i issued by a process p_i remains possible ($v \oplus \mathsf{op}_i \in L$) even if a process $p_k \neq p_i$ previously performed an append operation op_k (wherever op_k appears in the trace v).

- The set Q is the set of query operations, each query being a function from the trace language L to a set of application-dependent values. A $\mathsf{query} \in Q$, returns a view of the global content of the ledger as specified by the trace it operates on. Two arbitrary queries issued by two (possibly different) processes will in general return different results, but if the two processes have experienced sequences of operations that correspond to the same trace in L, their queries will return the same value.

The independence relation I expresses the fact that the content of a PC-ledger does not depend on the interleaving of the operations of different processes. Language L, on the other hand, specifies which contents (traces) are valid and through which operations In particular different applications may define L as a more or less constrained subset of $\mathbb{M}(\Sigma, I)$, as long as the forward-acceptability property holds.

Illustration. Let us consider money transfer. As previously suggested, this problem can be captured by a PC-ledger specification where the transfer operations by a process p_i define A_i, the invocation of a balance operation is a query, and L is defined as the set of traces of the transfer operations that produce positive balances only. As we can see, the money transfers issued by a process p_i are seen in the same order by all the processes, while money transfers issued by different processes may be seen in different orders. We observe that the corresponding language L satisfies forward acceptability because a transfer operation issued by a process p_i cannot invalidate an outgoing transfer from a different process p_j.

2.3   From a Specification to Executions

Now that we have defined what is a PC-ledger, we can explain how to make it "live" by defining what is an execution of it on top of an asynchronous crash-prone message-passing system. We do this with the help of the following definitions.

Histories. (While different, the following definitions are close to the ones used in [1])

- The *local history* of a process p_i is the sequence E_i of the append and query operations it has executed. If p_i executed op1 before op2 we write op1 $\rightarrow_i$ op2 ($\rightarrow_i$ is called process order).
- A history H is a set of local histories, one per process, $H = (E_1, \cdots, E_n)$.
- Given a history H and a process p_i, let $\widehat{H_i} = (\widehat{E_1}, \cdots, \widehat{E_n})$ such that
  - $\widehat{E_i} = E_i$.
  - $\widehat{E_j} = E_j \setminus Q_j$ for $j \neq i$ where Q_j denotes the set of queries issued by p_j.

Sequential Execution. A sequential execution SE is a sequence of triplets $SE = (e_x)_x$ where $e_x = (\text{op}, val, i)$, meaning that process p_i invoked $\text{op} \in A_i \cup Q$, with val being the returned value for $\text{op} \in Q$ and $val = \bot$ for $\text{op} \in A_i$.

Let $\text{proj}(SE, \Sigma)$ denote the sequence of append operations in SE, and $[\text{proj}(SE, \Sigma)]_I$ the equivalence class of $\text{proj}(SE, \Sigma)$ under the independence relation $\overset{I}{\sim}$ (defined in Sect. 2.2).

A sequential execution SE is *legal* if:

- The sequence of append operations is such that $[\text{proj}(SE, \Sigma)]_I \in L$.
- The value returned by a query depends only on the sequence of appends that precede it in SE.

Serializations

- A serialization S of a history H, is a legal sequential execution which contains all operations in H and respects all process orders $(\rightarrow_i)_{1 \leq i \leq n}$.
- Given a history H and a process p_i, a *local* serialization S_i is a serialization of $\widehat{H_i}$.

Distributed PC-Ledger Object: Definition. Given a PC-ledger specification $((A_i)_i, L, Q)$, a distributed PC-ledger object is a distributed object whose histories $H = (E_1, \cdots, E_n)$ verify the following properties:

- Any operation invoked by a correct process terminates.
- For any process p_i, there is a local serialization S_i of $\widehat{H_i}$.

3 An Algorithm Implementing a PC-Ledger

3.1 Reliable Broadcast

The algorithm that implements a PC-ledger assumes an underlying reliable broadcast communication abstraction. This abstraction provides the processes with two operations denoted r_broadcast() and r_deliver(). When a process invokes r_broadcast(m) (resp., r_deliver(m)), we say it r-broadcasts (resp., r-delivers) the message m. Reliable broadcast is defined by the following properties.

- RB-Validity. If a process p_i r-delivers a message m from a process p_j, then the process p_j r-broadcast m.
- RB-Integrity. Assuming all the messages are different, no process r-delivers twice the same message.
- RB-Termination-1. If a correct process r-broadcasts a message m, it r-delivers it.
- RB-Termination-2. If a correct process r-delivers a message m, all correct processes r-deliver m.

Validity and Integrity are safety properties. Validity relates the outputs to the inputs. Integrity states there is no duplication. The termination properties state that all correct processes r-deliver the same set M of messages, and this set includes all the messages they r-broadcast. Moreover a faulty process r-delivers a subset of M.

Using the technique "first forward and only then deliver", reliable broadcast is easy to implement on top of a point-to-point fully connected network. When a process invokes r_broadcast(m), it sends m to all the processes, and then r-delivers it to itself. When a process receives a message for the first time, it first forwards it to the other processes and only then r-delivers it locally. When a process receives a copy of a message it has already received, it discards it. Algorithms implementing reliable broadcast with additional qualities of service are described in [9, 16].

3.2 Local Data Structures

It is assumed that all the processes know the alphabet Σ (operations) and the language L defining the PC-ledger. The symbol $\oplus$ is used to explicitly denote the concatenation of an element at the end of a sequence. The symbol ϵ denotes the empty sequence. Since L is a trace language, we usually omit the equivalence class notation $[\cdot]_I$ for readability's sake when the context is clear, for instance writing $s \in L$ to mean $[s]_I \in L$ if $s \in \Sigma^*$ is a sequence.

The messages APPLY r-broadcast by the processes contain four fields: the index of the sender process, its sequence number, the identifier of the specific append operation issued by the sender (*opname*), and the parameters of this append operation (*param*).

Each process manages the following local variables.

- sn_i is an sequence number (initialized to 0) used by p_i to identify the messages it r-broadcasts.
- $del_i[1..n]$ is an array of sequence numbers (each initialized to 0. The entry $del_i[j]$ contains the greatest sequence number of the messages p_i has r-delivered from p_j.
- seq_i is the sequence of operations which locally represents the PC-ledger object, as seen by p_i. Its initial value is the empty sequence ϵ.

---

init: $sn_i \leftarrow 0$; $seq_i \leftarrow \emptyset$; $del_i[1..n] \leftarrow [0, \cdots, 0]$.

operation query() **is**   % query() is any operation of type Q
(01) $res \leftarrow$ query(seq_i); return(res).

operation append(*opname*, *param*) **is**   % $\langle opname, param \rangle$ is any operation $\in A_i$
(02) $sn_i \leftarrow sn_i + 1$;
(03)   r_broadcast APPLY(*opname*, *param*, sn_i, i);
(04)   wait ($del_i[i] = sn_i$);
(05)   return().

when APPLY(*opname*, *param*, sn, j) **is** r_delivered **do**
(06)   wait($sn = del_i[j] + 1) \wedge (seq_i \oplus \langle opname, param \rangle \in L$);
(07) $seq_i \leftarrow seq_i \oplus \langle opname, param \rangle$;
(08) $del_i[j] \leftarrow del_i[j] + 1$.

Algorithm 1: An algorithm implementing a PC-ledger (code for p_i)

Algorithm 1 is pretty simple. When a process p_i invokes an operation query(), it locally applies it to its local representation of the PC-ledger seq_i and returns the corresponding result (line 01). By construction, a process p_i only appends operations $\langle opname, param \rangle$ (lines 02–05) that *(i)* belong to the set A_i (the operations it is allowed to use), and *(ii)* are acceptable in p_i's current ledger representation seq_i, namely the concatenation of $\langle opname, param \rangle$ to seq_i remains in the trace language, $[seq_i \oplus \langle opname, param \rangle]_I \in L$. When process p_i invokes append(*opname*, *param*), with the pair $\langle opname, param \rangle \in A_i$, it first increases sn_i and r-broadcasts the message APPLY(*opname*, *param*, sn_i, i) to all the processes (including itself, lines 02–03).

When p_i r-delivers a message APPLY(*opname*, *param*, sn, j), it waits (line 06) until it has processed the previous append from p_j, and this new append satisfies the forward acceptability property. When this occurs, p_i adds this append to seq_i (line 07) and accordingly updates $del_i[j]$ (line 08).

4   Proof of the Algorithm

Notation. Considering an invocation op_j of append() by a process p_j, before(op_j) denotes all append() invocations by p_j that precede op_j.

Lemma 1. *Any invocation of an operation by a process that does not crash terminates.*

Proof. Let us first observe that any invocation of an operation query() by a correct process trivially terminates.

As far a the operation append() is concerned, let us assume (by contradiction) that some invocation op_j of an append invocation issued by a correct process p_j never returns. Since p_j is correct, by RB-Termination-1 p_j eventually r-delivers the message m_{op_j} corresponding to op_j. Let sn be the sequence number associated with op_j. Let us observe that all the append invocations issued by p_j with a sequence number smaller than sn have terminated (otherwise p_j could not have issued an operation with sequence number sn) and have therefore been processed at lines 07 and 08. It follows that the predicate $sn = del_j[j] + 1$ (line 06) is satisfied when m_{op_j} is r-delivered.

Let us note seq_j^0 the value of seq_j when op_j is invoked by p_j, and seq_j^1 its value when m_{op_j} is r-delivered by p_j. By assumption, we have $seq_j^0 \oplus op_j \in L$, since no nodes is Byzantine (for the sake of conciseness, we equate here op_j with its associated $\langle opname, param \rangle$ pair). Because all these invocations have already been processed by p_j when op_j is invoked, we have before(op_j) $\subseteq$ mset(seq_j^0). Because p_j does not perform any additional append() invocation after op_j (since by assumption op_j never returns), we also have mset(seq_j^1) $=$ mset(seq_j^0) $\cup$ A' for some $A' \subseteq \Sigma$ that fulfills $A' \cap A_j = \emptyset$. By recursively applying the *forward acceptability* property (defined in Sect. 2.2), this implies that $seq_j^1 \oplus op_j \in L$, and therefore that the second predicate at line 06 is also satisfied when m_{op_j} is r-delivered, leading to the execution of line 08, and the termination of append() at line 04.

$\square_{Lemma\ 1}$

Notations

- Let op_j^{sn} denote the append operation issued by p_j with sequence number sn. Hence the message APPLY($opname, param, sn, j$) is associated with this operation.
- A process p_i locally *processes* the operation op_j^{sn} when, after it r-delivered the message APPLY($opname, param, sn, j$), it executes the lines 07–08.
- If a message APPLY($opname, param, sn, j$) be is r-delivered by a correct process, we say it is *successful*. It follows from the RB-Termination properties that all the operations append() invoked by the correct processes give rise to successful APPLY messages.

Lemma 2. *If a process p_i processes op_j^k, any correct process processes it.*

Proof. Let us assume that op_j^k is the sth operation processed by p_i. We prove the lemma by induction on s. Let us note seq_i^{op} and del_i^{op} the values of seq_i and del_i at line 06 when the wait statement becomes true for op_j^k at p_i, and op_j^k is selected by p_i to be added to its ledger.

For $s = 1$, seq_i^{op} is the empty sequence ϵ, and $del_i^{\mathsf{op}}[j] = 0$ (since p_i has not processed any operation yet from any process). As the wait statement has just become true, we therefore have $k = del_i^{\mathsf{op}}[j] + 1 = 1$, and $seq_i^{\mathsf{op}} \oplus \mathsf{op}_j^k = \epsilon \oplus \mathsf{op}_j^k = \mathsf{op}_j^k \in L$ (since p_i is not Byzantine).

Let us consider a correct process p_ℓ. Due to the RB-Termination-2 property, p_ℓ eventually r-delivers the message $m_j^k = \text{APPLY}(opname, param, k, j)$ from p_j associated with op_j^k. Let us write seq_ℓ^{op} and del_ℓ^{op} the values of seq_ℓ and del_ℓ at line 06 just after m_j^k has been r-delivered.

By the RB-Integrity property, this is the first (and only) time p_ℓ r-delivers m_j^k, which implies $del_\ell[j]$ has not yet taken the value $k = 1$, and by monotony that $del_\ell^{\mathsf{op}}[j] = 0$. $del_\ell^{\mathsf{op}}[j] = 0$ implies that p_ℓ has not processed any operation from p_j yet, and therefore that $\mathsf{mset}(seq_\ell^{\mathsf{op}}) \subseteq \bigcup_{k' \neq j} A_k$. This last inclusion and the fact that $\mathsf{op}_j^k \in L$ (see above) implies by recursively using the *forward acceptability* property (Sect. 2.2) that $seq_\ell \oplus \mathsf{op}_j^k \in L$, and with $k = 1 = del_\ell^{\mathsf{op}}[j] + 1$ that the wait statement is immediately verified by p_ℓ, and that op_j^k is processed by p_ℓ, concluding the proof for $s = 1$.

Let us now assume that the property is true up to a value $s - 1 > 0$. When the wait statement becomes true for op_j^k at p_i, we have $del_i^{\mathsf{op}}[j] = k - 1$, implying p_i has already processed all the earlier operations $\mathsf{before}(\mathsf{op}_j^k) = \{\mathsf{op}_j^{k'}\}_{k' < k}$ issued by p_j, but has not yet processed any additional operation from p_j. As a consequence $\mathsf{mset}(seq_i^{\mathsf{op}}) \cap A_j = \mathsf{before}(\mathsf{op}_j^k)$. Furthermore, as earlier, we also have $seq_i^{\mathsf{op}} \oplus \mathsf{op}_j^k \in L$.

Let us consider a correct process p_ℓ. As earlier, p_ℓ eventually r-delivers the message m_j^k. Let us assume the condition of the wait statement at line 06 never becomes true for op_j^k, and op_j^k is never processed by p_ℓ. By induction hypothesis, since p_i has already processed all the operations in $\mathsf{mset}(seq_i^{\mathsf{op}})$, and $|\mathsf{mset}(seq_i^{\mathsf{op}})| = s - 1$, we know that p_ℓ also eventually processes all the operations in $\mathsf{mset}(seq_i^{\mathsf{op}})$, and at some point we have $\mathsf{mset}(seq_i^{\mathsf{op}}) \subseteq \mathsf{mset}(seq_\ell)$, and therefore $\mathsf{before}(\mathsf{op}_j^k) \subseteq \mathsf{mset}(seq_\ell)$, and $del_\ell[j] \geq k - 1$. Furthermore, since p_ℓ never processes op_j^k, we also have $del_\ell[j] < k$, and hence $del_\ell[j] = k - 1$, which implies $\mathsf{mset}(seq_\ell^{\mathsf{op}}) \cap A_j = \mathsf{before}(\mathsf{op}_j^k)$. As a result, there exists some set $A' \subseteq \bigcup_{k' \neq j} A_k$, such that $\mathsf{mset}(seq_\ell) = \mathsf{mset}(seq_i^{\mathsf{op}}) \cup A'$. Using $seq_i^{\mathsf{op}} \oplus \mathsf{op}_j^k \in L$, we can recursively apply the *forward acceptability* property, leading to $seq_\ell \oplus \mathsf{op}_j^k \in L$, meaning the wait statement eventually becomes true for op_j^k at p_ℓ, which contradicts the fact it is never processes by p_ℓ, and concludes the proof. $\square_{Lemma\ 2}$

Theorem 1. *Algorithm* 1 *implements a PC-ledger.*

Proof. The fact that the operations issued by the correct processes terminate follows from Lemma 1. So, the rest of the proof concerns the safety properties of a PC-ledger, namely: for any process p_i, there is a serialization S_i of $\widehat{H_i}$ (i.e. a legal sequential execution that is equivalent to $\widehat{H_i}$ from p_i's viewpoint).

Considering a process p_i, let us first recall the definition of $\widehat{H_i}$, namely $\widehat{H_i} = (\widehat{E_1}, \cdots, \widehat{E_n})$ such that $\widehat{E_i}$ is the local history of p_i, and, for each $j \neq i$, $\widehat{E_j}$ is the local history of p_j including only its append operations.

From Lemma 2 and the fact that the messages APPLY() are processed in their sending order (from a local point-to-point point of view) and in agreement with the forward acceptability property (from a global point of view), it follows that the append operations issued by any process p_j are added to seq_i in the order they have been invoked by p_j. Moreover, the queries issued by p_i are on the value of seq_i at the query time. It follows that the corresponding sequence of append issued by the processes and the query operations issued by p_i is a serialization S_i of $\widehat{H_i} = (\widehat{E_1}, ..., \widehat{E_n})$. In particular, as the value returned by each query issued by p_i depends on all the append operations that precede it in S_i, S_i is a legal sequential execution equivalent to $\widehat{H_i}$ from p_i's viewpoint. $\square_{Theorem\ 1}$

5   Conclusion

Considering asynchronous message-passing systems in which any number of processes may commit crash failures, this article has introduced the notion of a ledger where append operations from distinct processes are commutative while operations from a same processes are not (hence the name *PC-ledger* where PC stands for Process-Commutative).

After the formal definition of such objects, the article has shown how these objects can be implemented in asynchronous crash-prone distributed systems. On an application point of view, as already noticed, it is interesting to notice that, while money transfers from a given user are not commutative (in order to prevent double spending from occurring), money transfers from different users are commutative, and consequently money-transfer ledgers belong to the family of PC-ledgers.

Interestingly enough, this article has also shown the study of the class of applications where, while the operations of each process are not commutative, the operations issued by distinct processes are, can be based on sane foundations, namely sequences known as Mazurkiewicz's traces.

This paper has shown that the *trace languages*-based approach allows us to cope with the net effect produced by adversaries such as asynchrony and process crashes. So, and last but not least, a far from being trivial problem concerns the adversarial context defined by asynchrony and Byzantine process failures [11]. Can the proposed trace languages-based approach be used to address such a stronger non-deterministic context?

Acknowledgments. This work was partially supported by the French ANR project ByBLoS (ANR-20-CE25-0002-01) devoted the modular design of building blocks for large-scale fault-tolerant multi-users applications. The authors want thank the referees for their constructive comments.

A   Exercise: From a PC-Ledger to a Distributed PC-State Machine

Some applications do not require to keep the full history saved in a ledger object. In this case, it is easy to replace the sequence seq_i by a local variable $state_i$ which represents the current state of the ledger as know by p_i. The resulting PC-state machine Algorithm 2 is trivially obtained from Algorithm 1. The function $\delta()$ is the transition function of the corresponding state machine, which is assumed to return $\perp$ in case a transition is not allowed.

---

init: $sn_i \leftarrow 0$; $state_i \leftarrow$ initial value of the state machine; $del_i[1..n] \leftarrow [0, \cdots, 0]$.

operation query() **is** %   % query() is any operation of type Q
(01) $res \leftarrow$ query($state_i$); return(res).

operation append($opname, param$) **is** %   % $\langle opname, param \rangle$ is any operation A_i
(02) $sn_i \leftarrow sn_i + 1$;
(03)   r_broadcast APPLY($opname, param, sn_i, i$);
(04)   **wait** ($del_i[i] = sn_i$);
(05)   return().

when APPLY($opname, param, sn, j$) **is** r_delivered **do**
(06)   **wait**($sn = del_i[j] + 1) \wedge (\delta(seq_i, \langle opname, param \rangle) \neq \perp)$;
(07) $state_i \leftarrow \delta(seq_i, \langle opname, param \rangle)$;
(08) $del_i[j] \leftarrow del_i[j] + 1$.

Algorithm 2: From a PC-ledger to a PC-state machine (code for p_i)

References

1. Ahamad, M., Neiger, G., Burns, J.E., Hutto, P.W., Kohli, P.: Causal memory: definitions, implementation and programming. Distrib. Comput. **9**, 37–49 (1995). https://doi.org/10.1007/BF01784241
2. Auvolat, A., Frey, D., Raynal, M., Taïani, F.: Money transfer made simple. In: Bulletin of the European Association of Theoretical Computer Science (EATCS), vol. 132, pp. 22–43 (2020)
3. Cholvi, V., Fernández Anta, A., Georgiou, Ch., Nicolaou, N.C., Raynal, M., Russo, A.: Byzantine-tolerant distributed grow-only sets: specification and applications. In: Proceedings of the 4th International Symposium on Foundations and Applications of Blockchain (FBA 2021). LIPICS OASIS Series (2021)

4. Collins, D., et al.: Online payments by merely broadcasting messages. In: Proceedings of the 50th IEEE/IFIP International Conference on Dependable Systems (DSN 2020), pp. 26–38. IEEE Press (2020)
5. Fernández Anta, A., Konwar, M.K., Georgiou, Ch., Nicolaou, N.C.: Formalizing and implementing distributed ledger objects. SIGACT News **49**(2), 58–76 (2018)
6. Guerraoui, R., Kuznetsov, P., Monti, M., Pavlovic, M., Seredinschi, D.A.: The consensus number of a cryptocurrency. In: Proceedings of the 38th ACM Symposium on Principles of Distributed Computing (PODC 2019), pp. 307–316. ACM Press (2019)
7. Gupta, S.: A non-consensus based decentralized financial transaction processing model with support for efficient auditing. Master thesis, Arizona State University (2016). 83 p
8. Gupta, S., Hellings, J., Sadoghi, M.: Fault-Tolerant Distributed Transactions on Blockchains. Morgan & Claypool Series on Synthesis Lectures on Data Management, vol. 64 (2021). 248 p
9. Hadzilacos, V., Toueg, S.: A modular approach to fault-tolerant broadcasts and related problems. Technical report 94–1425, Cornell University (1994). 83 p
10. Li, Ch., Porto, D., Clement, A., Gehrke, J., Preguiça, N.M., Rodrigues, R.: Making geo-replicated systems fast as possible, consistent when necessary. In: Proceedings of the 10th Symposium on Operating Systems Design and Implementation (OSDI 2012), pp. 265–278. USENIX Association (2012)
11. Lamport, L., Shostack, R., Pease, M.: The Byzantine generals problem. ACM Trans. Program. Lang. Syst. **4**(3), 382–401 (1982)
12. Michael, M.M., Vechev, M.T., Saraswat, S.A.: Idempotent work stealing. In: Proceedings of the 14th ACM SIGPLAN Symposium on Principles and Practice of Parallel Programming (PPOPP 2009), pp. 45–54. ACM Press (2009)
13. Mazurkiewicz, A.: Trace theory. In: Brauer, W., Reisig, W., Rozenberg, G. (eds.) ACPN 1986. LNCS, vol. 255, pp. 278–324. Springer, Heidelberg (1987). https://doi.org/10.1007/3-540-17906-2_30
14. Nakamoto, S.: Bitcoin: a peer-to-peer electronic cash system (2008). https://bitcoin.org/bitcoin.pdf
15. Petit, A.: Recognizable trace languages, distributed automata and the distribution problem. Acta Informatica **30**(1), 89–101 (1993). https://doi.org/10.1007/BF01200264
16. Raynal, M.: Fault-Tolerant Message-Passing Distributed Systems: An Algorithmic Approach. Springer, Heidelberg (2018). https://doi.org/10.1007/978-3-319-94141-7, ISBN 978-3-319-94140-0, 550 p.
17. Riesen, A.: Satoshi Nakamoto and the financial crisis of 2008. https://andrewriesen.me/2017/12/18/2017-12-18-satoshi-nakamoto-and-the-financial-crisis-of-2008/
18. Shapiro, M., Preguiça, N., Baquero, C., Zawirski, M.: Conflict-free replicated data types. In: Défago, X., Petit, F., Villain, V. (eds.) SSS 2011. LNCS, vol. 6976, pp. 386–400. Springer, Heidelberg (2011). https://doi.org/10.1007/978-3-642-24550-3_29

Design and Implementation of Highly Scalable Quantifiable Data Structures

Victor Cook⬤, Christina Peterson$^{(\boxtimes)}$⬤, Zachary Painter⬤,
and Damian Dechev⬤

University of Central Florida, Orlando, FL 32816, USA
{victor.cook,clp8199,zacharypainter}@knights.ucf.edu
damian.dechev@ucf.edu

Abstract. Architectural imperatives due to the slowing of Moore's Law, the broad acceptance of relaxed semantics and the $O(n!)$ worst case verification complexity of generating sequential histories motivate a new approach to concurrent correctness. *Quantifiability* is proposed as a novel correctness condition that models a system in vector space to launch a new mathematical analysis of concurrency. Analysis is facilitated with linear algebra, better supported and of much more efficient time complexity than traditional combinatorial methods. In this paper, we design and implement a quantifiable stack (QStack) and queue (QQueue) and present results showing that quantifiable data structures are highly scalable through use of relaxed semantics, an explicit implementation trade-off permitted by quantifiability. We present a technique for proving that a data structure is quantifiable and apply this technique to show that the QStack is quantifiably correct.

Keywords: Concurrent correctness · Multicore performance · Relaxed semantics

1 Introduction

There are a number of correctness conditions for concurrent systems [1,2,12,13, 17,22]. The difference between the correctness conditions resides in the allowable method call orderings in a history based on the thread interleavings. Serializability [22] places no constraints on the method call order. Sequential consistency [17] requires that each method call takes effect in program order, i.e., all methods called by the same thread respect call order. Linearizability [13] requires that each method call takes effect at some instant, referred to as a *linearization point*, between its invocation and response; i.e. each method call takes effect in *real-time order*. Method calls that deviate from real-time order are considered to have *relaxed semantics*.

These correctness conditions require a concurrent history to be equivalent to a sequential history. While this way of defining correctness enables concurrent programs to be reasoned about as if they were sequential programs [9,15], it

© Springer Nature Switzerland AG 2021
V. Malyshkin (Ed.): PaCT 2021, LNCS 12942, pp. 371–385, 2021.
https://doi.org/10.1007/978-3-030-86359-3_28

imposes several inevitable limitations on a concurrent system. Such limitations include 1) requiring the specification of a concurrent system to be described as if it were a sequential system, 2) restricting the method calls to respect data structure semantics and to be ordered in a way that satisfies the correctness condition, leading to performance bottlenecks, and 3) burdening correctness verification with a worst-case time complexity of $O(n!)$ to compute the sequential histories for each of the possible interleavings of n concurrent method calls.

Quantifiability [8] is proposed as a new correctness condition that does not require reference to a sequential history. Quantifiability eliminates the need for demonstrating equivalence to sequential histories by evaluating correctness of a concurrent history based solely on the outcome of the method calls. Additionally, quantifiability requires *conservation* of method calls; that is, method calls do not fail unless explicitly cancelled. The elimination of constraints such as method call ordering and data structure semantics as permitted by quantifiability enables concurrent histories to be represented in vector space. The vector space model elegantly allows the verification of concurrent correctness based on the outcomes of method calls using linear algebra and holds the potential for many practical uses, including extension to traditional correctness conditions [23].

In this paper, we present the design and implementation of a quantifiable stack (QStack) and quantifiable queue (QQueue). The performance evaluation of the QStack and QQueue demonstrates that these data structures obtain significantly higher scalability when compared to the state-of-the-art linearizable counterparts. The QStack and QQueue achieve high scalability by leveraging the relaxed semantics permitted by quantifiability to avoid contention. Additionally, the QStack and QQueue obtain high throughput by implementing the method call conservation principle of quantifiability. We present a technique for proving that a data structure is quantifiably correct. We apply this technique to prove that the QStack is quantifiably correct.

Contributions to the field are:

1. We present a technique for proving that a data structure is quantifiably correct.
2. We implement a quantifiably correct concurrent stack (QStack) and queue (QQueue).
3. We present a performance evaluation which demonstrates that the QStack and QQueue achieve significantly higher scalability when compared to state-of-the-art linearizable counterparts.

2   Proving that a Data Structure is Quantifiably Correct

To prove that a concurrent data structure is quantifiability correct, it must be shown that each method call preserves atomicity (the method takes effect entirely or not at all), isolation (the method's effects are indivisible), and conservation (every method call either completes successfully, remains pending, or is explicitly cancelled). The techniques for proving atomicity in literature include Lipton's theory of reduction [18] for reasoning about sequences of statements that are indivisible, occurrence graphs that represent a single computation as a set of interdependent events [5], Wing's methodology [27] for demonstrating that a concurrent object's behavior exhibits its specification, and simulation mappings between the implementation and specification automata [7].

Since Lipton's approach [18] is focused on lock-based critical sections, occurrence graphs [5] do not model data structure semantics, and Wing's approach [27], simulation mappings [7], and formal proofs of linearizability [14] require reference to sequential histories, they are not sufficient for proofs of quantifiability. However, informal proofs of linearizability reason about correctness by identifying a single instruction for each method in which the method call takes effect, referred to as a *linearization point*. Proving that a data structure is quantifiably correct can be performed in a similar fashion by defining a *visibility point* for each method. A visibility point is a single instruction for a method in which the entire effects of the method call become visible. Unlike a linearization point, a visibility point does not need to occur at some instant between a method call's invocation and response.

Establishing a visibility point for a method demonstrates that its effects preserve atomicity and isolation, but it still remains to be shown that the method call's effects are conserved. A method call's effects are proven to be conserved by showing that it returns successfully or its pending request is stored in the data structure and will be fulfilled by a future method call. Additionally, statements must be provided for each method that prove that its invocation is guaranteed to fulfill a corresponding pending request if the elements of the data structure comprise only requests.

Let $\langle X.m(a^*)\ P \rangle$ denote a method call invocation of method m on object X with input arguments a^* by process P. An object's element, denoted x, contains the boolean field *request* to indicate if the element is a request to perform an operation on an element that does not yet exist in the object. Element x contains the *event* field to indicate the requested operation. Let $\langle X : t(r^*)\ P \rangle$ denote a method call response from object X with response values r^* by process P, where t is Ok if the response is successful, *Pending* if the response is pending, or an *Exception* if the response is unsuccessful. An *inverse method m'* of method m is a method such that applying m' immediately after m undoes the effects of m. Let m_{inv} be an inverse method of method m (e.g. *push* is an inverse for *pop* in a stack). A method call m is conserved if $\langle X : t(r^*)\ P \rangle$, $t = Ok \vee (\langle X : t(r^*)\ P \rangle,\ t = Pending \wedge (\exists x \in X,\ x.request = true \wedge x.event = \langle X.m(a^*)\ P \rangle))$. The proof for conservation of method calls requires demonstrating that 1) $\langle X : t(r^*)\ P \rangle$, $t = Ok$ (a method completes its operation) on the successful code path, 2) $\langle X : t(r^*)\ P \rangle$, $t = Pending \wedge (\exists x \in X,\ x.request = true \wedge x.event = \langle X.m(a^*)\ P \rangle)$ (a method's pending request is stored in the data structure) on the unsuccessful code path, and 3) $\forall x \in X$ such that $x.request = true \wedge \langle X.m_{inv}(a^*)\ P \rangle \wedge (\exists x \in X,\ x.request = true \wedge x.event = \langle X.m(a^*)\ P \rangle) \implies \langle X.m_{inv}(a^*)\ P \rangle\ \langle X : t(r^*)\ P \rangle, t = Ok \wedge \langle X.m(a^*)\ P \rangle\ \langle X : t(r^*)\ P \rangle, t = Ok$ (method m's pending request is fulfilled by inverse method m_{inv}).

3 Design of a Quantifiable Stack

The quantifiable stack (QStack) is designed to conserve method calls while avoiding contention wherever possible. Consider the state of a stack receiving two concurrent *push* operations. Assume a stack contains only Node 1. Two threads concurrently push Node 2(*a*) and Node 2(*b*). The state of the stack after both operations have completed is shown in Fig. 1. The order is one of two possibilities: 5, 7, 9, or 5, 9, 7. Based on this quantifiable implementation, either 7 or 9 are valid candidates for a *pop* operation.

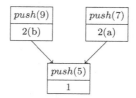

Fig. 1. Concurrent push representation

The QStack is structured as a doubly-linked tree of nodes. Two concurrent *push* method calls are both allowed to append their nodes to the data structure, forming a fork in the tree (Fig. 1). *Push* and *pop* are allowed to insert or remove nodes at any "leaf" node. To facilitate this design we add a descriptor pointer to each node in the stack. At the start of each operation, a thread creates a descriptor object with all the details necessary for an arbitrary thread to carry out the intended operation.

Algorithm 1. Stack: Definitions

```
1: Struct Node {              7: Struct Desc {
2: T value;                   8: T value;
3: Op op;                     9: Op op;
4: Node * nexts[];            10: bool active = true;
5: Node * prev;               11: };
6: };
```

Algorithm 1 contains type definitions for the QStack. *Node* contains the fields *value*, *op*, *nexts* and *prev*. The *value* field represents the abstract type being stored in the data structure. The *op* field identifies the node as either a pushed value or an unsatisfied pop operation. The *nexts* field is an array holding references to the children of the node, while *prev* contains a reference to its parent. *Descriptor* contains the *value* and *op* fields, as well as *active*. The *active* field designates whether the associated operation for the descriptor object is currently pending, or if the thread performing that operation has completed it. The stack data structure has a global array *tail*, which contains all leaf nodes in the tree. The stack data structure also has a global variable $forkRequest$ that is used to indicate that another branch should be added to a node in the stack and is initialized to null. The tree is initialized with a sentinel node in which the *active* flag is set to false.

Algorithm 2. Stack: Insert

1: **function** INSERT(**Node** * cur, **Node** * $elem$, **int** $index$)
2:     **Desc*** $d \leftarrow$ **new** Desc(v, op)
3:     **Node** * $curDesc \leftarrow cur.desc$
4:     **if** $curDesc.active = true$ **then**
5:         **return** $false$
6:     **if** $cur.desc.CAS(currDesc, d)$ **then**
7:         **if** $tail[index] \neq cur$ **then**
8:             $d.active \leftarrow false$
9:             **return** $false$
10:         **if** $cur.nexts.isEmpty()$ & $tail.count(cur) = 1$ **then**
11:             $elem.prev \leftarrow cur$
12:             $cur.nexts.add(elem)$
13:             $tail[index] \leftarrow elem$
14:             **Node** * $helperNode \leftarrow forkRequest$
15:             **if** $helperNode \neq null$ & $forkRequest.CAS(helperNode, null)$ **then**
16:                 **if** $helperNode.op = cur.op$ **then**
17:                     $helperNode.prev \leftarrow cur$
18:                     $cur.nexts.add(helperNode)$
19:                     initialize a new tail pointer and set it equal to $helperNode$
20:             $d.active \leftarrow false$
21:             **return** $true$
22:         **else**
23:             Remove dead branch
24:     $d.active \leftarrow false$
25:     **return** $false$

To conserve unsatisfied pops, we generalize the behaviour of *push* and *pop* operations with *insert* and *remove*. If a *pop* is made on an empty stack, we create a stack of waiting *pop* operations by calling *insert* and designating the inserted node as an unfulfilled *pop* operation. Similarly, if we call *push* on a stack that contains unsatisfied pops, we instead use *remove* to eliminate an unsatisfied *pop* operation, which then finally returns the value provided by the incoming *push*.

Algorithm 2 details the pseudocode for the *insert* operation. A node *cur* is passed in, which is expected to be a leaf node. In addition, *elem* is passed in, which is the node to be inserted. We check the descriptor of *cur* to see if another thread is already performing an operation at this node on line 4. If there is no pending operation, then we attempt to update the descriptor to point to our own descriptor on line 6. If this is successful, we check on line 10 if *cur* is a leaf node by ensuring *cur.nexts* is empty and that *tail* contains only 1 reference to *cur*. If it is not, that means that *cur* was previously a fork in the tree, but all nodes from one of the branches has been popped. In this case, we remove the index of the *tail* array corresponding to the empty branch, effectively removing the fork at *cur* from the tree. If *cur* is determined to be a leaf node on line 10, we are free to make modifications to *cur* without interference from other threads. In this case, *elem* is linked with *cur* and the tail pointer is updated. In our implementation, we choose to return *false* if a thread finds an active descriptor on line 4. This causes the thread to attempt to push at a different branch with less contention, which maximizes the number of threads working concurrently and minimizes contention. Alternatively, the thread could help the operation occurring at the node, but this would require the writes on lines 11, 12, and 13 to be done use CAS, and would likely increase contention.

The *remove* method is given by Algorithm 3. The *remove* method is similar to the *insert* method except that after the CAS on line 7, we check if *cur* is a leaf node before removing it from the tree.

Algorithm 3. Stack: Remove

```
1: function REMOVE(Node * cur, int index)
2:     Desc* d ← new Desc(op)
3:     Node * curDesc ← cur.desc
4:     Node * prev ← cur.prev
5:     if curDesc.active = true then
6:         return false
7:     if cur.desc.CAS(currDesc, d) & tail.count(cur) = 1 then
8:         if tail[index] ≠ cur then
9:             d.active ← false
10:            return false
11:        if cur.nexts.isEmpty() then
12:            v ← cur.value
13:            prev.nexts.remove(cur)
14:            tail[index] ← prev
15:            d.active ← false
16:            return true
17:        else
18:            Remove dead branch
19:     d.active ← false
20:     return false
```

Push and *pop* methods wrap these algorithms, as both operations need to be capable of inserting or removing a node depending on the state of the stack. Care should be taken that *push* only removes a node when the stack contains unsatisfied *pop* operations, while *pop* should only insert a node when the stack is empty, or already contains unsatisfied *pop* operations.

Algorithm 4 details the *push* method for the QStack. On line 2 we allocate a new node, and set the *value* and *op* field. Since a node may represent either a pushed value, or a waiting pop, we need to use *op* to designate the operation of the node. At line 7, we choose an index at which to try and add our node. The *tail* array contains all leaf nodes. The *getRandomIndex()* method avoids contention with other threads by choosing a random index.

If a thread is failing to make progress (line 17), we update the $forkRequest$ variable to contain the node for the delayed operation. When a successful $insert$ operation finds a non-null value in the $forkRequest$ variable on line 15, it inserts that node as a sibling to its own node. This creates a fork at the node cur, increasing the chance of success for future $insert$ operations.

If the node's operation is determined to be a pop on line 11, then the $push$ operation will fulfill the unsatisfied pop operation. Otherwise, the $push$ operation will proceed to insert its node into the stack. The pop method is given by Algorithm 5. Similar to push, a random index is selected on line 7 and the corresponding node is retrieved on line 8. If the node's operation is determined to be a $push$ on line 11 then the node is removed from the top of the stack. Otherwise, the stack is empty and the unsatisfied pop operation is inserted in the stack.

Algorithm 4. Stack: Push

```
 1: function PUSH(T v)
 2:     Node* elem ← new Node(v, PUSH)
 3:     bool ret ← false
 4:     int loops ← 0
 5:     while true do
 6:         loops + +
 7:         int index ← getRandomIndex()
 8:         Node * cur ← tail[index]
 9:         if cur = null then
10:             Continue
11:         if cur.op = POP then
12:             ret ← remove(cur, v)
13:         else
14:             ret ← insert(cur, elem, v)
15:         if ret then
16:             break
17:         if loops > FAIL_THRESHOLD & !forkRequest then
18:             forkRequest.CAS(null, cur)
19:             Break
```

Theorem 1. *The QStack is quantifiable.*

Proof. To prove that the QStack is quantifiable it must be shown that each of the methods preserve atomicity, isolation, and conservation. A visibility point is established for each of the methods that demonstrates that each method preserves atomicity and isolation.

Insert: The *insert* method creates a new descriptor on line 2, where the *active* field is initialized to true. When the CAS succeeds on line 6, any other thread that reads the descriptor on line 4 when calling *insert* (or line 5 of Algorithm 3 when calling *remove*) will observe that the *active* field is true and will continue from the beginning of the while loop on line 5 of Algorithm 4 when calling *push* (or line 5 of Algorithm 5 when calling *pop*). When the if statement on line 10 succeeds, the current thread sets the descriptor's *active* field to false on line 20.

Algorithm 5. Stack: Pop

```
 1: function POP(T &v)
 2:     Node* elem ← new Node(v, POP)
 3:     bool ret ← false
 4:     int loops ← 0
 5:     while true do
 6:         loops + +
 7:         int index ← getRandomIndex()
 8:         Node * cur ← tail[index]
 9:         if cur = null then
10:             Continue
11:         if cur.op = PUSH then
12:             ret ← remove(cur, &v)
13:         else
14:             ret ← insert(cur, elem, &v)
15:         if ret then
16:             v = cur.value
17:             break
18:         if loops > FAIL_THRESHOLD & !forkRequest then
19:             forkRequest.CAS(null, cur)
20:             Break
```

Since threads that were spinning due to the if statement on line 4 (or line 5 of Algorithm 3 when calling *remove*) are now able to observe the effects of the operation associated with the previous active descriptor, the visibility point for the *insert* method is line 20. Once an *insert* method call *m* becomes visible, a partial order is established such that the corresponding *push* or *pop* calling *m* is ordered before other concurrent *push* or *pop* method calls.

Remove: The *remove* method creates a new descriptor on line 2, where the *active* field is initialized to true. When the CAS succeeds on line 7, any other thread that reads the descriptor on line 5 when calling *remove* (or line 4 of Algorithm 2 when calling *insert*) will observe that the *active* field is true and will continue from the beginning of the while loop on line 5 of Algorithm 4 when calling *push* (or line 5 of Algorithm 5 when calling *pop*). When the if statement on line 11 succeeds, the current thread sets the descriptor's *active* field to false on line 15. Since threads that were spinning due to the if statement on line 5 (or line 4 of Algorithm 2 when calling *insert*) are now able to observe the effects of the operation associated with the previous active descriptor, the visibility point for the *remove* method is line 15. Once a *remove* method call *m* becomes visible, a partial order is established such that the corresponding *push* or *pop* calling *m* is ordered before other concurrent *push* or *pop* method calls.

Push: The *push* method accesses the node at a random tail index on line 8. If the operation of the node is a *pop*, then *remove* is called on line 12, so the visibility point is line 15 of Algorithm 3. Otherwise, *insert* is called on line 14, so the visibility point is line 20 of Algorithm 2.

Pop: The *pop* method accesses the node at a random tail index on line 8. If the operation of the node is a *push*, then *remove* is called on line 12, so the visibility point is line 15 of Algorithm 3. Otherwise, *insert* is called on line 14, so the visibility point is line 20 of Algorithm 2.

It now must be shown that the method calls are conserved. Since *insert* and *remove* are utility functions, only *push* and *pop* must be conserved.

Push: The *push* method checks if the operation of the node at the tail is a *pop* on line 11. If the check succeeds, then the *push* fulfills the unsatisfied *pop* by removing it from the stack at line 12. Otherwise, it proceeds with its own operation by calling *insert* at line 14. Since a *pop* request is guaranteed to be fulfilled if one exists due to the check on line 9, and *forkRequest* is updated on line 18 to the current node if the loop iterations exceeds the $FAIL\_THRESHOLD$, *push* satisfies method call conservation.

Pop: The *pop* method checks if the operation of the node at the tail is a *push* on line 11. If the check succeeds, then the *pop* proceeds with its own operation by removing it from the stack at line 12. Otherwise, it places its unfulfilled request by calling *insert* at line 14. Since a *pop* will only place a request if no nodes associated with a *push* operation exist in the stack due to the check on line 9, and *forkRequest* is updated on line 19 to the current node if the loop iterations exceeds the $FAIL\_THRESHOLD$, *pop* satisfies method call conservation.

4   Design of a Quantifiable Queue

The quantifiable queue (QQueue) is organized as an array of linked lists, each with a head and tail pointer. When a thread intends to perform an *enqueue* or *dequeue* operation, it first chooses an index of the array at random at which to base its operation. If performing an *enqueue*, the thread updates the tail's next pointer to the new node using CAS and then attempts to update the tail pointer to the new node using another CAS. In the case of a *dequeue*, the thread reads the value of the head and then updates the head to point to the head pointer's next field using CAS. Before any operation, a thread determines if the tail pointer is lagging behind by checking that $tail.next! = null$ and advances the tail pointer if this statement returns false. The array size is tunable based on the number of threads available. This modular approach used in the QQueue can be used to convert many types of lock-free containers into fast concurrent data structures.

Algorithm 6 contains the definitions for the QQueue. *Node* is similar to that of the QStack. The *op* field identifies the node as either a enqueued value or an unsatisfied *dequeue* operation. The queue data structure itself contains an array of *head* and *tail* references.

Algorithm 7 details the *enqueue* method for the QQueue. Similar to the QStack, we select a random index on line 5, and retrieve the corresponding node on line 6. The current *enqueue* operation will either fulfill an unsatisfied *dequeue* operation or insert a node as normal. The *remove* and *insert* operations can be any underlying, lock-free queue algorithm. In our experiments, we adapt the methods from the *enqueue* and *dequeue* methods of the MSQueue [20].

Algorithm 6. Queue: Definitions

```
1: Struct Node {
2: T value;
3: Op op;
4: Node * next;
5: };
6:
7: Struct Queue {
8: Node * head[];
9: Node * tail[];
10: };
```

Algorithm 7. Queue: Enqueue

```
1: function ENQUEUE(T v)
2:     Node* elem ← new Node(v, ENQUEUE)
3:     bool ret ← false
4:     while true do
5:         int index ← getRandomIndex()
6:         Node * cur ← tail[index]
7:         if cur.op = DEQUEUE then
8:             ret ← remove(cur, v)
9:         else
10:             ret ← insert(cur, elem, v)
11:         if ret then
12:             break
```

Algorithm 8. Queue: Dequeue

```
1: function DEQUEUE(T &v)
2:     Node* elem ← new Node(v, DEQUEUE)
3:     bool ret ← false
4:     while true do
5:         int index ← getRandomIndex()
6:         Node * cur ← tail[index]
7:         if cur.op = ENQUEUE then
8:             ret ← remove(cur, &v)
9:         else
10:             ret ← insert(cur, elem, &v)
11:         if ret then
12:             break
```

The *dequeue* method is given by Algorithm 8. It is identical to *enqueue* except for a key difference at line 7. In the case of a *dequeue*, we only remove a node if *cur* represents an enqueued value. If *cur* represents an unsatisfied *dequeue* operation, we insert our dequeue operation behind it to be satisfied later.

5   Quantifiability Applied to Other Types of Data Structures

Quantifiability is applicable to other abstract data types that deliver additional functionality beyond the standard producer/consumer methods provided by queues and stacks. Consider a reader method such as a *read* operation for a hashmap or a *contains* operation for a set. If the item to be read does not exist in the data structure, a *pending item* is created and placed in the data structure at the same location where the item to be read would be placed if it existed. If a pending item already exists for the item to be read, the reader method references this pending item. Once a producer method produces the item for which the pending item was created, the pending item is updated to a regular (non-pending) item. Since the reader methods hold a reference to this item, they may check the address when desired to determine if the item of interest is available to be read. A similar strategy can be utilized for writer methods.

6   Performance

A quantifiable stack (QStack) and a quantifiable queue (QQueue) are implemented to showcase the performance characteristics of quantifiable data structures. The QStack and QQueue were tested against the fastest available published work, along with classic examples. Stack results are shown in Fig. 2, and queue results in Fig. 3. The x-axis plots the number of threads available for each run. The y-axis plots method calls per microsecond. Plot line color and type show the different implementations.

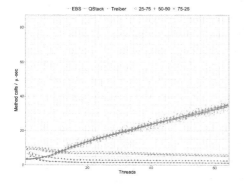

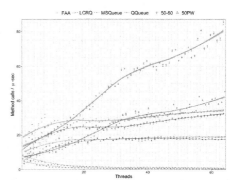

Fig. 2. QStack, EBS and Treiber stack.

Fig. 3. QQueue, FAA queue, LCRQ and MS queue.

Experiments were run on an AMD® EPYC® server of 2 GHz clock speed and 128 GB memory, with 32 cores delivering a maximum of 64 simultaneous multi-threads. The operating system is Ubuntu 18.04 LTS and code is compiled with gcc 7.3.0 using -O3 optimizations. The QStack was compared with the lock-free Elimination Backoff Stack (EBS) [3] and lock-free Treiber Stack [26]. The operations mix made little difference, but the Treiber Stack and EBS showed slightly higher performance at *push-pop* mixes of 25–75 because they quickly discard unsatisfied pop calls, returning *null*. As threads are added, Treiber drops off quickly due to contention. At five threads QStack overtakes Treiber and at 12 threads becomes faster than EBS. The salient result is that the QStack continues to scale, achieving over five times EBS performance with 64 threads. The other implementations consume resources to maintain order at microsecond scale instead of serving requests as quickly as possible with best efforts ordering. The testing methodology follows those used in the original EBS presentation, going from one to 64 threads with five million operations per thread. Memory is pre-allocated in the stack experiments, and for each run the program is restarted by a script to prevent the previous memory state from influencing the next run. The Boost library [4] is used to create a uniform random distribution of method calls based on the different mixes. Stack *push-pop* mixes of 25-75, 50-50 and 75-25 were tested for each implementation across all threads. Queue *enqueue-dequeue* mixes were temporal variations on a 50-50 mix. For both stack and queue, there were a minimum of 10 trials per thread per mix. The data was smoothed using the LOESS method as implemented in the ggplot2 library. Shaded areas indicate the 95% confidence limits for the lines.

The QQueue was compared with the lock-free LCRQ [21], the wait-free FAA queue [29] and the lock-free MS queue [19]. The LCRQ and FAA are the fastest queues in a recent benchmark framework with ACM verified code artifacts [28]. The MS queue is a classic like the Treiber stack. The framework uses only 50-50 mixes, one random (50-50) and one pairwise (50PW). The QQueue performs similarly to LCRQ until overtaking it at 14 threads, then overtaking the wait-free FAA queue at 18 threads. The FAA queue is exceptional as it performs as well or better than the alternative lock-free implementations. The TS-Queue [10] and the Multiqueue [24] are queues of interest published more recently than the FAA queue, but were not selected because verified code artifacts have not been published. Queue experiments follow the methodology of the Yang and Mellor-Crummey framework [28] and use the queue implementations provided in the source code. Memory allocation is dynamic within the framework. Benchmarks provided are two variations on 50-50 mixes, one random and the other pairwise. The different temporal distributions within the 50-50 mix have more influence on the results than different mixes (25-75, 50-50, 75-25) used in the stack experiments. In both the pairwise and random mixes, the QQueue continues to scale to the limit of hardware support, more than double the performance of FAA and LCRQ at 64 threads. The quantifiable containers continue to scale until all threads are employed, with slightly reduced slope in the simultaneous multi-threading region from 32 to 64 threads. Other implementations, including those that are linearizable with relaxed semantics, could maintain microscale order only at the cost of scalability. Furthermore, linearizability may cause unfairness where method calls are not conserved.

7   Related Work

Several data structure designs are presented in literature that reflect the principles of quantifiability. The concern of defining the behavior of partial methods when reaching an undefined object state is addressed by dual data structures [25]. Dual data structures are linearizable concurrent objects that hold reservations in addition to data to handle conditional semantics. The difference between dual data structures and quantifiable data structures is the allowable order that the requests may be fulfilled. The relaxed semantics of quantifiability provides an opportunity for performance gains over dual data structures.

Other data structure designs observe that contention can be reduced by allowing operations to be matched and eliminated if the combined effect does not change the abstract state of the data structure. The elimination backoff stack (EBS) [11] uses an elimination array where *push* and *pop* method calls are matched to each other within a short time delay if the main stack is suffering from contention. When operating on the central stack, the *pop* method is at risk of failing if the stack is empty. We note that if the elimination array timer were set to infinity, the elimination backoff stack would implement quantifiability's conservation principle, because all method calls would wait until they succeed.

The TS Queue [10] also relies on matching up method calls, enabling methods that would otherwise fail when reaching an undefined state of the queue to instead be fulfilled at a later time. In the TS Queue, rather than a global delay, there is a tunable parameter called *padding* added to different method calls. By setting an infinite time padding on all method calls, the TS Queue follows quantifiability's conservation principle. The EBS and TS-Queue share in common that they improve performance by using a window of time in which pending method calls are conserved until they can succeed.

Contention due to frequently accessed elements in a data structure can be further reduced by relaxing object semantics. The k-FIFO queue [16] maintains k segments each consisting of k slots implemented as either an array for a bounded queue or a list for an unbounded queue. This design enables up to k *enqueue* and *dequeue* operations to be performed in parallel and allows elements to be dequeued out-of-order up to a distance k. Quantifiability takes these relaxed semantics a step further by allowing method calls to occur out-of-order up to any arbitrary distance, leading to performance gains.

Interval-linearizability [6] has been defined for objects with no sequential specifications. The authors introduce the notion of an interval-sequential object which is specified by an automaton that is able to express any concurrency pattern of overlapping invocation of operations that may occur in an execution. While interval-linearizability is specifically intended for objects whose behavior must be defined in terms of input/output relations, quantifiability is intended for facilitating high-performance concurrent objects.

8   Conclusion

This paper presented the design and implementation of a quantifiable stack and queue. A technique for proving that a data structure is quantifiably correct is presented and applied to prove that the stack implementation is quantifiably correct. The relaxed semantics permitted by quantifiability allow for significant performance gains through contention avoidance in the implementation of concurrent systems. The performance evaluation showcases how the quantifiable stack and queue achieve substantially higher scalability than the state-of-the-art linearizable counterparts.

References

1. Afek, Y., Korland, G., Yanovsky, E.: Quasi-linearizability: relaxed consistency for improved concurrency. In: Lu, C., Masuzawa, T., Mosbah, M. (eds.) OPODIS 2010. LNCS, vol. 6490, pp. 395–410. Springer, Heidelberg (2010). https://doi.org/10.1007/978-3-642-17653-1_29
2. Aspnes, J., Herlihy, M., Shavit, N.: Counting networks. J. ACM (JACM) **41**(5), 1020–1048 (1994)
3. Bar-Nissan, G., Hendler, D., Suissa, A.: A dynamic elimination-combining stack algorithm. In: Fernàndez Anta, A., Lipari, G., Roy, M. (eds.) OPODIS 2011. LNCS, vol. 7109, pp. 544–561. Springer, Heidelberg (2011). https://doi.org/10.1007/978-3-642-25873-2_37
4. Beman Dawes, D.A., Rivera, R.: Boost C++ libraries, December 2018. https://www.boost.org/users/history/version_1_69_0.html. Accessed 15 Jan 2019
5. Best, E., Randell, B.: A formal model of atomicity in asynchronous systems. Acta Inform. **16**(1), 93–124 (1981)
6. Castañeda, A., Rajsbaum, S., Raynal, M.: Unifying concurrent objects and distributed tasks: interval-linearizability. J. ACM (JACM) **65**(6), 1–42 (2018)
7. Chockler, G., Lynch, N., Mitra, S., Tauber, J.: Proving atomicity: an assertional approach. In: Fraigniaud, P. (ed.) DISC 2005. LNCS, vol. 3724, pp. 152–168. Springer, Heidelberg (2005). https://doi.org/10.1007/11561927_13
8. Cook, V., Peterson, C., Painter, Z., Dechev, D.: Quantifiability: correctness of concurrent programs in vector space. In: 29th Euromicro International Conference on Parallel, Distributed and Network-Based Processing (2021)
9. Guttag, J.V., Horowitz, E., Musser, D.R.: Abstract data types and software validation. Commun. ACM **21**(12), 1048–1064 (1978)
10. Haas, A.: Fast concurrent data structures through timestamping. Ph.D. thesis, University of Salzburg, Salzburg, Austria (2015)
11. Hendler, D., Shavit, N., Yerushalmi, L.: A scalable lock-free stack algorithm. In: Proceedings of the Sixteenth Annual ACM Symposium on Parallelism in Algorithms and Architectures, SPAA 2004, pp. 206–215. ACM, New York (2004)
12. Herlihy, M., Shavit, N.: The Art of Multiprocessor Programming. Morgan Kaufmann (2012). Revised Reprint. ISBN 0123973375
13. Herlihy, M.P., Wing, J.M.: Linearizability: a correctness condition for concurrent objects. ACM Trans. Program. Lang. Syst. (TOPLAS) **12**(3), 463–492 (1990)
14. Herlihy, M.P., Wing, J.M.: Linearizability: a correctness condition for concurrent objects. ACM Trans. Program. Lang. Syst. **12**(3), 463–492 (1990)

15. Hoare, C.A.R.: Proof of correctness of data representations. In: Programming Methodology, pp. 269–281. Springer, New York (1978)
16. Kirsch, C.M., Lippautz, M., Payer, H.: Fast and scalable, lock-free k-FIFO queues. In: Malyshkin, V. (ed.) PaCT 2013. LNCS, vol. 7979, pp. 208–223. Springer, Heidelberg (2013). https://doi.org/10.1007/978-3-642-39958-9_18
17. Lamport, L.: How to make a multiprocessor computer that correctly executes multiprocess program. IEEE Trans. Comput. **28**(9), 690–691 (1979)
18. Lipton, R.J.: Reduction: a method of proving properties of parallel programs. Commun. ACM **18**(12), 717–721 (1975)
19. Michael, M.M., Scott, M.L.: Simple, fast, and practical Non-Blocking and blocking concurrent queue algorithms. Technical Report 600 (1995)
20. Michael, M.M., Scott, M.L.: Nonblocking algorithms and preemption-safe locking on multiprogrammed shared memory multiprocessors. J. Parallel Distrib. Comput. **51**(1), 1–26 (1998)
21. Morrison, A., Afek, Y.: Fast concurrent queues for x86 processors. In: Proceedings of the 18th ACM SIGPLAN Symposium on Principles and Practice of Parallel Programming (2013)
22. Papadimitriou, C.H.: The serializability of concurrent database updates. J. ACM (JACM) **26**(4), 631–653 (1979)
23. Peterson, C., Cook, V., Dechev, D.: Concurrent correctness in vector space. In: Henglein, F., Shoham, S., Vizel, Y. (eds.) VMCAI 2021. LNCS, vol. 12597, pp. 151–173. Springer, Cham (2021). https://doi.org/10.1007/978-3-030-67067-2_8
24. Rihani, H., Sanders, P., Dementiev, R.: Brief announcement: multiqueues: simple relaxed concurrent priority queues. In: Proceedings of the 27th ACM Symposium on Parallelism in Algorithms and Architectures, SPAA 2015, pp. 80–82. ACM, New York (2015). https://doi.org/10.1145/2755573.2755616. http://doi.acm.org/10.1145/2755573.2755616
25. Scherer, W.N., Scott, M.L.: Nonblocking concurrent data structures with condition synchronization. In: Guerraoui, R. (ed.) DISC 2004. LNCS, vol. 3274, pp. 174–187. Springer, Heidelberg (2004). https://doi.org/10.1007/978-3-540-30186-8_13
26. Treiber, R.K.: Systems programming: coping with parallelism. International Business Machines Incorporated, Thomas J. Watson Research Center, New York (1986)
27. Wing, J.M.: Verifying atomic data types. Int. J. Parallel Prog. **18**(5), 315–357 (1989)
28. Yang, C.: Fast wait free queue, October 2018. https://github.com/chaoran/fast-wait-free-queue. Accessed 5 Feb 2019
29. Yang, C., Mellor-Crummey, J.: A wait-free queue as fast as fetch-and-add. In: Proceedings of the 21st ACM SIGPLAN Symposium on Principles and Practice of Parallel Programming - PPoPP 2016, pp. 1–13. ACM Press, New York (2016)

Optimal Concurrency for List-Based Sets

Vitaly Aksenov[1]([✉]), Vincent Gramoli[2], Petr Kuznetsov[3], Di Shang[4],
and Srivatsan Ravi[5]

[1] ITMO University, St. Petersburg, Russia
[2] University of Sydney, Sydney, Australia
`vincent.gramoli@sydney.edu.au`
[3] LTCI, Télécom Paris, Institut Polytechnique de Paris, Palaiseau, France
`petr.kuznetsov@telecom-paris.fr`
[4] IBM, Sydney, Australia
[5] University of Southern California, Los Angeles, USA
`srivatsr@usc.edu`

Abstract. Designing an efficient concurrent data structure is a challenge that is not easy to meet. Intuitively, efficiency of an implementation is defined, in the first place, by its ability to process applied operations *in parallel*, without using unnecessary synchronization. As we show in this paper, even for a data structure as simple as a linked list used to implement the *set* type, the most efficient algorithms known so far are *not concurrency-optimal*: they may reject correct concurrent schedules. We propose a new algorithm for the list-based set based on a *value-aware try-lock* that we show to achieve *optimal* concurrency: it only rejects concurrent schedules that violate correctness of the implemented set type. We show that reaching this kind of optimality may be beneficial in practice. Our *concurrency-optimal* list-based set outperforms two state-of-the-art algorithms: the Lazy Linked List and the Harris-Michael List.

1 Introduction

Multicore applications require highly concurrent data structures. Yet, the very notion of concurrency is vaguely defined, to say the least. What do we mean by a "highly concurrent" data structure? Generally speaking, one could compare the concurrency of algorithms by running a game where the adversary decides on the schedules of shared memory accesses from different processes. At the end of the game, the more schedules the algorithm would accept without hampering high-level correctness, the more concurrent it would be. The algorithm that accepts all correct schedules would then be considered *concurrency-optimal* [1].

To illustrate the difficulty of optimizing concurrency, let us consider one of the most "concurrency-friendly" data structures [2]: the *sorted linked list* used to implement the *integer set* type. Since any modification on a linked list affects only a small number of contiguous list nodes, most of update operations on the list could, in principle, run concurrently without conflicts. For example, one of

© Springer Nature Switzerland AG 2021
V. Malyshkin (Ed.): PaCT 2021, LNCS 12942, pp. 386–401, 2021.
https://doi.org/10.1007/978-3-030-86359-3_29

the most efficient concurrent list-based set to date, the Lazy Linked List [3], achieves high concurrency by holding locks on only two consecutive nodes when updating, thus accepting *concurrent* modifications of non-contiguous nodes. The Lazy Linked List is known to outperform the Java variant [4] of the CAS-based Harris-Michael algorithm [5,6] under low contention because all its traversals, be they for read-only look-ups or for locating the nodes to be updated, are *wait-free, i.e.,* they ignore locks and logical deletion marks. As we show below, the Lazy Linked List implementation is however not *concurrency-optimal,* raising two questions: Is there a more concurrent list-based set algorithm? And if so, does higher concurrency induce an overhead that precludes higher performance?

The concurrency limitation of the Lazy Linked List is caused by the locking strategy of its update operations: both insert(v) and remove(v) traverse the structure until they find a node whose value is larger or equal to v, at which point they acquire locks on two consecutive nodes. Only then is the existence of the value v checked: if v is found (resp., not found), then the insertion (resp., removal) releases the locks and returns without modifying the structure. By modifying metadata during lock acquisition without necessarily modifying the structure itself, the Lazy Linked List over conservatively rejects certain correct schedules. To illustrate that the concurrency limitation of the Lazy Linked List may lead to poor scalability, consider Fig. 1 that depicts the performance of a 25-node Lazy Linked List (red curve) under a workload of 20% updates (insert/removals) and 80% contains on a 72-core machine. The list is comparatively small, hence all updates (even the failed insertions and removals) are likely to contend. We can see that when we increase the number of threads beyond 40, the performance drops significantly.

This observation suggests a desirable property that concurrent operations should conflict on metadata only when they conflict on data. To achieve this, we need to exploit the semantics of the high-level data type.[1]

Our main contribution is the Value-Based List (**VBL**), the most concurrent (in fact, *optimally* concurrent, as we formally prove) and probably the most efficient list-based set algorithm to date. It exploits the logical deletion technique of Harris-Michael that divides the removal of a node into a logical step (marking the node for deletion) and a physical step (unlinking the node from the list), and the wait-free traversal of the Lazy Linked List. In addition, our approach relies on a novel *value-aware* synchronization technique: first the lock, implemented using *compare-and-swap,* is taken, then the procedure checks whether the *value* in the next node has changed, if the validation is successful then the operation continues, otherwise, the operation restarts. Compared to the Lazy Linked List, this approach allows for the improvement of performance and even provides scalability in the highly contended cases (Fig. 1). We show that the resulting algorithm rejects a concurrent schedule only if otherwise the high-level correct-

---

[1] Note that this property refines the original notion of disjoint access parallelism (DAP) [7], trivially ensured by most linked-list implementations simply because all their operations "access" the *head* node and, thus, are allowed to conflict on the metadata.

ness of the implemented set type (linearizability [8]) is violated. Our algorithm is thus *concurrency-optimal* [1]: no correct list-based set algorithm can accept more schedules.[2]

The evaluation of VBL shows that achieving optimal concurrency in list-based set implementations does not necessarily result in a costly overhead, complementing the recent analysis of concurrency-optimality for tree-based dictionaries [9]. Extensive experiments on two x86-64 architectures machines, 72-way Intel machine and 64-way AMD machine, confirmed that VBL outperforms the state-of-the-art algorithms [3,4]. In particular, VBL outperforms the Lazy Linked List performance by 1.6× for 72 threads on the 20%-update workload of Fig. 1, which can be explained by the fact that our algorithm validates list data *before* locking, and not after. In addition, as our algorithm differs from Harris-Michael by avoiding meta-

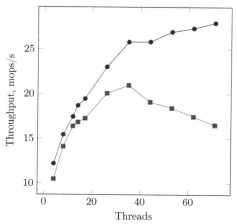

Fig. 1. The throughput of Lazy Linked List (red square curve) and VBL (blue circle curve). We consider the load with only 20% updates. Lazy Linked List behaves worse, as its operations potentially contend on meta-data even when they do not modify the data structure. (Color figure online)

data accesses during traversals, it outperforms it by up to 1.6× on read-only workloads. We report the performance of the Java variant of Harris-Michael list-based set with wait-free contains as presented in Shavit and Herlihy's book [4] and the Java optimised implementation with RTTI [3], and, in the technical report [10], on the performance of our own C++ translations of the Lazy algorithm (without memory management).

Roadmap. The rest of this paper is structured as follows. We present our methodology on modelling concurrency and prove the suboptimal concurrency of the Lazy and Harris-Michael linked lists in Sect. 2. In Sect. 3, we present our VBL list implementation. Section 4 presents the methodology for performance evaluation of concurrent list implementations and Sect. 5 presents a discussion of concurrency w.r.t list-based sets. The full proofs of linearizability and deadlock-freedom are deferred to the technical report [10]. Synchrobench benchmark suite [11] contains the code for all the lists considered in this paper.

---

[2] Here we adapt to list-based sets the notion of concurrency-optimality, introduced in [1] for generic search data structures.

2   Concurrency Analysis of List-Based Sets

2.1   Preliminaries

We consider a standard asynchronous shared-memory system, in which $n >$ 1 processes (or *threads* of computation) $p_1, \ldots, p_n$ communicate by applying operations on shared *objects*.

Sequential List-Based Set. An object of the *set* type stores a set of integer values, initially empty, and exports operations insert(v), remove(v), contains(v) where $v \in \mathbb{Z}$. The update operations, insert(v) and remove(v), return a boolean response, *true* if and only if v is absent (for insert(v)) or present (for remove(v)) in the list. After insert(v) is complete, v is present in the list, and after remove(v) is complete, v is absent from the list. The contains(v) returns a boolean *true* if and only if v is present in the list. The concurrent *set* implementations considered in this paper are based on a specific *sequential* one. The implementation, denoted *LL*, stores set elements in a *sorted linked list*, where each list node has a *next* field pointing to the successor node. Initially, the *next* field of the *head* node points to *tail*; *head* (resp. *tail*) is initialized with values $-\infty$ (resp., $+\infty$) that is smaller (resp., greater) than any other value in the list. We follow natural sequential implementations of operations insert, remove, and contains presented in detail in [10].

Executions. An *event* of a process p_i (we also say a *step* of p_i) is an invocation or response of an operation performed by p_i on a high-level object (in this paper, a set) implementation, or a *primitive* applied by p_i to a base object b along with its response. A *configuration* specifies the value of each base object and the state of each process. The *initial configuration* is the configuration in which all base objects have their initial values and all processes are in their initial states. An *execution fragment* is a (finite or infinite) sequence of events. An *execution* of an implementation I is an execution fragment where, starting from the initial configuration, each event is issued according to I and each response of a primitive matches the state of b resulting from all preceding events.

A *high-level history* $\tilde{H}$ of an execution α is the subsequence of α consisting of all invocations and responses of (high-level) operations.

Let $\alpha|p_i$ (resp. $H|p_i$) denote the subsequence of an execution α (resp. a histiry H) restricted to the events of process p_i. Executions α and α' (resp. histories H and H') are *equivalent* if for every process p_i, $\alpha|p_i = \alpha'|p_i$ (resp. $H|p_i = H'|p_i$). An operation π *precedes* another operation π' in an execution α (resp. history H), denoted $\pi \to_\alpha \pi'$ (resp., $\pi \to_H \pi'$) if the response of π occurs before the invocation of π' in α (resp. H). Two operations are *concurrent* if neither precedes the other.

An execution (resp. history) is *sequential* if it has no concurrent operations. An operation is *complete* in α if the invocation event is followed by a *matching* response; otherwise, it is *incomplete* in α. Execution α is *complete* if every operation is complete in α.

High-Level Histories and Linearizability. A complete high-level history $\tilde{H}$ is *linearizable* with respect to an object type τ if there exists a sequential high-level history S equivalent to $\tilde{H}$ such that (1) $\rightarrow_{\tilde{H}} \subseteq \rightarrow_S$ and (2) S *is consistent with the sequential specification of type* τ. Now a high-level history $\tilde{H}$ is linearizable if it can be *completed* (by adding matching responses to a subset of incomplete operations in $\tilde{H}$ and removing the rest) to a linearizable high-level history [8].

2.2   Concurrency as Admissible Schedules of Sequential Code

Schedules. Informally, a *schedule* of a list-based set algorithm specifies the order in which concurrent high-level operations access the list nodes. Consider the *sequential* implementation, LL, of operations insert, remove and contains. Suppose that we treat this implementation as a *concurrent* one, i.e., simply run it in a concurrent environment, without introducing any synchronization mechanisms, and let § denote the set of the resulting executions, we call them *schedules*. Of course, some schedules in § will not be linearizable. For example, concurrent inserts operating on the same list nodes may result in "lost updates": an inserted element disappears from the list due to a concurrent insert operation. But, intuitively, as no synchronization primitives are used, this (incorrect) implementation is as concurrent as it can get.

We measure the concurrency properties of a linearizable list-based set via its ability to accept all *correct* schedules in §. Intuitively, a schedule is correct if it respects the sequential implementation LL *locally*, i.e., no operation in it can distinguish the schedule from a sequential one. Furthermore, the schedule must be linearizable, even when we consider its extension in which all update operations are completed and followed with a contains(v) for any $v \in \mathbb{Z}$. Let us denote this extension of schedule σ by $\bar{\sigma}(v)$.

Given a schedule σ and an operation π, let $\sigma|\pi$ denote the subsequence of σ consisting of all steps of π.

Definition 1 (Correct schedules). *We say that a schedule σ of a concurrent list-based set implementation is* locally serializable *(with respect to the sequential implementation of list-based set LL) if for each of its operations π, there exists a sequential schedule S of LL such that $\sigma|\pi = S|\pi$. We say that a schedule is* correct *if (1) σ is locally serializable (with respect to LL), (2) for all $v \in \mathbb{Z}$, $\bar{\sigma}(v)$ is linearizable (with respect to the* set *type).*

Note that the last condition is necessary for filtering out schedules with "lost updates". Consider, for example a schedule in which insert(1) and insert(2) are applied to the initial empty set. Imagine that they first both read *head*, then both read *tail*, then both perform writes on the *head.next* and complete. The resulting schedule is, technically, linearizable and locally serializable but, obviously, not acceptable. However, in the schedule, one of the operations, say insert(1), overwrites the effect of the other one. Thus, if we extend the schedule with a complete execution of contains(2), the only possible response it may give is *false* which obviously does not produce a linearizable high-level history.

Note also that, as linearizability is a safety property [12], if $\bar{\sigma}(v)$ is linearizable, σ is linearizable too. (In the following we omit mentioning set and *LL* when we talk about local serializability and linearizability).

Concurrency-Optimality. A concurrent list-based set generally follows *LL*: every high-level operation, insert, remove, or contains, *reads* the list nodes, one after the other, until the desired fragment of the list is located. The update operation (insert or remove) then *writes*, to the *next* field of one of the nodes, the address of a new node (if it is insert) or the address of the node that follows the removed node in the list (if it is remove). Note that the (sequential) write can be implemented using a *CAS* primitive [5].

Let α denote an execution of a concurrent implementation of a list-based set. We define the *schedule σ exported by* α as the subsequence of α consisting of reads, writes and node creation events (corresponding to the sequential implementation *LL*) of operations insert, remove and contains that "take effect". Intuitively, taking effect means that they affect the outcome of some operation. The exact way an execution α is mapped to the corresponding schedule σ is implementation specific.

An implementation I *accepts* a schedule σ if there exists an execution of I that exports σ.

Definition 2 (Concurrency-optimality). *An implementation is* concurrency-optimal *if it accepts every correct schedule.*

2.3 Concurrency Analysis of the Lazy and Harris-Michael Linked Lists

In this section, we show that even state-of-the-art implementations of the list-based set, namely, the Lazy Linked List and the Harris-Michael Linked list are suboptimal w.r.t exploiting concurrency. We show that each of these two algorithms rejects some correct schedules of the list-based set.

Lazy Linked List. In this deadlock-free algorithm [3], the list is traversed in the wait-free manner and the locks are taken by update operations only when the desired interval of the list is located. A remove operation first marks a node for *logical* deletion and then *physically* unlinks it from the list. To take care of conflicting updates, the locked nodes are *validated*, which involves checking if they are not logically deleted. If validation fails, the traversal is repeated. The schedule of an execution of this algorithm is naturally derived by considering only the *last* traversal of an operation.

Figure 2 illustrates how the post-locking validation strategy employed by the Lazy Linked List makes it concurrency sub-optimal. As explained in the introduction, the insert operation of the Lazy Linked List acquires the lock on the nodes it writes to, prior to the check of the node's state.

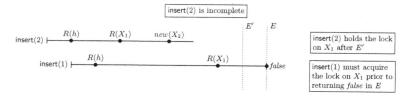

Fig. 2. A schedule rejected by the Lazy Linked List; initial list state is $\{X_1\}$ that stores value 1; $R(X_1)$ refers to reads of both *val* and *next* fields; $new(X_2)$ creates a new node storing value 2

Fig. 3. A schedule rejected by the Harris-Michael Linked List; the initial state of the list is $\{X_2, X_3, X_4\}$; each X_i stores value i; note that not all schedules are depicted for succinctness.

One can immediately see that the Lazy Linked List is not concurrency optimal. Indeed, consider the schedule depicted in Fig. 2. Two operations, insert(1) and insert(2) are concurrently applied to the list containing a single node X_1 storing value 1. Both operations first read h, the head of the list, then operation insert(2) reads node X_1 and creates a new node, X_2, storing 2. Immediately after that, operation insert(1) reads X_1 and returns *false*.

The schedule is correct: it is linearizable and locally serializable. However, it cannot be accepted by the Lazy Linked List, as insert(1) needs a lock on X_1 previously acquired by insert(2). Thus, the implementation is concurrency suboptimal: an operation may engage in synchronization mechanisms even if it is not going to update the list.

Harris-Michael Linked List. Like the Lazy Linked List, the *lock-free* Harris-Michael algorithm (cf. [4, Chap. 9]) separates logical deletion of a node from its physical removal (both steps use *CAS* primitives). If a CAS associated with logical deletion fails, the operation is restarted. Unlike the Lazy Linked List, however, if the physical removal fails (e.g., a concurrent update performed a successful CAS on the preceding node) the operation completes, and unlinking the logically deleted node from the list is then left for a future operation. Every update operation, as it traverses the list, attempts to physically remove such nodes. If the attempt fails, the operation is restarted. The delegation of physical removals to future operations is crucial for lock-freedom: an operation may only be restarted if there is a concurrent operation that took effect, i.e., global progress is made. But, as we show below, this delegation precludes some legitimate schedules.

Strictly speaking, this algorithm is not locally serializable with respect to the sequential implementation *LL*. Indeed, if a remove operation completes after logical deletion, we may not be able to map its steps to a write to a next field of the preceding node without "over-writing" a concurrent update. Therefore, for the sake of concurrency analysis, we consider a variant of LL in which remove operations only remove nodes *logically* and physical removals are put to the traversal procedure of future update operations. Now to define the schedule incurred by an execution of the algorithm, we consider the read and write steps that are part of the last traversal of an operation, node creation steps by insert operations, and successful logical deletions by remove operations. However, the Harris-Michael Linked List is not concurrency-optimal even with respect to this adjusted sequential specification.

Consider the schedule depicted in Fig. 3. Two operations, insert(1) and remove(2) are concurrently applied to the list containing three nodes, X_2, X_3 and X_4, storing values 2, 3 and 4, respectively. Note that operation remove(2) marks node X_2 for deletion but does not remove it physically by unlinking it from h. (Here we omit steps that are not important for the illustration.) Note that so far the schedule is accepted by the Harris-Michael algorithm: an earlier update of h by operation insert(1) causes the corresponding CAS primitive performed on h by remove(2) to fail.

After the operation completes, we schedule two concurrent operations, insert(4) and insert(3). Suppose that the two operations concurrently read head, X_1 and X_2. As they both witness X_2 to be marked for logical deletion, they both will try to physically remove it by modifying the next field of X_1. We let insert(3) to do it first and complete by reading X_3 and returning *false*. In the schedule depicted in Fig. 3, insert(4) also writes to X_1, and then successfully reads X_3 and X_4, and returns *false*. However, in the execution of the Harris-Michael algorithm, the attempt of insert(4) to physically remove X_2 will fail, causing it to restart traversing the list from the head. Thus, this schedule cannot be accepted.

3   The VBL List

In this section, we address the challenges of extracting maximum concurrency from list-based sets and present our VBL list. As we have shown in the previous section, an update in the Lazy Linked List acquires locks on nodes it is about to modify prior to checking the node's state. Thus, it may reject a correct schedule in which the operation does not modify the list. The schedule rejected by the Harris-Michael Linked List (Fig. 3) is a bit more intricate: it exploits the fact that Harris-Michael List involves *helping* which in turn induces additional synchronization steps leading to rejection of correct schedules.

Deriving a concurrency-optimal list requires introducing *value-based* node validation along with the logical-deletion technique. This observation inspired our *value-aware try-lock*.

3.1   Value-Aware Try-Lock

The class Node now contains the fields: (i) val for the value of the node; (ii) next providing a reference to the next node in the list; (iii) a boolean deleted to indicate a node to be marked for deletion and (iv) a lock to indicate a mutex associated with the node.

The value-aware try-lock supports the following operations:

(1) lockNextAt(Node node) first acquires the lock on the invoked node, checks if the node is marked for deletion or if the next field does not point to the node passed as an argument, then releases the lock and returns *false*; otherwise, the operation returns *true*.
(2) lockNextAtValue(V val) acquires the lock on the invoked node, checks if the node is marked for deletion or if the value of the next node is not val, then releases the lock and returns *false*; otherwise returns *true*.

3.2   VBL List

We now describe our VBL implementation. The list is initialized with 2 nodes: *head* (storing the minimum sentinel value) and *tail* (storing the maximum value), *head.next* stores the pointer to *tail*, both *deleted* flags are set to *false*. The pseudocode is presented in Fig. 1.

Contains. The contains(v) algorithm starts from the *head* node and follows *next* pointers until it finds a node with the value that is equal to or bigger than v. Then, the algorithm simply compares the value in the found node with v.

Inserting a Node. The algorithm of insert(v) starts with the traversal (Line 24) to find a pair of nodes ⟨*prev, curr*⟩ such that *prev.val* is less than v and *curr.val* is equal to or bigger than v. The traversal is simple: it starts from *head* and traverses the list in a wait-free manner until it finds the desired nodes. If *curr.val* is equal to v (Line 25) then there is no need to insert. Otherwise, the new node with value v should be between *prev* and *curr*. We create a node with value v (Lines 26–27). Then, the algorithm locks *prev* and checks that it still can insert the node correctly (Line 28): *prev.next* still equals to *curr* and *prev* is not marked as deleted. If both of these conditions are satisfied, the new node can be linked. Otherwise, it cannot: the correctness of the algorithm (namely, linearizability) would be violated; so the operation restarts from the traversal (Line 24). Note that to improve the performance, the algorithm starts the traversal not from *head* but from *prev*.

Removing a Node. The algorithm of remove(v) follows the lines of insert(v): first it finds the desired pair of nodes ⟨*prev, curr*⟩. If *curr.val* is not equal to v then there is nothing to remove (Line 36). Otherwise, the algorithm has to remove the node with value v. At first, it takes the lock on *prev* and checks two conditions (Line 39): *prev.next.val* equals to v and *prev* is not marked as deleted. The first condition ensures concurrency-optimality by taking care of the scenario described above: one could have removed and inserted v while the thread was asleep. The second condition is necessary to guarantee correctness, i.e., the node *next* is not linked to *deleted* node, which might result in a "lost update" scenario. If any of the conditions is violated, the algorithm restarts from

Algorithm 1. VBL list

```
 1: Shared variables:
 2:    head.val ← −∞
 3:    tail.val ← +∞
 4:    head.next ← tail
 5:    head.deleted ← false
 6:    tail.deleted ← false
 7:    head.lock ← new Lock()
 8:    tail.lock ← new Lock()

 9: contains(v):
10:    curr ← head
11:    while curr.val < v do
12:        curr ← curr.next
13:    return curr.val = v

14: waitfreeTraversal(v, prev):
15:    if prev.deleted then
16:        prev ← head
17:    curr ← prev.next
18:    while curr.val < v do
19:        prev ← curr
20:        curr ← curr.next
21:    return ⟨prev, curr⟩
```

```
22: insert(v):
23:    prev ← head
24:    ⟨prev, curr⟩ ← waitfreeTraversal(v, prev)
25:    if curr.val = v then return false
26:    newNode.val ← v
27:    newNode.next ← curr
28:    if not prev.lockNextAt(curr) then
29:        goto Line 24
30:    prev.next ← newNode
31:    prev.lock.unlock()
32:    return true

33: remove(v):
34:    prev ← head
35:    ⟨prev, curr⟩ ← waitfreeTraversal(v, prev)
36:    if curr.val ≠ v then
37:        return false
38:    next ← curr.next
39:    if not prev.lockNextAtValue(v) then goto Line 35
40:    curr = prev.next
41:    if not curr.lockNextAt(next) then
42:        prev.unlock()
43:        goto Line 35
44:    curr.deleted ← true
45:    prev.next ← curr.next
46:    curr.lock.unlock()
47:    prev.lock.unlock()
48:    return true
```

Line 35. Then, the algorithm takes the lock on $curr = prev.next$ and checks a condition $curr.next$ equals to $next$ in Line 41 (note that the second condition is satisfied by the lock on $prev$ as $curr$ is not marked as deleted). This condition ensures correctness: otherwise, the link $next$ to $prev$ will be incorrect. If it is not satisfied, the algorithm restarts from Line 35. Afterwards, the algorithm sets $curr.deleted$ to $true$ (Line 44) and unlinks $curr$ (Line 45).

Correctness. We show that the VBL list accepts only correct schedules of the list-based set. We then show that the VBL list accepts *every* correct schedule of the list-based set, thus establishing its concurrency-optimality.

Theorem 1. *Every schedule of the VBL list is linearizable w.r.t the set.*

The full proof is deferred to the companion technical report. Observe that the only nontrivial case to analyse for proving deadlock-freedom is the execution of the update operations. Suppose that an update operation π fails to return a matching response after taking infinitely many steps. However, this means that there exists a concurrent insert or remove that successfully acquires its locks and completes its operation, thus implying progress for at least one correct process.

Theorem 2. *The* VBL *implementation accepts only correct list-based set schedules locally serializable (wrt LL).*

Concurrency-Optimality. We prove that the VBL accepts every correct interleaving of the sequential code. The goal is to show that any finite schedule rejected by our algorithm is not correct. Recall that a correct schedule σ is locally serializable and, when extended with all its update operations completed and contains(v), for any $v \in \mathbb{Z}$, we obtain a linearizable schedule.

Note that given a correct schedule, we can define the contents of the list from the order of the schedule's write operations. For each node that has ever been created in this schedule, we derive the resulting state of its *next* field from the last write in the schedule. Since in a correct schedule each new node is first created and then linked to the list, we can reconstruct the *state of the list* by iteratively traversing it, starting from the *head*.

Theorem 3 (Optimality). VBL *implementation accepts all correct schedules.*

4   Experimental Evaluation

Experimental Setup. In this section, we compare the performance of our solution to two state-of-the-art list-based set algorithms written in different languages (Java and C++) and on two multicore machines from different manufacturers: A 4-socket Intel Xeon Gold 6150 2.7 GHz server (Intel) with 18 cores per socket (yielding 72 cores in total), 512 Gb of RAM, running Debian 9.9. This machine has OpenJDK 11.0.3; A 4-socket AMD Opteron 6276 2.3 GHz server (AMD) with 16 cores per socket (yielding 64 cores in total), running Ubuntu 14.04. This machine has OpenJDK 1.8.0_222 (We delegate the AMD results to the tech report).

Concurrent List Implementations. We compared our VBL algorithm (VBL) to the lock-based Lazy Linked List (Lazy) [3] and Harris-Michael's non-blocking list (Harris-Michael) [5,6] with its wait-free and RTTI optimization suggested by Heller et al. [3] using the Synchrobench benchmark suite [11]. To compare these algorithms on the same ground we primarily used Java as it is agnostic of the underlying set up. The evaluation of the C++ implementations of these algorithms is deferred to the companion technical report [10]. The code of the implementations is part of Synchrobench at https://github.com/gramoli/synchrobench.

Experimental Methodology. We considered the following parameters:

- **Workloads.** Each workload distribution is characterized by the percent $x\%$ of update operations. This means that the list will be requested to make $(100 - x)\%$ of contains calls, $x/2\%$ of insert calls and $x/2\%$ of remove calls. We considered three different workload distribution: 0%, 20%, and 100%. Percentages 0% and 100% were chosen as the extreme workloads, while 20% update ratio corresponds to the standard load on databases. Each operation contains, insert, and remove chooses its argument uniformly at random from the fixed key range.
- **List size.** On the workloads described above, the size of the list depends on the range from which the operations take the arguments. Under the described workload the size of the list is approximately equal to the half of the key range. We consider four different key ranges: 50, 200, $2 \cdot 10^3$, and $2 \cdot 10^4$. To ensure consistent results we pre-populated the list: each element is present with probability $\frac{1}{2}$.
- **Degree of contention.** This depends on the number of cores in a machine. We take enough points to reason about the behavior of the curves.

Results. We run experiments for each workload 5 times for 5 s with a warm-up of 5 s. Figure 4 contains the results on Intel machine. Our new list algorithm outperforms both Harris-Michael's and the Lazy Linked List algorithms, and remains scalable except for the situation with very high contention, i.e., high update ratio with small range. We find this behavior normal at least in our case, since the processes contend to get the cache-lines in exclusive mode and this traffic becomes the dominant factor of performance in the execution.

Comparison Against Harris-Michael. Harris-Michael's algorithm in general scales well and performs well under high contention. Even though the three algorithms feature the wait-free contains, our original implementation of the Harris-Michael's contains was slower than the other two. The reason is the extra indirection needed when reading the *next* pointer in the combined *pointer-plus-boolean* structure. To avoid reading an extra field when fetching the Java AtomicMarkableReference we implemented the run-time type identification (RTTI) variant with two subclasses that inherit from a parent node class and that represent the marked and unmarked states of the node as previously suggested [3]. This optimization requires, on the one hand, that a remove casts the subclass instance to the parent class to create a corresponding node in the marked state. It allows, on the other hand, the traversal to simply check the mark of each node by simply invoking instanceof on it to check the subclass the node instantiates. As we see, Harris-Michael's algorithm has very efficient updates because it only uses CAS, however it spends much longer on list traversals.

Comparison Against the Lazy Linked List. The Lazy Linked List has almost the same performance as our algorithm under low contention because both algorithms share the same wait-free list traversal with zero overhead (as the sequential code does) and for the updates, when there is no interference

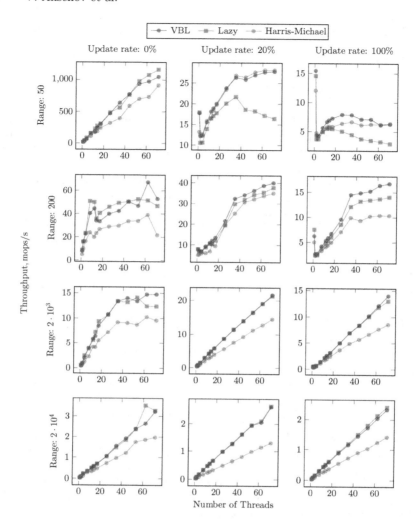

Fig. 4. Evaluation on Intel

from concurrent operations, the difference between the two algorithms becomes negligible. The difference comes back however as the contention grows. The Lazy Linked List performance drops significantly due to its intense lock competition (as briefly explained in Sect. 1). By contrast, there are several features in our implementation that reduce significantly the amount of contention on the locks. We observed a tremendous increase in execution time for the Lazy Linked List because of the contention on locks.

5   Related Work and Concluding Remarks

List-Based Sets. Heller *et al.* [3] proposed the Lazy Linked List and mentioned the option of validating prior to locking, and using a single lock within an insert. One of the reasons why our implementation is faster than the Lazy Linked List is the use of a novel value-aware try-lock mechanism that allows validating before acquiring the lock.

Harris [5] proposed a non-blocking linked list algorithm that splits the removal of a node into two atomic steps: a logical deletion that marks the node and a physical removal that unlinks the node from the list. Michael [6] proposed advanced memory reclamation algorithms for the algorithm of Harris. In our implementation, we rely on Java's garbage collector for memory reclamation [13]. We believe that our implementation could outperform Michael's variant for the same reason it outperforms Harris' one because it does not combine the logical deletion mark with the next pointer of a node but separates metadata (logical deletion and versions) from the structural data (check [4] for variants of these list-based sets). Fomitchev and Ruppert [14] proposed a lock-free linked list where nodes have a backlink field that allows to backtrack in the list in case a conflict is detected instead of restarting from the beginning of the list. Its *contains* operation also helps remove marked nodes from the list. Gibson and Gramoli [15] proposed the *selfish linked list*, as a more efficient variant of this approach with the same amortized complexity, relying on wait-free *contains* operations. These algorithms are, however, not concurrency-optimal: schedule constructions similar to those outlined for the Harris-Michael and Lazy linked lists apply here.

Concurrency Metrics. Sets of accepted schedules are commonly used as a metric of concurrency provided by a shared-memory implementation. For static database transactions, Kung and Papadimitriou [16] use the metric to capture the parallelism of a locking scheme. While acknowledging that the metric is theoretical, they insist that it may have "practical significance as well, if the schedulers in question have relatively small scheduling times as compared with waiting and execution times". Herlihy [17] employed the metric from [16] to compare various optimistic and pessimistic synchronization techniques using commutativity [18] of operations constituting high-level transactions. A synchronization technique is implicitly considered in [17] as highly concurrent, namely "optimal", if no other technique accepts more schedules. In contrast to [16,17], we focus here on a *dynamic* model where the scheduler cannot use the prior knowledge of all the shared addresses to be accessed.

Optimal concurrency, originally introduced in [1], can also be seen as a variant of metrics like *permissiveness* [19] and *input acceptance* [20] defined for transactional memory. The concurrency framework considered in this paper though is independent of the synchronization technique and, thus, more general. Our notion of local seriazability, also introduced in [1], is also reminiscent to the notion of *local linearizability* [21].

Concurrent interleavings of *sequential* code has been used as a base-line for evaluating performance of search data structures [22]. Defining *optimal concurrency* as the ability of accepting *all* correct interleavings has been originally proposed and used to compare concurrency properties of optimistic and pessimistic techniques in [1].

The Case for Concurrency-Optimal Data Structures. Intuitively, the ability of an implementation to successfully process interleaving steps of concurrent threads is an appealing property that should be met by performance gains.

In this paper, we support this intuition by presenting a concurrency-optimal list-based set that outperforms (less concurrent) state-of-the-art algorithms. Does the claim also hold for other data structures? We believe that generalizations of linked lists, such as skip-lists or tree-based dictionaries, may allow for optimizations similar to the ones proposed in this paper. The recently proposed concurrency-optimal tree-based dictionary [9] justifies this belief. This work presents the opportunity to construct a rigorous methodology for deriving concurrency-optimal data structures that also perform well.

Also, there is an interesting intermingling between progress conditions, concurrency properties, and performance. For example, the Harris-Michael algorithm is superior with respect to both the Lazy Linked List and **VBL** in terms of progress (lock-freedom is a strictly stronger progress condition than deadlock-freedom). However, as we observe, this superiority does not necessarily imply better performance. Improving concurrency seems to provide more performance benefits than boosting liveness. Relating concurrency and progress in concurrent data structures remains an interesting research direction.

References

1. Gramoli, V., Kuznetsov, P., Ravi, S.: In the search for optimal concurrency. In: Suomela, J. (ed.) SIROCCO 2016. LNCS, vol. 9988, pp. 143–158. Springer, Cham (2016). https://doi.org/10.1007/978-3-319-48314-6_10
2. Sutter, H.: Choose concurrency-friendly data structures. Dr. Dobb's J. (2008)
3. Heller, S., Herlihy, M., Luchangco, V., Moir, M., Scherer, W.N., Shavit, N.: A lazy concurrent list-based set algorithm. In: Anderson, J.H., Prencipe, G., Wattenhofer, R. (eds.) OPODIS 2005. LNCS, vol. 3974, pp. 3–16. Springer, Heidelberg (2006). https://doi.org/10.1007/11795490_3
4. Herlihy, M., Shavit, N.: The Art of Multiprocessor Programming. Morgan Kaufmann (2012)
5. Harris, T.L.: A pragmatic implementation of non-blocking linked-lists. In: Welch, J. (ed.) DISC 2001. LNCS, vol. 2180, pp. 300–314. Springer, Heidelberg (2001). https://doi.org/10.1007/3-540-45414-4_21
6. Michael, M.M.: High performance dynamic lock-free hash tables and list-based sets. In: SPAA, pp. 73–82 (2002)
7. Guerraoui, R., Kapalka, M.: Principles of Transactional Memory, Synthesis Lectures on Distributed Computing Theory. Morgan and Claypool (2010)
8. Herlihy, M., Wing, J.M.: Linearizability: a correctness condition for concurrent objects. ACM Trans. Program. Lang. Syst. **12**(3), 463–492 (1990)

9. Aksenov, V., Gramoli, V., Kuznetsov, P., Malova, A., Ravi, S.: A concurrency-optimal binary search tree. In: Rivera, F.F., Pena, T.F., Cabaleiro, J.C. (eds.) Euro-Par 2017. LNCS, vol. 10417, pp. 580–593. Springer, Cham (2017). https://doi.org/10.1007/978-3-319-64203-1_42
10. Aksenov, V., Gramoli, V., Kuznetsov, P., Ravi, S., Shang, D.: A concurrency-optimal list-based set. CoRR abs/1502.01633 (2021)
11. Gramoli, V.: More than you ever wanted to know about synchronization: synchrobench, measuring the impact of the synchronization on concurrent algorithms. In: PPoPP, pp. 1–10 (2015)
12. Lynch, N.A.: Distributed Algorithms. Morgan Kaufmann (1996)
13. Sun Microsystems: Memory Management in the Java HotSpot Virtual Machine, April 2006. http://www.oracle.com/technetwork/java/javase/memorymanagement-whitepaper-150215.pdf
14. Fomitchev, M., Ruppert, E.: Lock-free linked lists and skip lists. In: PODC, pp. 50–59 (2004)
15. Gibson, J., Gramoli, V.: Why non-blocking operations should be selfish. In: Moses, Y. (ed.) DISC 2015. LNCS, vol. 9363, pp. 200–214. Springer, Heidelberg (2015). https://doi.org/10.1007/978-3-662-48653-5_14
16. Kung, H.T., Papadimitriou, C.H.: An optimality theory of concurrency control for databases. In: SIGMOD, pp. 116–126 (1979)
17. Herlihy, M.: Apologizing versus asking permission: optimistic concurrency control for abstract data types. ACM Trans. Database Syst. 15(1), 96–124 (1990)
18. Weihl, W.E.: Commutativity-based concurrency control for abstract data types. IEEE Trans. Comput. 37(12), 1488–1505 (1988)
19. Guerraoui, R., Henzinger, T.A., Singh, V.: Permissiveness in transactional memories. In: Taubenfeld, G. (ed.) DISC 2008. LNCS, vol. 5218, pp. 305–319. Springer, Heidelberg (2008). https://doi.org/10.1007/978-3-540-87779-0_21
20. Gramoli, V., Harmanci, D., Felber, P.: On the input acceptance of transactional memory. Parallel Process. Lett. 20(1), 31–50 (2010)
21. Haas, A., et al.: Local linearizability for concurrent container-type data structures. In: CONCUR 2016, vol. 59, pp. 6:1–6:15 (2016)
22. David, T., Guerraoui, R., Trigonakis, V.: Asynchronized concurrency: the secret to scaling concurrent search data structures. In: ASPLOS, pp. 631–644 (2015)

Mobile Agents Operating on a Grid: Some Non-conventional Issues in Parallel Computing

Fabrizio Luccio and Linda Pagli$^{(\boxtimes)}$

Department of Informatics, University of Pisa, Pisa, Italy
{luccio,pagli}@unipi.it

Abstract. This paper is a contribution to studying the parallel operation of moving agents on a grid where some targets have to be reached and attended, as a strong generalization and improvement of a previous results. In particular the well-known problems of *cops and robber* and *sparce sensor networks* are considered as applicative examples of this new computing scheme. We assume that the set of targets may change in time with new targets arising while previous ones are being taken care of. An evolving scheme is considered where all targets needing attention are finally served, or their number may never end but is constantly kept under a fixed limit. Mathematical expressions for the number of agents and their processing time are given as a function of various parameters of the problems, and their relation is studied under the concepts of *work* and *speed-up* inherited from parallel processing.

Keywords: Grid · Agent · Sensor · Cop · Robber · Travel time · Computation time · Work · Speed-up

1 Rules of the Game

Mobile agents operating on a two-dimensional grid have been extensively studied as a mathematical processing model in a variety of settings and applications. In the present work we focus on the two problems of harm detection in the so-called cops and robber chase, and of sparse sensor networks management via mules. In both cases mobile *agents* (cops or mules) travel on a grid for reaching specific *target* nodes (robbers or sensors) where an operation is needed as contrasting a danger or attending a sensor. The agents follow a parallel processing paradigm where several agents operate concurrently.

There exists an extremely rich literature on these subjects where the operations are conducted on a grid, see for example the classical review [1] on the cops and robber problem, the implications of using more cops than necessary [2], and the concurrent operations of several cops and robbers [6]; and the recent survey

Partially supported by MIUR of Italy under Grant 20174LF3T8 AHeAD: efficient Algorithms for Harnessing networked Data.

V. Malyshkin (Ed.): PaCT 2021, LNCS 12942, pp. 402–410, 2021.
https://doi.org/10.1007/978-3-030-86359-3_30

on sensor network applications [9]. But, surprisingly, some specific character-istics of this type of precessing, and of the problems arising, have never been considered. In particular we study:

1. The division of the processing time in two phases with different character-istics, namely the *travel time* of an agent to reach its destination, and the succeeding *processing time*.
2. The influence on the overall process of the *relative numbers* of agents and targets. Inheriting some concept from parallel processing [5], we define the *work* w_k of a process carried out by k agents in time t_k as $w_k = k \cdot t_k$, and the *speed-up* between the actions of j over $i < j$ agents as w_i/w_j. If $w_i/w_j > 1$ the speed-up is said to be *super-linear*, a case that may occur only in special circumstances, e.g. see [7].
3. The evolution of the system if the set of targets needing attention changes continuously in time, with new targets arising while previous targets are being taken care of. This phenomenon will be treated in a continuously evolving scheme where the number of targets needing attention may never end but is constantly kept under a fixed limit.
4. (less important) For a better balancing of the agent movements we work on a grid bounded by a *diamond* of *side n*, wth n odd; all the well-known results and algorithms reported in the literature with a rectangular boundary are valid with obvious minor transformations.

2   Several Cops Capture Several Robbers

The standard cops and robber problem is typically aimed at finding the mini-mal number of cops needed to reach a moving robber on a graph, contextually studying the movements involved, see [1]. Cops and robber move alternatively, one edge a time, until a cop reaches the robber's vertex. The problem has been thoroughly studied on rectangular grids.

Here we assume that k cops and l robbers are present, with arbitrary k and l. All move on a diamond grid, and all robbers must be reached by a cop. To the best of our knowledge this situation has been studied only in [4] with $k \geq l \geq 1$, for deciding if the capture of all robbers is possible on a digraph. Our approach, aim, and graph traveled, are completely different. In particular we refer to the capture of more than one robber, judging the possible advantage of increasing the number of cops.

We may assume that the cops start in the central node of the diamond as the most favourable position to reach any robber that may appear and move around according to the problem rules to avoid being reached. The optimal strategy for all actors is the one already known for the problem on the rectangular grid, with alternative moves of a pair of cops against one robber until this is pushed to the diamond boundary in a *siege* condition (e.g., see [6]), to be inevitably reached at the next step. This situation is depicted in Fig. 1. It can be easily shown that the standard algorithm allows the capture in $(n + 1)/2$ moves on the diamond, which is also a lower bound on the capture time.

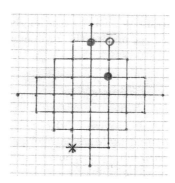

Fig. 1. A diamond with $n = 9$. Two cops (black dots) are initially placed in the center of the diamond and push the robber (white dot) to the farthest *siege* shown, on the diamond boundary. The robber manages to escape as far as possible, still it cannot avoid being pushed in a siege node at a distance $(n + 1)/2$ from the center, in at most $(n + 1)/2$ moves. The star indicates the position of a next robber, as far as possible from the cops.

Now take arbitrary values k and l, with k even as cops chase robbers in pairs. To start consider $k = 2$ and $l = 2$, with the two robbers captured one after the other. To delay the chase as long as possible, the first robber will move to the siege shown in Fig. 1, and the second will be captured in the starred vertex. The total chase *time* $t_{k,l} = t_{2,2}$ is given by $(n + 1)/2$ moves to catch the first robber, plus $n - 1$ moves for reaching the second robber from the previous siege position, with a total of $t_{2,2} = (3n - 1)/2$ and work $w_{2,2} = 3n - 1$.

If four cops chase the same two robbers, $k = 4$ and $l = 2$, two pairs of cops push the robbers into two sieges in parallel, in total time $t_{4,2} = (n + 1)/2$ and work $w_{4,2} = 2n + 2$. We then observe a *remarkable super-linear speed-up* of $(3n - 1)/(2n + 2)$ that tends to $3/2$ for increasing grid dimension n. This advantage with using more cops depends on the necessity or the smaller team to regain a better position in the grid before attacking the second robber.

We are in front of a new situation in parallel processing, where the running time of an algorithm includes a part for getting to the data before a real processing stage takes place, that in the cops and robber problem consists in invading the robber's cell. This latter time is assumed to be null. If we add a capturing time τ once the robber is reached, we must rewrite $t_{2,2} = (3n - 1)/2 + 2\tau$, $w_{2,2} = 3n - 1 + 4\tau$, and $t_{4,2} = (n + 1)/2 + \tau$, $w_{4,2} = 2n + 2 + 4\tau$, with a lower speed-up than before, that in fact tends to 1 if the value of τ prevails over n.

With similar computation we find $t_{2,3} = (5n - 3)/2 + 3\tau$ and $t_{4,3} = n + 1 + 2\tau$; $t_{2,4} = (7n - 5)/2 + 4\tau$ and $t_{4,4} = (3n - 1)/2 + 2\tau$; $t_{2,5} = (9n - 7)/2 + 5\tau$ and $t_{4,5} = 2n + 3\tau$, and so on. We can prove that, for $l \geq 2$, the general law for $k = 2$ versus $k = 4$ is the following:

$$t_{2,l} = ((2l - 1)n - 2l + 3)/2 + l\tau, \tag{1.1}$$

$$t_{4,l} = ((l-1)n - l + 3)/2 + \lceil l/2 \rceil \tau, \text{ for } l \text{ even} \tag{1.2}$$

$$t_{4,l} = ((l-1)n - l + 5)/2 + \lceil l/2 \rceil \tau, \text{ for } l \text{ odd} \tag{1.3}$$

where a penalty occurs in $t_{4,l}$ for l odd (see (1.2) and (1.3)) due to the chase of four cops on a single robber in one of the rounds (a companion phenomenon may cause *slow-down* in parallel computation, see [5]).

We note that the speed up $w_{2,l}/w_{4,l}$ tends to $(n(2l-1)+2l\tau)/(n(2l-2)+2l\tau)$, so it is always super-linear although its value decreases for increasing l and τ.

General results similar to the ones of relations (1.1), (1.2), and (1.3) can be proved for arbitrary values of k, l. We do not insist on this as our goal was to enlighten this kind of effect before passing to a more intriguing situation.

3   Chasing a Continuous Stream of Invaders

The cops and robber problem can be seen an a paradigm of ridding of harmful moving invaders a region often represented as a two-dimensional grid. The invaders try to resist as long as possible and the longest chase inevitably ends on the region boundary.

Despite the very rich literature on this subject, nobody seems to have extended these studies to the practical case where new invaders (i.e., robbers) appear while the patrolling agents (the cops) are chasing the original ones. Now k is a constant and l is a function of time. A new paradigm for this case must be set up, with many possible variations two of which are considered below. While the standard chasing rules will be maintained, a new relation among the values of k and l is going to be a key parameter of any chasing algorithm.

3.1   Herd Immunity from Invaders

The first computing scheme assumes that the cops must reach a time in which all the robbers appeared thus far have been captured, so that a "herd immunity" is attained for the grid. Let us assume that the appearance law for the robbers takes one of the two forms:

$$l = l_0 + \lambda t, \tag{2.1}$$

$$l = l_0 + \lambda n t, \tag{2.2}$$

where t is the total time to get the immunity, l_0 (the initial number of robbers) and λ are constant, and the incoming number of new robbers is proportional only to the time t, relation (2.1); or to the time t and to the grid side n, relation (2.2), assuming that the intruders come into the region through its boundary. Clearly other cases could be considered.

Referring to the standard algorithm with pairs of cops chasing one robber, the time t can be evaluated noting that the k cops initially chase $k/2$ robbers in $(n+1)/2$ steps; then $k/2$ pairs of cops chase the remaining $l - k/2$ robbers in $(l - k/2)/(k/2)$ consecutive stages requiring $(n-1)$ steps each. That is: $t = (n+1)/2 + (n-1)(l-k/2)/(k/2)$. Substituting the value (2.1) for l, and grouping the terms containing t, we easily find :

$$t = (2l_0(n-1) - k(n-3)/2) / (k - 2\lambda(n-1)), \text{ for } \lambda t \text{ new robbers.} \quad (3.1)$$

This is an interesting expression. First note that the denominator must be greater than zero, that is, once n and λ are fixed a minimum number of cops $k > 2\lambda(n-1)$ is established for getting the herd immunity, with the time sharply increasing for k approaching $2\lambda(n-1)$. Second, the nominator is positive for $l_0 > k(n-3)/4(n-1)$ to avoid that the herd immunity is trivially reached in the chasing phase to the l_0 robbers.

Finally relation (3.1) has the structure $(A - kB)/(k - C)$ with $A > kB$ and $k > C$. If αk cops are used instead of k, with $\alpha > 1$, the works for k and αk are respectively $w_k = (kA - k^2B)/(k - C)$ and $w_{\alpha k} = (\alpha kA - \alpha^2 k^2 B)/(\alpha k - C)$. With easy computation we find a speed-up $w_k/w_{\alpha k} > 1$, consistent with the relation $A > kB$.

If we substitute the value (2.2) for l in the expression for t we have:

$$t = (2l_0(n-1) - k(n-1)) / (k - 2\lambda n(n-1)), \text{ for } \lambda nt \text{ new robbers} \quad (3.2)$$

yielding results quite similar to the ones found for expression (3.1). In particular the condition $k > 2\lambda(n-1)$ holds and a super linear speed-up occurs as before. However we must impose the much stronger condition $k > 2\lambda n(n-1)$ on the denominator, implying that the number of cops must be proportional to the area (number of vertices) of the grid instead of to the length of its boundary.

If a capturing time τ is added in each parallel capture, expressions (3.1) and (3.2) are respectively rewritten as:

$$t = (2l_0(n-1+\tau) - k(n-3)/2) / (k - 2\lambda(n-1+\tau)) \quad (4.1)$$

$$t = (2l_0(n-1+\tau) - k(n-1)) / (k - 2\lambda n(n-1+\tau)), \quad (4.2)$$

with similar considerations on time of convergence and speed-up.

3.2   Controlling the Size of an Everlasting Invasion

A new relevant situation is tolerating an everlasting intruder invasion if the number of invaders is constantly kept under control. As an example, still consider the previous invader arrival laws (2.1) and (2.2) and assume that the number of active invaders be kept at its initial value l_0. The problem can be more easily solved than one may expect.

Let $l = l_0 + \lambda t$, and recall that $(n+1)/2$ steps are needed for capturing one or more intruders in parallel. Dividing the time in slots of $(n+1)/2$ steps, $\lambda(n+1)/2$ invaders will be captured by k cops, that is $\lambda(n+1)/2 = k/2$ since k cops can capture $k/2$ intruders in one slot of time. We let:

$$k = \lambda(n+1) \quad (5.1)$$

to remain with exactly l_0 active invaders at the end of the time slot; and the process will go on forever in the succeeding time slots. Similarly, under the arrival law $l = l_0 + \lambda nt$ we have:

$$k = \lambda n(n+1) \quad (5.2)$$

that requires a much higher number of cops for maintaining the number l_0 of active intruders.

In both cases a number of cops smaller than the ones in (5.1), (5.2) will cause the number of active intruders to constantly increase, and a greater number of cops will bring the number of active intruders to an end as studied in the previous sub-section.

If a capturing time τ is added in each parallel capture the time slot becomes $(n+1)/2 + \tau$ and relations (5.1), (5.2) respectively become $k = \lambda(n+1) + 2\tau$ and $k = \lambda n(n+1) + 2\tau$.

Note that the issue of speed-up as a function of the number k of cops has no relevance here since k is fixed.

4   Data Collection in a Sparse Sensor Network

A variety of applications have been recently developed for environmental monitoring, or for regional activity control in a broad sense, where the "region" may be described as a grid with a large network of inexpensive wireless *sensors* deployed in the nodes, e.g. see [10–12] and the survey [9]. The sensors have the role of collecting data on the surrounding environment to be communicated to a central *access point*, however they have enough power for transmitting just a prompt signal when they have data to deliver, while the data are to be collected with other means. Typically some specialised *agents* also called *mules*, e.g. robots, are sent back and forth from the access point to do the collecting job. This is the setting considered here.

Our sensor network scheme has many characteristics in common with the cops and robber setting, where agents stand for cops and sensors stand for robbers. We deploy the network on a diamond grid of side n where the sensors are uniformly scattered, while the access point and the agent warehouse are placed in the central node, see Fig. 2. Only the agents now move along the grid edges while the sensors are steady.

Excluding the center, the number of grid nodes is $N = 2n^2 - 2n$, and the sum of all the node distances from the center is $2n(2n^2 - 3n + 1)/3$; so the average distance of a sensor from the access point is less than $2n/3$ and tends to this value for $n \to \infty$. Then $\beta = 4n/3$ will be taken as an upper bound for the *travel time* of an agent, that is the expected back and forth time for visiting a sensor and return to the access point.

As before let k be the number of agents. Let αN be the expected number of *active* sensors, meaning the ones requiring attention at a given moment, and let $\bar{t}$ be the expected time before a new batch of active sensor arise. Each subset of $\alpha N/k$ sensors will be assigned do a different agent.

In addition, in the present case we cannot ignore the *processing time* τ of an agent collecting data from a sensor: we consider the two cases:

$$\tau = \gamma \text{ constant,} \qquad (6.1)$$

$$\tau = \gamma t, \text{ with } \gamma \text{ constant and time t,} \qquad (6.2)$$

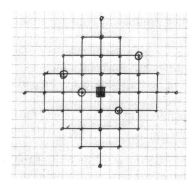

Fig. 2. A diamond grid with $n = 9$ with four sensors (white dot). The access point with agents warehouse is in the center (black square) and the sensors are at a distance 1, 2, 3, and 4 from it. The average distance of a node from the center is $2n/3$ for $n \to \infty$.

where in case (6.2) t is the elapsed time between the request of attention of a sensor and the moment in which an agent arrives, imagining the the sensor have been collecting data in between. The problem is now determining the number k of agents needed in the average to gather data continuously from the active sensors, with the number of the latter kept under control. More specifically, maintaining this number at αN. In this continuous stream of operations we assume that a sensor request of attention arrives when an agent has just initiated its return to the access point, so a time $\beta = 4n/3$ is required to the agent to reach a new sensor.

The following mathematical development must be taken with a bit of caution because time will be measured as a number of agent steps while some of the numerical figures to be used, as α, β and γ, may not be integers. The results that will be found must then be merely considered as indicative of the influence of the different parameters, while getting precise values would imply the use of floor and ceiling rounding thereby complicating the computation without any real advantage for our present purpose.

Consider an agent visiting the $\alpha N/k$ active sensors assigned to it under condition (6.1). The travel time to reach the first sensor is $\beta = 4n/3$, followed by a processing time γ to collect the data. Then the agent travels back to the access point to deposit the data, and then travels again to the next active agent. The process goes on with the data of the i-th sensor processed after $i(\beta + \gamma)$ steps, so that the whole batch of active sensors is visited in time $t = \alpha N(\beta + \gamma)/k$. Imposing $t \leq \bar{t}$ and noting that $N < 2n^2$, we have:

$$k \geq 2\alpha n^2(4n + 3\gamma)/3\bar{t}. \tag{7.1}$$

with the integer ceiling of the above expression giving a number of agents that can do the job.

At a first glance relation (7.1) seems to be absurd, as the number k of agents appears to be of order n^3, that is, there should be more agents than grid nodes for $n \to \infty$. Obviously the result is justified on practical bases by the low value of α and the high value of $\bar{t}$. However an asymptotic study of this figures is not significant as it will be explained in the next section.

Under condition (6.2) the situation is more intriguing. The agent meets the first sensor after a travel time β and processes its data in time $\gamma\beta$, then leaving the sensor after $\beta(1+\gamma) = \beta\,\Gamma$ steps, where for simplicity we denote $1+\gamma$ by Γ. The second sensor is then reached in $\beta\,\Gamma + \beta = \beta\,(\Gamma+1)$ steps, and left after additional $\gamma\beta\,(\Gamma+1)$ steps, that is at time $\beta\,(\Gamma+1)(\gamma+1) = \beta\,(\Gamma^2+\Gamma)$. The process goes on, with the data of the i-th sensor processed after $\beta\,(\Gamma^i + \Gamma^{i-1} + \cdots + \Gamma)$ steps, and the whole batch of $l = \alpha N/k$ active sensors visited in total time $t = \beta\,(\Gamma^l + \Gamma^{l-1} + \ldots \Gamma) = \beta\,(\Gamma^l - \Gamma)/(\Gamma-1) = \beta\,((\gamma+1)^l - (\gamma+1))/\gamma$. Then we have $t < \beta\,(\gamma+1)^l/\gamma$, and with proper substitutions we pose :

$$t < 4n\,(\gamma+1)^{2\alpha n^2/k}/3\gamma \le \bar{t}.$$

By taking the logarithm base two on the two sides of the second inequality we have $(2\alpha n^2/k)\log(\gamma+1) + \log 4n - \log 3\gamma \le \log \bar{t}$, that is:

$$k \ge 2\alpha n^2 \log(\gamma+1)/(\log \bar{t} + \log 4n - \log 3\gamma). \tag{7.2}$$

Again the integer ceiling of the above expression gives a number of agents that can do the job, maybe cancelling from the denominator the values $\log 4n$ and $\log 3\gamma$ that are practically negligible compared with $\log \bar{t}$.

The comparison of Relation (7.2) with (7.1) shows an interesting variation in the value of k due to the presence of n^2 instead of n^3 in the nominator, and of $\log \bar{t}$ instead of $\bar{t}$ in the denominator.

The concept of parallel speed-up must be revisited in the present paradigm where the value of k computed in both cases (7.1), (7.2) depends on the value of $\bar{t}$. If this latter value is imposed a priori, k is uniquely determined. Then an interesting figure to take as the agents work is the product $k\bar{t}$, with the purpose of studying how this value is changed if $\bar{t}$ is differently fixed.

With relation (7.1) we approximately have $k\bar{t} = \Delta$ with Δ constant, implying that the speed-up is constantly equal to one for all values of $\bar{t}$. With relation (7.2) the situation is much more interesting since we approximately have $k \log \bar{t} = \Delta$. Using a larger number μk of agents, with $\mu > 1$, we have a very strongly super-linear speed-up: $w_k/w_{\mu k} = 2^\mu/\mu$.

5   Some Considerations and Possible Extensions

The present work is a stimulus to investigate the relation between the number of agents working on a region (a two-dimensional grid in our example) and the work they are called to perform: a basic concept in parallel computing but scarcely considered in the large literature of agents moving on a graph. Not much is known on this subject. We can just mention a study on a specific problem [8], and a general definition of the *cop throttling number* [3] defined as the sum of the agents number plus their moving time for capturing a robber. For the first time we also consider that the agents have to move *and* compute, where these operations obey different mathematical rules.

The mathematical structure of our resulting figures are different from the ones generally stated in algorithmic complexity. In particular the time required to complete the agent operations is not carried out in order of magnitude as usual, since the value of the multiplicative constants getting into the global computation of time cannot be ignored. Some of our results, particularly the ones presented in Sect. 4, must then be considered on different grounds.

Obvious developments of this research should be carried out, as considering different distributions and arrival laws of the targets; a deeper investigation of the agent moving laws and data processing rules; and, above all, extending our study to different families of graphs.

References

1. Bonato, A., Nowakovski, R.: The Game of Cops and Robbers on Graphs. American Mathematical Society, Providence, RI (2011)
2. Bonato, A., Pérez-Giménez, X., Prałat, P., Reiniger, B.: The game of overprescribed cops and robbers played on graphs. Graphs Combinatorics **33**(4), 801–815 (2017). https://doi.org/10.1007/s00373-017-1815-2
3. Breen, J., Brimkov, B., Carlson, J., Hogben, L., Perry, K.E., Reinhart, C.: Throttling for the game of cops and robbers on graphs. Discret. Math. **341**, 2418–2430 (2018)
4. Hahn, G., MacGillivray, G.: A note on k-cop, l-robber games on graphs. Discret. Math. **306**, 2492–2497 (2006)
5. Karp, R.M., Ramachandran, V.: A survey of parallel algorithms for shared memory machines. In: van Leeuwen, J. (ed.) Handbook of Theoretical Computer Science. Elsevier, Amsterdam, pp. 869–941 (1990)
6. Luccio, F., Pagli, L.: Cops and robber on grids and tori: basic algorithms and their extension to a large number of cops. J. Supercomput. (2021). https://doi.org/10.1007/s11227-021-03655-1
7. Luccio, F., Pagli, L., Pucci, G.: Three non conventional paradigms of parallel computation. In: Meyer, F., Monien, B., Rosenberg, A.L. (eds.) Nixdorf 1992. LNCS, vol. 678, pp. 166–175. Springer, Heidelberg (1993). https://doi.org/10.1007/3-540-56731-3_16
8. Luccio, F., Pagli, L.: More agents may decrease global work: a case in butterfly decontamination. Theoret. Comput. Sci. **655**, 41–57 (2016)
9. Mukherjee, A., et al.: A survey of unmanned aerial sensing solutions in precision agriculture. J. Netw. Comput. Appl. **148**, 102461 (2019)
10. Shah, R.C., et al.: Data MULESs: modeling and analysis of a three-tier architecture for sparse sensor networks. Elsevier Ad Hoc Netw. **1**, 215–233 (2003)
11. Tekdas, O., et al.: Using mobile robots to harvest data from sensor fields. IEEE Wirel. Commun. **16–1**, 22–28 (2009)
12. Trotta, A., et al.: BEE-DRONES?: ultra low-power monitoring systems based on unmanned aerial vehicles and wake-up radio ground sensors. Comput. Netw. **180**, 107425 (2020)

Computing Services

Parallel Computations in Integrated Environment of Engineering Modeling and Global Optimization

Victor Gergel⬛, Vladimir Grishagin(✉)⬛, Alexei Liniov⬛,
and Sergey Shumikhin⬛

Lobachevsky State University, Gagarin Avenue 23, 603950 Nizhni Novgorod, Russia
{gergel,vagris,alin}@unn.ru

Abstract. Contemporary software systems of computer-aided engineering do not contain efficient tools for optimization of designed models especially if such optimization requires finding out the global solution. For stating and solving problems of optimal choice arising in the course of model investigation they, as a rule, are combined with specialized software aimed inherently at corresponding classes of optimization statements. Among those the problems of global optimization are the most complicated ones and there exist few solvers for global optimization. In the paper the results of combining the engineering modeling system OpenFOAM and global optimization solver Globalizer as an integrated system of computations are described. This integrated system is oriented at multiprocessor architectures and implements parallel methods of global optimization. Solving applied optimization problems modeled by OpenFOAM and optimized by Globalizer confirms the perspective of the approach.

Keywords: Parallel computations · Engineering modeling · Global optimization

1 Introduction

Many contemporary technologies of studying and solving scientific and practical problems are based on application of computer systems of modeling oriented at wide classes of objects and processes to be modeled and investigated. For example, very powerful computer-aided engineering (CAE) system ANSYS covers a wide range of finite element analysis problems, provides diversified environments for building and analyzing models from GUI to create a model step by step manually up to using programming language APDL (Ansys Parametric Design Language). ANSYS is compatible with some CAD-systems, for instance, with SolidWorks, Autodesk Inventor and others. However, it is necessary to purchase an expensive license to work with the system, and this significantly limits its scope. Moreover, ANSYS allows parallelizing the computations but requires separate licenses for each computational node.

© Springer Nature Switzerland AG 2021
V. Malyshkin (Ed.): PaCT 2021, LNCS 12942, pp. 413–419, 2021.
https://doi.org/10.1007/978-3-030-86359-3_31

Another software for modeling and analyzing the problems of mechanics, hydro- and aerodynamics, etc., is the system OpenFOAM [1,2] that is an open source software and uses the GNU GPL (General Purpose License) license. One of the advantages of this system is the openness of the source code, which opens up the possibility of developing and building specific user models and freedom of integration with any programs, due to flexibility of use. In particular, there are rich possibilities for parallelizing the computational procedures using MPI. The results obtained in the system OpenFOAM are comparable to the results of ANSYS, but at the same time performing the correct modeling requires additional efforts as a payment for versatility and full access to the entire internal structure and settings.

An important direction of obtaining qualitative solutions when analyzing problems by means of modeling systems is the ability to obtain optimal (in one or another sense) parameters of the constructed model. Some of CAE-systems, including ANSYS and OpenFOAM, contain tools for simple optimization but for solving complicated time-consuming optimization problems they require, as a rule, interaction with external solvers of such the problems. As such the solvers the systems LGO (Lipschitz Global Optimization) [3], LINDO [4], MATLAB Global Optimization Toolkit [5], GlobSol [6], BARON (Branch-And-Reduce Optimization Navigator) [7], IOSO (Indirect Optimization on the basis of Self-Organization) [8], BOA (IBM Bayesian Optimization Accelerator) [9] and others (see, for instance, the comparative description of optimization solvers in the papers [10–12]). Most of them implement sequential optimization methods, but only a few packages contain capabilities for the parallelization of the optimization process (see [6,8]). Among the optimization solvers the supercomputer system Globalizer [13,14] is one of the most powerful systems (see the link https://github.com/sovrasov/ags_nlp_solver). It is oriented at complicated time-consuming global optimization problems and provides deep and efficient parallelization on heterogeneous high-performance computer architectures.

In any case, whatever any modeling and optimization systems, there exists a complicated problem of interaction between them, because the structures of the systems, their data formats and interfaces differ significantly. As a consequence, it is required to develop special software (wrapper) for their connection which can provide such the interaction including possible parallelization of computational process. This paper presents a version of such wrapper which is open for connecting to them both optimization solvers and applied modeling systems. For solving an applied problem the wrapper was configured for interaction of Globalizer and OpenFOAM in a parallel mode.

The rest of the paper is organized as follows. Section 2 is devoted to a brief description of the integrated system of Globalizer and OpenFOAM. Section 3 contains the results of applying the integrated system to optimization of a real object - a beam with complex geometry - and estimations of parallelization effectiveness. The last section concludes the paper.

2   OpenFOAM+Globalizer Integration

The system Globalizer is intended for numerical solving constrained multidimensional global optimization problems which can be described by the statement

$$f(y^*) = min\{f(y), y \in B, g_j(y) \leq 0, 1 \leq j \leq m\} \tag{1}$$

$$B = \{y \in R^N : a_j \leq y_j \leq b_j,\, 1 \leq j \leq N\} \tag{2}$$

The objective function $f(y)$ and constraints $g_j(y), 1 \leq j \leq m$ are supposed to satisfy over the box B the Lipschitz condition , i.e., for all $y', y'' \in B$

$$|f(y') - f(y'')| \leq L \left\| y' - y'' \right\|, \tag{3}$$

$$|g_j(y') - g_j(y'')| \leq L_j \left\| y' - y'' \right\|, 1 \leq j \leq m \tag{4}$$

where $\|\bullet\|$ denotes the Euclidean norm in the space R^N and the Lipschitz constants $L, L_1, ..., L_m$ are positive.

Solving the problem (1) consists in finding after a finite number k of trials (evaluations of objective function) an approximate solution y_k^* being close to the exact global minimizer y^*, for example, such that

$$\left\| y_k^* - y^* \right\| < \epsilon \tag{5}$$

where $\epsilon > 0$ is a predefined accuracy of the search.

Optimization problems under such assumptions are, as a rule, multiextremal and this circumstance requires the use of methods that are capable to find global solution. Moreover, the multiextremality leads to exponential growth of number of trials when the dimensionality increasing [15] that makes these problems very complicated and time-consuming for high dimensionalities. At the same time, the Lipschitz conditions (3), (4) allow one to obtain estimations (5) after finite number of trials that is impossible, for example, for continuous functions.

The algorithmic kernel includes information-statistical algorithms of global optimization [15,16] in combination with schemes of complexity reduction. The first of such schemes reduces the constrained optimization (1) to an equivalent problem of optimization in the box (2) on the base of the index method [15] that unlike the widely applied penalty method does not use any tuning coefficients. The next step in the fight against complexity consists in reducing dimensionality. Two approaches are used for decreasing dimension of problems to be solved. The first one is based on Peano mappings that transform the multidimensional box (2) onto one-dimensional interval and the multidimensional optimization problem to an equivalent univariate one [15,16]. In the system a family of mappings called multiple Peano curves are used which generates a family of one-dimensional problems solved in parallel. The second approach to dimensionality reduction applies the scheme of nested optimization [15,17] that replaces solving the problem (1) with solving a set of recursively connected univariate subproblems. In both the schemes the parallel information-statistical global optimization methods are taken for solving one-dimensional subproblems and they provide another one level of parallelism (so called parallelization by characteristics). Globalizer is open in relation to a module that will describe the problem (1) and perform computation of objective function $f(y)$ and constraints $g_j(y), 1 \le j \le m$. For this goal the system provides a header file describing the interface with the solver.

For integration of optimization solver and applied system modeling an object, a program-wrapper was developed which allows one to form the statement of optimization problem from parameters and characteristics of the object modeled by applied software and implements calculation of these characteristics at points chosen by solver during optimization. The wrapper is open for connecting on sides of both optimization solver and applied modeling system and in the current study has been configured for Globalizer and OpenFOAM. The general scheme of the integrated system is presented in Fig. 1.

To accelerate calculations, the wrapper has been designed with the possibility of using multiple computing nodes and running several computing processes of OpenFOAM (green boxes) simultaneously.

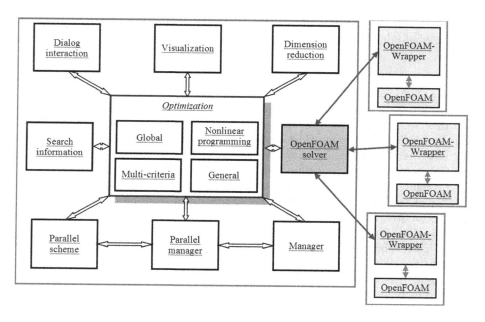

Fig. 1. Structure of integrated system

3   Optimizing the Beam Profile of Complex Geometry

The integrated system was applied for finding optimal parameters of objects modeled by OpenFOAM. Here the results for one of those - a beam of complex geometry - are presented.

The beam drawing is shown in Fig. 2.

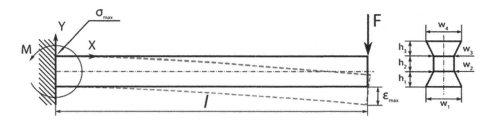

Fig. 2. Beam drawing

Parameters of the model are l - beam length, F - power load, heights h_1, h_2, h_3 and widths w_1, w_2, w_3, w_4. As characteristics of the beam, its weight W, maximum deformation MD and stress MS are considered.

The goal was to optimize the beam weight s.t. restrictions on maximum deformation and stress. In the problem the parameters l and F were fixed and equal to 1000 mm and 10000 Pa correspondingly, whereas the 7-dimensional vector of geometric parameters $y = (h_1, h_2, h_3, w_1, w_2, w_3, w_4)$ measured in millimeters was vector of optimization parameters. The statement of the optimization problem was as follows.

$$W(y^*) = min\{W(y), y \in B\} \tag{6}$$

$$B = \{y \in R^7 : 0 \le h_j \le 40, 1 \le j \le 3, 0 \le w_s \le 50, 1 \le s \le 4\} \tag{7}$$

s.t.

$$MD(y) \le 4,40e - 5\ m, MS(y) \le 3,30e + 9\ Pa. \tag{8}$$

The solution is presented in Tables 1 and 2.

Table 1. Results of optimization

Characteristic	Value
Weight	4,27 kg
Maximum deformation	3,82e−5 m
Maximum stress	3,14e+9 Pa

Table 2. Optimal parameters

Parameters	Values
h_1, h_2, h_3	(11.57, 12.04, 12.27) mm
w_1, w_2, w_3, w_4	(34.12, 10.02, 12.21, 30.21) mm

For estimation of parallelization efficiency of the integrated system the optimization was performed several times with different numbers of parallel threads. The achieved speed-ups are reflected in the Fig. 3. These results demonstrate perspective of parallelization when solving problems of the class in question.

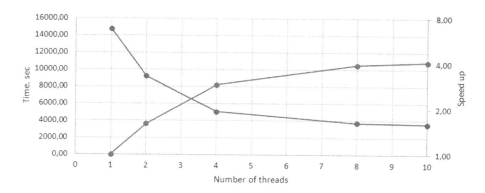

Fig. 3. Acceleration results

4 Conclusion

Integrated environment of the engineering modeling system OpenFOAM and global optimization solver Globalizer has been considered and its main characteristics and capabilities has been described. The environment enables multi-level parallelization both in modeling and optimizing parts. Results of solving an applied 7-dimensional constrained optimization problem where the objective function and constraints are modeled by OpenFOAM and optimization is performed by parallel algorithms of Globalizer have been presented. The results confirms the perspective of the approach based on integration modeling and optimizing software systems and efficiency of parallelization. The continuation of the research will consist in expanding the spectrum of optimization algorithms more oriented at different classes of modeled objects and in embedding the new tools of parallelization.

Acknowledgements. This work was supported by the Ministry of Science and Higher Education of the Russian Federation, project no. 0729-2020-0055, and by the Research and Education Mathematical Center, project no. 075-02-2020-1483/1.

References

1. Jasak, H.: OpenFOAM: open source CFD in research and industry. Int. J. Nav. Archit. Ocean Eng. **1**(2), 89–94 (2009)
2. Nóbrega, J.M., Jasak, H. (eds.): OpenFOAM®: Selected Papers of the 11th Workshop. Springer, Cham (2019). https://doi.org/10.1007/978-3-319-60846-4

3. Pintér, J.D.: Global Optimization in Action (Continuous and Lipschitz Optimization: Algorithms, Implementations and Applications). Kluwer Academic Publishers, Dordrecht (1996)
4. Lin, Y., Schrage, L.: The global solver in the LINDO API. Optim. Methods Softw. **24**(4–5), 657–668 (2009)
5. Venkataraman, P.: Applied Optimization with MATLAB Programming. Wiley, Hoboken (2009)
6. Kearfott, R.B.: GlobSol user guide. Optim. Methods Softw. **24**, 687–708 (2009)
7. Sahinidis, N.V.: BARON: a general purpose global optimization software package. J. Glob. Optim. **8**, 201–205 (1996). https://doi.org/10.1007/BF00138693
8. Egorov, I.N., Kretinin, G.V., Leshchenko, I.A., Kuptzov, S.V.: IOSO optimization toolkit - novel software to create better design. In: 9th AIAA/ISSMO Symposium on Multidisciplinary Analysis and Optimization (2002). https://www.iosotech.com/ru/ioso_nm.htm
9. Choudary, C.S.: Introducing IBM Bayesian Optimization Accelerator (2021). https://developer.ibm.com/components/ibm-power/blogs/boa-not-just-for-rocket-scientists/
10. Mongeau, M., Karsenty, H., Rouzé, V., Hiriart-Urruty, J.B.: Comparison of public-domain software for black box global optimization. Optim. Methods Softw. **13**, 203–226 (2000)
11. Pintér, J.D.: Software development for global optimization. In: Lectures on Global Optimization, vol. 55, pp. 183–204. Fields Institute Communications (2009)
12. Rios, L.M., Sahinidis, N.V.: Derivative-free optimization: a review of algorithms and comparison of software implementations. J. Glob. Optim. **56**, 1247–1293 (2013). https://doi.org/10.1007/s10898-012-9951-y
13. Gergel, V., Barkalov, K., Sysoyev, A.: Globalizer: a novel supercomputer software system for solving time-consuming global optimization problems. Numer. Algebra Control Optim. **8**, 47–62 (2018)
14. Gergel, V., Kozinov, E.: Efficient multicriterial optimization based on intensive reuse of search information. J. Glob. Optim. **71**(1), 73–90 (2018). https://doi.org/10.1007/s10898-018-0624-3
15. Strongin, R.G., Sergeyev, Y.D.: Global Optimization with Non-convex Constraints: Sequential and Parallel Algorithms. Kluwer Academic Publishers, Dordrecht (2000)
16. Sergeyev, Y.D., Strongin, R.G., Lera, D.: Introduction to Global Optimization Exploiting Space-Filling Curves. Springer, New York (2013). https://doi.org/10.1007/978-1-4614-8042-6
17. Grishagin, V., Israfilov, R., Sergeyev, Y.: Comparative efficiency of dimensionality reduction schemes in global optimization. In: AIP Conference Proceedings, vol. 1776, p. 060011 (2016)

Implementing Autonomic Internet of Things Ecosystems – Practical Considerations

Kumar Nalinaksh[1], Piotr Lewandowski[2], Maria Ganzha[2],
Marcin Paprzycki[2(✉)], Wiesław Pawłowski[3],
and Katarzyna Wasielewska-Michniewska[2]

[1] Warsaw Management University, Kawęczyńska 36, 03 -772 Warsaw, Poland
kumarnalinaksh21@gmail.com
[2] System Research Institute, Polish Academy of Sciences, Warsaw, Poland
{piotr.lewandowski,maria.ganzha,marcin.paprzycki,
katarzyna.wasielewska-michniewska}@ibspan.waw.pl
[3] University of Gdańsk, Wita Stwosza 57, 80-308 Gdańsk, Poland
wieslaw.pawlowski@ug.edu.pl

Abstract. Development of next generation Internet of Things ecosystems will require bringing in (semi-)autonomic behaviors. While the research on autonomic systems has a long tradition, the question arises, are there any "off-the-shelf" tools that can be used directly to implement autonomic solutions/components for IoT deployments. The objective of this contribution is to compare real-world-based, autonomy-related requirements derived from ASSIST-IoT project pilots with existing tools.

Keywords: Internet of Things · Autonomic systems · Self-* mechanisms

1 Introduction

The idea of autonomic systems can be traced back to early works in the discipline known as cybernetics [1]. However, the modern understanding of the concept arose from seminal work performed by IBM, within the scope of the autonomic computing initiative (ACI) [2]. Here, (and in later work [3]) IBM proposed four categories of, so called, "Self-*" properties, which were to capture main aspects for development of autonomic systems: (1) *Self-configuration*: automatic component configuration; (2) *Self-healing*: automatic fault discovery and correction; (3) *Self-optimization*: automatic resource monitoring and control to ensure optimal performance in accordance with specified requirements; (4) *Self-protection*: diligent detection and protection from random attacks. Later, seven Self-* properties have been proposed [4,5]. Let us leave aside the number and scope of Self-* properties and come back to them later.

To realize the Self-* mechanisms, the MAPE-K (Monitor, Analyze, Plan, Execute, Knowledge) loop was proposed [3]. In the MAPE-K autonomic loop

© Springer Nature Switzerland AG 2021
V. Malyshkin (Ed.): PaCT 2021, LNCS 12942, pp. 420–433, 2021.
https://doi.org/10.1007/978-3-030-86359-3_32

sensors gather data about the managed element, while actuators make modifications to it. Specifically, a manager tracks the state of an element and makes adjustments using the data gathered by the sensors. As part of its development work on the Autonomic Computing Toolkit, IBM created a prototype version of the MAPE-K loop, called the Autonomic Management Engine.

Independently, recent years are characterized by rapid developments in the area of the Internet of Things. Here, the main idea is to deploy sensors and actuators, connected using heterogeneous networking infrastructures (wireless and wired), to deliver novel services for the users. With the size of IoT ecosystem deployments reaching thousands of elements, it becomes clear that it is not going to be possible to "hand manage" them. In this context, recently, the European Commission requested research in the area of Self-adaptive, Self-aware and semi-autonomous IoT systems[1]. One of the projects that was funded as a result of this call is ASSIST-IoT[2]. This project is grounded in four pilots, and each one of them has specific needs for Self-* mechanisms. This leads to the question: can these needs be satisfied using existing solutions/tools? The aim of this work is to answer this question.

In this context we proceed as follows. In Sect. 2 identified Self-* needs of the ASSIST-IoT pilots are discussed. Next, in Sect. 3 we present known to us tools that can be used in context of implementation of Self-* mechanisms. We follow, in Sect. 4, with discussion on how the existing solutions address the identified needs. Section 5, summarizes our findings.

2 Autonomic Computing for the Real-World IoT

Results of the ASSIST-IoT project will be validated in four pilots: (1) port automation, (2) smart worker protection, and (3) cohesive vehicle monitoring and diagnostics. The latter one is divided into sub-pilots dealing with (3a) car engine monitoring, and (3b) car exterior monitoring. Let us now discuss which Self-* mechanisms have been identified in each pilot, during the requirements analysis phase of the project.

2.1   Port Automation Pilot

Owing to the high volume of TEUs (an inexact measure of cargo capacity that is frequently employed by port authorities) handled and the growing number of stopovers, the Malta Freeport Terminal (MFTL) is nearly at capacity, with almost constant congestion in the terminal area and sporadic disruptions having a significant effect on business operations. As a consequence, four main problems can occur: (1) longer vessel dwell periods; (2) increased berthing-wait-time; (3) vessels being moved to other terminals; and (4) increased wait and turn-around times of land-side vehicles, all of which contribute to increased environmental

---

[1] https://ec.europa.eu/info/funding-tenders/opportunities/portal/screen/opportuniti es/topic-details/ict-56-2020.

[2] https://assist-iot.eu/.

load, transportation inefficiency, and cost of efficient movements. To deal with the existing threats, using solutions provided by the ASSIST-IoT project, three business scenarios have been identified: (1) asset monitoring in the terminal yard, (2) automated container handling equipment cooperation, and (3) rubber-tired gantry remote control, with augmented reality assistance. All of those scenarios will need Self-* capabilities, that will work seamlessly in heterogeneous IoT environment.

Scenarios (1), (2) and (3) will need Self-inspection (sometimes called Self-awareness) to understand where particular assets are located and what is the current state of those assets. Self-healing and Self-diagnosis will also be impactful as they will allow to automatically detect issues, autonomously fix some of them or call for human operator as a last resort. (1) will also require Self-configuration capabilities so that new devices can easily connect and acquire required configuration (i.e. map of the port). (2) will additionally need Self-organization and Self-adaptation to autonomously carry on container handling via organizing work efficiently and by adapting to a changing port environment.

2.2   Smart Safety of Workers Pilot

Construction companies and relevant administration agencies, such as the European Agency for Safety and Health at Work, place a high emphasis on compliance with workplace safety and health standards and risk management at small or large, private or public construction projects. A vast number of people with varying degrees of knowledge and experience collaborate with each other, control equipment, and interface with heavy machinery on each building site, which is occupied by many subcontracted firms. Their experience, best practises, and risk management culture offer a layer of security for construction workers, but it does not ensure that all accidents could be avoided. Accidents will happen in a split second with no indications. Furthermore, unless appropriate monitoring mechanisms are in place, a potentially life-saving immediate intervention to an accident couldn't be feasible.

ASSIST-IoT solution will enable this pilot to collect accurate and appropriate data in order to produce intelligent insights for the protection of all people involved at every work site within a vast construction site. Such data and observations, along with the clear implementation of data security policies, will advance understanding and increase awareness about workplace safety, as well as lead to the digital transformation of construction processes that retains the employee at the leading edge. In this application area, the main goal of ASSIST-IoT is to prevent and detect common Occupational Safety and Health (OSH) hazards such as stress, exhaustion, overexposure to heat and ultraviolet rays, slips, trips, falls from heights, suspension injuries, lack of mobility due to loss of consciousness, collision with heavy equipment, entrapment and PPE misuse. The success of implementing this pilot test-bed would result in two key outcomes: better working conditions for thousands of workers and a clear return on investment (ROI) for the facility. This pilot has been divided into four business scenarios, (1) occupation safety and health monitoring; (2) fall arrest monitoring; (3) safe navigation; and (4) health and safety inspection support.

(1), (2), (3) and (4) will need Self-inspection to understand where particular events are taking place or if (and where) someone is accessing dangerous zone. Self-diagnosis is required so that whole system can autonomously detect any potential issues. (1) and (3) will also need Self-configuration to automatically connect upcoming devices and ensure that up-to-date configuration in dynamic construction site environment is available; (3) will need Self-adaptation so that in case of dynamically occurring risks, the safest route can be selected.

2.3   Cohesive Vehicle Monitoring and Diagnostics Pilot

Currently, ICT penetration in the automobile industry is just a fraction of what it should be, and it is mostly dominated by car manufactures. Because of high costs and bandwidth problems, communication between vehicle fleets and original equipment manufacturers is also restricted. Due to safety and security concerns, most IoT integration in vehicles programs struggle to incorporate data from various sources (e.g. industry data, environmental data, data from inside the car, historical vehicle maintenance data) and to obtain access to vehicle data. Although real-time operation of a moving vehicle creates safety risks and therefore prohibits full unrestricted access to the information and control firmware, there is no theoretical obstacle to trustworthy third parties having access to onboard sensor measurements for diagnostics and monitoring. Furthermore, no existing application or implementation incorporates and delivers automotive details to a customer in an immersive friendly atmosphere based on their position and relationship with the vehicle, avoiding recalls.

The use of the ASSIST-IoT reference architecture in this pilot will improve the automotive OEMs' ability to track the pollution standards of vehicles that are currently on the road in order to ensure that the fleet maintains certification limits over its lifespans. Monitoring fleet pollution levels allows for the prompt execution of corrective measures, if necessary, to return them to acceptable levels. There are two independent sections of this pilot: (1) a Ford initiative and (2) a TwoTronic initiative. The Ford initiative is divided into two business scenarios: (1) fleet in-service conformity verification; and (2) vehicle diagnostics; while the TwoTronic initiative deals with vehicle exterior condition inspection and documentation.

Scenarios (1) and (2) will need Self-learning to constantly improve their capabilities and Self-diagnosability to ensure that all components of the system provide realistic measurements. We also assumed that Self-configuration will be required to always be up-to-date with current requirements.

3   State-of-the-art in tools for Autonomic Computing

Let us now summarize the state-of-the-art in the area of tools that can be used to implement Self-* mechanisms. Most important factor taken into account was their out-of-the-box Self-* capability. Moreover, tools were selected on having publicly available (open source) code repositories that have recent updates indicting that these tools are currently under active development. Another factor was the potential to generalize particular tool to solve novel problems.

AMELIA: Analysable Models Inference. In [9] authors report on tracking IoT system trajectories, i.e. a series of latitude and longitude coordinate points mapped with respect to time [10], in a complex spatial context. This is combined with accessible space landmarks, to create graph-based spatial models. These are, in turn, analysed by the MAPE-K loop's Analyse feature, to search for goal and requirement violations, during system runtime. The project is available as a virtual environment, in which it can be run and the findings replicated as published. The authors run the simulations[3] using actual data sets derived from Taxis (IoT devices) and used city's landmarks as the graph's nodes. The project is primarily built on Python with MongoDB as it can resolve geo-spatial queries. Shell scripts are used to interface between the project and the operating system. If required, the project can be built and run locally with different parameters.

OCCI-Compliant Sensor Management. The Open Cloud Computing Interface (OCCI) specifies an API for managing large and diverse cloud services that is independent of the service provider. Various tools offer interfaces for identifying, initiating, and implementing modifications to complex cloud environments. The authors built an OCCI monitoring extension[4] in JAVA that offered managing the implementation and setup of monitoring sensors in the cloud [11]. In an OCCI-compliant runtime model, sensors and their monitoring results are described. This extension transforms the OCCI runtime model into a knowledge base that, when coupled with the other objects in the OCCI ecosystem, facilitates full control loops for Self-adaptation into cloud systems. The authors integrated the project with a real-world cloud infrastructure and included two sample scenario implementations for other researchers using the test environment to validate the project outcomes. A Hadoop cluster was implemented and dynamically scaled in both instances.

PiStarGODA-MDP: A Goal-Oriented Framework to Support Assurances Provision. A Self-adaptive system often works in a complex and partly unknown context, which introduces uncertainty that must be addressed in order for it to accomplish its objectives. Furthermore, several other types of uncertainties exist, and the causes of these uncertainties are not consistently resolved by current approaches in the Self Adaptive System (SAS) life cycle. This begets the question of how can the goals of a system that is subject to continuous uncertainties be guaranteed? Here, the authors proposed and implemented a goal-oriented assurance method that allows monitoring sources of uncertainty

---

[3] https://dsg.tuwien.ac.at/team/ctsigkanos/amelia/.
[4] https://gitlab.gwdg.de/rwm/de.ugoe.cs.rwm.mocci.

that arise during the design phase, or during runtime execution of a system [12]. The SAS is designed with the goals in mind, and the Self-adaptation occurs during the runtime. GORE (Goal-Oriented Specifications Engineering) is used for separating technological and non-technical criteria into clearly specified goals and justifications for how to accomplish them. These goal models are converted into reliability and cost parametric formulae using symbolic model checking, which are then used as runtime models to express the likelihood of SAS goals being reached. Based on the principle of feedback control, the controller continuously monitors the managed system's costs and reliability statuses, as well as contextual constraints, at runtime to address parameterized uncertainties. The runtime models are then used to assess (i) system's reliability and cost, and (ii) policy measures that should be activated to accomplish the goals, influencing SAS adaptation decisions. The authors evaluated their project's[5] approach using the Body Sensor Network (BSN) implemented in OpenDaVINCI[6] and were able to effectively provide guarantees for Self-adaptive systems' goals. JavaScript and Java were used in the project's development. Heroku hosts the pistarGODA modelling and analysis environment.

TRAPP: Traffic Reconfiguration Through Adaptive Participatory Planning. Traffic management is a difficult challenge from the standpoint of Self-adaptation because it is hard to prepare ahead with all potential scenarios and behaviours. Here, authors present a method for autonomous agents to collaborate in the absence of a centralised data collection and arbitrator [13]. TRAPP integrates the SUMO [14] and EPOS [15] frameworks. EPOS is a decentralised combinatorial optimization approach for multi-agent networks, while SUMO is a simulation environment for traffic dynamics. SUMO sends EPOS a list of potential routes for each vehicle, and EPOS generates the designated plan for each vehicle, which SUMO picks up and executes. The mechanism described above occurs on a regular basis. Periodical adaptation cycles are operated by the managing system, which, in accordance with the MAPE-K loop, monitor data, evaluate it for traffic issues or anomalies, schedule subsequent activities to adjust the way participatory preparation occurs, and eventually perform the adaptation actions by configuring EPOS accordingly. The revised configuration is used the next time EPOS is invoked. The authors run simulations[7] by deploying 600 cars in the city of Eichstatt, which has 1131 roads. Python and Jupiter notebook were used to create the project.

mRUBiS: Model-Based Architectural Self-healing and Self-optimization. Self-adaptive software is a restricted system that uses a feedback mechanism to adjust to changes in the real world. This mechanism is implemented by the adaptation engine, while the domain logic is realised by adaptable software and controlled by the engine. The authors came to the conclusion that

---

[5] https://github.com/lesunb/pistarGODA-MDP.
[6] https://github.com/se-research/OpenDaVINCI.
[7] https://github.com/iliasger/TRAPP.

there is no off-the-shelf product for designing, testing, and comparing model-based architectural Self-adaptation and hence they developed mRUBIs [17]. It simulates adaptable software and allows for "issues" to be injected into runtime models. This helps developers to test and compare different adaptation engine variants as well as validate the Self-adaptation and healing properties of the adaptation engine. The authors developed a generic modelling language called "CompArch" to interact with the project[8], while the project itself has been implemented in JAVA.

Lotus@Runtime: Tool for Runtime Monitoring and Verification of Self-adaptive Systems. Lotus@Runtime tracks execution traces provided by a Self Adaptive System and annotates the probability of occurrence of each system operation using a Labelled Transition System model [18]. In addition, the probabilistic model is used at runtime to check adaptability properties. A warning function built into the tool notifies the Self-adaptive device if a property is violated. These notifications are handled by ViolationHandler module that the user implements during planning phase. The project[9] is based over the existing LoTuS[10] project built in JAVA. The authors used Tele Assistance System (TAS) and Travel Planner Application (TPA) [19] for validating the project[11].

Intelligent Ensembles. Autonomous components are deployed in a physical world in smart cyber-physical systems (CPS) like smart cities, where they are supposed to collaborate with each other and also with humans. They must be capable of working together and adapt as a group to deal with unexpected circumstances. To address this problem, the authors applied Intelligent Ensembles. They're dynamic groups of components that are generated at runtime depending on the components' current state. Components are not capable of communicating with one another; rather, the ensemble is responsible for communication. The Intelligent Ensembles framework uses a declarative language called "EDL" for describing dynamic collaboration groups [20]. The project[12] is built over the Eclipse Modelling Framework and the Z3 SMT solver.

CrowdNav and RTX. The authors look at the issue of a crowdsourced navigation system (CrowdNav). It's a city traffic control system that gathers data from a variety of sources, such as cars and traffic signals, and then optimises traffic guidance. The authors solve this problem by interpreting and adapting the stream of data from the distributed system using Real-Time Experimentation (RTX) tool [21]. The project[13] is written in Python, configures Kafka and Spark

---

[8] https://github.com/thomas-vogel/mRUBiS.
[9] https://github.com/davimonteiro/lotus-runtime.
[10] https://github.com/lotus-tool/lotus-tool.
[11] https://drops.dagstuhl.de/opus/volltexte/2017/7145/.
[12] https://drops.dagstuhl.de/opus/volltexte/2017/7144/.
[13] https://drops.dagstuhl.de/opus/volltexte/2017/7143/.

and links them together. Its architecture is straightforward and restricted to the most relevant input and output parameters, with Big Data analytics guiding Self-adaptation based on a continuous stream of operational data from Crowd-Nav. To help in the assessment of different Self-adaptation strategies for dynamic large-scale distributed systems, the authors built a concrete model problem using CrowdNav and SUMO in this exemplar [14,16].

DeltaIoT: Self-adaptive Internet of Things. Wireless connectivity absorbs the majority of energy in a standard IoT unit, so developing reliable IoT systems is critical. Finding the correct network configurations, on the other hand, is difficult because IoT implementations are subject to a multitude of uncertainties, such as traffic load fluctuations and connectivity interruption. Self-adaptation enables hand-tuning or over-provisioning of network settings to be automated.A feedback loop is installed on top of the network to track and measure the motes and the environment, allowing the IoT system to adapt autonomously. The DeltaIoT project[14] consists of an offline simulator and a physical setup of 25 mobile nodes which can be remotely controlled for field testing. The IoT system is installed on the KU Leuven Computer Science Department's property. DeltaIoT [22] is the very first Self-adaptation research project to have both a simulator and a physical system for testing. DeltaIoT is used in Self-adaptation studies. It allows researchers to test and compare emerging Self-adaptation approaches, techniques, and resources in the IoT. The WebService Engine is a user interface for inspecting and controlling the Internet of Things system. A WSDL file is used to describe this interface. Just one person may do Self-adaptation at a time, hence accessibility to the web service is restricted.

TAS: Tele Assistance System. TAS [23] was created with the help of the Research Service Platform (ReSeP)[15]. ReSeP is built upon the Service-Oriented Architecture (SOA) principles using JAVA. The tool is an example of a service-based system (SBS). It gives preventive care to chronic patients in their own homes. TAS makes use of sensors mounted in a wearable interface, and remote services from healthcare, pharmacy, and emergency response providers. Periodic samples of a patient's critical parameters are taken and exchanged with a medical service for study. The service may invoke a pharmacy service based on the review to distribute new medication to the patient or to change and upgrade the medication dose. Using ReSeP the authors defined two different adaptation policies and validated it with TAS. The first policy was to retry twice in case of service failure whereas the second policy selects an alternate service with similar cost and invokes it. The experiment found that the first policy kept the costs low but failed the reliability constraint while the second one passed successfully albeit with high costs.

---

[14] https://people.cs.kuleuven.be/~danny.weyns/software/DeltaIoT/.
[15] https://github.com/davimonteiro/resep.

DEECo: Dependable Emergent Ensembles of Components. The authors conclude that developing complex Self-adaptive smart CPS is a difficult challenge that current software engineering models and methods can only partially solve. The appropriate architecture of a smart CPS adopts a holistic view that considers the overall system goals, operating models that include system and climate uncertainties, and the communication models that are being used. To answer these issues the authors used DEECo [25]. It's a model and framework for creating sophisticated smart CPS. It also provides precise information about the consequences of adaptation techniques in complex smart CPS. The Java and C++ are included with the DEECo component model [24]. The C++ architecture is used for real-world deployment on embedded devices, like the STM32F4 MCU. Java i.e. JDEECo, on the other hand, is used for adaptation and autonomous components simulation. JDEECo simulates implementations of hybrid network environments, mixing IP networks and mobile/vehicle ad-hoc networks (MANETS/VANETS), as seen in current smart-* systems, by using the OMNeT extensions INET and MIXIM. The project[16] was created specifically for the purpose of developing and simulating dynamic Self-adaptive smart CPS. Authors used a smart parking scenario to validate its usage.

Znn.com Rainbow [26] is a framework for designing a system with Self-adaptive, run-time capabilities for monitoring, detecting, deciding, and acting on system optimization opportunities. Znn.com is an N-tier-style web-based client-server system. Rainbow uses the following guidelines to handle Znn.com's Self-adaptation at peak periods: (i) Changing the server pool size and (ii) switching between textual and multimedia response [27]. The project[17] has been built using several different languages such as PHP, Shell, Brainfuck, Awk, HTML and Perl etc.

Dragonfly. The authors [28] noted that when designing cyber-physical Systems, we often encounter defiant systems that can evolve and collaborate to achieve individual goals but struggle to achieve global goals when combined with other individual systems. They suggest an integration strategy for converting these defiant systems to also achieve the overall objectives. Dragonfly[18] is a drone simulator that allows users to simulate up to 400 drones at once. Simulations may be performed in both regular and unusual conditions. The wrappers implement the drones' adaptive behavior and enable runtime adaptation. The simulator is built using JAVA, AspectJ, HTML and Docker.

DARTSim. Cyber-Physical systems make use of Self-adaptive capabilities to autonomously manage uncertainties at the intersection of the cyber and physical worlds. Self-adaptation-based approaches face several challenges such as:

---

[16] https://github.com/d3scomp/JDEECo.
[17] https://github.com/cmu-able/znn.
[18] https://github.com/DragonflyDrone/Dragonfly.

(i) sensing errors while monitoring environment, (ii) not being able to adapt in time due to physical constraints etc., (iii) objectives that cannot be coupled together in a single utility matrix such as providing good service vs avoiding an accident. To evaluate and compare various Self-adaptation approaches aiming to address these unique challenges of smart CPS, DARTSim was created in 2019 [29]. DARTSim is a simulation of an autonomous team of unmanned aerial vehicles (UAVs) conducting a reconnaissance mission in a hostile and unfamiliar area. The squad must follow a predetermined path and pace when attempting to locate the targets. The lower it goes, the more likely it is to find targets, but also the more likely it is to be killed by threats. The high-level Self-adaptation decisions that the machine must make to achieve mission success are the subject of DARTSim. This sCPS has the "smartness" needed to conduct the task autonomously thanks to the adaptation manager who makes these decisions. When a mission detects at least half of the threats, it is considered effective. The project[19] outcome is available as a C++ library or via a TCP interface[20].

4   Needs vs Available Tools – Critical Analysis

Existing tools/platforms vary in application and abstraction level. Their range of capabilities varies from solving specific problem using specific type of IoT Device (TRAPP) to a high-level generic ones that need non-trivial amount of additional work to solve concrete tasks (e.g. OCCI-compliant, fully causal-connected runtime models supporting sensor management). In this context, let us consider existing tools and evaluate their potential to deliver Self-* mechanisms identified within the ASSIST-IoT pilots. In total, to satisfy all pilot requirements the following Self-capabilities were identified: (1) Self-inspection (or Self-awareness), (2) Self-diagnosis, (3) Self-healing, (4) Self-configuration, (5) Self-organization, (6) Self-adaptation, and (7) Self-learning.

When considering available solutions, we verified whether their public source code repositories were available and then focused on the fact whether the Self-* capabilities were available out of the box or with minimal additional work required. Following is a list of considered solutions with Self-* capabilities that they support:

AMELIA (1) Self-inspection, (6) Self-adaptation, (7) Self-learning
OCCI (1) Self-inspection, (2) Self-diagnosis, (3) Self-healing, (6) Self-adaptation, (7) Self-learning
PiStarGODA-MDP (1) Self-inspection, (6) Self-adaptation, (7) Self-learning
TRAPP (6) Self-adaptation, (7) Self-learning.
mRUBiS (1) Self-inspection, (2) Self-diagnosis, (3) Self-healing, (5) Self-organization, (6) Self-adaptation, (7) Self-learning
Lotus@Runtime: (1) Self-inspection, (2) Self-diagnosis, (3) Self-healing, (6) Self-adaptation

---

[19] https://github.com/cps-sei/dartsim.
[20] https://hub.docker.com/r/gabrielmoreno/dartsim/.

Intelligent Ensembles (6) Self-adaptation
CrowdNav and RTX (6) Self-adaptation, (7) Self-learning
DeltaIoT (1) Self-inspection, (2) Self-diagnosis, (5) Self-organization, (6) Self-adaptation
TAS: (1) Self-inspection, (5) Self-organization, (6) Self-adaptation
DEECo (1) Self-inspection, (5) Self-organization, (6) Self-adaptation
Znn.com (1) Self-inspection, (6) Self-adaptation
Dragonfly (1) Self-inspection, (6) Self-adaptation
DARTSim (1) Self-inspection, (6) Self-adaptation

In summary, it is easy to observe that there is no available solution capable of running heterogeneous Self-* IoT deployments that satisfies needs of all pilots considered by ASSIST-IoT. Particularly, Self-configuration seems to be a missing component. If no tool is able to provide common abstraction for detecting and automatically connecting and configuring various devices that are present in an IoT ecosystem it will be hard to imagine a widespread adoption of IoT-based solutions. As a general note, most projects followed a very high-level approach to Self-*, leaving implementation of components below MAPE-K loop (or analogous solution) to the user. This is understandable, as most of them were not designed with IoT deployments in mind, yet widespread adoption needs to be preceded by developing a well-rounded solution that answers the common Self-* problems on a more concrete level. There is a set of Self-* enabled tools that focus on selected problems (for example TRAPP is specific to a car traffic management) but they are very hard to generalize to conveniently handle as diverse scenarios as worker safety, coordination between port machinery and detect defects in car exhaustion system. Those expectations might sound very ambitious, yet this is a general trend in Software Engineering, where Cloud-based solutions abstracted away many of the hard problems to the point of few clicks in web-based UI. We predict that the same is required in IoT based environments.

5   Concluding Remarks

The aim of this work is to consider how existing autonomic computing solutions match actual needs of Internet of Things deployments. Proceeding in this directions we have, first, outlined requirements related to autonomic computing, in 4 real-world pilots, grounding the work to be completed in the ASSIST-IoT project. Second, we have summarized state-of-the-art of existing ready-to-use tools that are claimed to support implementation of autonomic systems. Finally, we have matched the two, and critically analysed the results.

Overall, we conclude that there is no solution available that can address all challenges that have been identified in ASSIST-IoT, in the context of applying Self-* in considered business scenarios and use cases. The existing solutions would need to be adapted and combined to cover the set of required features. Additionally, some of them would need to be verified for their adaptability and performance in heterogeneous IoT ecosystems that are very ambitious target environment, for the technological solutions. We foresee that ASSIST-IoT will not only give opportunity to verify a set of approaches proposed so far, in a real-life deployments, but will also advance state-of-the-art in Self-* systems in terms of providing Self-* capabilities for IoT-centric environments.

Acknowledgment. Work of Maria Ganzha, Piotr Lewandowski, Marcin Paprzycki, Wiesław Pawłowski and Katarzyna Wasielewska-Michniewska was sponsored by the ASSIST-IoT project, which received funding from the EU's Horizon 2020 research and innovation program under grant agreement No. 957258.

References

1. Cybernetics: Or Control and Communication in the Animal and the Machine, 2nd revised ed. 1961. Hermann & Cie–MIT Press, Paris–Cambridge (1948). ISBN 978-0-262-73009-9
2. Kephart, J.O., Chess, D.M.: The vision of autonomic computing. Computer **36**(1), 41–50 (2003). https://doi.org/10.1109/MC.2003.1160055
3. IBM: An Architectural Blueprint for Autonomic Computing, IBM White Paper (2005). https://www-03.ibm.com/autonomic/pdfs/AC%20Blueprint%20White %20Paper%20V7.pdf
4. Poslad, S.: Autonomous systems and artificial life. In: Ubiquitous Computing Smart Devices, Smart Environments and Smart Interaction, pp. 317–341. Wiley (2009). ISBN 978-0-470-03560-3
5. Nami, M.R., Sharifi, M.: A survey of autonomic computing systems. In: Shi, Z., Shimohara, K., Feng, D. (eds.) IIP 2006. IIFIP, vol. 228, pp. 101–110. Springer, Boston, MA (2006). https://doi.org/10.1007/978-0-387-44641-7_11
6. What is autonomic computing?, 1 August 2018 OpenMind. https://www. bbvaopenmind.com/en/technology/digital-world/what-is-autonomic-computing/. Accessed 15 May 2021
7. Kramer, J., Magee, J.: Self-managed systems: an architectural challenge. In: Future of Software Engineering (FOSE 2007), pp. 259–268 (2007). https://doi.org/10. 1109/FOSE.2007.19
8. Sarma, S., et al.: Cyberphysical-System-on-Chip (CPSoC): a self-aware mpsoc paradigm with cross-layer virtual sensing and actuation. In: Proceedings of the 2015 Design, Automation & Test in Europe Conference & Exhibition, pp. 625–628 (2015). EDA Consortium
9. Tsigkanos, C., Nenzi, L., Loreti, M., Garriga, M., Dustdar, S., Ghezzi, C.: Inferring analyzable models from trajectories of spatially-distributed internet of things. In: 2019 IEEE/ACM 14th International Symposium on Software Engineering for Adaptive and Self-Managing Systems (SEAMS), 2019, pp. 100–106 (2019). https:// doi.org/10.1109/SEAMS.2019.00021
10. Zheng, Y., Zhou, X. (eds.): Computing with Spatial Trajectories. Springer Science & Business Media, Heidelberg (2011)
11. Erbel, J., Brand, T., Giese, H., Grabowski, J.: OCCI-compliant, fully causal-connected architecture runtime models supporting sensor management. In: IEEE/ACM 14th International Symposium on Software Engineering for Adaptive and Self-Managing Systems (SEAMS), pp. 188–194. Montreal, QC, Canada (2019). https://doi.org/10.1109/SEAMS.2019.00032

12. Félix Solano, G., Diniz Caldas, R., Nunes Rodrigues, G., Vogel, T., Pelliccione, P.: Taming uncertainty in the assurance process of self-adaptive systems: a goal-oriented approach. In: IEEE/ACM 14th International Symposium on Software Engineering for Adaptive and Self-Managing Systems (SEAMS), pp. 89–99. Montreal, QC, Canada (2019). https://doi.org/10.1109/SEAMS.2019.00020
13. Gerostathopoulos, I., Pournaras, E.: TRAPPed in traffic? A self-adaptive framework for decentralized traffic optimization. In: IEEE/ACM 14th International Symposium on Software Engineering for Adaptive and Self-Managing Systems (SEAMS), pp. 32–38. Montreal, QC, Canada (2019). https://doi.org/10.1109/SEAMS.2019.00014
14. Lopez, P.A., et al.: Microscopic traffic simulation using sumo. In: The 21st IEEE International Conference on Intelligent Transportation Systems. IEEE (2018). https://elib.dlr.de/124092/
15. Pournaras, E., Pilgerstorfer, P., Asikis, T.: Decentralized collective learning for self-managed sharing economies. ACM Trans. Auton. Adapt. Syst. (TAAS) 13(2), 10 (2018)
16. Chen, C., Liu, Y., Kreiss, S., Alahi, A.: Crowd-robot interaction: crowd-aware robot navigation with attention-based deep reinforcement learning. In: 2019 International Conference on Robotics and Automation (ICRA), pp. 6015–6022 (2019)
17. Vogel, T.: MRUBiS: an exemplar for model-based architectural self-healing and self-optimization. In: Proceedings of the 13th International Conference on Software Engineering for Adaptive and Self-Managing Systems (SEAMS 2018), pp. 101–107. Association for Computing Machinery, New York, NY, USA (2018). https://doi.org/10.1145/3194133.3194161
18. Barbosa, D.M., Lima, R.G.D.M., Maia, P.H.M., Costa, E.: Lotus@Runtime: a tool for runtime monitoring and verification of self-adaptive systems. In: 2017 IEEE/ACM 12th International Symposium on Software Engineering for Adaptive and Self-Managing Systems (SEAMS), 2017, pp. 24–30 (2017). https://doi.org/10.1109/SEAMS.2017.18
19. Zeng, L., Benatallah, B., Ngu, A.H., Dumas, M., Kalagnanam, J., Chang, H.: QoS-aware middleware for Web services composition. IEEE Trans. Softw. Eng. 30(5), 311–327 (2004). https://doi.org/10.1109/TSE.2004.11
20. Krijt, F., Jiracek, Z., Bures, T., Hnetynka, P., Gerostathopoulos, I.: Intelligent ensembles - a declarative group description language and java framework. In: 2017 IEEE/ACM 12th International Symposium on Software Engineering for Adaptive and Self-Managing Systems (SEAMS), 2017, pp. 116–122 (2017) https://doi.org/10.1109/SEAMS.2017.17
21. Schmid, S., Gerostathopoulos, I., Prehofer, C., Bures, T.: Self-adaptation based on big data analytics: a model problem and tool. In: 2017 IEEE/ACM 12th International Symposium on Software Engineering for Adaptive and Self-Managing Systems (SEAMS), 2017, pp. 102–108 (2017). https://doi.org/10.1109/SEAMS.2017.20
22. Iftikhar, M.U., Ramachandran, G.S., Bollansée, P., Weyns, D., Hughes, D.: DeltaIoT: a self-adaptive internet of things exemplar. In: 2017 IEEE/ACM 12th International Symposium on Software Engineering for Adaptive and Self-Managing Systems (SEAMS), 2017, pp. 76–82 (2017). https://doi.org/10.1109/SEAMS.2017.21
23. Weyns, D., Calinescu, R.: Tele assistance: a self-adaptive service-based system exemplar. In: 2015 IEEE/ACM 10th International Symposium on Software Engineering for Adaptive and Self-Managing Systems, 2015, pp. 88–92 (2015). https://doi.org/10.1109/SEAMS.2015.27

24. Keznikl, J., Bureš, T., Plášil, F., Kit, M.: Towards dependable emergent ensembles of components: the DEECo component model. In: Joint Working IEEE/IFIP Conference on Software Architecture and European Conference on Software Architecture, pp. 249–252 (2012). https://doi.org/10.1109/WICSA-ECSA.212.39
25. Bures, T., Gerostathopoulos, I., Hnetynka, P., Keznikl, J., Kit, M., Plasil, F.: DEECo: an ensemble-based component system. In: Proceedings of CBSE 2013, pp. 81–90. Vancouver, Canada (2013)
26. Cheng, S.-W.: Rainbow: Cost-Effective Software Architecture-Based Self-Adaptation. Ph.D. Dissertation, TR CMUISR-08-113, Carnegie Mellon University School of Computer Science, May 2008
27. Cheng, S.W., Garlan, D., Schmerl, B.: Evaluating the effectiveness of the Rainbow self-adaptive system. In: 2009 ICSE Workshop on Software Engineering for Adaptive and Self-Managing Systems, Vancouver, BC, Canada, 2009, pp. 132–141 (2009). https://doi.org/10.1109/SEAMS.2009.5069082
28. Maia, P.H., Vieira, L., Chagas, M., Yu, Y., Zisman, A., Nuseibeh, B.: Dragonfly: a tool for simulating self-adaptive drone behaviours. In: IEEE/ACM 14th International Symposium on Software Engineering for Adaptive and Self-Managing Systems (SEAMS), pp. 107–113. Montreal, QC, Canada (2019). https://doi.org/10.1109/SEAMS.2019.00022
29. Moreno, G., Kinneer, C., Pandey, A., Garlan, D.: DARTSim: an exemplar for evaluation and comparison of self-adaptation approaches for smart cyber-physical systems. In: 2019 IEEE/ACM 14th International Symposium on Software Engineering for Adaptive and Self-Managing Systems (SEAMS) (2019)

Information-Analytical System to Support the Solution of Compute-Intensive Problems of Mathematical Physics on Supercomputers

Yury Zagorulko[1]([✉]) [iD], Galina Zagorulko[1] [iD], Alexey Snytnikov[2,4] [iD],
Boris Glinskiy[3] [iD], and Vladimir Shestakov[1] [iD]

[1] A.P. Ershov Institute of Informatics Systems of Siberian Branch of the Russian Academy of
Sciences, Novosibirsk, Russia
zagor@iis.nsk.su
[2] Institute of Automation and Electrometry of Siberian Branch of the Russian
Academy of Sciences, Novosibirsk, Russia
[3] Institute of Computational Mathematics and Mathematical Geophysics of Siberian
Branch of the Russian Academy of Sciences, Novosibirsk, Russia
[4] Novosibirsk State Technical University, Novosibirsk, Russia

Abstract. The paper presents an approach to the development of an information-analytical system to support the solution of compute-intensive problems of mathematical physics on supercomputers. The basis of this system is a knowledge base built on the basis of the problem domain ontology. This system provides effective information and analytical support to users thanks to detailed systematized descriptions of (a) the methods and algorithms designed for solving problems on a supercomputer, (b) software components implementing parallel algorithms and fragments of a parallel code, and (c) parallel architectures and devices used in them. Moreover, the system contains information about publications and information resources on this subject. These capabilities saves considerably the time required for mastering the methods for solving compute-intensive problems of mathematical physics on supercomputers since all the necessary information is structured and collected in one place, namely, in the knowledge base of the information-analytical system.

Keywords: Supercomputers · Mathematical physics · Intelligent support · Ontology · Information-analytical system

1 Introduction

Nowadays, there are many high-performance supercomputers which allow solving the problems of great computational complexity. In particular, supercomputers are widely used in the mathematical modeling of various physical phenomena and processes. However, researchers rarely use modern supercomputers to solve their problems. The reason for this is that the researchers are not familiar with the supercomputer architecture and

© Springer Nature Switzerland AG 2021
V. Malyshkin (Ed.): PaCT 2021, LNCS 12942, pp. 434–444, 2021.
https://doi.org/10.1007/978-3-030-86359-3_33

specifics of implementing the algorithms designed for solving their problems on super-computers. Hence, it becomes urgent to create the means of intelligent support (IS) for solving the compute-intensive problems of mathematical physics on supercomputers [1].

We believe that the basis of this intelligent support should be the knowledge about the problem domain (PD) under consideration, including information about the methods and algorithms for solving the problems of mathematical physics and their implementations, and the experience of solving these problems on supercomputers, presented in techniques (manuals and textbooks), expert rules, and software components.

The first step to organizing the IS is providing information and analytical support to the user, that is providing the user with convenient access to the information about all the available methods and algorithms for solving problems on a supercomputer, about the capabilities and limitations of each of them, and about their implementation characteristics. An information and analytical Internet resource built on the ontology of the problem domain could become a means of such support.

This paper is devoted to the development of a system providing the user with information and analytical support for solving compute-intensive problems of mathematical physics on supercomputers.

Section 2 presents a conceptual scheme of intelligent support for solving compute-intensive problems on supercomputers. Section 3 describes the knowledge base of the system proposed. Section 4 contains information about the implementation of information-analytical Internet resource. Section 5 provides an overview of current information resources that support solving tasks on supercomputers. The Conclusion summarizes the intermediate results of the design and implementation of the information-analytical system and outlines plans for the future.

2   Conceptual Scheme of Intelligent Support for Solving Compute-Intensive Problems

The main idea of intelligent support for solving compute-intensive problems of mathematical physics on supercomputers is to use the knowledge about this problem domain presented in the form of ontologies [2] and the experience of solving these problems presented as techniques (manuals and textbooks), expert rules and implemented software components (fragments of parallel code).

Figure 1 shows a conceptual scheme of intelligent support for solving compute-intensive problems designer for the client-server architecture.

The lower part of Fig. 1 shows the information-analytical system implemented on the client side, and the upper part depicts the supercomputer system acting as a server and used for solving the problem. (Note that in this scheme ordinary personal computers can be used as a client.)

The information-analytical system (IAS) includes the following components (see Fig. 1): a library of software components (SC library), a simulation module (SM), a knowledge base (KB), an information-analytical Internet resource (IAIR) and an expert system (ES).

A knowledge base is the central component of the IAS since it is used by an information-analytical Internet resource and an expert system. The KB includes the ontology of the problem domain "Solving compute-intensive problems of mathematical physics on supercomputers", containing, in particular, the descriptions of computational methods, parallel algorithms, parallel architectures and technologies, and ontologies of several subject domain built on the basis of the first ontology and supplementing it with entities from specific subject domains, as well as a set of expert rules (inference rules) expanding the possibilities of logical inference on ontologies.

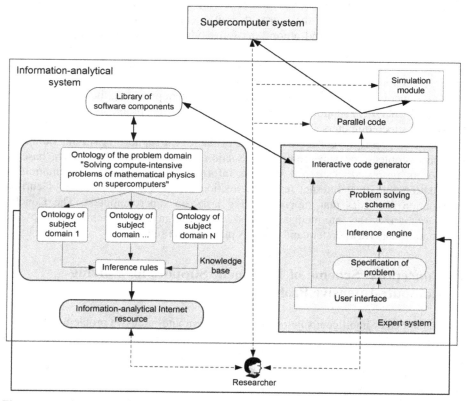

Fig. 1. Conceptual scheme of intelligent support for solving compute-intensive problems on supercomputers.

The ontology-based IAIR is designed to provide the user with information support in solving compute-intensive problems. It provides comprehensive information about the available methods and algorithms, capabilities and limitations of each of them, characteristics of their implementations. This resource is equipped with an advanced user web interface providing content-based access to this kind of information (i.e. convenient ontology-driven navigation through the IAIR content and search using the terms of the problem and subject domains).

The IAIR users have access to the information necessary to solve their problems: a set of available methods and algorithms, descriptions of the features of their numerical implementations, descriptions of the tools available for creating a parallel code, and descriptions of available supercomputer architectures and programming features for these architectures. Thanks to these capabilities, the IAIR user can considerably save on the time required for in-depth acquaintance with the problem domain: all the necessary information is well structured and collected in one place.

The expert system is designed to assist the user in building a parallel code to solve his/her problem on a supercomputer. In addition to the knowledge base, the expert system includes a user interface (UI), an inference engine (solver), and an interactive code generation module.

The ES user interface is primarily intended to specify the problem the user wishes to solve. Using the UI interactive capabilities, the user first selects from the drop-down lists formed on the basis of the ontology an area of interest (currently, it is Astrophysics, Geophysics or Plasma Physics) and a class of problems (for example, modeling a gas discharge in a plasma). Then, the system invites him to specify the problem in more detail, introducing parameters such as the dimension of the problem, the accuracy of the solution, static/dynamic nature, geometry of the computational domain, boundary conditions, and others. The user can either select the values of these parameters from the drop-down lists, also formed on the basis of the ontology, or set them through the data entry forms (in the case of numerical values).

The inference engine using the ontology, inference rules and specification of the problem builds an optimal scheme for solving it.

The interactive code generator supports the creation of a parallel code that solves the problem. This module substitutes the corresponding code fragments from the library of software components (SC) into the scheme for solving the problem. If there is no suitable component in the SC library, the user can substitute it himself, taking it from a standard library or writing a new one.

The SC library includes code fragments implementing the necessary algorithms executed on a supercomputer. The software components are provided with unified specifications, and can thus be integrated into the common code.

The simulation module [3] evaluates the scalability of the resulting code. This module allows the user, basing on the studying of the code behavior with different numbers of cores in model time, to choose an optimal number of computational cores for implementing the code.

3 The Knowledge Base

The core of the knowledge base is the problem domain ontology "Solving compute-intensive problems of mathematical physics on supercomputers," which is described in detail in [1, 4]. It includes concepts typical for any scientific domain, such as *Problem, Object of Research, Research Method,* and *Branch of Science.* Specific to the problem domain under consideration are such entities as *Physical Object* and *Physical Phenomena, Fundamental Laws of Nature, Physical Model, Mathematical Model, Equation System, Numerical Method, Parallel Algorithm, Target Architecture, Parallel Programming Technology,* and *Software Product.*

In addition to the aforesaid basic concepts of the problem domain and relations between them, the ontology described includes classes of a scientific activity ontology serving to present additional information about the problem domain. For these purposes, we use the classes *Branch of Science, Activity, Publication, Event, Person, Organization, Information resource, Geographic Location*, and relations between them.

The knowledge base can include ontologies of several subject domains. For the IAS described, we have developed ontologies for such areas of mathematical physics as *Astrophysics, Geophysics*, and *Plasma physics*.

All the ontologies are presented in the OWL language [5]. Descriptions of ontologies include descriptions of classes and their properties (T-Box), i.e. class attributes (Data Properties) and binary relations (Object Properties) between class objects, as well as class instances (individuals) forming the content of the knowledge base (A-Box) from specific problems, methods, algorithms, software components and elements of parallel architectures.

When developing ontologies, we used ontology design patterns. They are documented descriptions of proven solutions to the typical problems of ontological modeling [6]. They are created to streamline and facilitate the ontology building process and to help the developers avoid some highly repetitive errors of ontological modeling. The patterns used relate to the so-called content patterns [7], defining schemes for describing the key concepts of the subject domains under consideration. Let us consider, as an example, the use of patterns to develop the ontology of plasma physics. Figure 2 presents a pattern for describing numerical methods.

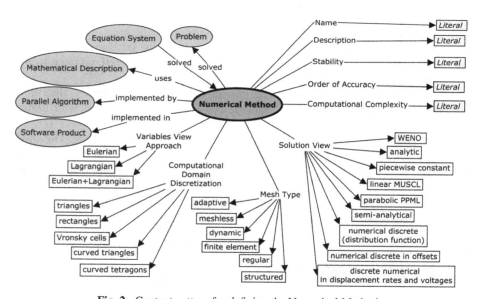

Fig. 2. Content pattern for defining the Numerical Method.

From a meaningful point of view, this pattern is a semantic neighborhood of the central concept, which in this case is the *Numerical method* class. For this class, properties (attributes and relations) are defined. Attributes are represented as Data Properties for the properties whose values are of a standard data type (*Name, Description, Stability, Absolute Accuracy, Order of Accuracy, Computational Complexity*) or as Object Properties for the properties with values from an enumerated data type (*Solution View, Mesh Type, Computational domain Discretization, Variables View Approach*). Relations define the links between the objects of the class considered and the objects of other classes and are represented as Object Properties.

As it was said above, the knowledge base also contains expert rules (inference rules) allowing, basing on the problem specification, selecting the software components for generating a code and determining the architecture on which it will be executed, and vice versa. To describe the expert rules, the SWRL (Semantic Web Rule Language) [8] is used.

To construct ontologies and inference rules, we used the ontology editor Protégé 5.2 [9]. Inference in the OWL ontologies is carried out on the basis of the axioms and inference rules specified in the ontology by means of one of the reasoners (Pellet, FaCT++, HermiT).

4   Implementation of Information-Analytical Internet Resource

Based on the ontology described above, an information-analytical Internet resource was developed to support the solution of compute-intensive problems of mathematical physics on a supercomputer. Figure 3 shows the page of the web interface of this IAIR. On the left side of this page, the concepts of ontology organized in a hierarchy in a general-particular relationship are presented. When you select a class, a list of objects of the selected class and objects of its descendant classes are displayed in the central part of the page. When you select an object from the list, you can see the description of the properties of this object (values of its attributes and relations with other objects) in a tabular or graphical form.

Figure 3 presents the page with a description of the Numerical *Particle-in-Cell Method*. From the description, you can learn that this method can be used to solve problems in plasma physics, namely, to simulate certain phenomena and processes occurring in plasma. This method uses the *Distribution* (of particles in the cells) *function* and helps to solve the *Vlasov equation* and *Boltzmann equation*. The page also shows the main properties of the *Particle-in-Cell Method*, such as *Order of Accuracy, Computational Complexity, Stability, Solution View, Mesh Type* and others.

To create the resource, we used the technology for the development of intelligent scientific Internet resources [10] providing a shell of an intelligent scientific Internet resource, a set of basic ontologies and technique for constructing an ontology from the basic ontologies. The technique involves the development of a system of ontology concepts (T-Box) using the Protégé editor. Populating the ontology by specific objects and their properties (A-Box) can be performed both in the same Protégé editor and in a specialized data editor [11], which is part of the shell of the Internet resource.

Using this data editor, you can create, edit and delete the objects of ontology classes and relations between them. Note that the data editor works under the control of an ontology, which makes it possible not only to facilitate significantly the correct data entry, but also to ensure their logical integrity.

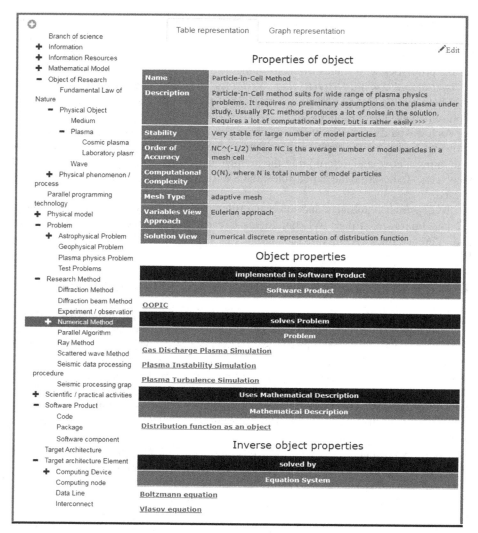

Fig. 3. The IAIR to support the solution of compute-intensive problems of mathematical physics on supercomputers.

When creating a new object, the class required is selected in the visualized hierarchy of ontology classes. After that, according to the description of the class presented in the ontology, the data editor creates a form that includes the fields for entering the values of the object's attributes and its relations with other objects already specified in the ontology (in the A-Box part). The types of these relations and classes of these objects are defined by the corresponding ontology relations (represented in the T-Box). Based on this knowledge and the current state of the A-Box, the data editor provides the user with a list of objects with which the object being created can be connected by this relation. The user only has to select the desired object from this list. When specifying the attributes of an object, the user can either set their values manually, or select from a list of possible values (if a set of values is defined for this attribute in the ontology).

Note that in fact the data editor supports the use of the content patterns presented in Sect. 3.

5 Related Works

At the moment, there are several information resources helping the users solve their problems on supercomputers. The most important of these are AlgoWiki [12], parallel.ru [13], HPCwire [14] and the HPRC website [15].

The information resource AlgoWiki [12] is positioned as an open Internet-encyclopedia on the properties of algorithms and peculiarities of their implementation on various software and hardware platforms, ranging from mobile to exascale supercomputer systems. The stated goal of the AlgoWiki resource is to provide a comprehensive description of each algorithm, which will help assess its potential in relation to a specific computing platform. To do this, AlgoWiki provides a description for each algorithm, as well as a description of its parameters required for both sequential and parallel numerical implementation, indicating the most time-consuming parts. In addition, AlgoWiki provides references to ready-made packaged solutions.

The parallel.ru web resource [13] is aimed at informing the users about the events taking place in the supercomputer community (new technologies, software products and tools for developing parallel code, conferences, new supercomputers, etc.). In addition, this resource provides information and consulting services in high performance computing, such as training in parallel programming technologies, designing and configuring cluster computing systems, providing computing resources for the real computing of varying complexity and intensity, etc.

HPCwire [14] is positioned by its creators as the #1 news and information resource covering the fastest computers in the world and the people who run them. This resource is aimed at professionals in the field of science, technology and business interested in high performance and data-intensive computing. HPCwire covers many different topics, from the late-breaking news and emerging technologies in the HPC, to new trends, expert analysis, and exclusive features.

There are a number of other resources that also provide users with access to high-performance clusters and train the users to work on them. All this is done, as a rule, on a paid basis. So, for example, using the site of the group of high performance research computing (HPRC group) from the University of Texas [15], you can access three clusters with a total peak performance of 947 TF and high-performance storage of 13.5 PB. The HPRC group website also provides consultations, technical documentation and training for the users of these resources.

The resources presented above systematize information related to supercomputers and high-performance computing without using ontologies, which significantly reduces their capabilities, both in terms of representing knowledge and data, and ease of access to them. Representation of such information in the form of an ontology makes it possible to provide not only effective content-based access to it, but also support the user in choosing optimal algorithms and parallel architectures when solving his applied problems due to the possibilities of logical inference on ontology.

Note that there are a number of projects in which ontologies are also used to increase the efficiency of using computing resources and to support the users solving their problems on them.

For example, the authors of paper [16] propose to use ontologies for the formal description of Grid resources. The authors developed an ontology for the Data Mining knowledge area, which makes it possible to simplify the development of Grid applications focused on identifying knowledge from data. The ontology offers an expert in this area a reference model for various data analysis tasks, methodologies and software that has been developed to solve this problem, and helps the user in choosing the optimal solution.

Paper [17] describes an approach using the ontology of the Grid resource metadata and appropriate inference mechanisms, which provide a more efficient management of the Grid resources.

It is also worth noting the fragmented programming technology and its supporting system LuNA [18] designed for the automatic construction of parallel programs. This system is based on the concept of active knowledge, which is close to ontology, i.e. automatic or semi-automatic conversion of a set of expert knowledge about the subject area into a parallel program, possessing different non-functional properties and working correctly on a supercomputer. The LuNA system provides the language enabling you to describe the algorithm for solving a problem in the form of a set of information-dependent tasks, which allows them to be executed in parallel, dynamically redistributed across the nodes of a multicomputer, providing dynamic load balancing on the nodes.

However, the LuNA system supports the user at the level of software implementation and, in part, at the level of binding to the supercomputer architecture [2], but not at the level of constructing a mathematical model.

6   Conclusion

The paper presents an approach to the development of an information-analytical system designed to support the solution of compute-intensive problems of mathematical physics on supercomputers.

It is important to note that the system being developed focuses primarily on supporting the researchers who cannot solve their professional problems neither on personal computers as they lack computing power nor on supercomputers as the researchers lack of proper knowledge and experience. Therefore, this system provides the user with a detailed systematized descriptions of parallel algorithms and architectures and offers him content-based access to the structured descriptions of available software components implementing parallel algorithms and fragments of the parallel code, to the descriptions of available parallel architectures and devices used in them, as well as to the publications and information resources relevant to this problem domain.

Note that all this information is structured and systematized on the basis of an ontology developed by specialists in ontological modeling together with specialists in high performance computing, which ensures its high quality.

In addition, the information-analytical system includes an expert system that has to help the user to build a parallel code solving his problem on a supercomputer.

The system is under active development. The purpose of the authors of the system is to make it practical and useful. As noted above, the knowledge base plays an important role in the users' intelligent support; therefore, we pay special attention to its development. At the moment, the ontology contains about 150 concepts (classes of ontology) and 600 objects including several dozen classes of problems, numerical methods, program codes, and publications on this topic. The knowledge base also includes a set of dozens of expert rules, which are constantly updated.

Acknowledgment. This work is financially supported by the Russian Foundation for Basic Research (Grants no. 19-07-00085 and no. 19-07-00762).

References

1. Zagorulko, G., Zagorulko, Y., Glinskiy, B., Sapetina, A.: Ontological approach to providing intelligent support for solving compute-intensive problems on supercomputers. In: Kuznetsov, S.O., Panov, A.I. (eds.) RCAI 2019. CCIS, vol. 1093, pp. 363–375. Springer, Cham (2019). https://doi.org/10.1007/978-3-030-30763-9_30
2. Sharman, R., Kishore, R., Ramesh, R. (eds.): Ontologies: A Handbook of Principles, Concepts and Applications in Information Systems. Springer, New York (2007)
3. Podkorytov, D., Rodionov, A., Choo, H.: Agent-based simulation system AGNES for networks modeling: review and researching. In: Proceedings of the 6th International Conference on Ubiquitous Information Management and Communication (ACM ICUIMC 2012), Paper 115. ACM (2012)
4. Sapetina, A., Glinskiy, B., Zagorulko, G.: Content of ontology for solving compute-intensive problems of the cosmic plasma hydrodynamics. In: Journal of Physics: Conference Series, vol. 1640, p. 012013 (2020). https://doi.org/10.1088/1742-6596/1640/1/012019
5. Antoniou, G., van Harmelen, F.: Web ontology language: OWL. In: Staab, S., Studer, R. (eds.) Handbook on Ontologies, pp. 67–92. Springer, Heidelberg (2004). https://doi.org/10.1007/978-3-540-24750-0_4
6. Gangemi, A., Presutti, V.: Ontology design patterns. In: Staab, S., Studer, R. (eds.) Handbook on Ontologies. IHIS, pp. 221–243. Springer, Heidelberg (2009). https://doi.org/10.1007/978-3-540-92673-3_10
7. Zagorulko, Y., Borovikova, O., Zagorulko, G.: Development of ontologies of scientific subject domains using ontology design patterns. In: Kalinichenko, L., Manolopoulos, Y., Malkov, O., Skvortsov, N., Stupnikov, S., Sukhomlin, V. (eds.) DAMDID/RCDL 2017. CCIS, vol. 822, pp. 141–156. Springer, Cham (2018). https://doi.org/10.1007/978-3-319-96553-6_11
8. SWRL: A semantic web rule language combining OWL and RuleML. http://www.w3.org/Submission/SWRL/. Accessed 05 Apr 2021
9. Protégé. https://protege.stanford.edu. Accessed 05 Apr 2021
10. Zagorulko, Y., Zagorulko, G.: Ontology-based technology for development of intelligent scientific internet resources. In: Fujita, H., Guizzi, G. (eds.) SoMeT 2015. CCIS, vol. 532, pp. 227–241. Springer, Cham (2015). https://doi.org/10.1007/978-3-319-22689-7_17
11. Zagorulko, Y., Borovikova, O., Zagorulko, G.: Methodology for the development of ontologies for thematic intelligent scientific Internet resources. In: Proceedings of the 2nd Russian-Pacific Conference on Computer Technology and Applications (RPC), pp. 194–198 (2017)

12. AlgoWiki: Open Encyclopedia of Algorithm Properties. https://algowiki-project.org/ru/. Accessed 05 Apr 2021
13. Parallel.ru. https://parallel.ru. Accessed 05 Apr 2021
14. HPCwire website. https://www.hpcwire.com/. Accessed 05 Apr 2021
15. HPRC website. https://hprc.tamu.edu/. Accessed 05 Apr 2021
16. Cannataro, M., Comito, C.: A data mining ontology for grid programming. In: Proceedings of 1st International Workshop on Semantics in Peer-To-Peer and Grid Computing (In Conjunction with WWW 2003), Budapest, Hungry, pp. 113–134 (2003)
17. Amarnath, B.R., Somasundaram, T.S., Ellappan, V.M., Buyya, R.: Ontology-based grid resource management. Softw. Pract. Exp. **39**(17), 1419–1438 (2009)
18. Malyshkin, V., Akhmed-Zaki, D., Perepelkin, V.: Parallel programs execution optimization using behavior control in LuNA system. J. Supercomput. (2021). https://doi.org/10.1007/s11227-021-03654-2

The Web Platform for Storing Biotechnologically Significant Properties of Bacterial Strains

Aleksey M. Mukhin[1,2(✉)] ⓘ, Fedor V. Kazantsev[1,2,3] ⓘ,
Alexandra I. Klimenko[1,2,3] ⓘ, Tatiana N. Lakhova[1,2] ⓘ, Pavel S. Demenkov[1,2] ⓘ,
and Sergey A. Lashin[1,2,3] ⓘ

[1] Kurchatov Genomics Center Institute of Cytology and Genetics SB RAS,
10, Lavrentiev Avenue, Novosibirsk 630090, Russia
mukhin@bionet.nsc.ru
[2] Institute of Cytology and Genetics SB RAS, 10, Lavrentiev Avenue,
Novosibirsk 630090, Russia
[3] Novosibirsk State University, 2, Pirogov street, Novosibirsk 630090, Russia

Abstract. Current biology tasks are impracticable without bioinformatic data processing. Information technologies and the newest computers provide the ability to automatically execute algorithms on an extensive data set and store either strong- or weak-structured data. A well-designed architecture of such data warehouses increases the reproducibility of investigations. However, it is challenging to create a data schema that aids fast search of properties in such warehouses. This paper describes the method and its implementation for storing and processing microbiological and bioinformatical data. The web platform stores genomes in FASTA format, genome annotations in table files that indicate gene coordinates in the genomes, structural and mathematical models to compare different strains and predict new properties.

Keywords: Database · Process · Web · Application · Architecture · Microbial data · Genome data

1 Introduction

The biotechnology industry companies demand a large amount of data sources to create new or improve existing technology lines. Thus, the bioresource collections creation and their improvement are essential tasks for advances in biotechnology.

Several bioresource collections are used to describe phenotypical properties of biological entries [1–3]. It is necessary to keep in mind that such collections contain both phenotypic and genomic data. In the paper, we present a web platform for storing biotechnologically significant properties of bacterial strains. These properties include phenotypical traits, environment conditions, growth rates, substrate consumption/production rates, genomic data and others. These properties have different types and this factor contributes into the complexity of searching procedure. It is important to collect the information on all the steps of bacterial strain gathering and evaluation. We divide the storing data into several levels:

© Springer Nature Switzerland AG 2021
V. Malyshkin (Ed.): PaCT 2021, LNCS 12942, pp. 445–450, 2021.
https://doi.org/10.1007/978-3-030-86359-3_34

- The first level is bacteria isolation information—place of isolation, environment conditions, and other characteristics.
- The second level is laboratory experiment data to determine the phenotype and preparation of strains for genome sequencing, analyzing a transcriptome (identification and quantification of RNA molecules), analyzing a proteome (identification proteins and their complexes).
- The third level is bioinformatical data.
- The fourth level is the data that are derivatives of existing data.

The information on levels 1–3 may take up to 1 GBytes per one strain. The derivative data amount on the 4th level may increase dramatically. The expected collection size is about over 10000 bacteria strains. The data should be stored in a qualitative manner to provide more reliable analysis and investigation reproducibility [4].

Some principles from life science were proposed for the construction of well-structured systems like biocollections [5]:

- Findable – (meta)data is uniquely and persistently identifiable. Should have basic machine-readable descriptive metadata;
- Accessible – data is reachable and accessible by humans and machines using standard formats and protocols;
- Interoperable – (meta)data is machine-readable and annotated with resolvable vocabularies/ontologies;
- Reusable – (meta)data is sufficiently well-described to allow (semi)automated integration with other compatible data sources.

To follow these principles, we should provide requirements to the system and implement them. Some software products (such as openBIS [6] and SEEK [7]) are implemented following these principles, however, they focus on storing of multi-project data The openBIS stores metadata in the PostgreSQL database for saving the hierarchy of the stored data. For storing data, it uses an additional service that works with the controller service. It cannot index values and properties from data to aid the global search. Metadata in SEEK has a tabular type, but some knowledge and data cannot be described in this format and it is rather inconvenient for bioinformatics pipelines. We have taken into account the features of these systems and the FAIR principles to develop the web platform and database for a biological research of biotechnologically significant bacterial strains.

2   Web-Platform

2.1   Architecture

At the moment, the web platform architecture has four modules:

1. The data store is a network resource for storing the weak- and nonstructured files. A JSON scheme indexes the files.

2. The metadata server is a database for storing structured bacterial strain properties. This module is designed to describe the bacterial strain, index the results, and group results related to the single bacterial strain.
3. The web service core handles requests from the user interface and data processing scripts by REST API protocol.
4. The user interface provides typical data representation formats.

2.2 The Data Store

The data store is the Network File System (NFS) resource. We developed the storing structure, which represents bacterial strain data levels (described above). Following the structure, the folder is created for each bacterial strain, and all results (genomic data, annotations, models) are formed in a tree structure by subfolders.

5. raw_data – short nucleotide sequences (reads) presented as text in the system (FASTQ format), which are generated by a laboratory sequencer.
6. assembled_genome – contains an assembled genome, which is presented as a text (FASTA format). Also, this folder stores quality statistics. We use SPAdes [8] for assembling.
7. annotations – contains annotations for assembled genome. This data describes biological properties from a genome. The folder has the following subfolders:

    a. Full-genome annotation files, which determine gene, locus, tRNA locations. They are produced by PROKKA toolbox [9].
    b. A list of found metabolic pathways in KEGG [10] and MetaCyc [11] terms.
    c. Enzyme Commission numbers, which define a set of enzymes encoded in the genome.
    d. A "BLAST" folder stores comparison results with Silva [12] database to determine homologous genes by BLAST algorithm. This analysis can determine taxonomic identity of the genome, and it has an essential value for searching.
    e. structural_models – it describes structural (graph) models. There are lists of predicted proteins in UniProt terms, genes in Entres Gene terms, Gene Ontology terms, and metabolic pathways in KEGG terms. These results are generated from the ANDSystem toolbox [13]. They can be presented as interaction graphs, so it is possible to inspect the strain metabolic pathways and compare them on different strains.

8. math_models – it stores mathematical models generated using PROKKA and ANDSystem result files. With these models, it is possible to perform computational analysis, in particular, to modify parameters to model gene engineering experiments.

We performed each analytic steps using several toolboxes and databases. To combine the execution of steps, we use the Bash scripts as analytical pipelines.

For structuring the result files for each section, the data store has annotation JSON files. This file is readable for a human and a machine as well. The description of a strain is built by these JSON files. These files should have the reserved name "result.json" and have the following structure:

```
{
«NAME»: «Short name the page of the results» (ex: Gene
network...)
«description»: «result description» (ex: which methods)
«objects»: [
    {
       «name»: «name», (Gene network of the process N)
       «description»: «text», (Description of the process)
       «paths»: [ (Paths to files )
         {
            «name»: «filename»,
            «description»: «text»,
            «type»: «text» (format file)
            «path: «path to file» (path to the file from the
current folder)
         },
         ...
    ]
  }
]
}
```

2.3 The Metadata Server

The metadata server is designed for storing metainformation about bacterial strains – a link between stored data divided into groups and additional information for improving search. The server was developed as a PostgreSQL database.

- Strain table (strain) – it is a shortened description of a strain (name, a path to the folder, environment conditions). A primary key is a strain number from the data store.
- Taxon table (taxon) – it is a table, which stores taxonomic identity of each strain. There are three columns:

  - strain_id – strain identification
  - taxon_level – level of taxonomic identity (number)
  - taxon_name – the name of taxon on "taxon_level"

- genomes – it is a table, which stores a path to genomes
- Annotation – it is a table with an annotation of a genome. It is presented as a GFF table for each strain
- Genes – list of genes generated from gene network reconstruction
- Proteins – list of proteins generated from gene network reconstruction
- Gene Ontology network (go_net)

- strain_id - strain identification
- go_id – gene ontology identification
- go_net_xml – gene network of Gene Ontology, XML text

• KEGG metabolic network (keg_net)

- strain_id – strain identification
- go_id – gene ontology identification
- kegg_net_xml – KEGG metabolic network, XML text

• Math_models - mathematical models in SBML format (without gene regulation).

2.4 The Web-Service Core

The web service was developed in Java language with Spring Boot framework. This module's task is to link the data store with the metadata server and to provide REST API interfaces for users. The web service uses Hibernate library to work with Database Management System (DBMS) and Spring-integration library to connect the data store by SFTP protocol. Users can interact with web service by developing scripts in Python, Bash, R, and other programming languages.

2.5 The Graphical User Interface

The web application is developed using the standard web technology stack (HTML, CSS, JavaScript) and Vue.JS framework to construct the multi-component application. This application interacts with the web service by REST API protocol exchanging JSON objects as messages and results.

3 Conclusion

Presented web platform for storing biotechnologically significant properties of bacterial strains contains the information on more than 600 bacterial strains (total size > 1TB). The platform provides links to genomes and processing results. The annotation JSON scheme, the metadata server and relation tables are implemented.

Currently, we are working to improve the database structure and result visualization. The bioinformatics researchers of Kurchatov Genomics Center are working with the developed web platform (e.g., searching for new candidates for super production).

Funding. The work was funded by the Kurchatov Genomic Center of the Institute of Cytology and Genetics of Siberian Branch of the Russian Academy of Sciences (Novosibirsk, Russia) according to the agreement with the Ministry of Education and Science RF, No. 075-15-2019-1662.

References

1. VIR Institute. History, scientific word by most important accessions. Passport database of the VIR's plant genetic recources collection – VIR Institute. History, scientific word by most important accessions. Passport database of the VIR's plant genetic recources collection. https://www.vir.nw.ru/en/. Accessed 07 Apr 2021

2. Lashin, S.A., et al.: An integrated information system on bioresource collections of the FASO of Russia. Vavilovskii Zhurnal Genet. Selektsii **22**(3), 386–393 (2018). https://doi.org/10.18699/VJ18.360

3. Guralnick, R.P., Zermoglio, P.F., Wieczorek, J., LaFrance, R., Bloom, D., Russell, L.: The importance of digitized biocollections as a source of trait data and a new VertNet resource. Database **2016**, baw158 (2016). https://doi.org/10.1093/database/baw158

4. Smith, V.S., Blagoderov, V.: Bringing collections out of the dark. ZooKeys **209**(209), 1–6 (2012). https://doi.org/10.3897/zookeys.209.3699

5. Wilkinson, M.D., et al.: Comment: the FAIR guiding principles for scientific data management and stewardship. Sci. Data **3**(1), 1–9 (2016). https://doi.org/10.1038/sdata.2016.18

6. Bauch, A., et al.: OpenBIS: a flexible framework for managing and analyzing complex data in biology research. BMC Bioinform. **12**(1), 1–19 (2011). https://doi.org/10.1186/1471-2105-12-468

7. Wolstencroft, K., et al.: SEEK: a systems biology data and model management platform. BMC Syst. Biol. **9**(1), 1–12 (2015). https://doi.org/10.1186/s12918-015-0174-y

8. Bankevich, A., et al.: SPAdes: a new genome assembly algorithm and its applications to single-cell sequencing. J. Comput. Biol. **19**(5), 455–477 (2012). https://doi.org/10.1089/cmb.2012.0021

9. Seemann, T.: Prokka: rapid prokaryotic genome annotation. Bioinformatics **30**(14), 2068–2069 (2014). https://doi.org/10.1093/bioinformatics/btu153

10. Kanehisa, M., et al.: KEGG for linking genomes to life and the environment. Nucleic Acids Res. **36**(SUPPL. 1), D480–D484 (2008). https://doi.org/10.1093/nar/gkm882

11. Karp, P.D., Riley, M., Paley, S.M., Pellegrini-Toole, A.: The MetaCyc database. Nucleic Acids Res **30**(1), 59–61 (2002). https://doi.org/10.1093/nar/30.1.59

12. Quast, C., et al.: The SILVA ribosomal RNA gene database project: improved data processing and web-based tools. Nucleic Acids Res. **41**(D1), D590–D596 (2013). https://doi.org/10.1093/nar/gks1219

13. Ivanisenko, V.A., Demenkov, P.S., Ivanisenko, T.V., Mishchenko, E.L., Saik, O.V.: A new version of the ANDSystem tool for automatic extraction of knowledge from scientific publications with expanded functionality for reconstruction of associative gene networks by considering tissue-specific gene expression. BMC Bioinform. **20**(S1), 34 (2019). https://doi.org/10.1186/s12859-018-2567-6

Cellular Automata

Minimal Covering of the Space
by Domino Tiles

Rolf Hoffmann[1]([✉]), Dominique Désérable[2], and Franciszek Seredyński[3]

[1] Technische Universität Darmstadt, Darmstadt, Germany
hoffmann@informatik.tu-darmstadt.de
[2] Institut National des Sciences Appliquées, Rennes, France
domidese@gmail.com
[3] Department of Mathematics and Natural Sciences,
Cardinal Stefan Wyszynski University, Warsaw, Poland
f.seredynski@uksw.edu.pl

Abstract. The objective is to find a Cellular Automata (CA) rule that is able to cover a *2d* array of cells by a minimum number of so-called "Domino Tiles". The aimed patterns are called *min patterns*. Two probabilistic CA rules were designed using templates, small matching patterns. For each of the 12 domino tile pixels a template is declared. If no template is matching then a noise is injected in order to drive the evolution to a valid (full covering) pattern. The First Rule shows the basic mechanism of searching coverings. It evolves very fast stable sub–optimal coverings, starting from a random configuration. The Second Rule is designed in a way that it can find min patterns with a high expected value. The longer the evolution time, the more probably a min pattern appears.

Keywords: Covering problem · Tilings · Matching templates ·
Probabilistic cellular automata · Asynchronous updating

1 Introduction

Our goal is to find a covering of the 2D space by a minimum of so-called *domino tiles* using Cellular Automata (CA). Our problem is one of the diverse covering problems [1] and it is related to the NP-complete *vertex cover problem* introduced by Hakimi [2] in 1965. A vertex cover of an undirected graph is a subset of its vertices such that for every edge (u, v) of the graph, either u or v is in the vertex cover. This means that all vertices are fully connected/reachable through the network of edges defined by the set of nodes of the vertex cover.

A minimum cover is a vertex cover which has the smallest number of vertices for a given graph. This covering problem is closely related to the general dominating set problem in graph theory [3]. Hakimi proposed a solution method based on Boolean functions, later integer linear programming [4], branch-and-bound, genetic algorithm, local search [5], and learning automata [6] were used, among others. Other related problems are the *Location Set Covering Problem* [7] and the *Central Facilities Location Problem* [8]. For covering problems there are a

© Springer Nature Switzerland AG 2021
V. Malyshkin (Ed.): PaCT 2021, LNCS 12942, pp. 453–465, 2021.
https://doi.org/10.1007/978-3-030-86359-3_35

lot of applications, in physics, chemistry, engineering, economy, urban planning, network design, etc. For example, finding a minimum vertex cover in a network corresponds to locating an optimal set of nodes on which to place controllers such that they can monitor the data going through every link in the network.

Our problem is to cover the *2d*-grid space by domino tiles. It is different from the classical cover problem because sophisticated constraints have to be taken into account, namely two adjacent cells (vertices) of the vertex cover have to be used pairwise (forming a domino, the two kernel pixels of a domino tile), and dominoes are not allowed to touch each other. The mapping of our problem to classical algorithms appears not to be straight forward, therefore it was not further investigated in this paper.

Our approach is to treat this special domino covering problem as a pattern formation problem using non-overlapping or partially overlapping tiles. Concerning tiling problems we cite [9]: *"If we have as many copies as we like of a finite set of shapes, can we fill a given region with them? Tiling problems are potentially very hard ... For finite regions, this problem becomes NP-complete ... In fact, tiling problems can be difficult even for small sets of very simple tiles, such as polyominoes ..."*. Our problem is important as it is a general optimization problem minimizing the cost (number of invested tiles) and it maybe applied to spin-systems where the spins try to keep a maximal distance of each other but are not allowed to exceed a certain range.

For the problem of forming a *Domino Pattern* we yielded already good results by using a probabilistic CA rule [10, 12]. There the number of dominoes was maximized by using overlapping tiles. We want to follow the same general approach, but now the problem is different because the number of tiles has to be minimized rather than to be maximized. In [11] a related approach was taken in order to cover the space by sensor tiles which is useful to solve the wireless-sensor-network covering problem. In this paper we incorporate new ideas, like the injection of asymmetric noise (more zeroes than ones). Our work was also inspired by *Parallel Substitution Algorithms* [13].

In Sect. 2, the domino covering problem is stated. In Sect. 3, two probabilistic CA rules are designed and tested. The First Rule shows the basic concept. The Second Rule finds min patterns, and its performance is evaluated for different field sizes. Conclusions are given in Sect. 4.

2   Problem Statement

A given 2d cell array (also called *field*) shall be covered by so-called *domino tiles* (Fig. 1(a)). A tile consists of square elements that we call "pixels". We do not call them "cells" in order to distinguish them from the cells of the CA field to be covered.

A *domino tile* of size 3×4 consists of two edge-to-edge connected black pixels (the *kernel*, the *true* domino) and 10 surrounding white pixels (the *hull*). For short we will use the term "domino" for a *domino tile* and also for a *true domino* when misunderstanding is implausible. Two types of dominoes are distinguished, the horizontal oriented domino (D_H) and the vertical oriented (D_V).

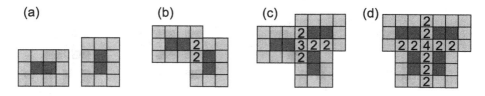

Fig. 1. (a) Horizontal and vertical domino tile, (b) two cells of two domino hulls are overlapping, marked by *2*, (c) the cell marked by *3* is the overlap of three domino hulls, (d) a case with *4* overlapping hull cells.

A pattern is a field where the cells are colored black or white. A *domino pattern* is a pattern that contains domino tiles (at least one). A "valid" domino pattern consists of domino tiles that completely cover the given field. We may imagine that the pixels of the tiles paint the cells of the field. So a valid domino pattern does not contain uncovered/unpainted cells.

The objective is to find a CA rule that can form a *valid domino pattern* that covers a *2d* field by a minimum number d_{min} of domino tiles. We call such a pattern *min pattern*, in contrast to a *max pattern* that contains a maximum number d_{max} of tiles.

We will design a CA rule for a square array of $N = n \times n$ cells with state values $\in \{0,1\}$. It is enclosed by border cells of constant value 0. So the whole array is of size $(n+2) \times (n+2)$. In our graphical representations, color white or green represents value 0, and color black or blue represents value 1.

It is allowed –and often necessary for a valid solution– that hull pixels of different domino tiles overlap with each other. For a valid pattern we do not require that border cells are covered, but they may be covered by hull pixels. The possible levels of overlapping, ranging from $v = 2$ to $v = 4$, are displayed in Fig. 1(b–c–d).

The number of dominoes is denoted as $d = d_H + d_V$, where d_H is the number of horizontal dominoes and d_V is the number of vertical dominoes. A further requirement could be that the number of domino types should be equal (or almost equal) (*balanced pattern*): $d_H = d_V$ if d_{min} is even, and $d_H = d_V \pm 1$ if d_{min} is odd, where $d_{min}(n)$ is the number of dominoes of a valid min pattern.

To summarize it briefly, we want to solve the **Minimal domino covering problem:** Find a valid coverage with a minimum number of domino tiles. The domino tiles are allowed to overlap and to cover border cells. By contrast, in our former work on dominoes we addressed the **Maximal domino covering problem:** Find a valid coverage with a maximum number of domino tiles.

Some valid 7×7 patterns are depicted in Fig. 2. We search for *min patterns* such as pattern (a) that covers all the 49 cells in the square by 5 dominoes. We can observe there some cells with overlap level $v = 2$. Another option is the search for max patterns like (g, h) that was already investigated in [10,12]. For 7×7, the minimal number is $d_{min} = 5$ and the maximal number is $d_{max} = 10$. In this example, all border cells in the patterns (a) and (d) are not covered, whereas all of them are covered in (g) and (h). Note that if we shift the dominoes in pattern

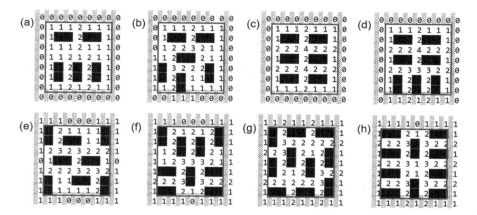

Fig. 2. Samples for valid (full covering) patterns of 7×7 fields. (a) Min pattern (with a minimal number of dominoes). (g, h) Max pattern (with a maximal number of dominoes). The number of dominoes is 5 for (a), 6 for (b, c), 7 for (d), 8 for (e), 9 for (f), and 10 for (g, h). The patterns (a, b, d, e, f) are balanced (difference between the number of horizontal and vertical dominoes is at most 1). No balanced max pattern exists (to be proved). – The numbers (aligned to the right) specify the cover level, also depicted in the border cells.

(a) and (d) in direction to the borders we can yield patterns with cover level $v = 1$ only.

The Number of Dominoes in Min Patterns. A formula giving the *maximal* number of dominoes that can be placed in an $n \times n$ square was already presented in [12]. We have no formula yet for the *minimal* number of dominoes that can cover the square without gaps. For some n it seems to be easy to find a min pattern, for instance 3×4 tiles can cover a 12×12 field without overlap. But for other field sizes it is not straightforward. One idea is to partition the space into sub-fields, cover them separately, and then join them. For fields up to 12×12 the numbers of min/max dominoes are:

n	2	3	4	5	6	7	8	9	10	11	12
d_{min}	1	1	2	4	4	5	6	9	10	12	12
d_{max}	1	2	4	6	8	10	13	16	20	24	28

3 Design of the CA Rules

Cell State. We use the cell state $z = (s, h)$, where $s \in \{0, 1\}$ is the *main state* used to represent the pattern and h is the *hit value* that stores the number of template hits defined later in more detail.

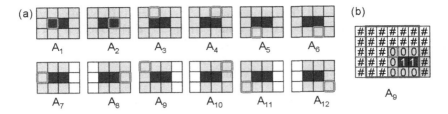

Fig. 3. (a) The 12 templates of the horizontal domino tile. The value of the template center (marked, the so-called reference pixel) is used for cell updating if all other template pixel values match with the corresponding cell values of the current configuration. (b) Template A_9 represented as an array with Don't Care (#). The templates size was reduced from (5×7) to a matching window of size (5×5).

Updating. In our previous work with related problems [10–12] we achieved good results by the use of a probabilistic rule and asynchronous updating. So we will follow this line. Nevertheless, several experiments with synchronous updating were carried out. The used probabilistic rules could also form domino patterns, but the time needed was longer. Furthermore, it was more difficult to drive the evolution to a min or max pattern.

Several tests showed that the later defined rules were robust against different updating orders, even the common fixed sequential order. Therefore the problem could effectively be mapped onto a multi-processor system by choosing updating sequences that minimize the communication between processors. A detailed analysis is a topic for further research.

3.1 Templates

The idea is to modify the current CA configuration step by step using *templates* such that domino patterns evolve out of an initial random configuration.

Templates are local matching patterns that can be seen as another class of tiles (we may call them *template tiles*). The insight is that a valid domino pattern is a field than can totally be covered by non-overlapping or overlapping template tiles without gaps.

We could find the template tiles by inspection (sliding window) of a given set of valid domino patterns. However we can derive them directly from the domino tiles (Fig. 3(a)) in a systematic way. For each of the 12 tile pixels i (marked in red, carrying the domino tile pixel value $dval(i)$), a template A_i is defined. A template can be seen as a copy of the tile, shifted in space in a way that the pixel i corresponds to the center of the template.

In the computation we represent a *template* A_i as an array of size $(a' \times b')$ of pixels, where $a' = 2a - 1$, $b' = 2b - 1$, and $(a \times b)$ is the size of the tile (its bounding box). Our horizontal tile is of size (3×4), thus their templates are of size (5×7) maximal (larger because of shifting). The pixels within a template are identified by relative coordinates $(\Delta x, \Delta y)$, where $\Delta y \in \{-(a-1), \ldots, a-1\}$,

$\Delta x \in \{-(b-1), \ldots, b-1\}$. The center pixel $(\Delta x, \Delta y) = (0,0)$ is called "*reference pixel*". Each template pixel carries a value $val(A_i, \Delta x, \Delta y) \in \{0, 1, \#\}$. The value of the reference pixel is called "*reference value*", $refval(A_i) = val(A_i, 0, 0) \in \{0, 1\}$. Its value is equal to the red marked value of the corresponding tile pixel, $refval(A_i) = dval(i)$. The reference value is

$refval(A_i) = 1$ for $i = 1, 2, 13, 14.$ (value 1 of the domino kernel pixels)
$refval(A_i) = 0$ for $i = 3..12, 15..24.$ (value 0 of the domino hull pixels).

The symbol $\#$ represents "Don't Care", meaning that a pixel with such a value is not used for matching (or does not exist, in another interpretation). Pixels with a value 0 or 1 are *valid* pixels, their values are equal to the values derived from the original tile. Some templates can be embedded into arrays smaller than $(a' \times b')$ when they have "$\#$" at their borders. Note that the valid pixels are asymmetrically distributed in a template because they are the result from shifting a tile.

Many of these templates are similar under mirroring, which can facilitate an implementation. For the vertical domino a corresponding set of 12 templates is defined by 90° rotation $(A_{13} - A_{24})$.

The templates $A_7 - A_{12}$ show white pixels that are not used because the template size (for the later described matching process) was restricted to (5×5). As an example, the reduced template A_9 is marked in Fig. 3(b) by the blue square. The implementation with these incomplete templates worked very well, but further investigations are necessary to prove to which extent templates can be incomplete.

We need also to define the term "*neighborhood template*" that is later used in the matching procedure. The neighborhood template A_i^* is the template A_i in which the reference value is set to $\#$, in order to exclude the reference pixel from the matching process.

3.2   Hit Value

Prior to the actual rule (changing the main state s) the hit value h is determined. At the current site (x, y) all neighborhood templates are tested and the number of hits is stored.

The neighborhood templates A_i^* are tested against the corresponding cell neighbors $B^*(x, y)$ in the current (5×5)–window with its center at (x, y). Thereby the marked reference position $(\Delta x, \Delta y) = (0, 0)$ of a neighborhood template is aligned with the center of the window.

It is possible that several neighborhood templates match (then tiles are overlapping), but there can be no conflicts because then all templates have the same reference value as derived from the tile. As no conflicts can arise, the sequence of testing the templates does not matter.

The hit value $h(x, y)$ is given by:

- 0, if no neighborhood template matches within the cell's neighborhood B^*. This means that cell (x, y) is not covered by a tile pixel.

- 1, if exactly one neighborhood template A_i^* matches with $refval = 0$.
- 2 – 4, if such a number of neighborhood templates A_i^* match with $refval = 0$. This means that 2 – 4 domino tiles are overlapping.
- 100, if exactly one neighborhood template $A_i \in \{A_1, A_2, A_{13}, A_{14}\}$ matches with $refval = 1$. The number 100 was arbitrarily chosen in order to differentiate such a hit from the others. The reason is, that the whole gained information (hit number and $refval$) has to be available in the subsequent main rule. So the hit value is a sort of encoding of the number of matches in combination with the reference value.

Remark. The hit number $h(x, y)$ holds the actual value after matching with all the neighborhood templates. Because of the random sequential updating scheme, the h-values in the (x, y)–neighborhood may not be up-to-date and may carry depreciated values from a former micro time-step evaluation. Nevertheless, the h-values correspond mainly to the cover levels v, especially when the pattern becomes more stable. This inaccuracy introduces some additional small noise which can even speed-up the evolution. And when the pattern becomes stable, the hit number is equal to the cover level, $\forall (x, y) : h(x, y) = v(x, y)$.

3.3   Processing Scheme

All cells are sequentially computed and updated at every time-step t. A new generation at time–step $t + 1$ is then declared after N cell updates (sub-steps) during the compute interval between t and $t + 1$. The processing scheme is:

1. The next cell at position (x, y) (according to an arranged random or deterministic order) is selected for computing. Unless otherwise stated, we will use a random order which is re-computed for each new time-step.
2. The new hit value $h'(x, y)$ (number of template matches) is computed and immediately updated ($h \leftarrow h'$). It is then used in the following main rule to compute the new main state s'. For our later defined rules it is not really necessary to store the hit value, rather it is a temporary information used to evaluate the main rule. Nevertheless, it is useful to store the hit value for more extraordinary rules and for debugging. In addition, the hit value is close to the cover level, especially when the pattern is not very noisy and converges to a stable pattern.
3. The main rule $f = s'(x, y)$ is evaluated defining the new main state taking into account the main states of cell (x, y) and its neighbors, the actual hit value $h(x, y)$ and the center value ($refval$) of a matching template in the case of a hit. A more detailed explanation is given later. Then the main state is immediately updated ($s \leftarrow s'$).

3.4   The First Rule

The main working principle is shown by the *First Rule*. It yields valid non-optimal domino patterns with a different number of dominoes ranging between minimum (very seldom) and maximum (seldom). Then following, the *Second, Minimizing Rule* is designed that can produce min patterns with a high expectation.

The Sub-Rule A of the First Rule. The main basic working principle will be shown by the basic rule A. We assume that the hit value $h(x, y)$ was already computed at the site (x, y) just before. The basic rule A is:

$$s'(x, y) = \begin{cases} s(x, y) & \textbf{default} & (a) \\ 0 & \textbf{if } (h > 0) \textbf{ and } (h \neq 100) & (b1) \\ 1 & \textbf{if } (h = 100) & (b2) \\ random \in \{0, 1\} \text{ with probability } \pi_0 \textbf{ if } h = 0 & (c) \end{cases}.$$

The new state is set to 0 if the hit value is $h = 1, 2, 3, 4$ (b1). This means at least one neighborhood template matches with reference (center) value 0. The new state is set to 1 if the hit value is $h = 100$ (b2). This means a match of a neighborhood template with reference value 1. In other words, if a complete template match is found, the cell's state remains unchanged, and if only the reference value in the center is wrong, it is adjusted. Noise is injected, if there is no hit (c), in order to further the evolution.

We can observe four classes of patterns during an evolution.

- **class I** (valid stable). The dominoes are covering the square totally without gaps and the reached pattern is stable.
- **class II** (partially stable). The dominoes are covering the field not totally with at least one gap. Gap cells are toggling their state values $(0 \leftrightarrow 1)$ due to the injection of noise that never ends. However, such patterns consist of dominoes which do not change (neither position, nor orientation or number). As the gaps do not disappear through the noise, we may interpret such a situation as a live-lock.
- **class III** (valid transient). Such a pattern is valid but not stable. It appears and disappears during an evolution.
- **class IV** (invalid transient). Such a pattern is invalid and not stable. It appears and disappears during an evolution. It can be totally noisy or partially noisy showing some domino tiles.

When we test Sub-Rule A on square fields we observe a fast convergence to class I and class II patterns only. Rule A was tested on a (7×7)–square with 1000 runs under the time limit $T_{limit} = 200$ with different probabilities. The number of evolved dominoes allowing up to 4 gaps was:

dominoes	5	6	7	8	9	10		
frequency [1/1000]	1	13	219	510	236	21	$\pi_0 = 1$	$t_{avrg} = 5.47$
frequency [1/1000]	0	18	255	512	202	13	$\pi_0 = 0.5$	$t_{avrg} = 6.47$
frequency [1/1000]	0	20	234	522	213	11	$\pi_0 = 0.25$	$t_{avrg} = 9.79$

The higher the probability π_0 the faster the aimed patterns evolve. Therefore we may choose the highest possible probability $\pi_0 = 1.0$. The number of dominoes lies between minimum (very seldom) and maximum (seldom). For all the three examples the average number of evolved dominoes d_{avrg} was close to

8. An evolved pattern remains stable or partially stable because the rule detects complete tiles everywhere except for gaps. So there will be noise injected only at gaps which does not influence the already found dominoes [12] if the gaps are isolated. As we aim at patterns without gaps, the following additional rule can solve this problem.

The Sub-Rule B of the First Rule. We define an additional sub-rule B that will turn class II patterns into class I patterns. Analyzing the class II patterns of Rule A we observe uncovered toggling gaps with hit value $h = 0$. The idea is to disseminate the "gap information" to the cells in the von-Neumann neighborhood. If a cell in the neighborhood detects a hit-zero cell, it will produce additional noise in order to drive the evolution to a stable pattern without gaps. Thereby already found dominoes in a class II pattern can be destroyed.

The sub-rule B is:

$$s''(x,y) = \begin{cases} s'(x,y) & \text{default} \\ random \in \{0,1\} \; with \; probability \; \pi_1 \; \textbf{if} \; \exists h(x \pm 1, y \pm 1) = 0 \end{cases}$$

Test of the First Rule (A Followed by B). The whole rule was tested on a (7×7)–square with 1000 runs under the time limit $T_{limit} = 200$. Several probabilities were checked, best convergence was obtained for $\pi_0 = 1.0$ and $\pi_1 = 0.15$. All patterns reached stability (class I). However, no min pattern with 5 dominoes evolved. The distribution was:

dominoes	5	6	7	8	9	10
frequency [1/1000]	0	1	24	311	586	78

The average number of dominoes evaluates to $d_{avrg} = 8.71$. The average needed time was $t_{avrg} = 22.1$ ($min \; 2 - max \; 189$).

We can conclude that this rule evolves very fast valid domino patterns, but it is very unlikely that a min pattern appears. Therefore we needed to design a modified Second Rule, that can evolve min patterns with a high expectated value.

3.5   The Second Rule: Minimizing the Number of Dominoes

The Second Rule shall evolve min patterns with a high probability. The designed rule consists of Sub-Rule C followed by Sub-Rule D.

Sub-Rule C. Rule C is a modified version of Rule A. The idea is to inject more "white" than "black" noise when there is an uncovered cell ($h = 0$). Why? Because min patterns contain much more white (0) than black (1) colored cells. A min pattern with d dominoes contains approximately $10d$ white and $2d$ black cells, the ratio as it is within a single domino tile. This means that a min pattern contains near to $1/5$ black cells.

The sub-rule C is:

$$s'(x, y) = \begin{cases} s(x, y) & \textbf{default} \\ 1 & \textit{with probability } \pi_{00} \textbf{ if } (h = 0) \textbf{ and } (s = 0) \\ 0 & \textit{with probability } \pi_{01} \textbf{ if } (h = 0) \textbf{ and } (s = 1) \end{cases}$$

This rule injects black noise with probability π_{00} if the uncovered cell is white, and injects white noise with probability π_{01} if the uncovered cell is black. The chosen probabilities (for best results) were $\pi_{00} = 0.1$ and $\pi_{01} = 0.9$. At the moment we cannot explain properly why the best working ratio $1/9$ is higher than the ratio $1/5$ prompted before.

Note that Rule C is a more general form of Rule A. Therefore it can emulate rule A by setting $\pi_{00} = \pi_{01} = 0.5\pi_0$. As Rule A alone is not effective enough to form min patterns, the additional sub-rule D was defined.

Sub-Rule D. The idea for this rule is to inject noise where there is a high overlap. Several attempts were made of reacting on that condition with noise, like using different probabilities for different hit values, or determining the hit density in a local window. The result was, that a relatively simple condition (together with rule C) was very effective, although more complex conditions may slightly improve the performance.

The rule D is:

$$s''(x, y) = \begin{cases} s'(x, y) & \textbf{default} \\ 1 & \textit{with probability } \pi_2 \textbf{ if } (h = 2) \end{cases}$$

Test of the Second Rule (C Followed by D). The whole rule was tested on a (7×7)–square with 1000 runs under the time limit $T_{limit} = 1000$. The best performing probabilities found were used, $\pi_2 = 0.07$ and $\pi_{00} = 0.1$ and $\pi_{01} = 0.9$. The evolved patterns belong to class I (valid stable) and class III (valid transient). 35 out of 1000 with 5 or 6 dominoes are stable, the remaining are transient. 136 min patterns were found, 22 of them were stable and 114 were transient.

The distribution of the found patterns is:

dominoes	5	6	7	8	9	10
frequency [1/1000]	136	695	166	3	0	0

The average number of dominoes evaluates to $d_{avrg} = 6.03$. The average time needed was $t_{avrg} = 358.2$ ($min\ 2 - max\ 993$). Compared to the First Rule, the Second Rule is able to evolve min patterns with a high expectation rate (13.6% for this test). Comparing the time, the Second Rule needs $359.2/23.16 = 15.5$ times more to find a target pattern within the range given by the corresponding average number of dominoes $d_{avrg} = 6.03$.

Figure 4 shows the evolution of a stable min pattern with 5 dominoes. During the evolution, the first valid pattern with 7 dominoes appears at $t = 26$, it is a

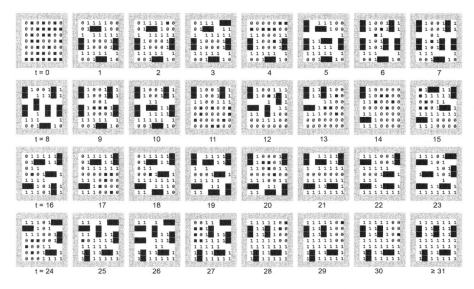

Fig. 4. Example of an evolution of a 7×7 min domino pattern. The number of appearing dominoes is fluctuating. Valid patterns contain no cells with cover level $v = 0$. The first valid pattern appears at t = 26 but it is not minimal. The second valid pattern appears at t = 31, it is minimal and stable. Coloring: gray (border), 0 ($s = 0, v = 0$), 1 ($s = 0, v = 1$), blank ($s = 0, v > 1$), full black square (domino part, $s = 1, v = 1$), small black square ($s = 1, v = 0$).

transient. Then at $t = 31$ a min pattern appears that is stable. No more noise is injected because the cover level is $v = 1$ everywhere.

The percentage of finding an optimal min pattern can be increased by increasing the time limit T_{limit} as the following table shows.

dominoes	5	6	7	8	9	10	
$freq.$ [1/1000]	136	695	166	3	0	0	$T_{limit} = 1000, N_{run} = 1000$
$freq.$ [1/1000]	501	497	1	0	0	0	$T_{limit} = 5000, N_{run} = 1000$
$freq.$ [1/100]	80	20	0	0	0	0	$T_{limit} = 20000, N_{run} = 100$

3.6 Performance for Other Field Sizes

The Second Rule was also tested on other field sizes under different time limits (Table 1). For sizes up to (5×5) all runs yielded optimal min patterns. For larger fields, min patterns were found among others with a high expectation rate. The average number of dominoes is d_{avrg}. Only min pattern were found for $n \leq 5$, then $d_{avrg} = d_{min}$ yields. For $n \geq 6$ not only min patterns were found but the difference $d_{avrg} - d_{min}$ is small. This means that the probability to find a min pattern is high. The last column gives the distribution of the dominoes with a certain number. In the case 11×11, no min pattern with 12 dominoes (0×12)

464 R. Hoffmann et al.

Table 1. Simulation results (average number of dominoes, average time) for different field sizes under certain time limits. The average number of evolved dominoes reaches the minimum or comes close to it.

N = n x n	d min	number runs	T Limit	d avrg	t avrg	distribution of dominoes
2 x 2	1	100	20	1.00	1.8	100x1
3 x 3	1	100	300	1.00	32.6	100x1
4 x 4	2	100	1500	2.00	171.9	100x2
5 x 5	4	100	200	4.00	24.2	100x4
6 x 6	4	100	10 000	4.09	290	91x4 + 9x5
7 x 7	5	100	20 000	5.20	406	80x5 + 20x6
8 x 8	6	100	50 000	7.09	1 531	7x6 + 77x7 +16x8
9 x 9	9	100	50 000	9.06	1 161	92x9 + 8x10
10 x 10	10	100	100 000	11.35	3 438	4x10 + 57x11 + 39x12
11 x 11	12	100	200 000	14.4	77 138	0x12 + 6x13 + 48x14 + 46x15

was found under this time-limit. For larger filed sizes the effort to find an optimal min pattern increases more and more. To find an optimum for large n needs a distributed implementation on a parallel computing system. Further work has to be done to find the time-complexity in theory and through more experiments. Another idea is to split the problem into sub-problems and then join the partial solutions (divide and conquer).

4 Conclusion

We designed two CA rules to find sub-optimal and optimal min domino patterns. The first rule evolves very fast stable valid patterns, with an average number of dominoes lying between minimum and maximum. The underlying design principle is methodical and based on a set of templates derived from all pixels of a domino tile. The second rule injects asymmetric noise, cells are colored more often white than black, and this rule tries to alter cells with a high overlap. It can find min pattern with a high probability although the time may exceed the available processing resources. In future work, the CA rules could be mapped onto parallel processing systems, compared to vertex cover algorithms, or a divide-and-conquer strategy could be considered.

References

1. Snyder, L.V.: Covering Problems. In: Foundations of location analysis, 2011, pp. 109–135. Springer, Boston, MA (2011)
2. Hakimi, S.L.: Optimum distribution of switching centers in a communication network and some related graph theoretic problems. Op. Res. **13**, 462–475 (1965)
3. Haynes, T.W., Hedetniemi, S.T., Slater, P.J.: Fundamentals of domination in graphs, Pure and applied mathematics 208, Dekker (1998)

4. Gomes, F.C., Meneses, C.N., Pardalos, P.M., Viana, G.V.R.: Experimental analysis of approximation algorithms for the vertex cover and set covering problems. Comput. Op. Res. **33**, 3520–3534 (2006)
5. Richter, S., Helmert, M., Gretton, C.: A stochastic local search approach to vertex cover. In: Hertzberg, J., Beetz, M., Englert, R. (eds.) KI 2007. LNCS (LNAI), vol. 4667, pp. 412–426. Springer, Heidelberg (2007). https://doi.org/10.1007/978-3-540-74565-5_31
6. Mousavian, A., Rezvanian, A., Meybodi, M.R.: Cellular learning automata based algorithm for solving minimum vertex cover problem. In: 22nd Iranian Conference on Electrical Engineering (ICEE), pp. 996–1000 (2014)
7. Church, R.L., ReVelle, C.S.: Theoretical and computational links between the p-median, location set-covering, and the maximal covering location problem. Geogr. Anal. **8**(4), 406–415 (1976)
8. Mehrez, A.: Facility Location Problems, Review, Description, and Analysis. Geogr. Res. Forum **8**, 113–129 (2016)
9. Moore, C., Robson, J.M.: Hard tiling problems with simple tiles. Discret. Comput. Geom. **26**(4), 573–590 (2001)
10. Hoffmann, R., Désérable, D., Seredyński, F.: A probabilistic cellular automata rule forming domino patterns. In: Malyshkin, V. (ed.): PaCT 2019. LNCS, vol. 11657. Springer, Cham (2019). https://doi.org/10.1007/978-3-030-25636-4
11. Hoffmann, R., Seredyński, F.: Covering the space with sensor tiles. In: Gwizdałła, T.M., Manzoni, L., Sirakoulis, G.C., Bandini, S., Podlaski, K. (eds.) Cellular Automata. ACRI 2020. Lecture Notes in Computer Science, vol. 12599, pp. 156–168. Springer, Cham (2021). https://doi.org/10.1007/978-3-030-69480-7_16
12. Hoffmann, R., Désérable, D., Seredyński, F.: A cellular automata rule placing a maximal number of dominoes in the square and diamond. J. Supercomput. **77**(8), 9069–9087 (2021). https://doi.org/10.1007/s11227-020-03549-8
13. Achasova, S., Bandman, O., Markova, V., Piskunov, S.: Parallel Substitution Algorithm: Theory and Application. World Scientific, Singapore, New Jersey, London, Hong Kong (1994)

Application of the Generalized Extremal Optimization and Sandpile Model in Search for the Airborne Contaminant Source

Miroslaw Szaban[1]([✉])(ID), Anna Wawrzynczak[1,2](ID), Monika Berendt-Marchel[1](ID), and Lukasz Marchel[1]

[1] Institute of Computer Science, Siedlce University
of Natural Sciences and Humanities, Siedlce, Poland
{miroslaw.szaban,awawrzynczak,monika.berendt-marchel}@uph.edu.pl
[2] National Centre for Nuclear Research, Otwock, Poland

Abstract. In this paper, the Generalized Extremal Optimization (GEO) algorithm is combined with the Sandpile model to localize the airborne contaminant source based on the contaminant concentration's spatial distribution. The GEO algorithm scans the proposed model's solution space to find the contamination source by comparing the Sandpile model output with the contaminant distribution over the considered area. The comparison is made by evaluating the assessment function considering the differences between the distribution of the sand grains from the Sandpile model and contaminant concentrations reported by the sensor network monitoring the considered area. The evolution of the sand grains in the Sandpile model is realized by the cellular automata cells. The proposed GEO-Sandpile localization model efficiency is verified using the synthetic contaminant concentration data generated by the Gaussian dispersion model: conducted test cases presented in this paper covered the various wind directions, and release source positions. Obtained results support the statement that the proposed algorithm can, with acceptable accuracy, localize the contaminant source based only on the sparse-point concentrations of the released substance.

Keywords: Sandpile model · Generalized Extremal Optimization (GEO) · Gaussian dispersion model · Airborne contaminant · Cellular automata

1 Introduction

Emissions and storage of toxic materials pose a constant risk of releasing them into the atmosphere, threatening human health and the environment. The most dangerous are cases when the dangerous level of the contaminant of unknown origin is detected. Knowledge about the release source coordinates and estimated release rate allows in a short time to undertake appropriate steps to

V. Malyshkin (Ed.): PaCT 2021, LNCS 12942, pp. 466–478, 2021.
https://doi.org/10.1007/978-3-030-86359-3_36

prevent further contamination from spreading into the atmosphere. In recent years, many scientists have been dealing with locating the source of contamination in open areas. Existing algorithms that can cope with the task can be divided into two categories. The first ones are based on the backward approach and are dedicated to the open areas or a continental-scale problem. The second are based on the forward approach. In this case, the appropriate dispersion model parameters are sampled (among them source location) to chose the one giving the smallest distance measure between the model outputs and sensors measurement in the considered spatial domain. Such an inverse problem has no unique analytical solution but might be analyzed with probabilistic frameworks, as the Bayesian approach, where all searched quantities are modeled as random variables. In [6,17,31] authors presented the reconstruction of the airborne contaminant source utilizing the Bayesian approach in conjunction with Markov Chain Monte Carlo, Sequential Monte Carlo, and Approximate Bayesian Computation algorithm. A comprehensive literature review of past works on solutions of the inverse problem for atmospheric contaminant releases can be found in (e.g.[14]). The offered algorithms require multiple runs of the appropriate dispersion algorithm simulating contaminants' transport in the atmosphere. Unfortunately, many models are computationally expensive, and a single simulation can take many minutes, hours, or even days. This is problematic because, in real-life situations, it is crucial to quickly estimate the most probable location of the contamination source based solely on the sensor network's concentration data. Recently, artificial neural networks have been proposed to simulate contaminant transport in the urbanized area [29,30].

This study proposes a new tool for predicting the source of contamination based on the registered substance concentrations. This approach is based on applying the Sandpile model, mapping a given problem area by sand falling within a limited space. The idea is to check does the Sandpile model can be applied to simulate the airborne contaminant transport. The Generalized Extremal Optimization (GEO) algorithm is used to indicate the point of dropping the sand grains as an algorithm capable of dealing with any combination of model variables. In the proposed model, the GEO algorithm searches the solution space to find the contamination source. The Sandpile model was used to evaluate the assessment function, simulating the spread of contamination in the studied area. The proposed algorithm's operation was verified using the synthetic data of the contaminant concentration over the simulation domain. The synthetic spatial distribution of contaminant was generated using the Gaussian dispersion model (e.g.[34]). The Gaussian plume model is the most common air pollution model. It is based on a simple formula describing the three-dimensional concentration field generated by a point source under stationary meteorological and emission conditions. Despite many simplifications, the Gaussian plume model is used up to now by many researchers.

The paper is organized as follows. The following section presents the Sandpile model and the idea of its application. Section 3 outlines the concept of the GEO algorithm. In Sect. 4, the cellular automata (CA) for the simulation model is introduced. The simulation model is widely described in Sect. 5. Section 6 brings closer the Gaussian dispersion model used to generate the synthetic data. Results are described in Sect. 7. The last section concludes the paper and presents the future prospects.

2   Sandpile Model

The Sandpile model is a prototypical deterministic model for studying self-organizing criticality. In this model, proposed by Bak, Tang, and Wiesenfeld in 1987 [4,7,16], the steady-state (critical state) is collapsing at some point in time. The simplest Sandpile model starts with a single column configuration. Then, in each step, if a column has at least two more grains than its right-hand neighbor, it passes one grain. It has been proved [13,18] that this model converges only to one configuration, in which the evolution rule cannot be applied in any column. This configuration is called a fixed point. All possible arrangements obtained from the columns' initial configuration by using the evolution rule are characterized in the space of a two-dimensional grid [10].

When describing the Sandpile model, a simple mental model with rice can be used [11]. Let us consider a pile of sand on a small table. Dropping another grain on the pile may cause avalanches that slide down the slopes of the pile. The dynamics of the resulting avalanche in such a situation depend on the steepness of the slope. During the avalanche, the sand will rest somewhere on the table. If the avalanche continues, some of the grains are falling down the edge of the table. If one grain is added to the pile, on average, it increased the slope. In the long perspective of grain-spreading, the slope evolves to a critical state. At that moment, a single grain dropped into a pile determines a massive avalanche. This thought experiment suggests that the critical condition is very prone to stimuli because a small change (internal or external) can have a significant effect [15].

In the first stage, the class of graphs $G = (V \cup \{s\}, E)$ on which the model is based is defined. G must be finite, unoriented, connected, and loopless. It can have multiple edges and include a distinctive vertex called a terminal vertex. The set of graphs is marked as G. The notation $u \sim v$ is used to denote the neighborhood of G, e.g., $u, v \in E$. The configuration of the Sandpile G model is the vector $\eta = (\eta_\nu, \nu \in V) \in Z_+^{|V|}$. The number η_ν represents the number of sand grains present at the vertex ν in the configuration η. When this number exceeds a certain threshold, the top is said to become unstable and descends (see, Fig. 1), giving one grain of sand to each of its neighbors. The probability of a neighbor obtaining a grain from an unstable vertex is selected from the interval $p \in (0, 1)$. The terminal apex plays a special role, as it can accept an infinite number of grains and never fall [7]. When the stack's local slope due to the difference in height between the top and the adjacent vertex exceeds the local threshold, the grains redistribute, allowing one grain to fall to another top. It is done by reducing the height of one pile and increasing the height of adjacent piles. At the next iteration of sand pouring, the local threshold is reselected. This procedure is repeated until all the vertices are stable, at which point a new

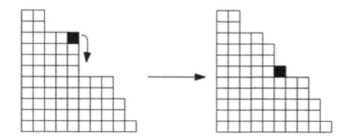

Fig. 1. The avalanche operation of the Sandpile model. Dropping seeds create a slope; after crossing the threshold equal to 3, the slope is falling (avalanche effect) [10].

grain will be introduced on the left side. The configuration $\eta = (\eta_\nu, \nu \in V)$ is stable when $\eta_\nu \leq d^G(\nu)$, where $d^G(\nu)$ is the degree of vertex ν in G [1,7].

In our paper, we used an asynchronous sequential approach because seeds dropping in one point could induce the avalanche, which generates changes in several points recursively to the stable state of the SandPile Model in CA. During the avalanche, the sand seeds go down with direction according to avalanche direction. However, when seeds may go down in a few direction, then the direction is selected with probability proportional to the slope's height in such direction.

The Sandpile model is a model that can be easily modified and adapted to the problem under study. There are one- or two-dimensional variations with open, closed, or infinite boundaries. This model is widely used in such fields as physics, economics, mathematics, theoretical computer sciences, etc. [2,5,19,21,23]. It has also been used for cellular automata, information systems, calculating earthquakes [3], studies of river sediments [24], the spread of forest fires [22], studies of the Earth's magnetosphere [8], or in studies of precipitation distribution [1] and also to diffusion problem (like Lattice-Gas, Lattice-Boltzmann, etc.) [9].

3   GEO Algorithm

Fabiano Luis de Soussa and Fernando Manuel Ramos proposed in 2002 a Generalized Extremal Optimization (GEO) [25] algorithm designed to operate on bit strings capable of dealing with any combination of continuous-discrete or integer variables. GEO algorithm can be used in a multi-modal space, discontinuous, and subject to any constraints. It was proposed as a stochastic method developed to solve optimization problems in complex design spaces. The GEO algorithm is a type of evolutionary algorithm which, like a genetic algorithm or simulated annealing, is a global search metaheuristic but has only one free parameter to adjust. It is based on the natural evolution model and was developed specifically for the use in complex optimization problems [12,26,27,33]. GEO is based on the theory of self-organizing criticality to explain complex systems' behavior in fields such as biology, geology, economics, and others. This theory assumes that large interactive systems evolve naturally to the critical point. A single change of one system element generates the so-called avalanches that can reach any number of elements. The power law describes the probability distribution of the size of avalanches s in the form: $P(s) \sim s^{-\tau}$, where $\tau > 0$. It means that the likelihood of smaller avalanches is greater than that of larger avalanches. However, avalanches of the entire system's size may occur with high probability [26, 27].

Self-organizing criticality can explain features of systems such as natural evolution. It can be proved by a simple model assuming that the selected species are placed next to each other on a line with periodic boundary conditions. The ends of these lines are joined in a circle. Each species has a randomly assigned suitability number with an even distribution range of $(0, 1)$. The species least adapted (lowest suitability value) is forced into a mutation and then assigned a new random number. The change in the adaptation of the weakest species affects the neighbors' efficiency, as they have also assigned new random numbers, even if well adapted. After several iterations, the system evolves to the critical point where all species are beneficial above the established critical threshold. However, the dynamics of such a system ultimately cause the number of species to fall below the critical threshold in avalanches. The reason for this is the worst species mutation in each iteration. As evolution progresses, the species being poorly adapted are forced to evolve. In this case, starting with a population with an even, random distribution of adaptation, the system will evolve and eventually reach a situation where all species will be useful above a certain threshold. An avalanche occurs when one or more species are below a critical threshold. An avalanche's size is the number of species that fall below the threshold between two iterations where all species are above that threshold [26]. Each species in the GEO algorithm is represented by a fragment of the sequence representing the entire ecosystem of the species. In genetic algorithms, variables are encoded in a chromosome-like chain. Each variable is encoded in binary code, and all variables form a string of finite lengths. In GEO, there is no population of strings (unlike the genetic algorithm), and a single bit-string represents the individual. Each bit in the string has an assigned efficiency number representing the level of adaptability of that bit according to the gain or loss resulting from the value of the objective function after bit mutation (inversion) [25, 26].

4   Two-Dimensional Cellular Automata for Localization Model

CAs and their potential to efficiently perform complex computations are described by S. Wolfram in [32]. In this paper, the two-dimensional CA is considered. CA is a rectangular grid of $X \times Y$ cells, each of which can take on k possible states. After determining the initial states of all cells (i.e., the initial configuration of a CA), each cell changes its state according to a rule - transition function TF, which depends on the states of cells in a neighborhood around it. In this paper, the finite CA with the stable boundary conditions is used. The CA cells transition is done asynchronously. As a transition function in this paper, the Sandpile model was used, which was set at the CA grid. The evolution of the sands grains in the Sandpile model was realized by the CA cells.

Assuming, that the contaminant distribution domain is of size $[0, 10000] \times [0, 10000]$ (in meters) in CA, the data space should be mapped from $[0, 10000] \times [0, 10000]$ into the grid of $X \times Y$ cells. In this paper the $X \times X$ was assumed for simplicity with X = 100.

5   The GEO-Sandpile Localization Model

The proposed localization model joins the Sandpile model set on the cellular automata and GEO algorithm to search for the drop point of the sand grains (the source), which approximates contaminant concentration data reported by the sensors distributed over a considered domain. As a result, GEO algorithm localizes the source of the airborne contamination. The steps of the applied methodology are shown in the pseudocode of the proposed algorithm and described as follows:

BEGIN

- Load the values of contaminant concentrations in the positions of the sensors.
- Load the parameters of the experiment. (*The parameter τ, generation number*).
- Create the initial GEO individual (ind_B). (*The individual symbolizes a potential solution to the problem; thus, it represents the position of the source hazardous substance release in a two-dimensional space. First, each of the two components is mapped to a binary string, and then the strings are combined into one string which is a binary representation of the individual (ind_B)*).
- Rate the fitness of the ind_B. (*The individual is rated using a fitness function based on the difference in value between the Sandpile model results and the contaminant concentrations in the positions of the sensors*).
  WHILE *The number of generations of GEO algorithm is realized* DO
  {
  - Create the neighbors:
    WHILE *Each of the genes do not be reversed* DO
    {
      * Create the GEO neighbour individual ind_{Ni}. (*The neighbor individual i is created by the $i - th$ bit value inversion, for each bit in binary sequence of ind_B*).
      * Rate the fitness each of the ind_{Ni}.
    }
  - Create ranking r_i of bits based on descending value of the fitness function for every ind_{Ni}.
  - Count the probabilities ($p_i = r_i^{-\tau}$) for inversion (mutation) of bits based on created rank r_i. (*τ is the parameter of the GEO*).
  - Mutate the bits of the individual ind_B according to the probabilities p_i.
  - Save the best individual ind_B.

}
- The solution of the GEO algorithm represents the best obtained individual ind_B.

END

The fitness function of GEO algorithm is computed as the sum of differences between the results of the Sandpile model and the contaminant concentrations in sensors locations. The difference between these values at each point is calculated according to the formula (e.g., [28]):

$$f(C_j^M, C_j^E) = \sum_{j=1}^{N} [\log(C_j^M) - \log(C_j^E)]^2 \tag{1}$$

where: C_j^M is the value of concentration in the $j-th$ sensor location, C_j^E is the value of concentration estimated by the Sandpile model in the $j-th$ sensor location, N is the number of all sensors. If the Sandpile model's result equals 0, then it is assumed to be $1e^{-200}$ to allow logarithm calculation. The final evaluation function is the sum of the differences from all considered points (positions of the sensors). The evaluation function tends to the minimum because the smaller the value of the obtained difference, the better the mapping of the Sandpile model with the target sensors' concentrations.

6   Verification of the GEO-Sandpile Model Effectiveness in the Localization of the Contaminant Source

6.1   Generation of Testing Data

In this chapter, the GEO-Sandpile model effectiveness in practical application, i.e., localization of the airborne contaminant source, is presented. The synthetic data of the contaminant concentration over the simulation domain were generated using the Gaussian dispersion model (e.g. [34]) to verify the proposed algorithm's operation. The Gaussian plume model is the most common air pollution model. It is based on a simple formula describing the three-dimensional concentration field generated by a point source under stationary meteorological and emission conditions. Despite many simplifications, the Gaussian plume model is used up to now by many researchers. In this model for uniform steady wind conditions, the concentration $C(\tilde{x}, \tilde{y}, z)$ of the emission (in micrograms per cubic meter) at any point $\tilde{x}$ meters downwind of the source, $\tilde{y}$ meters laterally from the centerline of the plume, and z meters above ground level can be written as follows:

$$C(\tilde{x}, \tilde{y}, z) = \frac{Q}{2\pi\sigma_y\sigma_z U} \exp[-\frac{1}{2}(\frac{\tilde{y}}{\sigma_y})^2] \times \{\exp[-\frac{1}{2}(\frac{z-H}{\sigma_z})^2] + \exp[-\frac{1}{2}(\frac{z+H}{\sigma_z})^2]\}, \tag{2}$$

where U is the wind speed directed along x axis, Q is the emission rate or the source strength, and H is the effective height of the release equal to the sum of the release height and plume rise ($H = \tilde{H} + h$). In the Eq. 2 σ_y and σ_z are the standard deviations of concentration distribution in the crosswind and vertical direction and depends on $\tilde{x}$. These two parameters were defined empirically for different stability conditions by Pasquill and Gifford (e.g. [34]).

The Gaussian dispersion plume model was used to generate a map of contaminant spread in a given area. We restrict the diffusion to the stability class C in an urban area (Pasquill type stability for the rural area). The sample distribution of the contaminant within the considered domain presents the right panel in Fig. 2.

6.2   Test Cases Assumptions

Described in detail in Sect. 5 the GEO-Sandpile model was examined in the sense of its efficiency in localizing the contamination source using the synthetic sensors data. These data were generated by the Gaussian plume model for an urbanized area (Sect. 6.1). The test domain was the square $10\,km \times 10\,km$. The contaminant source was placed in the position $(2000\,m, 5000\,m)$ and $5\,m$ above the ground, within the domain. The emission rate was varying from $5000\,\frac{g}{s}$ to $50000\,\frac{g}{s}$. The wind speed from the following set $\{3\,\frac{m}{s}, 5\,\frac{m}{s}, 7\,\frac{m}{s}, 10\,\frac{m}{s}\}$ with four wind directions parallel to both axes were considered. The sensors were placed $2.5\,m$ above the ground in a regular grid 100×100, mapped on the whole square domain $10000\,m \times 10000\,m$. The concentrations reported by the sensors grid have been passed to the GEO algorithm to verify if it can find airborne contaminant sources within the domain. The GEO algorithm was using the Sandpile model working on the applied two-dimensional CA with size 100×100. The larger size of CA, the more accurate results we could obtain, assuming a denser data grid. The number of sand grains dropped in this model were from set $\{10^4, 10^5, 10^7, 10^8\}$. The number of generations applied in the GEO algorithm was equal to 100.

7   Results of the GEO-Sandpile Localization Model

7.1   Results for Various Wind Speed Test Cases

The research's initial software was designed in Java within the master thesis [20]. The first stage of the research was to determine the optimal value of τ, the only parameter in the original GEO algorithm. The τ has a crucial role in GEO algorithm behavior and its effects on the mutation probability. A too high value of τ may cause the search space exploration in a deterministic way. On the other hand, too little value can lead to a completely random search of space. For this reason, determining the appropriate value of this parameter is extremely important. In the presented study, this value was selected from the interval $(0, 3)$ with a step equal to 0.1.

The conducted experiments revealed that wind speed assumed in the test domain affects the optimal value of τ in the GEO algorithm. Table 1 presents the GEO-Sandpile localization model results for the tests assuming various wind

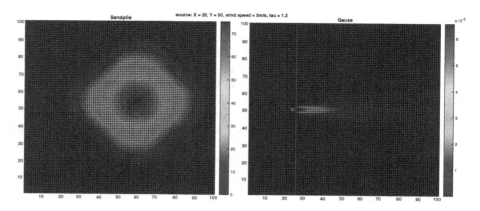

Fig. 2. The heat map of distribution of 10^5 sand grains in Sandplie model selected by GEO with $\tau = 1, 2$ (left panel) for the contaminant concentration distribution obtained by the Gaussian model (right panel) with the setup: wind speed $3\,\frac{m}{s}$ directed along x axis, emission rate $5000\,\frac{g}{s}$ and target source position $(20, 50)$.

speeds. As we can see with wind speed changes, the best value τ for this localization model is also changing. For the high wind speeds, the τ is equal to 1.4, while for lower speeds, the τ decreases. For selected values of τ we can analyze found by the GEO-Sandpile model contaminant source positions. GEO is searching for the most suited distribution of grains in the Sandpile model to contaminant distribution reported by the sensors network. So, GEO founds the actual spatial distribution of the contaminant and its center of mass. However, it cannot consider the gradual blowing off the contamination along the wind direction correspondingly to its speed. Consequently, the location of the target contaminant source and contaminant spatial distribution center is different and depends on wind parameters. To predict the source location, we should analyze the center of distribution (found with the use of GEO algorithm), presented in the third column in Table 1, and slope length for dropped sand grains (presented in the fourth column in Table 1). The target source position can be assessed by shifting the center of mass of the spatial Sandpile distribution obtained from GEO by the vector equal to slope length in the direction opposite to the wind direction. The last column in Table 1 presents the contaminant source position's final assessment. As we can see (Table 1), the source positions are the most correctly predicted for the test case study with wind speed equal to $3\,\frac{m}{s}$ and $\tau = 1.2$. For this case, the target source position was $(20, 50)$ while the predicted one is $(21, 53)$. This test case setup was selected as a reference setup in the presented subsequent analysis. The heat map for this is presented in Fig. 2. The right panel presents the Gaussian dispersion model's contaminant distribution being the input data for the GEO-Sandpile localization model. The left panel presents the grains' spatial distribution in the best Sandpile model configuration. The differences between both figures are obvious, but our aim is not to simulate the contaminant distribution but to find the source position, which is done quite

Table 1. GEO-Sandpile localization algorithm results for various wind speed scenarios with corresponding optimal values of τ. The wind direction was along the x-axis. Sand grains dropped in this model were equal to 10^5.

Experiment parameters	Target source position	Best GEO solution for Sandpile model	Slope length for best GEO score	Predicted source position
Wind speed 3 $\frac{m}{s}$, $\tau = 1.2$,	**(20, 50)**	(58, 53)	37	**(21, 53)**
Wind speed 5 $\frac{m}{s}$, $\tau = 1.2$,	**(20, 50)**	(58, 46)	37	**(21, 46)**
Wind speed 7 $\frac{m}{s}$, $\tau = 1.4$,	**(20, 50)**	(58, 52)	37	**(21, 52)**
Wind speed 10 $\frac{m}{s}$, $\tau = 1.4$,	**(20, 50)**	(62, 42)	37	**(25, 42)**

Table 2. GEO-Sandpile localization algorithm results for various wind directions and target contaminant source positions. The wind speed was $3 \frac{m}{s}$, emission rate $5000\frac{g}{s}$. The number sand grains in Sandplie model was equal to 10^5. The GEO parameter $\tau = 1.2$.

Wind direction	Target source positionl	Best GEO solution for Sandpile model	Slope length for best GEO score	Predicted source position
Along x-axis	**(20, 50)**	(58, 53)	37	**(21, 53)**
Opposite x-axis	**(80, 50)**	(42, 49)	39	**(81, 49)**
Along y-axis	**(50, 20)**	(44, 62)	37	**(44, 25)**
Opposite y-axis	**(50, 80)**	(51, 37)	39	**(51, 76)**

correctly. Similar results were obtained for the test case with wind speed equal to $7 \frac{m}{s}$ and $\tau = 1.4$, where the source position was predicted in $(21, 52)$ position, while the target position was $(20, 50)$. Assuming the size of the test domain (100×100), the prediction accuracy is acceptable.

7.2   Results for Various Wind Directions and Target Source Positions

The next stage of the study was to analyze the sensitivity of the GEO-Sandpile localization model to the changes in the wind direction and various target source positions. The wind's four directions were considered: along the x-axis, opposite x-axis, along the y-axis, and opposite y-axis. In all cases the wind speed was equal to $3 \frac{m}{s}$, emission rate was $5000 \frac{g}{s}$. We have also tested the localization algorithm against various target contamination source positions. The selection of the results presents Table 2. Analysis of the Table 2 allows concluding that in all considered test cases, the proposed GEO-Sandpile localization model predicts the contaminant source location with reasonably high precision.

8 Conclusions an Future Works

This paper proposes the Generalized Extremal Optimization algorithm combined with the Sandpile model for airborne contaminant source localization. The proposed algorithm input data are the contaminant concentrations reported by the sensors over a considered domain. The output is the assessment of the contaminant source position within the domain. The testing contaminant concentration data were generated with the use of the Gaussian dispersion model.

Conducted tests confirmed that the Sandpile model might be used as a model simulating the transport of airborne contaminants with some accuracy. Moreover, presented test cases verified the proposed GEO-Sandpile localization model efficiency assuming various wind directions, speed and contaminant source position within the domain. For the tested wind speed scenarios, the optimal value of τ parameter in the GEO algorithm was estimated, enabling accurate prediction of the airborne contaminant source. Conducted studies revealed that the proposed GEO-Sandpile localization model works well for different wind conditions (direction and speed) and various target contaminant source locations. Obtained results allow stating that the proposed algorithm can be successfully used in different optimization problems like predicting the location of the source of the airborne toxin.

The current research stage allows estimating only the contaminant source position. In a real threat situation, assessing the released substance's quantity is also essential for the emergency responders. Thus, future work will focus on algorithm testing to formulate and prove the relationship between contaminant emission rate and the number of dropped grains in the Sandpile model for a more accurate prediction of the contamination source.

References

1. Aegerter, C.M.: A sandpile model for the distribution of rainfall? Phys. A **319**, 1–10 (2003)
2. Bak, P.: How Nature Works: The Science of Self-Organized Criticality, 1st edn. Springer, Heidelberg (1999)
3. Bak, P., Tang, C.: Earthquakes as a self-organized critical phenomenon. J. Geophys. Res. **94**(B11), 15635–15637 (1989)
4. Bak, P., Tang, C., Wiesenfeld, K.: Self-organized criticality: an explanation of the 1/f noise. Phys. Rev. Lett. **59**(4), 381–384 (1987)
5. Bjorner, A., Lovász, L., Shor, W.: Chip-firing games on graphs. Eur. J. Combin. **12**, 283–291 (1991)
6. Borysiewicz, M., Wawrzynczak, A., Kopka, P.: Stochastic algorithm for estimation of the model's unknown parameters via Bayesian inference. In: Proceedings of the Federated Conference on Computer Science and Information Systems, pp. 501–508. IEEE Press, Wroclaw (2012). ISBN 978-83-60810-51-4
7. Chan, Y., Marckert, J.F., Selig, T.: A natural stochastic extension of the sandpile model on a graph. J. Combin. Theory Ser. A **120**, 1913–1928 (2013)
8. Chapman, S.C., Dendy, R.O., Rowlands, G.: A sandpile model with dual scaling regimes for laboratory, space and astrophysical plasmas. Phys. Plasmas **6**, 4169 (1999)

9. Désérable, D., Dupont, P., Hellou, M., Kamali-Bernard, S.: Cellular automata in complex matter. Complex Syst. **20**(1), 67–91 (2011)
10. Formenti, E., Pham, T.V., Duong Phan, T.H., Thu, T.: Fixed-point forms of the parallel symmetric sandpile model. Theoret. Comput. Sci. **533**, 1–14 (2014)
11. Frette, V., Christensen, K., Malthe-Sørenssen, A., Feder, J., Jøssang, T., Meakin, P.: Avalanche dynamics in a pile of rice. Nature **379**, 49–52 (1996)
12. Galski, R.L., de Sousa, F.L., Ramos, F.M., Muraoka, I.: Spacecraft thermal design with the genrelized extremal optimization algorithm. In: Inverse Problems, Design and Optimization Symposium, Brazil (2004)
13. Goles, E., Morvan, M., Phan, H.D.: Sandpiles and order structure of integer partitions. Discrete Appl. Math. **117**, 51–64 (2002)
14. Hutchinson, M., Oh, H., Chen, W.H.: A review of source term estimation methods for atmospheric dispersion events using static or mobile sensors. Inform. Fusion **36**, 130–148 (2017)
15. Hesse, J., Gross, T.: Self-organized criticality as a fundamental property of neural systems. Front. Syst. Neurosci. **8**, 166 (2014)
16. Bhaumik, H., Santra, S.B.: Stochastic sandpile model on small-world networks: scaling and crossover. Phys. A: Stat. Mech. Appl. **511**, 258–370 (2018)
17. Kopka, P., Wawrzynczak, A.: Framework for stochastic identification of atmospheric contamination source in an urban area. Atmos. Environ. **195**, 63–77 (2018)
18. Latapy, M., Mataci, R., Morvan, M., Phan, H.D.: Structure of some sandpiles model. Theoret. Comput. Sci. **262**, 525–556 (2001)
19. Latapy, M., Phan, H.D.: The lattice structure of chip firing games. Phys. D **155**, 69–82 (2000)
20. Marchel, L.: Zastosowanie modelu 'Sandpile' w procesie optymalizacji. Master thesis, supervisor: Szaban M. (2020). (in Polish)
21. Parsaeifard, B., Moghimi-Araghi, S.: Controlling cost in sandpile models through local adjustment of drive. Phys. A **534**, 122185 (2019)
22. Ricotta, C., Avena, G., Marchetti, M.: The flaming sandpile: self-organized criticality and wildfires. Ecol. Model. **119**(1), 73–77 (1999)
23. Rossin, D., Cori, R.: On the sandpile group of dual graphs. Eur. J. Combin. **21**(4), 447–459 (2000)
24. Rothman, D., Grotzinger, J., Flemings, P.: Scaling in turbidite deposition. J. Sediment. Res. **64**(1a), 59–67 (1994)
25. De Sousa, F.L., Ramos, F.M., Paglione, P., Girardi, R.M.: New stochastic algorithm for design optimization. AIAA J. **41**(9), 1808–1818 (2003)
26. De Sousa, F.L., Vlassov, V., Ramos, F.M.: Generalized extremal optimization: an application in heat pipe design. Appl. Math. Modell. **28**, 911–931 (2004)
27. De Sousa, F.L., Ramos, F.M., Soeiro F.J.C.P., Silva Neto, A.J.: Application of the generalized extremal optimizayion algorithm to an inverse radiative transfer problem. In: Proceedings of the 5th International Conference on Inverse Problems in Engineering: Theory and Practice, Cambridge, UK, 11–15 July 2005 (2005)
28. Wawrzynczak, A., Danko, J., Borysiewicz, M.: Lokalizacja zrodla zanieczyszczen atmosferycznych za pomoca algorytmu roju czastek. Acta Scientiarium Polonorum Adm. Locorum **13**(4), 71–91 (2014). (in Polish)
29. Wawrzynczak, A., Berendt-Marchel, M.: Can the artificial neural network be applied to estimate the atmospheric contaminant transport? In: Dimov, I., Fidanova, S. (eds.) HPC 2019. SCI, vol. 902, pp. 132–142. Springer, Cham (2021). https://doi.org/10.1007/978-3-030-55347-0_12. ISBN 978-3-030-55346-3

30. Wawrzynczak, A., Berendt-Marchel, M.: Computation of the airborne contaminant transport in urban area by the artificial neural network. In: Krzhizhanovskaya, V.V., et al. (eds.) ICCS 2020, Part II. LNCS, vol. 12138, pp. 401–413. Springer, Cham (2020). https://doi.org/10.1007/978-3-030-50417-5_30. ISBN 978-3-030-50416-8

31. Wawrzynczak, A., Kopka, P., Borysiewicz, M.: Sequential Monte Carlo in Bayesian assessment of contaminant source localization based on the sensors concentration measurements. In: Wyrzykowski, R., Dongarra, J., Karczewski, K., Waśniewski, J. (eds.) PPAM 2013. LNCS, vol. 8385, pp. 407–417. Springer, Heidelberg (2014). https://doi.org/10.1007/978-3-642-55195-6_38

32. Wolfram, S.: A New Kind of Science. Wolfram Media (2002)

33. Xie, D., Luo, Z., Yu, F.: The computing of the optimal power consumption of semi-track air-cushion vehicle using hybrid generalized extremal optimization. Appl. Math. Model. **33**, 2831–2844 (2009)

34. Zannetti, P.: Gaussian models. In: Zannetti, P. (ed.) Air Pollution Modeling, pp. 141–183. Springer, Boston (1990). https://doi.org/10.1007/978-1-4757-4465-1_7

Author Index

Printed in the United States
by Baker & Taylor Publisher Services